MEMS Reference Shelf

MEMS Reference Shelf

Series Editors:

Stephen D. Senturia
Professor of Electrical Engineering, Emeritus
Massachusetts Institute of Technology
Cambridge, Massachusetts

Roger T. Howe
Department of Electrical Engineering
Stanford University
Stanford, California

Antonio J. Ricco
Small Satellite Division
NASA Ames Research Center
Moffett Field, California

For other titles in this series, go to:
www.springer.com/series/7724

Jeffrey H. Lang
Editor

Multi-Wafer Rotating MEMS Machines

Turbines, Generators, and Engines

Springer

Editor
Jeffrey H. Lang
Department of Electrical Engineering &
Computer Science
Massachusetts Institute of Technology
77 Massachusetts Ave.
Cambridge MA 02139-4307
USA
lang@mit.edu

ISSN 1936-4407
ISBN 978-0-387-77746-7 e-ISBN 978-0-387-77747-4
DOI 10.1007/978-0-387-77747-4
Springer New York Dordrecht Heidelberg London

Library of Congress Control Number: 2009933947

Printed on acid-free paper

Springer is part of Springer Science+Business Media (www.springer.com)

Preface

This book is an outgrowth of the MIT Micro Engine Project. This project began at the Massachusetts Institute of Technology in the Fall of 1995, and later expanded through collaborations with the Georgia Institute of Technology, the Clark Atlanta University, and the University of Maryland at College Park. By the time the project ended in the Summer of 2008, almost 13 years later, it had supported over one hundred people including faculty, staff, post doctoral researchers, and many students.

The overall objective of the Micro Engine Project was to develop a small but power-dense gas turbine generator based on MEMS fabrication technologies. Thus, the project sought to develop a fuel-burning jet engine that would drive an electric generator to produce electric power for general purpose use. Along the way, the project would advance the science and engineering of many disciplines from the MEMS perspective. The purpose of this book is to pull together the results of that study in a compact, and hopefully integrated, manner.

The Micro Engine Project was by its very nature a highly multi-disciplinary project pursuing advances in materials, structures, fabrication, combustion, heat transfer, turbomachinery, bearings and electromechanics, all at the MEMS scale. Most of these critical disciplines are discussed in this book. To begin, Chapters 1 and 2 lay the foundation for the project, both in terms of its practical objectives, and its scientific and engineering rationale. Chapter 3 then discusses materials, structures, and packaging. Chapters 4 and 5 focus on MEMS fabrication, with an emphasis on technologies specifically required to make a device as complex as a microengine. Chapter 4 in particular focuses on silicon-related fabrication, while Chapter 5 focuses on electroplating magnetic components, such as windings and cores, into a silicon superstructure. Chapter 6 presents our development of very-high-speed air bearings. Chapter 7 presents our work on thermofluidics and turbomachinery. Chapter 8 studies electromechanics as required to develop an MEMS electric generator. Finally, Chapter 9 focuses on combustion at the MEMS scale.

It has been my great pleasure to work with the contributors to this book for the last 13 years and beyond, and the many other faculty, staff, postdocs, and students involved in the project over those years. I am grateful for their enthusiastic hard work, from the beginning of the project through the completion of this book. Without that effort this book and the work described herein would never have materialized.

Cambridge, MA Jeffrey H. Lang
March 2009

Contents

Contributors

Mark G. Allen School of Electrical and Computer Engineering, Georgia Institute of Technology, 791 Atlantic Dr., Atlanta, GA 30332-0269, USA

David P. Arnold Department of Electrical and Computer Engineering, University of Florida, 229 Benton Hall, Gainesville, FL 32611-6200, USA

Sauparna Das Department of Electrical Engineering and Computer Science, Massachusetts Institute of Technology, Bldg 10-050, 77 Massachusetts Avenue, Cambridge, MA 02139, USA

Alan H. Epstein Department of Aeronautics and Astronautics, Massachusetts Institute of Technology, Bldg 31-265, 77 Massachusetts Avenue, Cambridge, MA 02139, USA

Stuart A. Jacobson Department of Aeronautics and Astronautics, Massachusetts Institute of Technology, Bldg 31-269, 77 Massachusetts Avenue, Cambridge, MA 02139, USA

Jeffrey H. Lang Department of Electrical Engineering and Computer Science, Massachusetts Institute of Technology, Bldg 10-176, 77 Massachusetts Avenue, Cambridge, MA 02139, USA

Hanqing Li Microsystems Technology Laboratories, Massachusetts Institute of Technology, Bldg 31-261E, 77 Massachusetts Avenue, Cambridge, MA 02139, USA

Martin Schmidt Department of Electrical Engineering and Computer Science, Massachusetts Institute of Technology, Bldg 3-240, 77 Massachusetts Avenue, Cambridge, MA 02139, USA

Zoltán S. Spakovszky Gas Turbine Laboratory, Massachusetts Institute of Technology, Bldg 31-265, 77 Massachusetts Avenue, Cambridge, MA 02139, USA

C.M. Spadaccini Center for Micro and Nano Technology, Engineering Technologies Division, 7000 East Avenue, Livermore, CA 94550, USA

S. Mark Spearing School of Engineering Sciences, University of Southampton, University Road, Southampton SO17 1BJ, UK

I.A. Waitz Department of Aeronautics and Astronautics, Massachusetts Institute of Technology, Bldg 33-207, 77 Massachusetts Avenue, Cambridge, MA 02139, USA

Chapter 1
Introduction to PowerMEMS

Alan H. Epstein

The term "PowerMEMS" was first suggested by Epstein and Senturia [1] in 1996 to describe microsystems which generated power or pumped heat. Their primary interest was micro-electro-mechanical systems (MEMS) heat engines, specifically Brayton cycles (such as gas turbines) and subsystems thereof. The promise identified was microsystems whose power densities equaled or exceeded those of the more familiar large-scale devices. Since that time, the PowerMEMS system has evolved into a broader concept which includes other conventional thermodynamic cycles, new heat engine concepts which may only be attractive at microscale, energy harvesting schemes, and micro-fuel cells – all based on MEMS approaches.

Thermodynamic cycles and their heat engine embodiments – Rankine, Brayton, Otto, Diesel, etc. – were at the heart of the industrial revolution. Such engines can convert chemical energy into heat and then into mechanical work. This mechanical energy may be used directly for applications such as vehicle propulsion or fluid pumping or converted into electric power. Propulsion is one goal of MEMS jet engines and rocket engines, while MEMS gas turbine, steam, and internal combustion engines are being developed to power electric generators for electric power production. With electrical, mechanical or thermal power input, micro heat engines can also pump heat as in familiar conventional heat-pumps, chillers, and refrigerators. PowerMEMS also encompasses heat engines which convert chemical energy directly to electrical energy, often using the thermoelectric effect. These devices may be intended primarily as power generators or may have a different primary function such as fuel reforming for fuel cells, with the power generation or scavenging being a secondary consideration. PowerMEMS also encompasses devices which convert mechanical energy directly into electric power such as self-charging electric watches and heel strike power generators.

The proliferation of small, portable electronics – computers, digital assistants, cell phones, GPS receivers, etc. – requires compact energy supplies. Increasingly,

A.H. Epstein (✉)
Department of Aeronautics and Astronautics, Massachusetts Institute of Technology, Bldg 31-265, 77 Massachusetts Avenue, Cambridge, MA 02139, USA

J.H. Lang (ed.), *Multi-Wafer Rotating MEMS Machines*, MEMS Reference Shelf,
DOI 10.1007/978-0-387-77747-4_1, © Springer Science+Business Media, LLC 2009

these electronics demand energy supplies whose energy and power density exceed that of the best batteries available today. The lack of adequate energy sources has motivated considerable effort on energy efficient approaches to electronics design. Also, the continuing advance in microelectronics permits the shrinking of electronic subsystems of mobile devices such as small animal-sized ground robots and air vehicles. These small, and in some cases very small, mobile systems require increasingly compact power and propulsion. Thus, the interest in PowerMEMS results from both an applications pull and a technology push. The primary motivations for the work described in this book were that of (1) portable electric power and (2) propulsion for very small aerial vehicles, but the approaches and technologies apply well to other applications discussed above.

1.1 Compact Portable Power Considerations

Energy supplies for portable electronics must satisfy many requirements. These are often in conflict for a specific technology so that design compromise is usual. Energy can be stored in a variety of manners – chemical, mechanical (e.g., springs), fluid (pressurized gas), potential (e.g., a liquid in a gravity field), or nuclear. This discussion will focus on chemical energy since at this time chemical energy density greatly exceeds that of other known non-nuclear sources. Such requirements included the following considerations.

Energy Density – A primary metric is the mass and volume required to hold the needed energy. This includes the primary energy source (the chemistry), the structure needed to contain it (fuel tank, pressure vessel, etc.), and the power converter (engine, fuel cell, etc.). The relative size of the energy chemistry versus that of the power converter varies greatly depending upon the peak power needed, total energy required (average power times the duration), and the energy source. An energy harvester is all power converter, a battery could be considered all chemistry (although practical batteries are now closer to 60–80%).

Power Density – Equally important is the mass and volume needed to generate the power (energy per unit time) the power system is required to deliver. Peak power required may size the power conversion system. For example, batteries designed for very long lives but very low power applications (such as for watches) now have twice the energy density of high peak power batteries (such as for power tools). Similarly for mechanical generators, given a fixed total system weight budget (generator plus fuel), a larger output power generator will weigh more with this weight displacing that of fuel. Again, the system designer must trade power density against energy density.

Efficiency – The fraction of the chemical energy consumed converted to electrical energy delivered to the reduces the gross energy density. Clearly, the size of the primary energy supply needed scales inversely with the efficiency. Energy not delivered to the user ends up as heat that must be rejected from the package. For some devices, rejection of this heat is a primary design driver, influencing energy system size, weight, and complexity.

Safety – It is important to consider safety since an apple-sized battery contains more chemical energy than a hand grenade. In addition to explosion and fire, concerns of toxicity, thermal injury, and environmental impact are important. For example, batteries for unmanned aeronautical and munitions applications can be formulated with more than twice the energy density considered safe for industrial or retail use.

Environment – Consideration of the storage and operating environment may drive the energy/power system design. Ambient temperature requirements from –40°C to +50°C are typical for outdoor applications. This is a large temperature range for common battery chemistries. At the high temperature end, safety requirements reduce energy density. At low temperatures, most batteries do not deliver much power. Indeed, experienced hikers and soldiers often sleep with their batteries in their sleeping bags to keep the batteries warm in cold climates. Systems which use liquid water, such as some fuel cells, must include special provisions for storage, startup, and operation in sub-freezing conditions. Heat engines produce less power as the temperature rises both due to a drop in the mass of air flowing through the engine and due to a reduction in efficiency. For both engines and many fuel cells, altitude of operation is an important consideration since these air-breathing systems depend on the atmosphere for their oxygen. Oxygen density drops off rapidly with increasing altitude – down by about 30% at 3,000 m. Unless turbocharging or a similar approach is used, power output will drop concomitantly. Clearly, air-breathing devices will not work underwater. Contamination from the ingestion of dust, aerosols (a problem in marine environments), and rain or snow can be a concern. Environmental challenges also include mechanical considerations such as orientation (right side up – upside down), vibration (jogging, vehicle motion), and shock (dropping a device on a hard floor from 2 m can result in loads of 5,000 g).

The point of the above discussion is that while energy and power density are obvious requirements for portable energy supplies, there are many other considerations that influence the performance and utility of a particular technology. Many of these considerations are application specific, implying that the optimal solution may be as well.

1.2 Chemical Energy Sources

Table 1.1 compares several chemical energy source in terms of both the gross gravimetric energy in the chemistry and an estimate of the net energy output including assumptions about the efficiency of current and near term conversion processes. Lithium battery chemistry has about the same theoretical energy density as TNT. Two current practical implementations of lithium battery chemistries are shown,

Table 1.1 Chemical energy sources

Energy source	Potential[a] W h/kg	Practical W h/kg[b] based on conversion device efficiency
Lithium battery	1,400	175 ($LiSO_2$) 300 ($LiSOCl_2$)
TNT	1,400	NA
Sugar	4,700	NA
Methanol	6,200	1,500–3,100
Diesel fuel	13,200	1,320–5,000
Hydrogen	33,000	1,150–23,000

[a] *Based on enthalpy* (after R. Paur).
[b] (*System conversion efficiency*) × (*Energy available in fuel*).

lithium sulfur dioxide ($LiSO_2$) and lithium thionyl chloride ($LiSOCL_2$). While $LiSOCL_2$ batteries can have twice or more than the energy density of $LiSO_2$, current implementations are not considered safe enough for commercial applications or even military or aerospace ones that put people at risk. Both are onetime use primary batteries. Rechargeable batteries formulations have 30–50% less energy density. Batteries are important because they widely available and inexpensive, so they represent a baseline to which all portable power source PowerMEMS concepts must compare favorably to be of value.

The energy density of both lithium batteries and explosives is low compared to other chemical energy sources, one reason being that they contain both fuel and oxidizer while the other chemistries use atmospheric oxygen as the oxidizer which therefore need not be carried by the power source (unless used underwater or in space).

Hydrogen has the highest chemical energy density but since it is a gas at standard conditions it must be contained. For current approaches, the containment mass considerably outweighs the hydrogen it holds. Storage methods available include: as a metal hydride, as a high pressure gas, and as a liquid. In all three cases, the net energy density accounting for containment is considerably less than that of liquid hydrocarbons fuels such as diesel or methanol. Sugar, a biological "fuel," has an energy density about a third of that of diesel.

Fuel energy density information is plotted in Fig. 1.1 in terms of gravimetric energy density versus volumetric energy density. It can be seen in the left graph that all three forms of contained hydrogen have an energy density about equivalent to that of the potential of lithium battery chemistry, and all are about an order of magnitude below that of diesel fuel, which has about the highest density of any known liquid fuel.

The right plot of Fig. 1.1 shows the same data but now accounts for the assumed efficiency of the power converters (but not their mass or volume). Current lithium primary batteries deliver about an order of magnitude less power than their theoretical potential. Proton-exchange-membrane (PEM) fuels cells run on hydrogen and operate at about 50% net efficiency. Current direct methanol fuel cells (DMFC) are

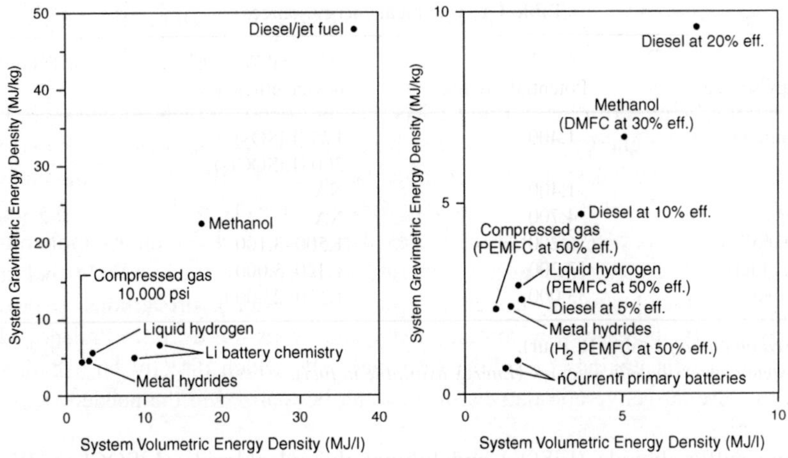

Fig. 1.1 *Left*: Energy density of common portable power fuels. *Right*: Net energy density accounting for converter efficiency but not converter mass and volume

about 30% efficient. Portable diesel generators below 10 kW are 20–30% efficient with efficiency increasing with size.

Figure 1.2 is the data of Fig. 1.1 replotted on a larger scale along with the efficiencies of large scale energy conversion systems in production. Large gas turbines are now 45–55% efficient. Modern, large (100+MW) combined cycle, gas turbine-steam power plants are the most efficient at 60%. Large diesels are in the 40–50% range, while automotive engines are closer to 30%. The best fuel cell systems

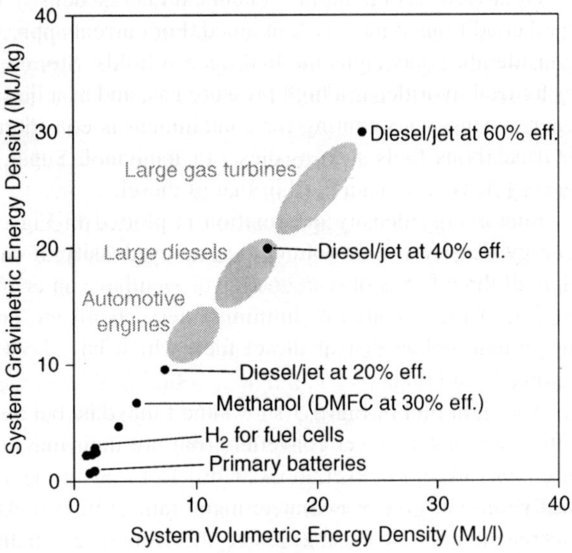

Fig. 1.2 Net fuel energy densities including conversion efficiency and efficiency of current conventional-sized devices

today are about 50% efficient. In all cases, the effective system efficiency can be improved if the waste heat is captured and used for other purposes such as heating or cooling.

One implication of Figs. 1.1 and 1.2 is that hydrocarbon fuels burned in air have 20–30 times the energy density of the best current lithium chemistry-based batteries, so that fuelled systems need only be modestly efficient to compete well with batteries. Another is that the very low net density of hydrogen (hydrogen plus container) renders it unattractive as a portable fuel for many applications. Thus, portable converters that must operate on hydrogen, such as PEM fuel cells, will likely manufacture hydrogen in situ from denser hydrocarbons, a process which greatly increases device complexity and drops the overall efficiency. This means that heat engines or solid oxide fuel cells which directly burn hydrocarbon fuel are attractive alternatives, if they can be realized in the needed sizes and efficiencies.

1.3 Power System Considerations

Since the size and efficiency of power converters vary greatly depending upon the technology, and the energy density of the fuels are so different, the optimal approach is quite application specific. If the power requirements are large but energy requirements are modest, then the power converter dominates the overall system mass. In this case, a compact power converter is favored even if it is relatively inefficient and uses a fuel with relatively low energy density. Conversely, when the total energy requirements are large, a more efficient power converter using high energy density fuel is favored, even if the converter is relatively large.

An example of the trade between power density and energy density is illustrated in Fig. 1.3. Here the total energy supply mass is plotted against the total energy; it can supply for three technologies – $LiSO_2$ primary batteries, a DMFC operating at 20% over all efficiency, and a notional MEMS heat engine. All three approaches are sized to deliver 50 W. A BA5590 battery weighs about a kilogram and can output 170 W-h (about twice the energy of a large laptop computer battery). If more energy than 170 W-h is needed, then a second battery must be added, etc. This discreteness results in the staircase appearance of the battery curve in Fig. 1.3. At an equivalent 20% conversion efficiency, methanol has a higher energy density than a battery, so the DMFC line has shallower slope than that of the battery. (The fuel energy density shown on the plot reflects the combined effect of the system efficiency and methanol concentration.) The fuel cell has a zero fuel mass of 2 kg, so for these assumptions, a battery is a lighter solution if less than 170 W-h is needed. As the energy required increases, the fuel cell looks increasingly attractive. The mass of a 50 g notional micro heat engine power system running on diesel fuel is shown for two overall efficiencies, 5 and 10%. Because of the very high power density of a micro heat engine, it is always better than a battery. At 5% efficiency it is superior to the fuel cell if the energy requirement is less than about 2,000 W-h. At 10%

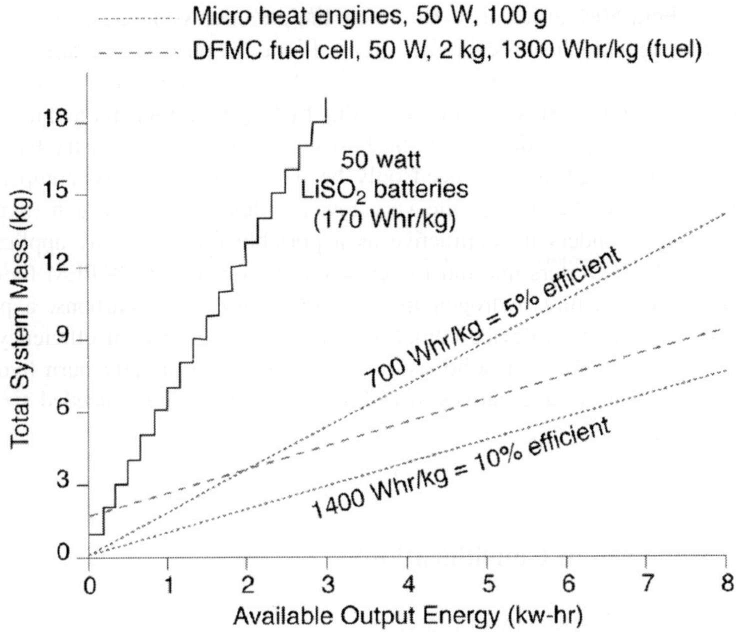

Fig. 1.3 Total mass for a 50 W power system as a function of total energy required for batteries, a direct methanol fuel cell (DMFC), and a notional micro heat engine using diesel fuel

efficiency, the micro heat engine is always better than the DMFC because the energy density advantage of the diesel fuel over methanol more than offsets the difference in efficiency.

The above example is for illustrative purposes only since the performance of a field ready DMFC or micro heat engine is projection only. However, several points can be made. First, the efficiency of a micro heat engine can be modest compared to that of the current conventional size variants shown in Fig. 1.2 for a micro solution to be a very attractive alternative to a battery in terms of size and mass. Second, the ability to use a liquid hydrocarbon, diesel in particular, is a powerful advantage compared to an approach requiring hydrogen. Third, the high power density of the micro heat engine compared to a fuel cell is an advantage when the total energy required is modest.

1.4 Conclusion

This chapter has outlined a motivation for powerMEMS, especially micro-heat engines. If their potential can be realized, they represent a very attractive alternative for many applications, compact power in particular. The engineering requirements for a MEMS gas turbine are discussed in Chapter 2. Subsequent chapters discuss the micro component technologies needed including structures and materials, bearings and rotor dynamics, fluid mechanics, combustion, electro-mechanics, and micro-fabrication. These chapters include theory, engineering design, fabrication, and test results.

Reference

1. Epstein, A.H., and Senturia, S.D., 1997, "Macro Power from Micro Machinery", *Science*, **276**, p. 1211

Chapter 2
System Design Considerations and Device Overview

Alan H. Epstein

2.1 Introduction

The primary objective of the engineering describe in this book is the realization of an MEMS gas turbine driven electric generator producing tens of watts of electric power. The same device with reduced electrical power extraction could produce tens a grams of thrust to propel very small air vehicles massing a few hundred grams. As a part of this effort, several MEMS engine subsystems were engineered and tested as stand alone devices, including: air-turbine driven generators, a motor driven gas compressor, combustion chambers, and turbochargers. Although challenging, useful, and complete in themselves, these devices and the requisite microfabrication technologies were engineered as part of the gas turbine engine's research and development program. The device design space, materials, and fabrication approaches were thus constrained to be consistent with the requirements of the gas turbine and within the bounds of realizable fabrication techniques. In this chapter, we describe the overall system design of the gas turbine and its subsidiary devices, presenting an overview of the MEMS devices and technologies which underlie them. The point of view presented in this chapter is that of a device designer rather than those described in greater detail in subsequent chapters. We start by reviewing the thermodynamics of a gas turbine, briefly describe the relevant mechanics considerations – including materials, fluids, and dynamics – which shape the design space, and then consider individual MEMS devices.

2.2 Thermodynamic and Scaling Considerations

Thermal power systems encompass a multitude of technical disciplines. The architecture of the overall system is determined by thermodynamics, while the design of the system's components is influenced by fluid and structural mechanics and by material, electrical, and fabrication concerns.

A.H. Epstein (✉)
Department of Aeronautics and Astronautics, Massachusetts Institute of Technology, Bldg 31-265, 77 Massachusetts Avenue, Cambridge, MA 02139, USA

J.H. Lang (ed.), *Multi-Wafer Rotating MEMS Machines*, MEMS Reference Shelf,
DOI 10.1007/978-0-387-77747-4_2,

A conventional, macroscopic gas turbine generator consists of a compressor, a combustion chamber, and a turbine driven by the combustion exhaust to power the compressor. The residual enthalpy in the exhaust stream provides thrust or can power a turbine to drive an electric generator. A macroscale gas turbine with a meter-diameter air intake area generates power on the order of 100 MW. Thus, tens of watts would be produced when such a device is scaled to millimeter size if the power per unit of air flow is maintained. When based on rotating machinery, such power density requires combustor exit temperatures of 1,200–1,600 K; rotor peripheral speeds of 300–600 m/s and thus rotating structures centrifugally stressed to several hundred MPa because the power density of both turbomachinery and electrical machines increase with the square of the speed, as does the rotor material centrifugal stress; low-friction bearings; and tight geometric tolerances and clearances between rotating and static parts to inhibit fluid leakage, the clearances in large engines are maintained at about one part in 2,000 of the diameter; and thermal isolation of the hot and cold sections.

These thermodynamic considerations are no different at micro- than at macroscale. But the physics and mechanics influencing the design of the components do change with scale, so that the optimal detailed designs can be quite different. Examples of these differences include the viscous forces in the fluid (larger at microscale), usable strength of materials (larger at microscale), surface area-to-volume ratios (larger at microscale), realizable electric field strength (higher at microscale), and manufacturing constraints (limited mainly to 2-D, planar geometries given current microfabrication technology). Chemical reaction times are invariant with size but most of the fluid residence time in gas turbine combustors is set by mixing considerations which do scale with size.

There are many thermodynamic and architectural design choices in a device as complex as a gas turbine engine. These involve tradeoffs among fabrication difficulty, structural design, heat transfer, and fluid mechanics. Given a primary goal of demonstrating that a high-power density MEMS heat engine is physically realizable, a simple-as-possible design philosophy is attractive, with performance traded for simplicity. For example, a recuperated cycle, which requires the addition of a heat exchanger transferring heat from the turbine exhaust to the compressor discharge fluid, offers many benefits including reduced fuel consumption and relaxed turbomachinery performance requirements, but introduces additional design and fabrication complexity.

How large should a "micro" engine be? A micron, a millimeter, a centimeter? Determination of the optimal size for such a device involves considerations of application requirements, fluid mechanics and combustion, manufacturing constraints, and economics. The requirements for many power production applications favor a larger engine size, 50–100 W. Viscous effects in the fluid and combustor residence time requirements also favor larger engine size. Current MEMS manufacturing technology places both upper and lower limits on engine size. The upper size limit is set mainly by precision etching depth capability, hundreds microns at this time. The lower limit is set by feature resolution and aspect ratio. Economic concerns include manufacturing yield and cost. A wafer of fixed size (say 200 mm diameter) would yield many more low-power engines than high-power engines at essentially the same manufacturing cost per wafer. (Note that the sum of the power produced by all of the engines on the wafer would remain about constant at 1–10 kW.) When commercialized, applications and market forces may establish a strong preference here. For the first demonstrations of a concept, a minimum technical risk approach is attractive. Analysis suggested that fluid mechanics are more difficult at smaller scales, so the largest size near the edge of then current microfabrication technology, 1–2 cm rotor diameter was selected.

Thermodynamic performance calculations indicate that the power per unit air flow from the configuration discussed below has the potential to reach 50–150 W/(g/s) of air flow (Fig. 2.1) . For a given rotor radius, the air flow rate is limited primarily by airfoil span as set by stress in the turbine blade roots. Calculations suggest that it might be possible to improve the specific work, fuel consumption, and air flow rate in later designs with recuperators to realize microengines with power outputs of as much as 50–100 W, power specific fuel consumption of 0.3–0.4 g/W-h, and thrust-to-weight ratios of 100:1. This level of specific fuel consumption approaches that of current small gas turbine engines but the thrust-to-weight ratio is 5–10 times better than that of the best aircraft engine. The extremely high thrust-to-weight ratio is simply a result of the so-called "cube–square law." All else being the same as the engine is scaled down linearly, the air flow and thus the power decreases with the intake area (the square of the linear size) while the weight decreases with the volume of the engine (the cube of the linear size), so that the power-to-weight ratio increases linearly as the engine size is reduced. Detailed calculations show that the actual scaling is not so dramatic, since the specific power is lower at the very small sizes [1]. A principal point is that a microheat engine is a different device than more familiar full-sized engines, with different weaknesses and different strengths.

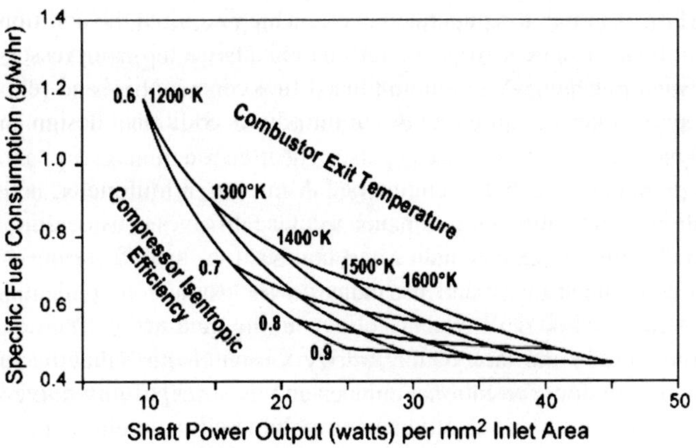

Fig. 2.1 Simple cycle gas turbine performance with H_2 fuel

2.2.1 Mechanics Scaling

While the thermodynamics are invariant down to this scale, the mechanics are not. The fluid mechanics, for example, are scale-dependent [2]. For example, viscous forces are more important at small scale. Turbomachinery pressure ratios of 2:1–4:1 per stage imply turbomachinery tip Mach numbers that are in the high subsonic or supersonic range. Airfoil chords on the order of a millimeter imply that a device with room temperature inflow, such as a compressor, will operate at Reynolds numbers (the ratio of inertial to viscous forces in a fluid) in the tens of thousands. With higher gas temperatures, turbines of similar size will operate at a Reynolds number of a few thousand. These are small values compared to the 10^5–10^6 Reynolds number range of large-scale turbomachinery and viscous losses will be concomitantly larger. But viscous losses make up only about a third of the total fluid loss in a high-speed turbomachine (3-D, tip leakage, and shock wave losses account for most of the rest) so that the decrease in machine efficiency with size is not so dramatic. The increased viscous forces also mean that fluid drag in small gaps and on rotating disks will be relatively higher, and indeed are much larger influences on overall machine performance. Unless gas flow passages are smaller than one micron, the fluid behavior can be represented as continuum flow so that molecular kinetics, Knudsen number considerations, are not important.

Heat transfer is another aspect of fluid mechanics in which microdevices operate in a different design space than large-scale machines. The fluid temperatures and velocities are the same but the viscous forces are larger, so the fluid film heat transfer coefficients are higher, by a factor of about 3. Not only is there more heat transfer to or from the structure, but also thermal conductance within the structure is higher due to the short length scale. Thus, temperature gradients within the structure are reduced. This is helpful in reducing thermal stress but makes thermal isolation challenging.

For structural mechanics, it is the change in material properties with length scale that is most important. Very small length scale influences both material properties and material selection. In a centimeter diameter engine, design features such as blade tips, fillets, orifices, seals, etc. may be only a few microns in size. Here, differences between mechanical design and material properties begin to blur. The scale is not so small (atomic lattice or dislocation core size) that continuum mechanics no longer applies. Thus, elastic, plastic, heat conduction, creep, and oxidation behaviors do not change, but fracture strength can differ. Material selection is influenced both by mechanical requirements and by fabrication constraints. For example, structure ceramics such as silicon carbide (SiC) and silicon nitride (Si_3N_4) have long been recognized as attractive candidates for gas turbine components due to their high strength, low density, and good oxidation resistance. Their use has been limited, however, by the lack of technology to manufacture flaw-free material in sizes large enough for conventional engines and by susceptibility to thermal shock. Shrinking engine size by three orders of magnitude virtually eliminates these problems. Indeed, semiconductor materials such as silicon have usable strength an order of magnitude better than conventional metals. This higher strength can be used to realize lighter structures, higher rotation speeds (and thus higher power densities) at constant geometry, or simplified geometry (and thus manufacturing) at constant peripheral speed.

2.3 Overview of an MEMS Gas Turbine Engine Design

The initial designs were directed at demonstrating bench-top operation of an MEMS-based gas turbine. This implies that, for a first demonstration, it would be expedient to trade engine performance for simplicity, especially fabrication simplicity. Most microfabrication technology has been developed for silicon. Since Si rapidly loses strength above 950 K, this becomes an upper limit to the turbine rotor temperature. But 950 K is too low a combustor exit temperature to close the engine cycle (i.e., produce net power) with the component efficiencies anticipated, so cooling is required for Si turbines. The simplest way to cool the turbine in a millimeter-sized machine is to eliminate the shaft, and thus conduct the turbine heat to the compressor, rejecting the heat to the compressor fluid. This has the great advantage of simplicity and the great disadvantage of lowering the pressure ratio of the now non-adiabatic compressor from about 4:1 to 2:1 with a concomitant decrease in cycle power output and efficiency. Hydrogen was chosen as the first fuel to simplify the combustor development. This expedient arrangement was referred to as the H_2 demo engine. It is a turbojet designed with the objective of demonstrating the concept of an MEMS gas turbine. It does not contain electrical machinery or controls.

One fundamental requirement is rotor-bearing technology compatible with microfabrication and capable of stable operation at a million rpm or more. The geometric and mechanical concept upon which the gas turbine was developed is illustrated in Fig. 2.2. This radial inflow air turbine consists of an outer array of nozzle guide vanes which accelerate the flow arranged around a 4 mm diameter

Fig. 2.2 A 4:1 pressure ratio, 4 mm rotor diameter radial inflow turbine stage

rotor which generates about 15 W of mechanical power when spinning at a peripheral speed of 250 m/s (1.2 Mrpm). The rotor is supported on airbearings at the center for axial forces and at the rotor periphery for radial forces. The device was manufactured by deep reactive ion etching (DRIE) the 400 μm tall airfoils and center thrust bearing from one side of a single crystal Si wafer and then freeing the rotor by etching the 12 μm wide journal bearing at the rotor periphery from the rear of a 600 μm thick wafer. Channels etched in capping wafers top and bottom (not shown) duct the air to turbine and the bearings. This turbine concept proved successful early in the research program so the basic arrangement – a thin disk with an air journal bearing on its periphery – was adopted for all of the devices in this program.

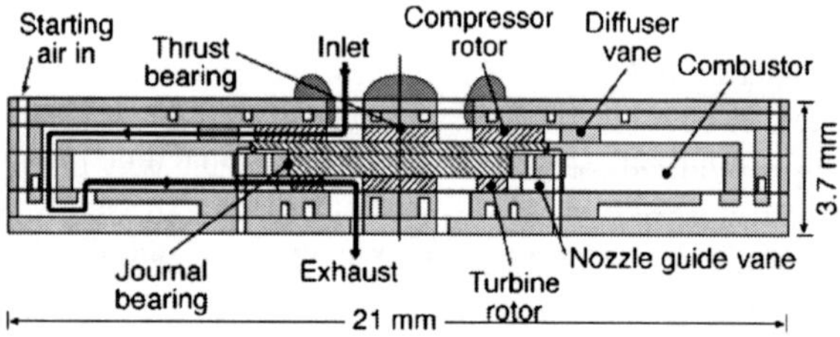

Fig. 2.3 H_2 demo engine with conduction-cooled turbine constructed from six silicon wafers

An elaboration of this geometry form the turbine and compressor of the initial H_2 demo engine design are shown in Fig. 2.3. The centrifugal compressor and radial turbine rotor diameters are 8 and 6 mm, respectively, in this first design (as fabrication capability was improved, later designs increased the component sizes). The compressor discharge air wraps around the outside of the combustor to cool the combustor walls, capturing the waste heat and so increasing the combustor efficiency while reducing the external package temperature. The rotor radial loads are supported on a journal bearing on the periphery of the compressor. Thrust bearings on the centerline and a thrust balance piston behind the compressor disk support the axial loads. The balance piston is the air source for the hydrostatic journal bearing pressurization. The thrust bearings and balance piston are supplied from external air sources. The design peripheral speed of the compressor is 500 m/s so that the rotation rate is 1.2 Mrpm. External air is used to start the machine. With 400 μm span airfoils, the unit was sized to pump about 0.36 g/s of air to produce 0.1 N of thrust or 17 W of shaft power. A cutaway engine chip is shown in Fig. 2.4 . In this particular engine design iteration, the airfoil span (height) is 225 μm and the disks are 300 μm thick.

Fig. 2.4 Cutaway H_2 demo gas turbine chip. Compressor blades are evident on the rotor disk

The following sections briefly elaborate on the component technologies of this engine design. It starts with a statement on microfabrication philosophy and then goes on to turbomachinery aerodynamic design, structures and materials, combustion, bearings and rotor dynamics, and controls and accessories. A system integration discussion then expands on the high-level tradeoffs which define the design space of an MEMS gas turbine engine.

An assumption underlying the design philosophy of these devices was that they were to be consistent with MEMS batch fabrication, with as few extensions to the state-of-the-art as possible. This constrains shapes to mainly prismatic or "extruded" geometries of constant height. Ongoing research with grayscale lithography suggests that smoothly variable etch depths (and thus airfoils of variable span) may be feasible in the near term [3]. Conceptually, more complex 3-D shapes can be constructed of multiple precision-aligned 2-D layers. But layering is expensive with current technology and 6 is considered a large number of precision-aligned layers for a microdevice. Since 3-D rotating machine geometries are difficult to realize, planar geometries are preferred. While 3-D shapes are difficult to microfabricate, in-plane 2-D geometric complexity is essentially free in manufacture since photolithography and etching processes an entire wafer at one time. These are much different manufacturing constraints than are common in the large-scale world so it is not surprising the optimal machine design may also be different.

2.4 Bearings and Rotor Dynamics

The mechanical design of gas turbine engines is dominated by the bearings and rotor dynamics considerations of high-speed rotating machinery. Micromachines are no different in this regard. As in all high-speed rotating machinery, the basic mechanical architecture of the device must be laid out so as to avoid rotor dynamic problems. The high peripheral speeds required by the fluid and electromechanics lead to rotor-bearing designs which are supercritical (operate above the natural resonant frequency of the rotor system), just as they often are in conventional gas turbines.

Key design requirements imposed by the rotor dynamics are that mechanical critical (resonant) frequencies lie outside the steady-state operating envelope, and that any critical frequencies that must be traversed during acceleration are of sufficiently low amplitude to avoid rotor rubbing or unacceptable vibrations. The bearings play an important role in the rotor dynamics since their location and dynamical properties (stiffness and damping) are a major determinant of the rotor dynamics. The bearings in turn must support the rotor against all radial and axial loads seen in service. In addition to the rotor dynamic forces, the bearing loads under normal operation include all the net pressure and electrical forces acting on the rotor as well as the weight of the rotor times the external accelerations imposed on the device. For aircraft engines, this is usually chosen as 9 g, but a small device dropped on a hard floor from 2 m experiences considerably larger peak accelerations. An additional requirement for portable equipment is that the rotor support be independent of device orientation. The bearing technology chosen must be compatible with the high temperatures in a gas turbine engine (or be protected within cooled compartments) and be compatible with the fabrication processes.

Early MEMS rotating machines were mainly microelectric motors or gear trains turning at significantly lower speeds and for shorter times than are of interest here, so these made do with dry friction bearings operating for limited periods. The higher speeds and longer lives desired for microheat engines require low-friction bearings. The very small size of these devices precludes the incorporation of commercially available rolling contact bearings. A microfabricated bearing solution is needed. Both electromagnetic and air bearings have been considered for this application.

Air bearings support their load on thin layers of pressurized gas. The simplest journal bearing is a cylindrical rotor within a close-fitting circular journal. This plane cylindrical geometry was the first approach adopted since it seemed the easiest to microfabricate. Gas bearings of this type can be categorized into two general classes which have differing load capacities and dynamical characteristics. When the gas pressure is supplied from an external source and the bearing support forces are not a first-order function of speed, the bearing is termed *hydrostatic*. When the bearing support forces are derived from the motion of the rotor, then the design is *hydrodynamic*. Since the MEMS gas turbines include air compressors, both approaches are applicable. Both can readily support the loads of machines in this size range and can be used at very high temperatures. The two types of bearings have differing load and dynamic characteristics. In hydrodynamic bearings, the load capacity increases with the speed since the film pressure supporting the rotor is generated by the rotor motion. This can be true for a hydrostatic bearing as well if the film pressure is increased with increasing rotor speed, for example, if the pressure is derived from an integral compressor as is the case with an engine or motor driven compressor.

The relevant physical parameters determining the bearing behavior are the length-to-diameter ratio (L/D); the journal gap-to-length ratio (g/L); and nondimensional forms of the peripheral Mach number of the rotor (a measure of compressibility), the Reynolds number, and the mass of the rotor. For a bearing supported on a hydrodynamic film, the load bearing capability scales inversely with $(g/D)^5$ which tend to dominate the design considerations [4].

The dynamical behavior of the rotor is of first-order concern because the high rotational speeds needed for high-power density by the turbo and electrical machinery require the rotor to operate at rotational frequencies several times the lowest radial resonant frequency of the bearing/rotor system. This complex issue is discussed in Chapter 6.

A rotor must be supported against axial as well as radial loads and so requires thrust bearings in addition to the radial bearings discussed above. Both hydrostatic and hydrodynamic approaches have been demonstrated. In either approach, the bearing must support the axial loads and remain stable. The devices described in this book have been designed for sub-critical thrust bearing operation so that bearing behavior traversing the critical frequency is not an issue.

Hydrostatic thrust bearings meter external air through supply orifices onto the bearing surface. The 400-μm-diameter thrust pad at the rotor center of the 4-mm-diameter turbine in Fig. 2.2 rotates relative to a stationary thrust bearing surface of similar diameter. The stationary bearing surface is perforated with a circular array of 12-μm-diameter nozzle orifices fed from a plenum which supplies the gas lubricating film between the bearing surfaces.

Because the area of the thrust bearings is only a small fraction of the disk area and because the disks may have significant net axial pressure and/or electromechanical forces on them, in many cases the thrust bearing do not have sufficient capacity to support the total axial load on the rotor. This is often the gas for large gas turbines as well. In these cases, an area of the disk is isolated and pressured so as to support the net steady axial load so that the thrust bearings can support the dynamics loads.

Bearings and rotor dynamics are discussed in more detail in Chapter 6.

2.5 Turbomachinery Fluid Mechanics

The turbomachine designs considered to date for MEMS engine applications have been centrifugal since this geometry is readily compatible with manufacturing techniques involving planar lithography. (It is also possible to manufacture single axial flow stages by using intrinsic stresses generated in the manufacturing process to warp what otherwise would be planar paddles into twisted blades, but such techniques have not been pursued for high-speed turbomachinery). In most ways, the fluid mechanics of microturbomachinery are similar to that of large-scale machines. For example, high-tip rotational speeds are needed to achieve high pressure ratios per stage. Micromachines are different in two significant ways: smaller Reynolds numbers (increased viscous forces in the fluid) and, currently, 2-D, prismatic geometry limitations. The low Reynolds numbers, 10^3–10^5, are simply a reflection of the small size, and place the designs in the laminar or transitional range. These values are low enough that it is difficult to diffuse the flow, either in a rotor or a stator, without separation. This implies that either most of the stage work must come from the centrifugal pressure change or that some separation must be tolerated. The design challenges introduced by the low Reynolds numbers are exacerbated by geometric restrictions imposed by current microfabrication technology. In particular, the fabrication constraint of constant passage height is a problem in these high-speed designs. High work on the fluid means large fluid density changes. In conventional centrifugal turbomachinery, density change is accommodated in compressors by contracting or in turbines by expanding the height of the flow path. However, conventional microfabrication technology is not amenable to tapering passage heights, so all devices built to date have a constant span. How these design challenges manifest themselves are somewhat different in compressors and turbines.

The engine cycle demands pressure ratios of 2–4, the higher the better. This implies that transonic tip Mach numbers and therefore rotor tip peripheral speeds in the 300–500 m/s range are needed. In simplified terms, the spinning rotor adds tangential kinetic energy to the fluid which the stator then slows, converting the energy into static pressure rise, a process known as diffusion. The flow wise distance over which this diffusion can occur reduces as the characteristic length scale is reduced. Excessively short diffusion passages result in loss of aerodynamic performance-excessive entropy generation, reduced flow per unit area. In conventional size machines the flow path contracts to control diffusion. Since this was not possible with established micromachining technology, the approach taken was to control diffusion in blade and vane passages by tailoring the airfoil thickness rather than the passage height as in the compressor in Fig. 2.5 [5, 6]. It should be noted that although the geometry is 2-D, the fluid flow is not.

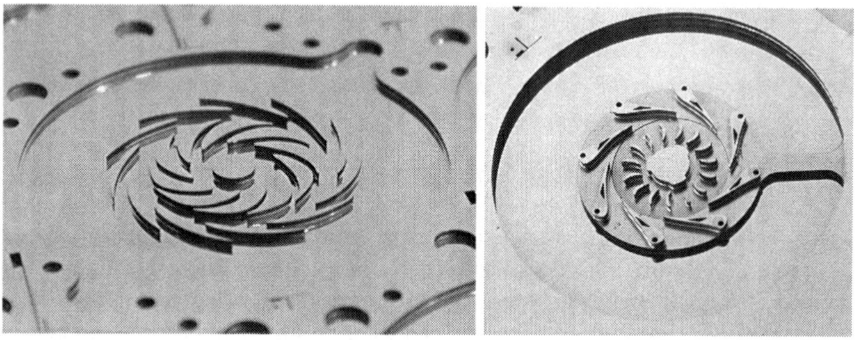

Fig. 2.5 Engine turbomachinery designed for 1.2 Mrpm. *Left* – 8 mm rotor diameter centrifugal compressor designed for 500 m/s tip speed. *Right* – the compressor driver, a radial inflow 6 mm rotor diameter engine turbine surrounded by an annular combustor volume

Tight clearances are usually highly desirable for turbomachinery aerodynamics in general to reduce leakage losses, but are a two-edged sword for the thick-bladed designs discussed above. Small tip clearance reduces leakage flow and its associated losses, but increases drag for designs in which the blade tip is at least as wide as the passage. The dimension of one microcompressor blade tested was a chord (flow-wise length) of about 1,000 μm and a span of 225 μm (Fig. 2.5 left). Thus the design tip clearance of 2 μm (set to avoid blade tip rubs) represents 0.2% of chord and 1% of span, percentage values similar to those of conventional large machines.

While the aerodynamic design of a microfabricated, centimeter-diameter radial inflow turbine shares many of the design challenges of a similar scale compressor, such as a constant airfoil span manufacturing constraint, the emphasis is different. Diffusion within the blade passages is not the dominant issue it is with the compressor, so the thick blade shapes are not attractive. The Reynolds numbers are lower, however, given increased viscosity of the high-temperature combustor exit fluid. A radial inflow engine turbine is shown in Fig. 2.5 right, surrounded by an annual combustor. The outer, stationary row of 8 turbine airfoils, the nozzle guide vanes, operate at a Re of 1,000–2,000 for millimeter-chord airfoils, 10^3 lower than large turbines.

Turbomachinery aerodynamics is discussed in detail in Chapter 7.

2.6 Structures and Materials

Structural considerations for the design of an MEMS gas turbine are in many ways similar to those of conventional engines. The design space is defined by the requirements of the thermodynamics (which require high stress and high temperatures), the properties of the materials, and the manufacturing capabilities. The material properties, in turn, are very much dependent on their processing. This section reviews materials selection, structural design features, high-temperature structures, analysis of such microstructures, and packaging (installation) technology.

2.6.1 Materials

Materials for gas turbine engines must exhibit high-specific strength (strength/ density) at high temperatures. High-temperature operation also requires creep and oxidation resistance. Other properties of interest include fracture toughness, modulus, and resistance to thermal shock. Table 2.1 compares the properties of metallic alloys used in gas turbines with those of single crystal silicon and silicon carbide.

Table 2.1 Comparison of gas turbine material Properties

	Ni-based super alloys	Titanium alloys	Micro SiC	Micro silicon
Centrifugal stress $\left[\sqrt{\sigma_f/\rho}\right]$ (m/s)	330	420	670	1,000
Thermal stress $\left[\alpha E/\sigma_{f/y}\right]$	2.7×10^{-3}	1.2×10^{-3}	1.1×10^{-3}	0.9×10^{-3}
Stiffness $\left[E/\rho\right]$ (MPa/Kg m^{-3})	~26	~25	~95	~70
Max temp (°C) (life limit)	~1000 (creep)	300 (strength)	~1500 (oxidation)	~600 (creep)

σ is fracture strength, ρ is density, α is coefficient of thermal expansion, E is modulus.

MEMS processing technologies are much more mature for silicon than for other materials so it is the first material an MEMS engineer considers. In terms of strength at temperature, single-crystal Si is the equal of common nickel alloys and, because it has only 1/3 the density, its specific strength is much higher, as illustrated in Fig. 2.6. It is quite oxidation-resistant and has thermal conductivity approaching that of copper, so it is resistant to thermal shock. On these grounds, it is an attractive material for gas turbine engines. However, at temperatures below about 900 K, Si is brittle, so usable strength is very much a function of the details of the processing. Structural life must be assessed with statistical methods.

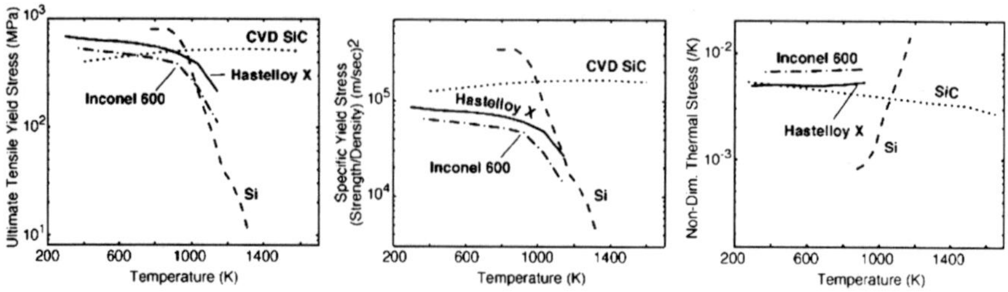

Fig. 2.6 Material properties important to gas turbine structures comparing nickel superalloys, silicon, and SiC

2.6.2 Structural Design Considerations

Structural design of an MEMS gas turbine has many of the same considerations as the design of large machines: basic engine layout is set by rotor dynamic considerations, centrifugal stress is the primary rotor load, stress concentrations must be avoided, and hot section life is creep- and oxidation-limited. Some large engine concerns do not exist at micron scale. For example, the microrotors are very stiff so that bending is not a concern and thermal stress from temperature gradients is not important at these sizes [7].

Large engines use standard tubing fittings and electrical connectors to pass fluids and electrical signals to the outside world. These do not exist at microscale. Most computer chips and MEMS devices do not require fluid connections and those that do operate not much above room temperature. For these applications, there are a variety of adhesives and polymer systems. A microgas turbine can have a surface temperature above 700 K and require fluid connection at pressures of 10 atm or more, so that high strength, high-temperature packaging approaches are needed. One approach which has proven successful is an adaptation of the hermetic package technology used for military electronics. This joins Kovar (a nickel alloy) tubing to silicon using glass as the bonding agent. The joining is done in a furnace above 1,100 K to melt the glass. Such joints can withstand pressures above 200 atm [8]. Another approach is to anodically bond glass layers to move the external fluid connections away from hot Si so that elastomer joints and O-rings can be used.

Materials, structural design and packaging are discussed in more detail in Chapter 3.

2.7 Combustion

The primary design requirements for gas turbine combustors include large temperature rise, high efficiency, low pressure drop, structural integrity, ignition, stability, and low emissions. These requirements are no different for a microcombustor which may flow less than 1 g/s of air than for a 100 kg/s large machine, but the implementation required to achieve them can be. Scaling considerations result in the power density of a microcombustor exceeding that of a large engine. However, the combustor volume relative to the rest of the microengine is much larger, by a factor of 40, than that of a large engine. The reasons for this scaling can be understood in reference to the basics of combustion science [9].

Combustion requires the mixing of fuel and air followed by chemical reaction. The time required to complete these processes is generally referred to as the required combustion residence time and effectively sets the minimum volume of the combustor for a given mass flow. The mixing time can scale with device size but the chemical reaction times do not. In a large engine, mixing may account for more than 90% of combustor residence time. One difference between large and microscale machines is the increased surface area-to-volume ratio at small sizes. This offers more area for catalysts; it also implies that microcombustors have proportionately larger heat losses. While combustor heat loss is negligible for large-scale engines, it is a dominant design factor at microscale since it can reduce the combustor efficiency and lower the reaction temperature. This narrows the flammability limits and decreases the kinetic rates. As an example, Fig. 2.7 [10] illustrates the viable design space for an H_2-fueled, 0.07 cc microcombustor as a function of equivalence ratio (fuel-to-air ratio relative to stoichiometric), the heat lost to the walls (Q), as constrained by flame stability, structure limits, and cycle requirement considerations. The design space shown permits a trade between heat loss and stoichiometry, which is especially important when burning hydrocarbons with narrower stoichiometry bounds.

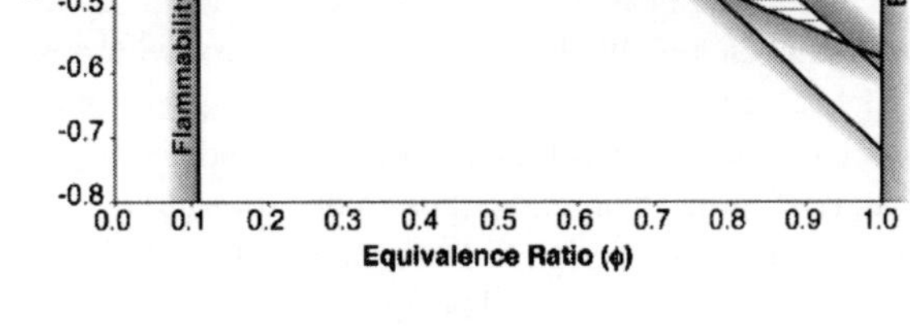

Fig. 2.7 Design space for a Silicon H_2 microcombustor

The design details are dependent on the fuel chosen. The design approach first taken was to separate the fuel–air mixing from the chemical reaction. This is accomplished by premixing the fuel with the compressor discharge air upstream of the combustor flame holders. This permits a reduction of the combustor residence time required by a factor of about 10 from the usual 5–10 ms. The disadvantage of this approach is a susceptibility to flashback from the combustor into the premix zone, which must be avoided. To expedite the demonstration of a microgas turbine engine, hydrogen was chosen as the initial fuel because of its wide flammability limits and fast reaction time. Hydrogen is particularly attractive because it will burn at equivalence ratios, ϕ, as low as 0.3 which yields adiabatic combustion temperatures below 1,500 K, facilitating the realization of simple premixed designs.

In the Si micromachined geometry of Fig. 2.3, to reduce heat losses through the walls and therefore to increase combustor efficiency, the combustor inlet air wraps around the outside of the 0.2 cc combustor before entering through flame holders in a reversed flow configuration. The Si liner in this case is conduction cooled and so can survive in a 2,000 K gas temperature environment. In this premixed approach, fuel is injected near the inlet of the upstream duct to allow time for fuel–air mixing without requiring additional combustor volume. This design takes advantage of microfabrication's ability to produce similar geometric features simultaneously, using 90 fuel injection ports, each 120 μm in diameter, to promote uniform fuel–air mixing. A hot wire loop provided ignition [11].

Combustion efficiencies approaching 100% with pressure drops of about 2–5% have been reported with H_2. These combustors have been operated at exit temperatures above 1,800 K. This is more than adequate combustor performance for a viable gas turbine engine.

Hydrocarbon fuels such as methane and propane have reaction rates only about 20% of those of H_2, requiring larger combustor volumes for the same heat release. They also must react closer to stoichiometric and therefore at higher temperatures, above 2,000 K. For gas-phase (homogeneous) combustion designs, this requires a multizone burner (stoichiometric zone followed by a dilution region) as used on most large gas turbines. Alternatively, heterogeneous reactions on the surface of a catalyst can widen the flammability limits and so reduce the combustion temperature. Both approaches have been demonstrated at microscale. Ethylene and propane have been burned in the H_2 combustors described above. The combustion efficiency with ethylene approached 90% while that for propane was closer to 60%.

Catalytic microcombustors have been produced by filling the combustor volume of the above geometries with platinum-coated nickel foam. For propane, the catalyst increased the heat release in the same volume by a factor of 4–5 compared to the propane–air results discussed above. Pressure drops through the foam are only 1–2% [12]. Presumably catalytic combustor performance can be improved by a better choice of catalyst (platinum was selected for H_2) and a geometry optimized for catalytic rather than gas-phase combustion.

A combustor designed for a standard liquid jet fuel, JP-8 (similar to kerosene), was tested by Peck [13]. The approach followed here was to pre-vaporize the liquid fuel using waste heat from the combustor, and then follow a prescription for gas-phase combustion as described above. Care was needed in the design of the vaporizer to avoid generating solid reside in the channels, the so-called coking common to many gas turbine applications.

Overall, experiments and calculations to date indicate that high-efficiency combustion systems can be engineered at microscale and achieve the heat release rate and efficiency needed for very small gas turbine engines.

2.8 Engine Controls and Accessories

All gas turbine engines require control systems to insure safe operation. Typically, the control system adjusts the fuel flow to deliver the requested power, and monitors engine operation to avoid unsafe conditions such as over-speed, over-temperature, or surge. Such a control system consists of sensors (speed, pressure, temperature, etc.), a feedback controller with a suitable set of control laws (now implemented in a digital computer), actuators such as a fuel control valve and compressor stability devices as needed (bleed valves, fences, variable stators). Engine accessories include an ignition system, fuel pump, lubrication system, and starter. All of this functionality is needed for a millimeter-scale MEMS gas turbine and all must fit within a microchip if the accessories are not to dwarf the engine. Following in the tradition of large engine development, the controls and accessories have received less attention to date than the major engine subsystems such as the compressor, turbine, and combustor but they are no less important to the ultimate success of the concept.

2.8.1 Engine Controls

The simplest engine control would consist of a single sensor feeding a digital controller which commands the fuel flow rate valves. The functional requirements for the valving and the sensors stem from the engine dynamics as represented in the control laws and from the engine environment. There is a very rich literature on MEMS sensors and valves, literally thousands of papers, and some units are commercially available. As for large jet engines, however, the combination of harsh environment, high frequency response, high accuracy, and high reliability means that sensors and actuators for MEMS-scale gas turbines must be specifically engineered for that environment.

2.8.2 Sensors

Large-scale gas turbines use compressor pressure ratio and/or rpm as the primary input to the fuel control system. Sensor selection for an MEMS engine is a trade among observability of the state (dynamical information represented by the sensed quantity), response time, difficulty of fabrication, and environmental compatibility. Sensing is complicated by the high rotational frequencies (1,000,000 rpm) and high temperatures (600 K at the compressor discharge) in a very small engine. In principle, sensors can be fabricated integrally with an engine or located remotely. A sensor remote from the gas path suffers reduced frequency response, which is already a challenge at the MEMS engine scale. In addition, it is difficult to be very remote from the gas path in a 0.5-mm-long engine. Integral sensors must withstand the high temperature of the gas path. Even for sensors fabricated in the cold sections, they must be capable of withstanding a wafer-processing environment that exceeds 1,000–1,400 K. This effectively precludes the use of low-temperature materials such as polymers and most metals in the device design. Integral sensors have several advantages. They can be very small and thus have high frequency response, and many can be fabricated in parallel for low-cost redundancy.

One integral solution was developed by Tang [14] who adapted a hot-film-type sensor to this application. Designed for placement on the wall above a rotor blade tip, this 50 μm square sensor is a heated, serpentine, polysilicon resistor positioned over a trench for thermal isolation (Fig. 2.8). The sensor and its leadouts are all polysilicon which is selectively doped to adjust its resistivity. Polysilicon has the advantage that it is compatible with most semiconductor fabrication techniques and can withstand high temperatures. Simulations confirmed by shock tube testing showed this approach to have sufficient sensitivity and frequency response to respond to the flow perturbations. With a total thickness of less than 1 μm, such sensors could be fabricated on the casing above the compressor blade tips. This type of resistor has also been shown to be usable as an igniter.

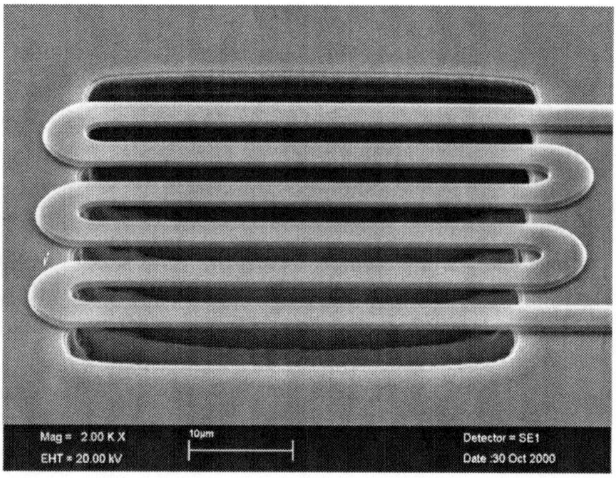

Fig. 2.8 A 50 μm sq polysilicon resistive sensor for engine temperature and rpm

2.8.3 Fuel Control Valves

Very small engines need very small fuel control valves. If integrated within the engine, the valve design must then be fully compatible with the fabrication and operating conditions of the gas turbine. This choice strongly constrains the valve design space. For example, the high processing and operating temperatures prohibit the use of polymers, so a hard valve seat must be used. The principal design requirements are flow rate, pressure, frequency response, very low power consumption and leakage, and high-temperature capability.

Yang et al. [15] developed MEMS fuel-metering valves for gaseous hydrocarbon fuels such as propane. The design is a simple silicon spring-mounted plunger opened by electrostatic forces and closed by a combination of the spring and fluid pressure forces (Fig. 2.9). The electrostatic approach has the advantages that very little power is needed to open a valve (40 nW) and the electrical materials (polysilicon) are compatible with high-temperature semiconductor fabrication technology. The 2-mm-square valve has a 1,000-μm-diameter plunger which rises 3 μm off the seat when actuated. The valve opens against 10 atm pressure and, when open, flows 35 sccm of N_2 at a pressure drop of 0.5 atm. Frequency response is several hundred Hertz. This design is on–off. Electric actuation-pressure scaling laws favor multiple small valves so the intent is to use a parallel array of 20 on–off valves, each with a capacity of 5% of the maximum fuel flow, to meter the fuel. All 20 valves would consume less than 1 mW total power and operate at temperatures approaching 1,000 K so they can be embedded in an engine chip. Many other arrangements are possible, such as logarithmic spacing of the valve orifice sizes to give finer fuel flow control.

Fig. 2.9 A 2 mm sq, high-temperature fuel control valve plunger on Si beam springs

2.8.4 Starter-Generator

Microelectrical machinery is required for power generation and electric starting, if desired. There is an extensive literature on microelectric motors, which is not reviewed here, but little work on generators. The requirements for the devices of interest here differ from previous work in that the power densities needed are at least two orders of magnitude greater than that of conventional-size and previous micromachines. Also, the thermal environment is much harsher. Integrating the electric machine within the engine offers the advantage of mechanical simplicity in that no additional bearings or structures are required over that needed for the fluid machinery. There is also a supply of cooling air available.

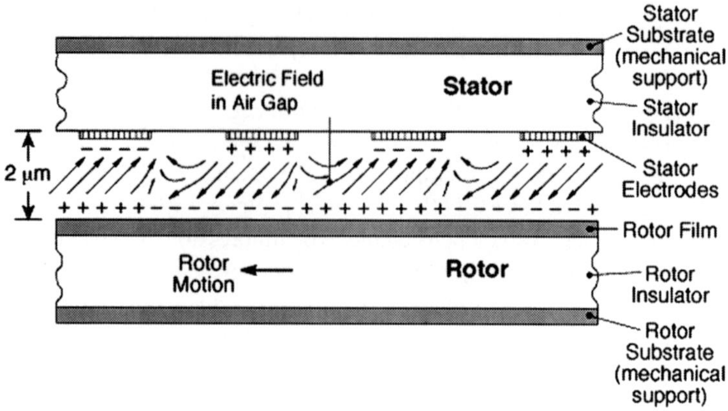

Fig. 2.10 Fields and charges in a microscale electric induction motor-generator

The operation of an electric induction machine can be understood with reference to Fig. 2.10 [16]. The machine consists of two components, a rotor and a stator. The rotor is comprised of a 5–20 μm thick insulator covered with a few microns of a poor conductor. The stator consists of a set of conductive radial electrodes supported by an insulator. A traveling electric potential is imposed on the stator electrodes with the aid of external electronics. The resulting rotating electric field then induces an image charge on the rotor. Depending on the relative phase between the motion of rotor charges (set by the rotor mechanical speed) and that of the stator field (set by the external electronics), the machine will operate as a motor, generator, or brake. Torque increases with the square of the electric field strength and frequency. The maximum electric field strength that an air gap can maintain without breakdown is a function of the gap dimension. In air, the breakdown field is a maximum at a gap of a few microns so that micromachines can potentially realize higher power density than large machines of the same design. Frequency is constrained by external electronics design and by fabrication constraints on the stator electrode geometry. Current technology is limited to about 300 V and 1–2 MHz.

Both electric and magnetic machine designs can be considered and, to first order, both approaches can yield about equivalent power densities. Since the magnetic machines are material property-limited at high temperature and because of the challenges of microfabricating magnetic materials (which are not compatible with standard semiconductor manufacturing techniques), electric designs were explored first. Power density scales with electric field strength squared, frequency, and rotational speed. The micromachinery of interest here operates at peripheral velocities 1–2 orders of magnitude higher than previously reported micromotors, and so yields concomitantly more power. Electric machines may be configured in many ways. Here an induction design was chosen since it requires neither electrical contact with the rotor nor knowledge of the rotor position.

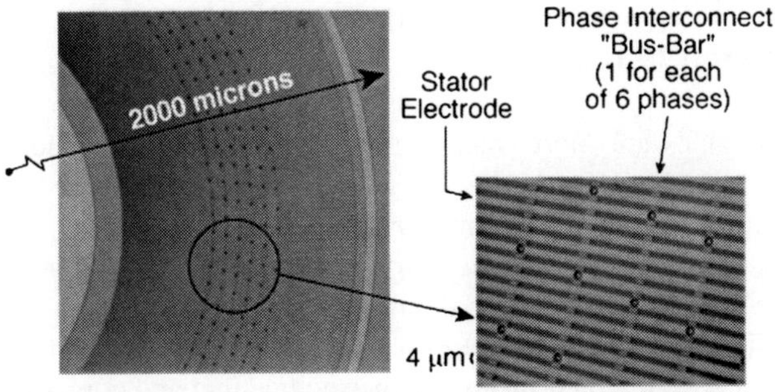

Fig. 2.11 131 pole, six-phase, 4 mm rotor diameter electric induction stator

This is consistent with a 6-mm-diameter machine producing about 10 W at a 3 μm air gap. A 4-mm-diameter, six-phase, 131-pole (786 electrodes, each 4 μm wide) stator for such a machine is shown in Fig. 2.11 [17]. Note that such an electric motor-generator occupies less than 20 μm thickness at the surface of the rotor and stator (mainly the insulator thickness). Thus the power density of this machine (excluding the external electronics) is many times that of a conventional magnetic motor-generator, on the order of 100 MW/m^3. Fréchette et al. [18] have reported a similar design run as an electric motor. The torque produced by these devices has agreed with theoretical predictions but high-power operation has yet to be reported.

To maximize power output, induction machines such as these require the spacing between the rotor and stator to be on the order of the stator pitch. The electrical torque produced scales with the square of the rotor–stator spacing, a few microns in this case. However, in these high-speed machines, the rotor periphery is at sonic velocity so the viscous drag in a gap of only 2–3 μ is extremely large. Indeed, this drag is the major loss mechanism for such an electrical machine. Thus, there is a basic design tradeoff for the electric motor-generator between power density and efficiency. Drag still makes up about half the total loss and limits the efficiency (shaft to net electrical) of these designs to 40–50%.

A magnetic induction machine has many fewer poles so that the optimum rotor–stator spacing is much larger (30–50 μm) and the drag concomitantly lower. Koser and Lang [19] designed such a machine based on the microplating technology developed by Park et al. [20]. The efficiency of this magnetic generator is calculated to approach 60%. It has the additional advantage that its external low frequency, low voltage electronics are easier to engineer and more compact than the high frequency, high voltage electronics of electric machines. Current stator materials are limited to only 500 K, however. Advanced materials may increase the operating temperature to 800 K. The rotor requires several hundred microns of iron for the magnetic path, which presents structural design challenges at 1,000,000 rpm. Magnetic machines will require careful thermal management when embedded within an MEMS gas turbine engine.

Chapters 5 and 8 discuss both electric and magnetic machines in considerable detail.

2.9 Engine Design Trades, Component Integration, and Design Evolution

Gas turbine engines are more than a set of components bolted together. Rather, a successful gas turbine is a highly integrated system engineered to meet specific requirements, often with artful compromises between conflicting demands of the fluid, thermo, structural, and manufacturing engineering. An MEMS gas turbine is no different in this respect. The two dominant design considerations are the fabrication complexity and the thermodynamic cycle requirements. The principal challenge is to arrive at a design which meets the thermodynamic and component functional requirements while staying within the realm of realizable micromachining technology.

For any gas turbine, maximizing net engine output power includes maximizing both power per unit mass flow (specific power) and mass flow. Specific power is sensitive to component efficiency and pressure ratio, especially at the low pressure ratios under consideration here, 2:1–4:1. High pressure ratio in a single-stage centrifugal machine implies high wheel speeds; 500 m/s peripheral speed was chosen as the maximum compatible with geometrically simple Si construction. This also requires bearings and rotor dynamics capable of such high-speed operation. High efficiency implies optimal airfoil design and tight clearances. Tight clearances, in turn, imply high manufacturing precision, careful design for centrifugal and thermal growth, and robustness to the occasional high-speed rub. The other key to high-specific power is high turbine inlet temperature. This requires high-temperature materials for the turbine or cooling or both. High mass flows require high through-flow Mach numbers and large flow areas. With these requirements in mind, let us examine how these trades influence the design of an MEMS gas turbine. Specifically, the next section discusses why the demo engine design in Figs. 2.3 and 2.4 is so configured. The principal geometric constraint on the design is that imposed by the current state-of-the-art in precision (micron-level accuracy) etch-depth capability of several hundred microns.

The engine design of Fig. 2.3 requires six wafers: two form the rotor (and the annular combustor volume), and two are needed on each side of the rotor for the hydrostatic thrust bearings and their associated plumbing. Replacing the hydrostatic bearings with hydrodynamic ones would eliminate one wafer. Adding a generator mounted on a rotating compressor tip shroud would add one wafer. Adding control valves would add an additional wafer above the compressor. Such a complete, self-contained (other than electronics) gas turbine generator could thus be built with seven or eight wafers, not including external electronics.

In a centrifugal compressor and turbine of geometry similar to Fig. 2.2, turning at 400–500 m/s tip speed, turbomachine design is constrained to a blade span of 400–500 μm by fabrication technology and bending stress at the blade roots. Given the blade height, air flow increases with rotor diameter. Rotor diameter is constrained by the number of engines required per wafer and by etching uniformity (if a design goal is to avoid the necessity of dynamically balancing the rotor). Etch uniformity scales inversely with rotor area and blade height (i.e., mass flow is constrained by etch uniformity technology). Another factor influencing uniformity is rotor/stator airfoil count. Etch rate is a function of the local geometry, so for best circumferential uniformity there should be an even multiple of rotor and stator airfoils. Such a design choice would be highly undesirable for airfoil long-term vibrational life. If the rotor is fabricated from more than one wafer and etched with multiple masks, then mask and wafer alignment is an issue which favors the fewest possible wafers and masks. Rotor diameters up to 20–40 mm with 400–1,000 μm blade spans, fabricated from 1 to 4 wafers, are consistent with current capabilities.

Bearing placement is another first-order concern. The bearings must support the load and remain in their stable operating regime. The load capability of air bearings scales with the bearing area. The bearing length is constrained by current etching technology to 300–500 μm (and widths of 10–25 μm), so load capacity scales with bearing diameter. For hydrostatic journal bearings of constant length, the optimum width of the bearing for maximum stable operating speed scales inversely with diameter. So as the bearing diameter is reduced, the air required to operate the bearing goes up (since the flow per unit area grows faster than the area shrinks) and the load capacity goes down. This implies that bearing diameter should be maximized, i.e., on the rotor diameter. Placing the bearing here also eliminates the need for a seal on the rotor, since in this case the leakage air is the air bearing fluid, but this approach increases bearing drag.

The cycle needs turbine inlet temperatures of at least 1,400–1,600 K to meet the application goals. These relatively high temperatures are needed due to the relatively high losses in the components and the secondary systems such as bearings. For example, the shaft mechanical losses (mainly bearing drag) are 4–8% of the shaft work for the design in Figs. 2.3 and 2.4, ten times that of large jet engines. These temperatures are above the 900–950 K maximum operating temperature of a single-crystal Si rotor. In the engine design of Fig. 2.3, the turbine rotor is conductively cooled through the compressor disk to the compressor air, which has a quite deleterious effect on the cycle performance and thus on the net engine output. Improving performance requires thermally isolating the turbine from the compressor which can be accomplished by hollowing the shaft between them. When this is done, the temperature capability of the turbine rotor must be increased. As for large engines, some combination of increased material capability and improved cooling is needed. Improved material capability can be provided by reinforcing the Si turbine disk with SiC, with the temperature capability increasing with fraction of SiC. So-called "film cooling" (it is really a film insulation) is the most effective cooling technique here since, in this case, the heat is prevented from entering the solid.

As with large engines, improving performance at the MEMS scale will require some combination of more complex designs, better materials, and improved manufacturing technology. For example, cycle pressure ratios much over 4:1–5:1 will require multiple spools (probably not concentric). Higher turbine inlet temperatures will require improved cooling and fabrication technology for such materials as silicon carbide, silicon nitride, and sapphire. Better compression system performance will require more attention to thermal isolation. These are all physically possible but will require investment in both the disciplinary and microfabrication technologies.

2.9.1 Electric Micro Motor-Compressor and Turbine-Generator

While much of the underlying motivation for the research described in this book was the realization of an MEMS gas turbine generator, the components of the gas turbine were developed separately and in various combinations form other useful MEMS devices. Two in particular are an electric motor driven air compressor and an air-turbine driven electric generator. The motor driven air compressor was embodied in a single thin disk, 4 mm diameter, 500 μm thick rotor contained within 5 bonded silicon layers (Fig. 2.12). The compressor is on one side of the rotor and was 4 mm diameter version of that shown in Fig. 2.5. On the opposite side of the rotor disk, an electric induction motor similar to that shown in Figs. 2.11–2.12 was fabricated. The rotor was supported on hydrostatic gas bearings similar to those described above for the gas turbine. On either side of the rotor disk layer, two additional Si layers were required for the air bearings and associated plumbing and for the electric motor stator and its lead outs. The design, construction, and testing of this device is described by Frechette et al. [21].

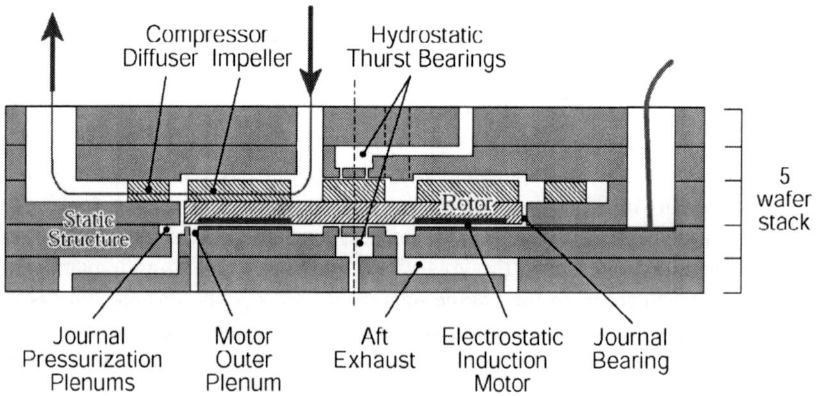

Fig. 2.12 Cross-section schematic of the five-wafer level motor-driven compressor

By changing the external electric circuit, the induction motor can be operated as an induction generator. With the compressor geometry replaced with that of the radial inflow turbine of Fig. 2.2, the motor-compressor became a turbine generator. This device is described in detail in Chapter 8.

2.10 Summary

This chapter has briefly described considerations underlying the design of the micro-components and disciplines described in the following chapters. The primary objective has been the engineering needed for an MEMS gas turbine engine producing tens of watts of electrical energy or tens of grams of thrust. Toward this goal, MEMS technologies for high-speed fluid mechanics and turbomachinery, rotor dynamics and bearings, materials and packaging, combustion, electric and magnetic rotating machinery, and microfabrication technologies have been developed with a consistent set of design and fabrication objectives and constraints. These topics are discussed at length in the following chapters.

References

1. Epstein, A.H., and Senturia, S.D., 1997, "Macro Power from Micro Machinery", *Science*, **276**, 1211.
2. Jacobson, S.A., 1998, "Aerothermal Challenges in the Design of a Microfabricated Gas Turbine Engine", AIAA 98-2545, 29th AIAA Fluid Dynamics Conference, Albuquerque, NM.
3. Waits, C.M., Modafe, A., and Ghodssi, R., 2003, "Investigation of Gray-Scale Technology for Large Area 3D Silicon MEMS Structures", *J. Micromech. Microeng.*, **13**, 170–177 PII: S0960-1317(03)52256-2
4. Breuer, K., Ehrich, F., Fréchette, L., Jacobson, S., Lin, C-C, Orr, D.J., Piekos, E., Savoulides, N., and Wong, C-W, 2003, "Challenges for Lubrication in High Speed MEMS," in *Nanotribology*, Hsu and Ying (eds), Kluwer Academic Press, New York.
5. Mehra, A., Jacobson, S.A., Tan, C.S., and Epstein, A.H., 1998, "Aerodynamic Design Considerations for the Turbomachinery of a Micro Gas Turbine Engine", presented at the 25th National and 1st International Conference on Fluid Mechanics and Power, New Delhi, India.
6. Mehra, A., 1997, "Computational Investigation and Design of Low Reynolds Number Micro-Turbomachinery," M.S. Thesis, MIT Department of Aeronautics and Astronautics .
7. Spearing, S.M., and Chen, K.S., 1997, "Micro-Gas Turbine Engine Materials and Structures", presented at 21st Annual Cocoa Beach Conference and Exposition on Composite, Advanced Ceramics, Materials and Structures .
8. Harrison, T.S., London, A.P., Spearing, S.M., 2001, "High Temperature, High Pressure Fluid Connections for Power Micro-Systems," Paper EE6.5, *Mat. Res. Soc. Symp. Proc.*, 654.
9. Waitz, I.A., Gautam, G., and Tzeng, Y.-S., 1998, "Combustors for Micro-Gas Turbine Engines," *ASME J. Fluids Eng.*, **120**, pp. 109–117.
10. Spadaccini, C.M., Mehra, A., Lee, J., Lukachko, S., Zhang, X., and Waitz, I.A., 2002, "High Power Density Silicon Combustion Systems for Micro Gas Turbine Engines," GT-2002-30082, ASME International Gas Turbine Institute TURBO EXPO '02, Amsterdam, The Netherlands.
11. Zhang, X., Mehra, A., Ayón, A.A., and Waitz, I.A., 2002, "Development of Polysilicon Igniters and Temperature Sensors for a Micro Gas Turbine Engine," *Technical Digest of MEMS 2002*, the 15th IEEE International Conference on Micro Electro Mechanical Systems, Las Vegas, NV, pp. 280–283.
12. Spadaccini, C.M., Zhang, X., Cadou, C.P., Miki, N., and Waitz, I.A., 2002, "Development of a Catalytic Silicon Micro-Combustor for Hydrocarbon-Fueled Power MEMS," *Technical Digest of MEMS 2002*, The 15th IEEE International Conference on Micro Electro Mechanical Systems, Las Vegas, NV, pp. 228–231.
13. Peck, J., 2008, "Development of a Liquid-Fuel Micro-Combustor", Ph.D. Thesis, MIT Department of Aeronautics and Astronautics .
14. Tang, D., 2001, "Rotor Speed Microsensor for the MIT Microengine," M.S. Thesis, MIT Department of Mechanical Engineering .
15. Yang, X., Holke, A., and Schmidt, M.A., 2002, "An Electrostatic, On/Off MEMS Valve for Gas Fuel Delivery of a Microengine," Solid-State Sensor, Actuator and Microsystems Workshop, Hilton Head Island, SC.

16. Nagle, S.F., and Lang, J.H., 1999, "A Micro-Scale Electric-Induction Machine for a Micro Gas Turbine Generator," 27th Annual Meeting of the Electrostatics Society of America .
17. Ghodssi, R., Fréchette, L.G., Nagle S.F., Zhang X., Ayón A.A., Senturia S.D., and Schmidt M.A., 1999, "Thick Buried Oxide in Silicon (TBOS): An Integrated Fabrication Technology for Multi-Stack Wafer-Bonded MEMS Processes," *Proceedings of the 1999 International Conference on Solid-State Sensors and Actuators*, Sendai, Japan, pp. 1456–1459.
18. Fréchette, L.G., Nagle, S.F., Ghodssi, R., Umans, S.D., Schmidt, M.A., and Lang, J.H., 2001, "An Electrostatic Induction Micromotor Supported on Gas-Lubricated Bearings," IEEE 14th International Micro Electro Mechanical Systems Conference, MEMS 2001, Interlaken, Switzerland.
19. Koser, H., and Lang, J. H., 2000, "Modeling a High Power Density MEMS Magnetic Induction Machine," *Proceedings of the Fourth International Conference On Modeling and Simulation of Microsystems*, Hilton Head, SC, pp. 286–289.
20. Park, J.W., Cros, F., Allen, M.G., 2002, "A Sacrificial Layer Approach to Highly Laminated Magnetic Cores," MEMS 2002 IEEE International Conference, Las Vegas, NV, pp. 380–383.
21. Frechette, L.G., Jacobson, S.A., Breuer, K.S., Ehrich, F.F., Ghodssi, R., Khanna, R., Wong, C.W., Zhang, X., Schmidt, M.A., Epstein, A.H., Feb 2005, "High-Speed Microfabricated Silicon Turbomachinery and Fluid Film Bearings," *IEEE J. Micromech. Syst.*, **14** (1),141–152.

Chapter 3
Materials, Structures and Packaging

S. Mark Spearing

3.1 Introduction

Realization of the Microengine and related devices is dependent on achieving four key requirements:

1. High peak cycle temperatures (1600–2500 K)
2. High peripheral speeds (400–600 m/s)
3. Effective bearings (air bearings for the microengine)
4. Reasonable individual component efficiencies.

Requirements (1) and (2) have direct implications for the choice of materials, processes and structural design while (3) and (4) result indirectly in constraints on these choices. (1) implies the use of refractory materials. (2) implies a high strength to density ratio for the material. (3) and (4) drive the design towards the selection of materials for which established processes exist, capable of achieving very tight dimensional tolerances in order to achieve the high-speed bearing performance as well as minimizing leakage flows within the engine. The highly integrated nature of the overall design requires high-fidelity prediction of the structural performance. As for any structural design, whether at the macro- or microscale this includes prediction of deformations (which affect bearing performance and leakage flows), prediction of failure and estimations of lifetime. In addition the development of effective packaging technologies for the microengine is highly dependent on material selection and structural design. In the course of pursuing the Microengine and other power MEMS projects, several recurrent themes emerged as particular challenges in the realm of materials and structures:

1. The effects of scale: quasi-fundamental, mechanism-dependent and indirect scaling [1, 2]

S.M. Spearing (✉)
School of Engineering Sciences, University of Southampton, University Road, Southampton, SO17 1BJ, UK

J.H. Lang (ed.), *Multi-Wafer Rotating MEMS Machines*, MEMS Reference Shelf,
DOI 10.1007/978-0-387-77747-4_3, © Springer Science+Business Media, LLC 2009

2. Characterization of materials at appropriate scale, geometry [3, 4]
3. Overall structural design process in uncharted design space: constitutive laws, probabilistic design, heat transfer [5]
4. Development of μfabricatable refractory structures [6, 7]
5. Insertion of functional materials including; piezoceramics, electrolytes, catalysts [8–10]
6. High precision deep etching of high-strength structures in inherently brittle material [11, 12]
7. Creation and characterization of high-strength wafer bonds [13, 14]
8. Engineering of thick (10–100 μm) high-integrity deposited layers (SiC, SiO_2) [15, 16]
9. Packaging, including considerations of CTE matching, joining, heat transfer [17, 18].

This set of considerations provides the motivation and guidance for this chapter. The organization is as follows: Section 3.2 focuses on materials selection considerations and discusses the effect of scale. Section 3.3 presents the approach used for designing rotating parts for low-temperature operation, including details the methods used for obtaining room temperature material properties. Section 3.4 introduces the considerations for high-temperature structural design and material and process selection. Section 3.5 presents the high-temperature structural analysis. Section 3.6 describes the approach and considerations for packaging of the Microengine. Section 3.7 provides concluding remarks.

In addition to the content of this chapter the interested reader is directed to the microengine annual reports, theses and published articles that underpin the work. These are referenced throughout the chapter, where appropriate.

3.2 The Effect of Scale and the Choice of Materials

3.2.1 Effect of Scale

One of the key distinguishing features of the Microengine is its small size. It is important to understand the implications of scale and how this modifies the design process over that used for macroscale systems [19]. Key factors that emerged during the course of the Microengine design were as follows:

1. The overall concept is enabled by favourable cube–square scaling (power, thrust, scale with area, mass with volume)
2. Stress levels are scale-independent. Throughout all aspects of the design of the microengine structures, continuum mechanics was found to be an appropriate analytical tool for assessing stress levels and deformations.
3. Structural configurations are dictated by fabrication and system considerations. In particular: (a) a reduced ability to tailor through-thickness dimensions for stress uniformity and (b) an increased sensitivity to deformations, which in turn affects the dimensional tolerances.

4. The materials available for use in the Microengine are brittle. This allows for high strength at small scale. Measured strengths well in excess of 1 GPa were routinely achieved.
5. Thermal shock, a problem encountered in designing macroscale turbomachinery components in ceramic materials, is a greatly reduced concern at the microscale. The low Biot number of the Microengine structures reduces the magnitude of temperature gradients and therefore the thermal stresses.
6. Surface effects (oxidation) potentially of increased concern for operational engines; however, the rates of oxidation were found to be sufficiently slow so as not to affect the development and testing of structures.
7. Creep and yield are largely unaffected (at 100–10 mm scale), although use of single crystal silicon introduces challenges, which are discussed later in this chapter.

3.2.2 Material Selection

Material selection for the Microengine was driven largely by the need to create a working device within a limited time scale. In order to achieve this, a conscious decision was taken to minimize programmatic risk associated with introducing new materials and developing new processes. As a result the first generation, demonstration engine was largely fabricated from silicon. Nevertheless the material indices for microfabricated silicon compare quite well with those for the metal alloys used in macroscale gas turbine power plant. Table 3.1 shows the relevant performance indices for turbomachinery [2]. The square root of the ratio of fracture (failure) strength to density provides an indication of the burst speed of the rotating components, which scales the mechanical power density of the device. The ratio of the Young's modulus to the density of the materials controls the deformation of the structure under centrifugal loading and the product of the Young's modulus and the thermal expansion coefficient provides a measure of the thermal stress that arises due to temperature gradients, which has implications for the dimensional tolerances,

Table 3.1 Selected performance indices and materials relevant to macro- and micro-gas turbines

	Ni-based superalloys	Titanium alloys	Macro (Micro) ceramics	Microfabricated silicon
Centrifugal strength σ_f/ρ] (m/s)	330	420	420 (670)	1000
Thermal strength [$\alpha E/\sigma_{f/y}$]	2.7×10^{-3}	1.2×10^{-3}	2.0×10^{-3} (1.1×10^{-3})	0.9×10^{-3}
Stiffness [E/ρ] (MPa/kgm^{-3})	~26	~25	~95	~70
Maximum temperature limit (°C)	~1000 (Creep)	~300 (Strength)	~1500 (Oxidation)	~600 (Creep)

particularly of the static structures. The relatively low density of silicon, its low thermal expansion coefficient and its high strength and stiffness make it an attractive material for rotating structures. Even its relatively poor temperature capability can be accommodated to some extent with careful design. However, the opportunity for introducing more refractory ceramic materials is clear from this table.

3.3 Design for Room Temperature Strength

3.3.1 Mechanical Testing

Given that silicon is a brittle material, the design of the microturbomachinery for room temperature operation must account for the stochastic aspects of the strength. Axisymmetric flexural specimens were preferred for this purpose, in which the specimen is supported around its edge and loaded centrally [20, 3]. This arrangement avoids extraneous flaws introduced at edges during specimen preparation affecting the strength. A modified microhardness indenter was found to be appropriate for this purpose. Specimens consisted of square die, 10 mm × 10 mm which were placed over a hole of diameter 5–8 mm. A load was applied centrally and the load at fracture recorded. Data such as that shown in Table 3.2 was obtained. Weibull analysis was applied to reduce the data, and obtain the Weibull reference stress and Weibull modulus to characterize the strength distribution. The reference strength correlates closely with the measured peak-to-peak roughness of the silicon surfaces.

Table 3.2 Data from planar biaxial flexure tests

	Mechanically ground (A)	Mechanically ground (B)	KOH etched	Chemically polished	STS DRIE
Sample size	19	30	25	4	20
Peak-to-peak surface roughness (μm)	~3	~1	~0.3	~0.1	~0.3
Reference strength, σ_0 (GPa)	1.2	2.2	3.5	>4	4.6
Weibull modulus, m	2.7	3.4–4.2	7.2–12	Not measurable	3.3

A particular concern for the design was the introduction of flaws by the deep reactive ion etching process. Additional surface roughness was observed at the intersections between horizontal surfaces and vertical etched walls. Since these locations also represent stress concentrations, locally reduced strength was of particular concern. To investigate this issue, a novel test specimen was developed: The Radiused-Hub Flexure Specimen (RHFS), which consists of a central hub defined

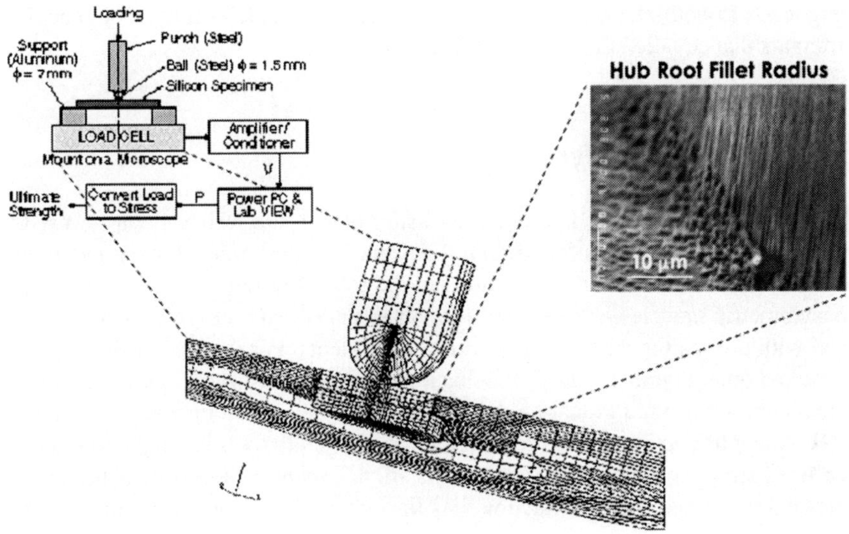

Fig. 3.1 RHFS strength test

by deep etching. The nominal stress concentration at the interface between the hub wall and the horizontal surface is calculated based on the nominal radius of the fillet and used to define a nominal local strength. A schematic of the specimen and details of a typical fillet are shown in Fig. 3.1. Data from such tests are shown in Table 3.3. The locally increased roughness significantly reduces the strength. The reference strength obtained from STS DRIE specimens drops from 4.6 GPa in Table 3.2 to 1.45 GPa in Table 3.3. However, use of a secondary, "smoothing" etch is shown to be very effective in recovering much of this strength loss. An isotropic SF_6 plasma etch recovers the local strength to 4 GPa. Such a strength recovering etch is essential if microengine rotating parts are to be able to achieve the high speeds required for power production. It is important to note that in all cases the Weibull modulus is low, indicating a large variability in the strengths. This is challenging for structural design.

The other significant materials characterization task for the room temperature performance was to assess the integrity of direct bonds between wafers. Strength-based and fracture-based tests were conducted to measure the bond strength and

Table 3.3 Data from radius hub flexure tests

	STS DRIE	DRIE + Wet isotropic etch	DRIE + SF_6 Plasma etch
Specimen size	20	16	18
Etch depth (μm)	NA	~1.8	~2.7
Reference strength, σ_0(GPa)	1.45	3.1	4.0
Weibull modulus, m	7.5	3.9-5.7	3~6

toughness. In both cases, it was determined that well-bonded wafers would achieve strengths that equalled that of monolithic silicon [21].

3.3.2 Structural Design

The mechanical strength data obtained using planar and RHFS's can be used to provide design allowables for structural design. Standard finite element techniques were used to obtain stress distributions in the rotor. For much of the rotor design, axisymmetric models were utilized in order to understand design tradeoffs associated with the scaling between individual components. For detailed design of blade roots and other features, 3-D global–local models were utilized. Figure 3.2 shows a stress contours from a global model and the local detail of a turbine blade trailing edge. The introduction of fillet radii at the blade and hub roots is critical to reducing the local stress concentrations at these locations. Such features are challenging to microfabricate and their production was the subject of a major experimental programme [10]. The stochastic properties of silicon were introduced into the design using the probabilistic design code CARES [5] which applies the experimentally determined Weibull probability density function at the integration points of the finite element model. This process allows for an assessment of the structural failure probability of the whole rotor. Results from such analyses can be presented graphically to allow design trades to be conducted. Figure 3.3 shows an example in which the effect of material reference strength and blade height on the failure probability is compared. As a practical matter, true probabilistic design was found to be unwieldy, so design was usually conducted to an allowable stress (typically 1 GPa) and then the failure probability was calculated to check the estimated reliability. The very low toughness of silicon ($\sim$0.9 MPa$\sqrt{\text{m}}$) implies that the failure probabilities can be heavily influenced by the introduction of flaws during processing and handling.

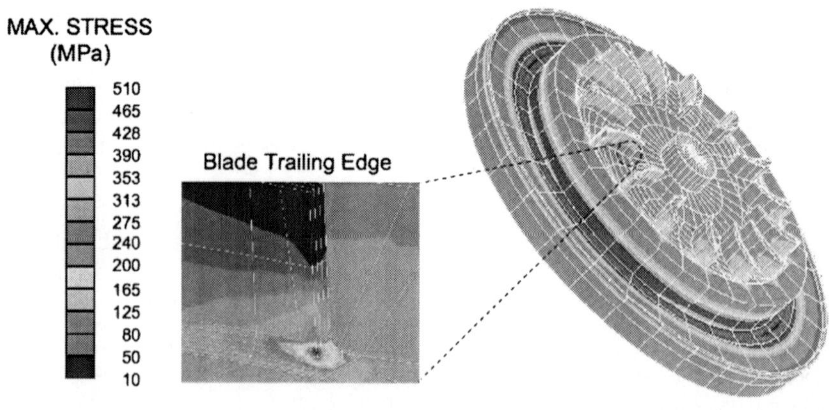

Fig. 3.2 Turbine stress contours obtained by finite element analysis

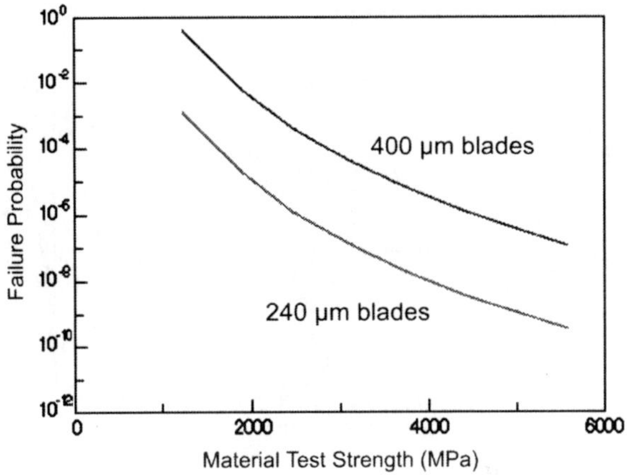

Fig. 3.3 Failure probabilities as a function of blade height and material test strength

A design stress of 1 GPa requires a maximum flaw size of less than 300 nm. This demands a conservative design approach and great care during fabrication.

The elastic deformation of the microengine rotor was assessed via the same axisymmetric and global finite element models used to assess strength. The critical aspects of the rotor design were the out of plane bending of the rotor and the radial expansion. Both deformations are of concern for bearing performance and the out of plane deflection must be controlled in order to minimize leakage flows and to avoid rubbing between the rotating and static structures. Such deformations need to be controlled to less than 1 μm. Options were explored to minimize the out of plane deformation of the rotor, via etching balancing features on the opposite side of the rotor to the blades. This is of most concern for the single-sided rotors used in the bearing test rig. In complete engine rotors, with turbine and compressor on opposite sides, the mass balance is automatically better and the bending stiffness greatly increased.

Structural dynamics analysis was also conducted, using the same 3-D finite element models. In all the fabricated rotors the blade natural frequencies were much higher than the rotor frequencies and the rigid body modes associated with the bearings. This is the result of the relatively low aspect ratio blades and the relatively high-specific stiffness of Si. If blade lengths were increased significantly, the natural frequencies would become more of a concern.

3.4 Design for High-Temperature Strength

Given the requirement for high-temperature operation of the turbine spool, design for high-temperature strength is of paramount importance. Silicon is not widely used at elevated temperatures; however, it has attracted some attention as a model system for experiments to understand the fundamentals of plasticity. Figure 3.4 shows literature data for the elevated temperature response of silicon. The pronounced

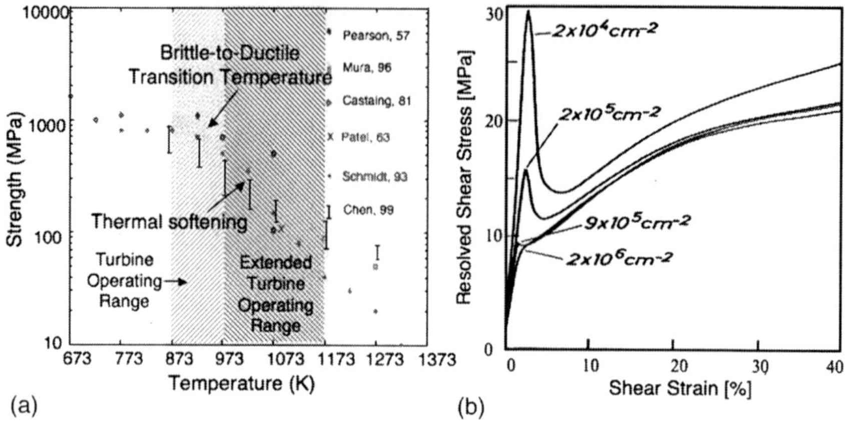

Fig. 3.4 High-temperature response of Si: (**a**) thermal softening and (**b**) strain softening (after [23])

reduction in strength at temperatures above 550°C and the highly non-linear stress–strain response are of particular concern. Experiments [22] conducted as part of the microengine project showed that there is a tendency for damage localization due to slip band formation, as shown in Figs. 3.5 and 3.6 for four-point bend specimens. In order to better understand the implications of this material behaviour for structural design, a constitutive model was developed based on continuum plasticity principles.

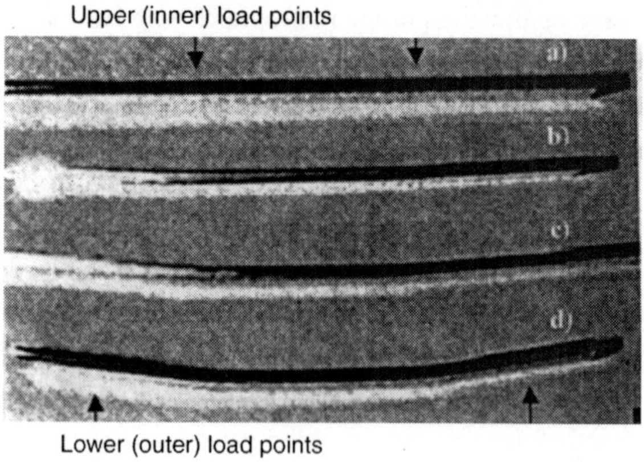

Fig. 3.5 Si four-point bend specimens after testing at elevated temperature. Note the pronounced deformation localization at the inner loading points

3.4.1 High-Temperature Material Model for Si

The material model was based on an existing isotropic plasticity model developed for metals [24, 25]. Two deformation mechanisms regimes are accounted for. At low dislocation densities, dislocation nucleation is considered to be the rate determining

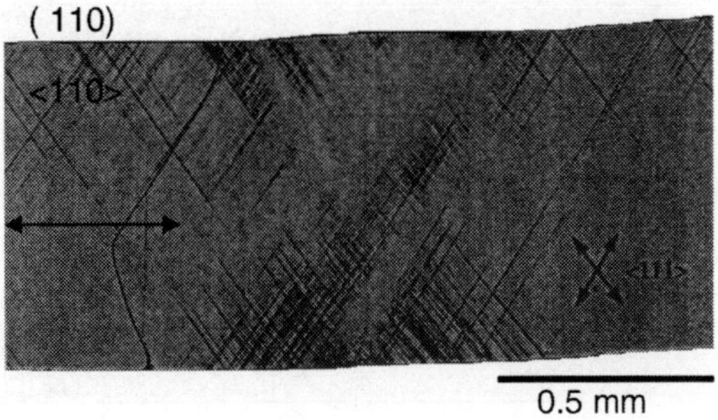

Fig. 3.6 Polished and etched cross-section of region under the inner loading points in Fig. 3.5 showing high concentration of slip bands

mechanisms. At higher dislocation densities, dislocation interactions are assumed to control deformation. Two internal state variables; the dislocation density and the internal shear resistance are used to characterize the material. Evolution equations are expressed in terms of stresses, strains and temperature. The model is largely empirical, requiring 15 parameters to be calibrated. However, the physical basis of the internal state variables ensures that there is some predictive capability. A complete description of the model is beyond the scope of this chapter, but can be found elsewhere [6].

The model was calibrated by uniaxial creep experiments using data such as that shown in Fig. 3.7. Cylindrical specimens of single crystal silicon 10 mm in diameter and 20 mm long were fabricated. These were tested in a conventional macroscale servohydraulic load frame with a clam-shell furnace surrounding the specimen. Data was obtained from tests conducted at two temperatures (600 and 800°C) and four stress levels at each temperature. The relatively low temperatures and short test durations ensured that surface oxidation of the specimens was not a concern.

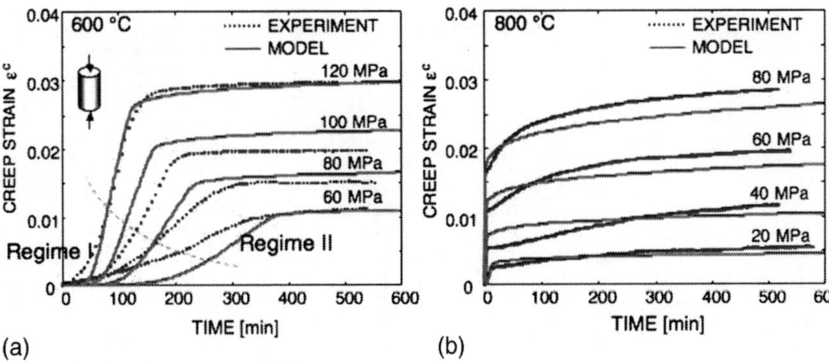

Fig. 3.7 Uniaxial creep data for silicon used to calibrate the constitutive model at (**a**) 600°C and (**b**) 800°C at four stress levels at each temperature

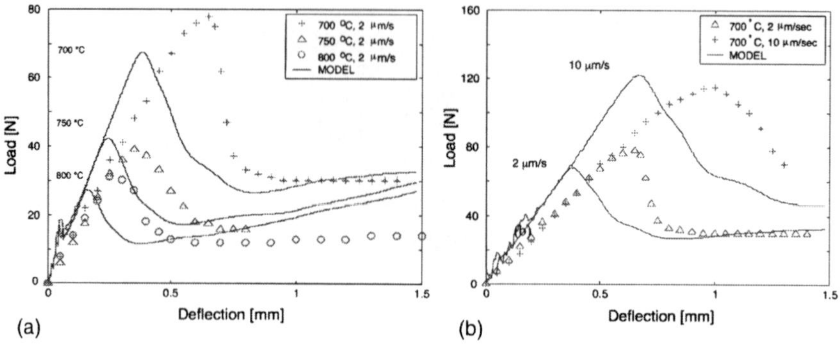

Fig. 3.8 Model predictions compared with four-point bend data for single crystal silicon, illustrating: (**a**) the effect of temperature and (**b**) the effect of strain rate

Once calibrated the model was validated by using it to predict the response in the monotonic loading of four-point bend tests. Data for specimens tested at three temperatures and two loading rates are shown in Fig. 3.8. The model provides a reasonable description of the data, thus confirming its validity for use in the structural design process. It is interesting to note that the model correctly predicts that the relatively compliant four-point bend specimen exhibits pronounced strain softening behaviour, even though this is not directly evident from the stiffer, uniaxial compression creep tests shown in Fig. 3.7.

Figure 3.9 shows data for four-point bend tests in which the loading was interrupted and either a constant displacement hold or a constant load hold were applied. It is clear that for both cases the peak load achieved is reduced below that found in constant rate loading. This illustrates the issues of load history and the requirement for a conservative design approach if silicon is to be used as the structural material for a microengine.

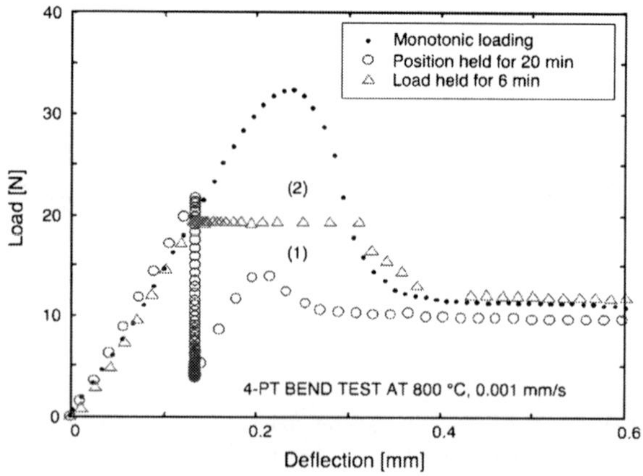

Fig. 3.9 Experimental results from (*1*) strain relaxation and (*2*) hold at load tests compared with the monotonic loading case for silicon specimens in four-point bending

3.5 Use of Silicon Carbide in Hybrid Structures

Given the temperature restrictions for the use of single crystal silicon, it is apparent that it is not an ideal material with which to construct high-performance microturbomachinery. However, it is important to understand why higher temperature materials are required. In the initial Si-based Microengine design, the key limitation on the performance is not the material temperature directly, rather the heat conduction between the turbine and the compressor. The overall engine efficiency is significantly reduced by the large heat flux from the turbine to the compressor. If thermal insulation can be incorporated between the turbine and the compressor, then higher efficiency will be achieved. However, this requires that the turbine operate at a higher temperature than silicon allows. To this end more refractory materials have been assessed for use.

3.5.1 Mechanical Considerations in SiC/Si Hybrid Structures

The tolerances on the design of the bearings, in terms of defining the rotating gap and the mass balance of the turbine/compressor spool place significant fabrication limitations on the material options available. Essentially, they stretch the limits of conventional microfabrication capabilities, and significantly exceed those achievable in more conventional machining, moulding or casting processes. This implies the continued use of lithographic patterning and etching or deposition processes. However, the candidate materials, chiefly refractory ceramics are not generally amenable to such processes. Etch rates are low, and the degree of anisotropy and control that can be achieved are very limited. With this background a hybrid approach was pursued that allowed the continued use of silicon microfabrication, but provided structural reinforcement using silicon carbide. A detailed description of this approach can be found elsewhere [6, 7, 26]. Figure 3.10 shows a schematic of a Microengine spool containing local silicon carbide reinforcement. Figure 3.11 shows finite element results which illustrate the structural reinforcement concept. The silicon carbide is much higher stiffness (460 GPa Young's modulus vs. 165 GPa for silicon) and retains its strength to temperatures in excess of 1600 K. This means that the silicon carbide core of the spool carries the majority of the centrifugal

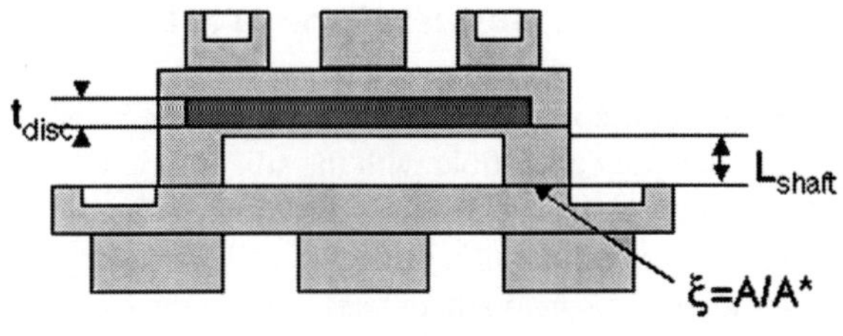

Fig. 3.10 Schematic of hybrid structure concept in which a silicon spool is reinforced with silicon carbide

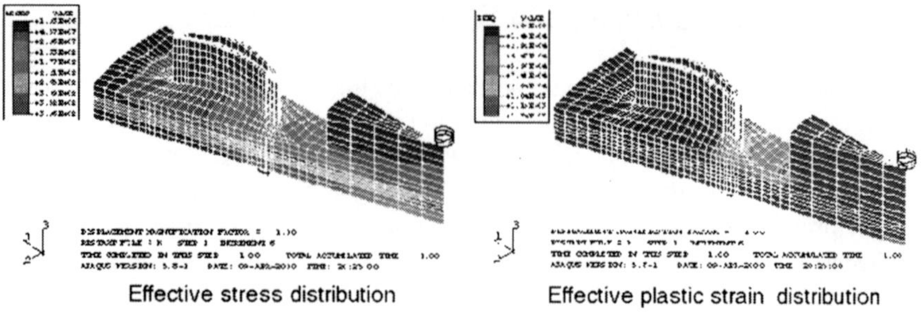

Fig. 3.11 Finite element stress contours and plastic strain contours showing the reduction in plastic strain resulting from the incorporation of silicon carbide

Fig. 3.12 Illustration of increased operating temperature arising from incorporation of silicon carbide

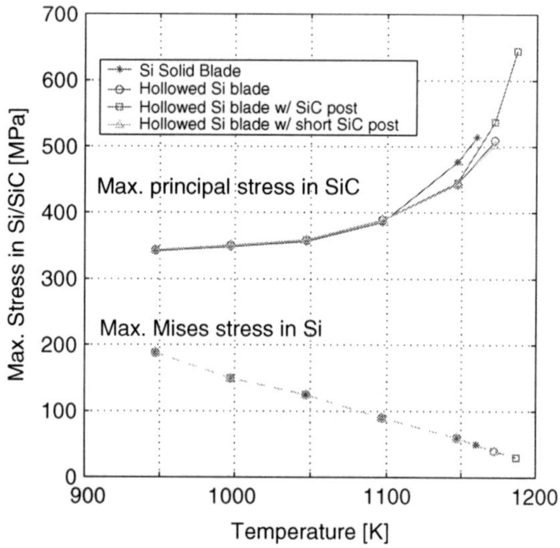

load of the spool. Figure 3.12 shows the effect on the turbine material temperature of including 30% through the thickness of silicon carbide. Operation up to 1150 K is achievable, which has a significant effect on overall engine efficiency, as it can cut the heat flux from the turbine to the compressor to ~5% of the all silicon case.

3.5.2 *Process Considerations in SiC/Si Hybrid Structures*

The use of silicon carbide reinforcement can be achieved with relative simple processes that are compatible with the silicon microfabrication used for the first generation microengines. These are illustrated in Fig. 3.13. There are four key steps. (1) an array of circular pits is etched in a silicon wafer. (2) the pits are filled with chemical vapour deposited silicon carbide. (3) the overburden of excess SiC is removed

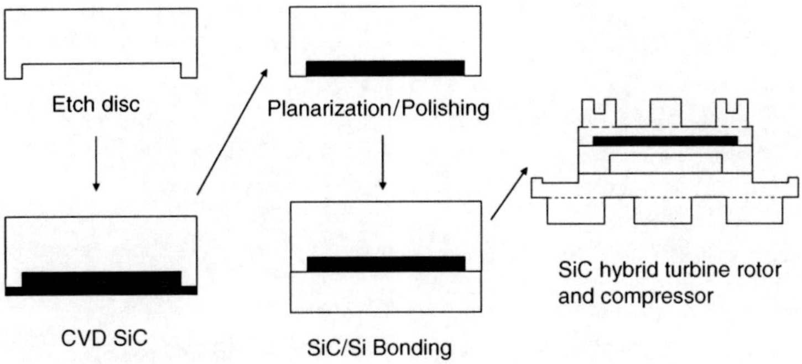

Fig. 3.13 Schematic process flow for creation of a Si/SiC hybrid spool

by grinding and polishing and (4) a second silicon wafer is bonded over the Si/SiC surface of the first wafer. The bonded wafer pair moves through the remainder of the process flow as though it were a monolithic silicon wafer. All the key dimensions (blades, bearings) are defined by silicon etching, which is well understood and can be controlled to the necessary tolerances.

The rationale behind this approach is that the silicon carbide provides significant structural reinforcement, while permitting most of the critical tolerances to be achieved in silicon, using well-established microfabrication processes. Of the four key process steps, deep etching of silicon is understood well and required little process development beyond that conducted for the rest of the project. Each of the other three key processes required significant process development and is discussed in turn in the following subsections.

3.5.2.1 Chemical Vapour Deposition of SiC

The key requirement for the SiC deposition is that it is deposited with very low levels of residual stress. The presence of residual stress causes wafer curvature, which makes subsequent processing difficult. A comprehensive study of the origins of residual stress in SiC has been conducted and can be found elsewhere [16]. SiC is deposited from a methyl-trichlorosilane precursor with a hydrogen carrier gas. Deposition occurs at low pressure and temperatures in excess of 1000°C. The deposited material has a highly columnar grain structure, as shown in Fig. 3.14. In addition there are multiple stacking faults visible within the grains. The columnar microstructure and the presence of stacking faults leads to the potential for high levels of residual stress.

The residual stress can be decoupled into two components: the thermal component (due to thermal expansion mismatch) and the "intrinsic" component due to the microstructure and growth processes. Wafer curvature measurements were used to characterize the residual stress as a function of deposition conditions. Independent measurements and accompanying analysis were used to separate the intrinsic and

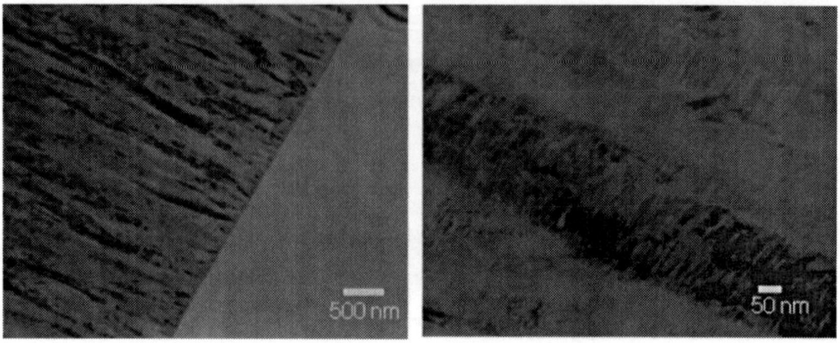

Fig. 3.14 Transmission electron micrographs of SiC grains on an Si substrate

Fig. 3.15 Variation of residual stress with deposition temperature for CVD SiC

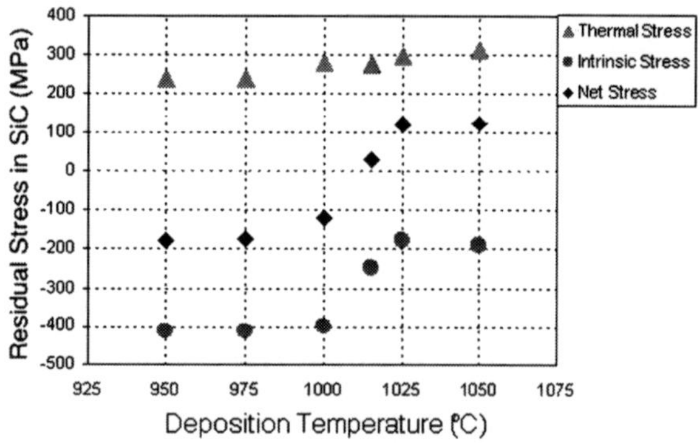

thermal components. Data was obtained as a function of process conditions and these are shown in Figs. 3.15 and 3.16. As can be seen, it is possible to control the process conditions to achieve near zero residual stress. This has proved to be a very robust process, with in excess of 50 wafers being successfully processed to date.

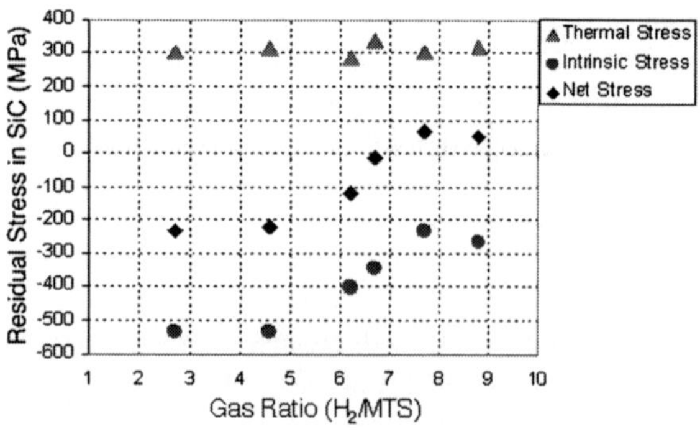

Fig. 3.16 Variation of residual stress with gas ratio for CVD SiC

3.5.2.2 Planarization of Si/SiC hybrid structures

A conventional diamond polishing procedure was used to generate smooth Si/SiC surfaces. A polished wafer is shown in Fig. 3.17. The SiC reinforcing disks are clearly visible.

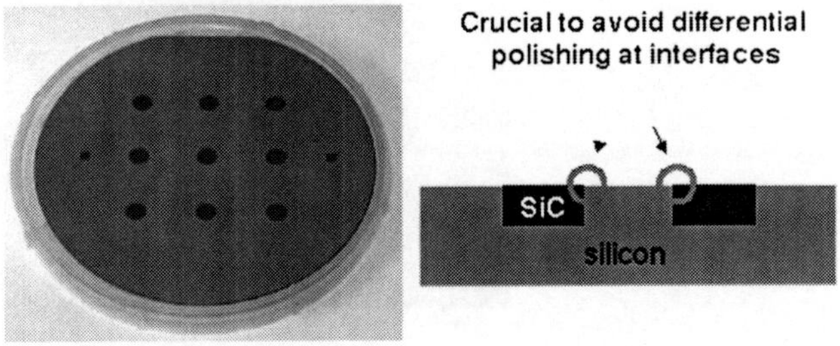

Fig. 3.17 Polished Si/SiC hybrid wafer

This process is still not sufficiently robust. Problems encountered include fracture of the wafers during polishing and the presence of height variations due to the hardness difference between the Si and SiC areas. This is an area that would require significant effort in order to realize an effective hybrid microengine rotor. Nevertheless considerable progress was made and some sufficiently smooth wafers were generated suitable for bonding.

3.5.2.3 Bonding of Si/SiC Hybrid Wafers

Initial attempts at bonding utilized identical procedures as for bonding pairs of Si wafers. However, the inadequacies of the polishing process and the occasional appearance of SiC nodules on the surfaces meant that this did not yield successful bonds. The process was modified by the inclusion of the deposition of a moderately thick layer of silicon oxide. The oxide layer overcoated any height variations in the Si/SiC layer. The oxide layer is readily planarized by chemical mechanical polishing (CMP) to yield a bondable interface. A cross-sectional micrograph of the resulting composite interface is shown in Fig. 3.18.

Mechanical testing was conducted which reveals that the silicon oxide layer is sufficiently thin that it has little effect on the structural integrity of the interface, at room temperature and at high temperature. Further work is required to refine this process so that high process yields are achieved; however, it appears to be feasible with the potential to achieve the necessary geometries and structural performance.

3.5.3 Mechanical Test Results

The concept of Si/SiC structures has been partially verified by testing Si/SiC beams in four-point bending. Results are shown in Figs. 3.19 and 3.20. The specimens were created by depositing SiC on Si and cutting them to size with a die saw.

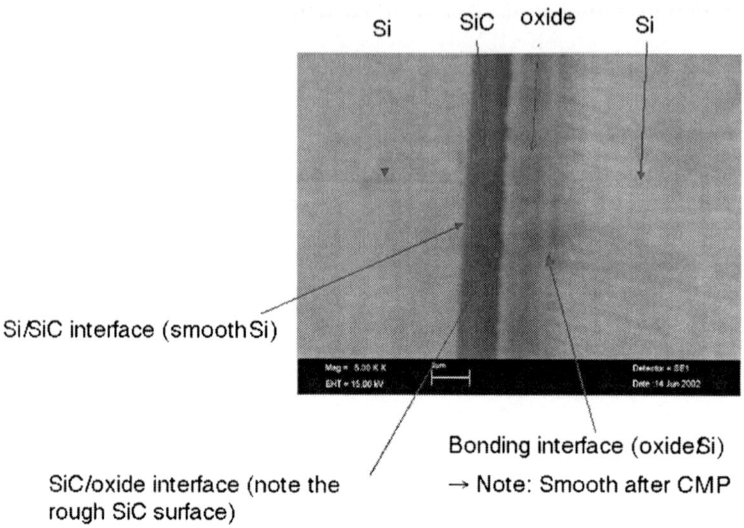

Fig. 3.18 Cross-section SEM of Si/SiC/SiO_2/Si interface layers

The specimens were then loaded in four-point bending. Figure 3.19 shows results for different SiC layer thicknesses. Figure 3.20 shows that strength is retained at temperatures as high as 900°C (1173 K). FE predictions have been made assuming the SiC remains elastic and the Si follows the constitutive behaviour described in Section 3.4.1. These predictions are superimposed on the data in Fig. 3.20. Good

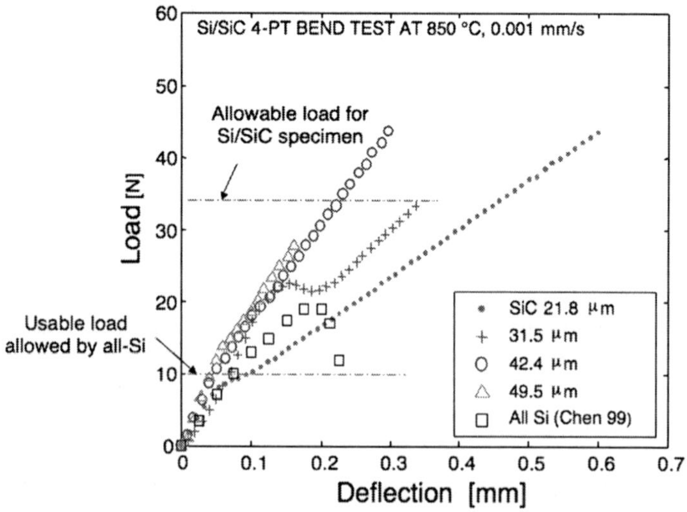

Fig. 3.19 Mechanical test results for Si/Sic hybrid beams in four-point bending

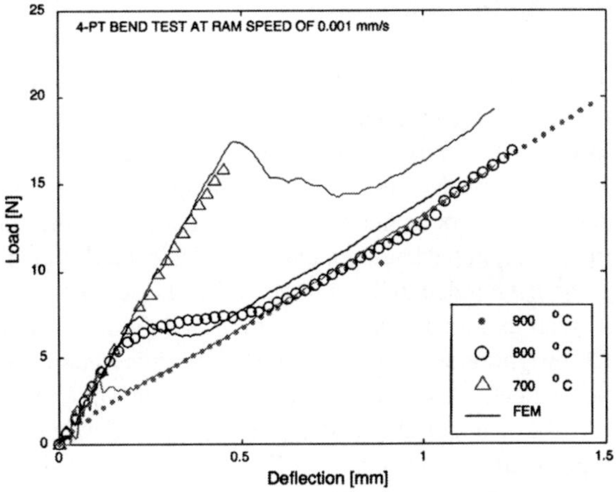

Fig. 3.20 Four-point bending test data and model predictions for SI/SIC beams tested at three temperatures

agreements between prediction and data are obtained, further validating the Si constitutive law. The SiC is assumed to be elastic with mechanical properties determined by micromechanical testing.

Having validated the concept and the modelling, the model has been used to guide trades in the design of actual microengine rotors. Figure 3.21 shows a design diagram which illuminates the trades between maintaining a stable bearing operation (by minimizing reduction of the journal bearing air gap) with variations in SiC thickness and strength. Such trades help to guide the effective determination of the division of programme resources between material and process development and advanced bearing concepts.

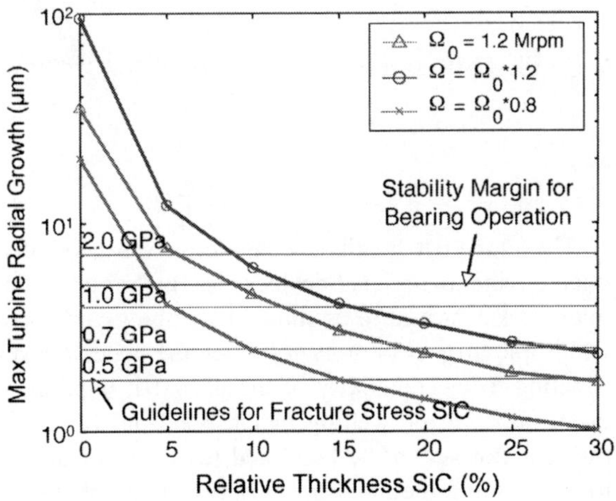

Fig. 3.21 Tradeoffs in the use of Si/SiC hybrid structures between the allowable radial turbine growth, the relative thickness of the SiC and the SiC fracture strength that can be obtained

3.6 Packaging

The packaging of the Microengine and other high-power MEMS is particularly demanding. The external surface temperatures are in excess of 500°C and combustible fuel must be delivered to the engine at the appropriate rate. In addition pressure taps, channels for thermocouple wires and optical access for the speed sensor needs to be achieved. For the majority of the microengine devices fabricated the principal packaging technology achieved fluidic connections using Kovar tubes, glass-frit bonded to the surface of the device. The tubes provided fluidic connections as well as mechanical and thermal isolation. For some devices, toward the end of the programme, anodic bonding of the silicon chip to a thermal-expansion matched borosilicate glass (Pyrex) support substrate was explored. The focus of this section is on the key materials, structures and process developments associated with achieving high-quality bonds between Kovar and silicon using a glass frit technology. Fuller descriptions of the technology used can be found elsewhere [27–29]. Figure 3.22 shows a microengine test device bonded to a fuel supply manifold. The Kovar tubes are visible around the periphery of the microengine die. Further tubes lead out from the manifold structure. Figure 3.23 shows a microcombustor test rig in operation, glowing red hot, with fluidic connections made by Kovar tubes. Figure 3.24 shows a close up of the glass frit bond between the Kovar tube and the Si surface.

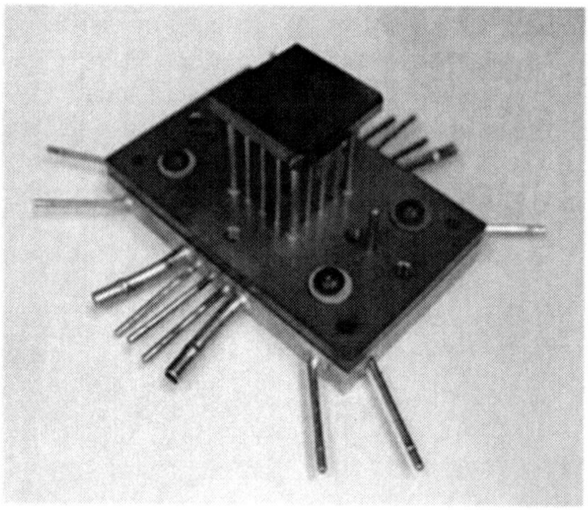

Fig. 3.22 Packaged microcombustor test structure, showing Kovar tubes

The glass frit bonding technology was generally found to be robust for the microengine and related devices. The higher pressures (>10 MPa) required for the microrocket proved to be more challenging and motivated a major research effort within the project in order to obtain the necessary reliability. The key issues contributing to the reliability of the glass frit bonding were: (a) the void content of the bonds, (b) the wetting of the bond on the silicon surface, (c) control of the residual stresses in the bond and (d) mechanical design, particularly with regard to establishing the interactions between neighbouring connections. Analytical and

Fig. 3.23 Microcombustor test structure in operation, showing connections made via Kovar tubes

Fig. 3.24 Detail of an individual tube-device connection, showing the glass frit

experimental studies were conducted to assess the efficacy of material, process and design choices. Finite element models, such as that shown in Fig. 3.25 , were developed to analyse the stress state in the joint, as a function of loading, geometry and material properties. Much of the evaluation of materials, processes and geometry were achieved using axial pull tests on single tube specimens, such as that shown in cross-section in Fig. 3.26. These allowed a straightforward evaluation of joint quality, allowing statistically significant data sets to be obtained, without incurring excessive fabrication expense. This approach was found to correlate satisfactorily with pressure loading of test specimens with multiple joints, although typically the failure pressures were lower that those which would be inferred from the stress levels in purely axial loading. The feasibility of joints formed between Kovar tubes and silicon MEMS dies was established by this means. Preforms of glass frits, available commercially for feedthroughs in microelectronic packaging were found to provide sufficient structural integrity for the microengine applications, with their

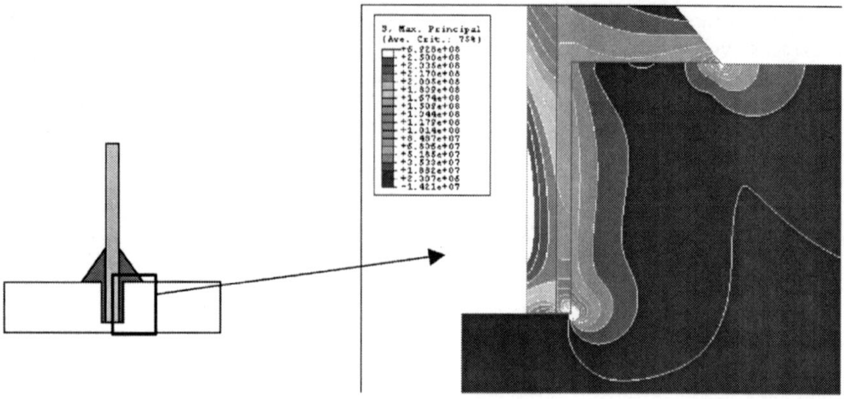

FIG. 3.25 Detail of stress contours from a finite element model of a fluidic connection, indicating regions of high stress at the tube base and edge of the glass frit

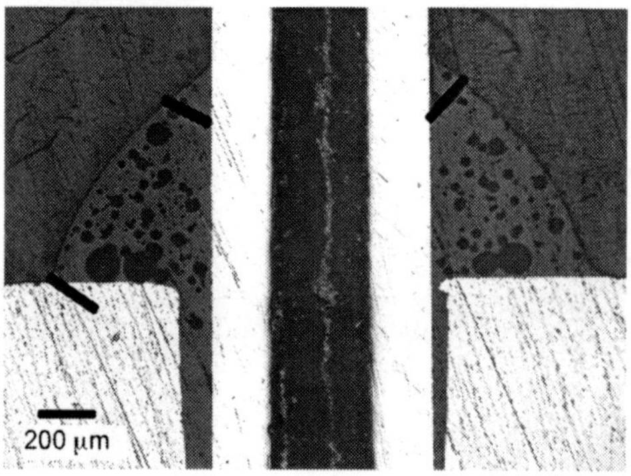

FIG. 3.26 Cross section of a fabricated joint showing the Kovar tube, glass frit and the silicon surface. Note the void content of the glass

relatively modest pressures (<1 MPa). The importance of a good thermal expansion match between the tube, silicon and the glass was established, and glass frit materials were chosen accordingly to minimize the thermal expansion mismatch over the entire temperature range. For high-pressure applications silicon sidewall bonding was found to be desirable as it maximized the stress transfer between the Si die and the tube in shear while minimizing the tensile stresses. The fracture loads in single joint specimens under axial tension and pressure loads can be estimated to first-order accuracy using linear elastic fracture mechanics, which was sufficient to evaluate materials, processes and design concepts. The load-carrying capability of joints in actual MEMS devices was generally found to be lower than that in single joint specimens. The differential pressures applied to joints in proximity and differences in joint configuration as manufactured are the principal causes of this discrepancy. Axial tension tests can be used to inspect joint quality, but the failure results cannot be used to predict the strength of the joint under pressure loading.

3.7 Summary

This chapter has provided an overview of materials, structural design and packaging issues associated with the MIT microengine concept. Key concepts covered have been provided as follows:

1. The use of Weibull statistics and micromechanical testing to provide design guidance for low-temperature structures.
2. The development of a high-temperature constitutive model for Si, and its validation against material test data.
3. The use of Si/SiC hybrid structures in order to increase the allowable operating temperatures while minimizing the deviation from Si processes.
4. The use of glass frit bonded Kovar tubes in order to provide fluidic connections to Microengine test devices.

These concepts have all had an impact beyond the immediate needs of the Microengine project. While there is much work to be done, the concept of utilizing high strength and in some cases refractory materials in MEMS has been demonstrated to a reasonably high technology readiness level. In particular the analytical tools required to evaluate and design structures utilizing such materials have been established. There is good potential for applications in commercial MEMS devices as well as in high-performance electronic packaging.

Acknowledgments The author would like to acknowledge all those who contributed to the Materials, Structures and Packaging tasks on the Microengine and related projects, including: Lallit Anand, Arturo Ayon, Shaun Berry, Kuo-Shen Chen, Dongwon Choi, Todd Harrison, Eugene Huang, Adam London, Hangqing Li, Kevin Lohner, Norihisa Miki, Bruno Miller, Hyung-Soo Moon, Anna Mracek, Erin Noonan, Yoav Peles, Christopher Protz, Dong-Jin Shim, Robert Shinavski, Christopher Spaddaccini, Hongwei Sun, Kevin Turner, Srikar Vengallatore, Douglas Walters, Wenjing Ye and Xin Zhang. SMS also is very grateful for the highly effective and rewarding collaborations across the whole Microengine project and for the various funding sources that made it all possible. He is especially thankful for the leadership and vision provided by Prof. Alan Epstein.

References

1. S. M. Spearing, "Materials Issues in MEMS" *Acta Mater.* **48**, 179–196, 2000
2. K.-S. Chen, "Micro-Engine Structural Design and Analysis", Ph.D Thesis, MIT, 1998
3. K.-S. Chen, A. Ayon, and S. M. Spearing, "Controlling and Testing the Fracture Strength of Silicon at the Mesoscale" *J. Am. Ceram. Soc.* **83**(6) 1476–1484, 2000
4. V. T. Srikar and S. M. Spearing. "Mechanical Test Methods for MEMS Devices" *Exp. Mech.* **43** (3) 228–237, 2003
5. K.-S. Chen, S. M. Spearing, and N. N. Nemeth, "Structural Design of a Silicon Micro-Turbo Generator" *AIAA J.* **39**(4) 720–728, 2001
6. H.-S. Moon, "Design of Si/SiC Hybrid Structures for a Micro-Engine" Ph.D Thesis, MIT 2002

7. H. S. Moon, D. Choi, and S. M. Spearing, "Development of Si/SiC Hybrid Structures for Elevated Temperature Micro-Turbomachinery" *J. MEMS*13(4) 676–687, 2004
8. V. T. Srikar, K. Turner, T.-Z. A. Ie, and S. M. Spearing, "Structural Design Considerations for Micromachined Solid Oxide Fuel Cells", *J. Power Sources* **125**(1), 62–69, 2004
9. C. D. Baertsch, K. F. Jensen, J. L. Hertz, H. L. Tuller, V. T. Srikar T. Vengallatore, S. M.Spearing, and M. A. Schmidt, "Fabrication and Structural Characterization of Self-Supporting Electrolyte Membranes for a Micro Solid-Oxide Fuel Cell" *J. Mater. Res.* **19**(9), 2604–2615 2004
10. K.-S. Chen, A. A. Ayon, X. Zhang, and S. M. Spearing, "Effect of Process Parameters on the Surface Morphology and Mechanical Performance of Silicon Structures after Deep Reactive Ion Etching (DRIE)" *J. Microelectromech. Syst.*, **11** (3) 264–275, 2002
11. A. A. Ayon, M. A. Schmidt, K. A. Lohner, K-S. Chen, S. M. Spearing, and H. Sawin "Deep Reactive Ion Etching of Silicon" *Mat. Res. Soc. Symp. Proc* **546**, 51–61, 1999
12. K.-S. Chen, A. A. Ayon, X. Zhang, S. M. Spearing, "Effect of Process Parameters on the Surface Morphology and Mechanical Performance of Silicon Structures after Deep Reactive Ion Etching (DRIE)" *J. Microelectromech. Syst.* **11**(3), 264–275, 2002
13. K. Turner and S. M. Spearing, "Modeling of Direct Wafer Bonding: Effect of Wafer Bow and Etch Patterns". *J. Appl. Phys.* **92**(12), 7658–7666, 2002
14. N. Miki, X. Zhang, R. Khanna, A. A. Ayón, D. Ward and S. M. Spearing "Multi-Stack Silicon-Direct Wafer Bonding for 3D MEMS Manufacturing" *Sens. Actuators, A, Physical* **103** (1–2), 194–201, 2003
15. X. Zhang, K.-S. Chen. and S. M. Spearing, "Thermal Mechanical Behavior of Thick PECVD Oxide Films for Power MEMS Applications," *Sens. Actuators, A, Physical*, **103** (1–2), 263–270, 2003
16. D. Choi, R. J. Shinavski, W. S. Steffier, and S. M. Spearing, "Residual Stress Control in Thick LPCVD Polycrystalline 3C SiC Coatings on Si Substrates" *J. Appl. Physics.* **97**(7), 074904, 2005
17. Y. Peles, V. T. Srikar, T. A. Harrison, C. Protz, A. Mracek, and S. M. Spearing, "Fluidic Packaging of Microengine and Microrocket Devices for High Pressure and High Temperature Operation" *J. Micro Electromech. Syst.* **13**(1), 31–40, 2004
18. D.-J Shim, H.-W Sun, V. T. Srikar, and S. M. Spearing, "Damage and Failure in Silicon-Glass-Metal Microfluidic Joints for High Pressure and Temperature MEMS Devices" *J. Microelectromech. Syst.***15** (1), 246–258, 2006
19. V. T. Srikar and S. M. Spearing. "Materials selection in Micromechanical Design: An Application of the Ashby Approach" *J. MicroElectromech. Syst.*, **12**(1), 3–10 2003
20. S. M. Spearing and K. S. Chen, "Micro Gas Turbine Engine Materials and Structures," *Ceramic Eng. Sci. Proc.* **18**(4), 11–18, 1997
21. A. Ayón, X. Zhang, K. Turner, D. Choi, B. Miller, S. F. Nagle, and S. M. Spearing, "Characterization of Low Temperature Silicon Wafer Bonding for Power MEMS Applications," *Sens. Actuators, A, Physical* **103**(1–2), 1–8, 2003
22. D. S. Walters and S. M. Spearing, "On the Flexural Creep of Single Crystal Silicon", *Scripta Materialia.* **42**, 769–774, 2000
23. K. Sumino, "Deformation Behavior of Silicon" *Metallur. Mater. Trans. A* **30A**, 1465–1479, 1999
24. L. Anand, "Constitutive Equations for the Rate-dependent Deformation of Metals at Elevated Temperatures" *J. Engine. Mater. Tech.* **104**, 12–17, 1982
25. S. Balasubramanian and L. Anand, "Elasto-viscoplastic Constitutive Equations for Polycrystalline fcc Materials at Low Homologous Temperatures" *J. Mech. Phys. Solid* **50**(1), 101–126, 2001
26. D. Choi, "Refractory Materials for Micro-Engine Applications" MIT, Ph.D Thesis 2004
27. T. Harrison, "Packaging technologies for Power MEMS devices" MIT. SM, 2000.

28. Y. Peles, V. T. Srikar, T. A. Harrison, C. Protz, A. Mracek, and S. M. Spearing, "Fluidic Packaging of Microengine and Microrocket Devices for High Pressure and High Temperature Operation" *J. Micro Electromech. Syst.*, **13**(1), 31–40, 2004
29. D.-J Shim, H.-W Sun, V. T. Srikar, and S. M. Spearing, "Damage and Failure in Silicon-Glass-Metal Microfluidic Joints for High Pressure and Temperature MEMS Devices" *J. Microelectromech. Syst.*.**15**(1), 246–258, 2006

Chapter 4
Microengine Fabrication

Hanqing Li and Martin Schmidt

4.1 Introduction

In this chapter, we will talk about the fabrication of the microengine devices using micro-electro-mechanical system (MEMS) technology. MEMS is an enabling technology based on the mature semiconductor fabrication technologies and has been widely used in making microscale devices such as sensors, actuators, RF communication devices, microfluidic devices, biology, and medical devices. The progress in MEMS technology, especially deep reactive ion etch (DRIE) and wafer bonding, in batch fabricating 3D structures at micrometer precision using Si wafers, an excellent structural material, gives birth to the concept of microengine devices for propulsion and power generation. The development of microengine devices also contributed in return to the understanding of the fundamentals of DRIE process [1–8] and improvement in multilayer bonding of heavily fabricated Si wafers [9–12].

We will first introduce various devices and their key components that have been built over the course of the microengine development. Then we will go over the basic processes before getting down to the major fabrications challenges unique to microengine devices. These challenges and our solutions to them will be addressed in more detail and a relatively complete process flow of the self-sustained engine (SSE), the last and most complicated microengine device will be presented. An outlook for further improvement in microengine fabrication technologies and a summary will bring an end to this chapter.

4.2 Microengine Devices

The goal of the microengine project is a portable battery charging system that burns regular liquid fuels to run a microscale gas turbine and generate a few watts of

H. Li (✉)
Microsystems Technology Laboratories, Massachusetts Institute of Technology, Bldg 31-261E, 77 Massachusetts Avenue, Cambridge, MA 02139, USA

J.H. Lang (ed.), *Multi-Wafer Rotating MEMS Machines*, MEMS Reference Shelf,
DOI 10.1007/978-0-387-77747-4_4,

electric power [13–15]. A microengine device is a complicated 3D structure which includes key components such as bearing systems, a free rotor, a combustion chamber, and a thermal control mechanism. The rotor itself consists of a compressor and a turbine and thermal isolation parts in between ideally, and it is suppose to rotate at a tip speed of a few hundred meter per second to generate power. Because of the high-speed rotation requirement, many of the dimensions in the devices have very tight tolerances that often exceed the specifications of the machines used, and this means that innovations are needed in fabricating each layer. The 3D structures require unprecedented stacking up of multiple heavily patterned Si wafers with high precision. Moreover, little is understood about the performance of the components of this microscale gas turbine device even though it is a stripped down design compared to the conventional gas turbines. Therefore, it is necessary to take a step-by-step approach in fabrication and system study of the microengine devices.

Figure 4.1 is a road map of the microengine devices. The first generation devices are a bearing rig [16–19] for the study of journal bearings and thrust bearings, and a combustor [20–23] for study of ignition and combustion in microscale devices. The second generation devices are turbocharger [24–26] and electric and magnetic generators [27, 28], the third generation devices are SSE, and the final goal is an integrated system of microengine and generator.

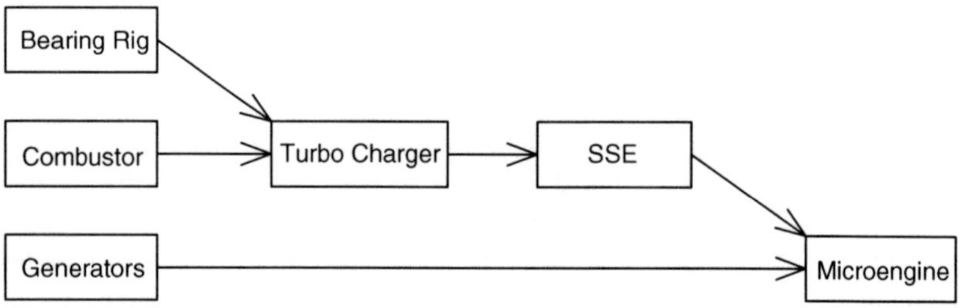

Fig. 4.1 Road map of microengine device development. The 1st generation devices include bearing rig, combustor and generators. They share the same basic structure. The 2nd generation turbo charger devices have a turbine and a compressor in their rotors. The 3rd generation SSE devices added a thermal shaft between the compressor and the turbine for high power efficiency. We have fabricated SSE devices and the generators separately. The ultimate goal is combine the SSE with a generator into a micro power generator

4.2.1 Bearing Rig

The purpose of the bearing rig device is to study the thrust bearing and journal bearing performance. A typical cross section view is shown in Fig. 4.2. It is a six-wafer structure. Layers 1, 2, 5 and 6 provide all the feeding channels to the rotor in layers 3 and 4. It should be point out that the main structure of various microengine devices differs mostly in the rotor layers, at least from a fabrication perspective. The

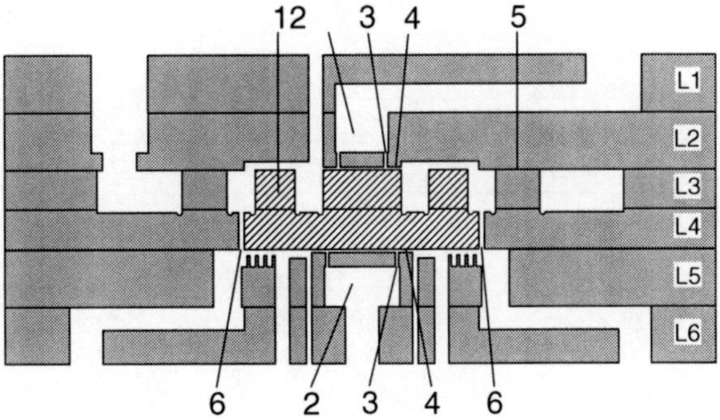

Fig. 4.2 Cross section drawing of the bearing rig. It has 6 layers, and the rotor is constructed by the blade layer L3 and journal bearing layer L4. The key components for fabrication are 1 rotor blades, 2 plenum, 3 nozzles, 4 thrust bearings, 5 stator blades, and 6 journal bearing

number of wafers and patterns in the rotor layers change from devices to device, and this, however, usually means major change in fabrication processes. The processes of the non-rotor layers remain more or less the same except for pattern change and sometimes tighter etch uniformity control when the devices grow larger.

The rotor in a bearing rig consists of one set of blades and a circular disk that form the inner diameter of the donut shape journal bearing which is typically 10–20 μm wide and about 300 μm deep. The small gaps, typically 5–6 μm, between the rotor and layer 2 and 5 define the thrust bearings which are fed through a ring of 10 μm nozzles in layer 2 and layer 5.

4.2.2 Combustor

A combustor is basically a bearing rig with a fixed rotor and its fabrication is easier than the bearing rig because one does not have to worry about making the rotor to tight specifications and releasing it before final assembly. However, the combustion study needs to have catalyst material integrated inside the device, therefore, making the bonding process more challenging due to the particles from the catalyst material. A special shadow wafer was made to help shield the bonding surfaces when inserting the catalyst material into the combustor.

4.2.3 Turbocharger

The turbocharger is the simplest form of a micro gas turbine. Its structure differs from a bearing rig in the rotor layers. As seen in the cross section schematic in Fig. 4.3, the rotor consists of a compressor in layer 3 and turbine in layer 4 that are bonded back to back. The compressor and the turbine have their own blades and journal bearings, and a wider running gap in the turbine depending on the bearing design.

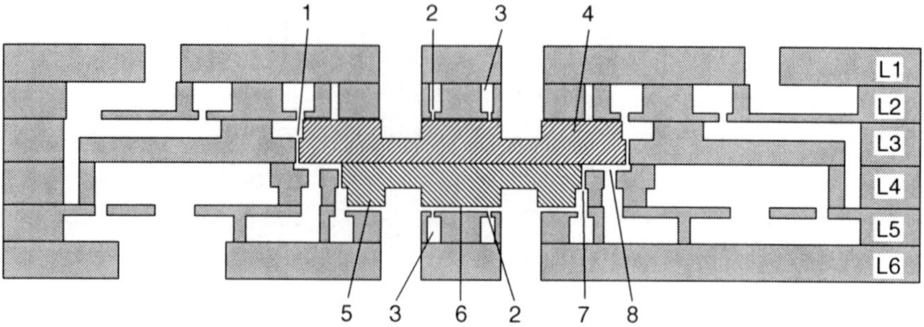

Fig. 4.3 Cross section drawing of a turbo charger. It has 6 layers. The compressor is in layer 3 (L3) and the turbine is in layer 4. The key components for fabrication are 1 journal bearing, 2 nozzles, 3 plenum, 4 compressor blades, 5 turbine blades, 6 thrust bearing, 7 running gap, and 8 seal clearance

In addition to the tight journal bearing and thruster bearing specifications as in the bearing rig, because of the increased number of rotor layers the fabrication of turbo charger has more prominent challenges in rotor balancing that is influenced by alignment between patterns and between wafers in the rotor layers, and the blade height uniformity. In this drawing the older version, where the rotor consists of two layers, is presented for clarity. A Newer version of the turbo charger that addressed the balancing issue will be described later.

4.2.4 Self-Sustained Engine (SSE)

The SSE is designed to minimize the thermal loss from the turbine to the compressor. The cross section of the SSE is shown in Fig. 4.4. Compared to the turbocharger, the SSE has an additional thermal shaft, 0.9 mm in diameter and 1.0 mm long, formed in layer 5 and layer 6 as the only connection between the high-temperature turbine and low-temperature compressor. Another change is in the blades which are now formed in a separate wafer, layer 3 and layer 8, instead of etched half way into a wafer (for better blade height uniformity).

As one can see with all these additions the SSE becomes a 10-layer devices. This makes the fabrication processes, especially the wafer bonding, even more challenging. The 10-layer structure requires 9 bonding steps, and all the small blades and shafts have to be perfectly bonded to make a good device. The detailed process flow for SSE will be described later.

The major fabrication challenges of microengine devices originate from the complexity of the devices and the need for high-speed rotation of the rotors. As described above, the microengine devices all have multilayer structure that requires silicon fusion bonding to assemble. Fusion bonding is sensitive to wafer quality, surface cleanness, processes, and patterns in the wafers. It is a non-trivial task to bond even

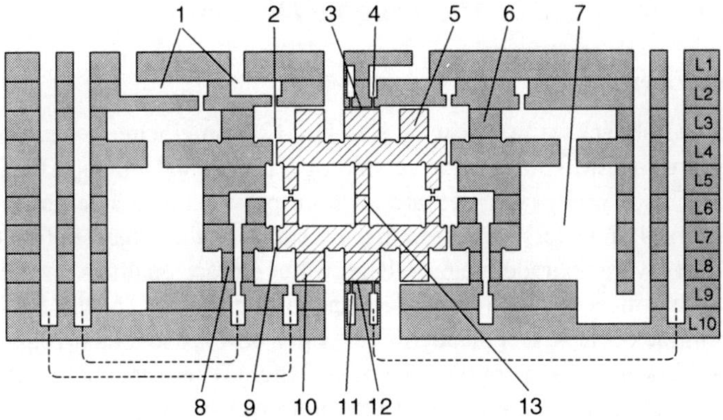

Fig. 4.4 Cross section drawing of a 10 layer self sustained engine (SSE). The layer numbers are labeled on the right of each layer as L1-L10. The 6 layer rotor is hatched for clarity. The components which are challenging to fabrication are marked: 1 forward plenum, 2 compressor journal bearing, 3 forward thrust bearing, 4 compressor nozzles, 5 compressor rotor blades, 6 compressor stator blades, 7 combustor chamber, 8 turbine stator blades, 9 turbine journal bearing, 10 turbine rotor blades, 11 turbine nozzles, 12 aft thrust bearing

two patterned wafers and the complexity grows exponentially as the number of layers increases.

For a microengine device to generate net power, the tip speed of the rotor should be about 500 m/s, which for example, is equivalent to over 2 million rpm for a device with rotors of 4 mm in diameter. Such high-speed rotation means extremely tight tolerance in many of the rotor-related specifications, especially laterally in the journal bearings precision for the small scale rotors and vertically in the blade height uniformity for balance control.

There exist other fabrication issues such as global variation in plenum depth uniformity and small nozzle etch, but generally the major microengine fabrication challenges fall into three categories: precision dimension control for the journal bearing, rotor balance that is affected in large part by the blade height uniformity, and multilayer assembly by Si fusion bonding. We call them the 3B (bearing, blades, and bonding) challenges for microengine fabrication. The fabrication team has put in a large amount of effort to address these issues. Before we get down to the solutions to these specific challenges of microengine devices, let us first review the basic MEMS processes routinely we routinely use.

4.3 Basic Fabrication Processes

This section lists the basic fabrication processes involved in microengine fabrication, starting with the materials used. Most of the processes described here are for general MEMS fabrication, but it also includes some techniques we developed in the course of microengine fabrication research.

4.3.1 Materials Used in Microengine Devices

4.3.1.1 Si Wafers

As mentioned above, we use Si wafers to build the microengine devices because of the matured MEMS fabrication technology for Si and its superb mechanical properties. The Si wafers we used are all 6 in. prime double side polished (DSP) N-type wafers. 4 in. wafers were used in early stages of the project before the clean-room facilities were upgraded to 6 in. High wafer surface qualities, the flatness and roughness, that meet the requirement of fusion bonding, are of the first priorities for microengine devices. Our standards for wafer bow are less than 10 μm for 4 in. diameter wafers and 15 μm for 6 in. diameter wafers. These are conservative empirical numbers due to the complexity of the multilayer bonding of patterned wafers. The reason to use N-type wafers is that they allow for easy IR inspection of bonded wafer stacks. When new wafers are received, the first thing we do is check the bonding quality by taking two wafers randomly, cleaning and bonding them according to our process.

The wafer thickness for our devices ranges from 330 to 1000 μm, as specified by the designs of the devices. The wafers that form the rotors in some devices require very small total thickness variation (TTV) across the wafer because the rotors have to be moved to a different location and the thrust bearing clearance is only a few micrometers. Large TTV could result in unacceptable variation in thrust bearing clearance. The TTV in the rotor wafers we used for these rotor wafers is 0.75 μm across the 6 in. diameter.

4.3.1.2 Glass Wafers

Glass wafers are used for packaging the microengine devices by anodic bonding. These glass wafers contain sodium ions which become mobile as temperature increases. Handling of the glass wafers must follow the clean room protocol strictly to avoid sodium contamination to CMOS equipments.

4.3.1.3 Thermal Oxide Films

The first process every wafer in our devices undergoes is thermal oxidation. The purpose is to protect the bonding surfaces from the various photolithography and etching processes. Sometimes thermal oxide is also used as a mask for the dry etch of Si. The protective oxide film is kept on throughout the processing of each wafer until right before fusion bonding when the wafers are cleaned and the protective oxide film is stripped by buffered oxide etcher (BOE) or HF. Silicon oxide is grown on the wafer by standard thermal oxidation. The thickness ranges from 0.1 to 2 μm according to the process flow design.

The thermal oxidation of Si wafer process flow is as follows:

(1) For processed wafers, especially the wafer that have undergone DRIE, an oxygen plasma ashing step is needed to remove the polymer residue from the DRIE

process. When a DRIE etched wafer is put in acetone, the polymer residue appears as floating fibers. Other normal cleaning processes such as piranha and RCA do not remove the polymer fiber properly. Failure to do ashing will result in masking by polymer residue in subsequent Si etch; usually in the form of random smooth-edged Si protrusions in the etched area and bad bonding. If the wafers are new, skip to RCA cleaning in step 4.

(2) Piranha clean the wafer to remove particles. Piranha is a mixture of 1 part of hydrogen peroxide, H_2O_2, and 4 parts of sulfuric acid, H_2SO_4. The wafers are usually cleaned in piranha for 10 min, followed by 5 min rinse in deionized (DI) water and 5 min spin dry at 2000 rpm.

(3) Strip the oxide by HF solutions or BOE. BOE is a 7:1 solution of ammonium hydroxide (NH4 OH) and 49% hydrofluoric acid (HF). One thing to remember about the use of BOE is to avoid over etch in BOE since BOE roughens the Si surface, although very slowly.

(4) RCA clean the wafer. The RCA process consists of two steps: 15 min in RC1, a 5:1:1 solution of DI water, hydrogen peroxide (H_2O_2) and ammonium hydroxide at 80°C, and 10 min in RC2, a 5:1:1 solution of DI water, H_2O_2, and hydro chloride acid (HCl) at 80°C, followed by DI water rinse and spin dry.

(5) Grow oxide in a diffusion tube at 1000–1050°C. The oxide thickness is determined by the oxidation time.

4.3.1.4 Deposited CVD Oxide Film

In some cases, we need a thick oxide film as nested mask for long DRIE etch. The thickness can be over 5 μm, way beyond the 2 μm limit of our thermal oxidation equipment. Chemical vapor deposition of oxide is our choice for the thick oxide. The nested mask process requires the deposited oxide behaves the same way in oxide etching chemicals, usually BOE, as the thermal oxide, i.e., having about 1:1 under cut with a proper mask. Test etching runs with dummy wafers are necessary with the CVD thick film process from our experience.

The wafers to be deposited are sometimes bonded stacks that are heavily processed. It is important to keep the wafer stacks clean and dry before transferring them to the CVD chamber that is in vacuum and at 400°C. Defective oxide could form on surfaces that are not properly cleaned. Cracking might occur in wafer stack if not completely dried, especially those with bonding defects on the edge because of trapped liquid therein.

4.3.1.5 Silicon Nitride (Si3N4) Film

Silicon nitride is used in the processing of our devices

(1) as a mask for BOE of thick oxide films since resist does not have very good adhesion to oxide in long-time BOE,
(2) for stress compensation to the thermal oxide film,
(3) as local oxidation barrier, and

(4) as protection of DRIE-etched side walls and bonding interfaces from HF or BOE oxide stripping.

The nitride is deposited in a low-pressure chemical vapor deposition (LPCVD) system. LPCVD nitride is Si rich and of low stress. It has the advantage of very conformal coverage with no pinholes. It covers completely even the vertical side walls of the etched features, and this feature can be used as protection of the side walls from erosion in later DRIE etching.

The wafer cleaning process before nitride deposition and CVD oxide deposition is similar to that for thermal oxidation:

(1) Oxygen plasma ashing, especially for those wafers that have been DRIE etched and might have polymer films residue,
(2) 10 min in piranha, a mixture of 1 part of hydrogen peroxide, H_2O_2, and 4 parts of sulfuric acid, H_2SO_4, followed by 5 min rinse in DI water and spin dry
(3) RCA cleaning: RCA is a two-step cleaning process. The first is 10 min dip in solution SC1, a 5:1:1 mixture of DI water, H_2O_2, and NH_3OH, heated to 80°C, followed by DI wafer dump rinse. The second step is 15 min dip in solution SC2, a 5:1:1 mixture of DI water, H_2O_2, and hydrochloric acid, HCl, heated to 80°C, followed by DI water dump rinse and spin dry.

4.3.2 Photolithography

The photolithography process, widely used in semiconductor microelectronics, is a major part of the fabrication processes of microengine devices. It is the first step of patterning wafers of our devices. The dimension control and the pattern quality are mainly determined by the quality of the photo process. Except for the thick resist, which is unique to MEMS technology, all the facilities and procedures are the same as standard semiconductor photolithography.

A typical photolithography process has the following steps

(1) Pre-cleaning of the wafers

The first thing is to choose the right wafers for the job. Check the wafer specifications such as thickness and bow, patterns already on the wafer, and type of films according to the design, and check the clean room temperature and humidity which is important to the photoresist adhesion on the wafers especially to thick photoresist. The humidity should be between 38 and 42% and the temperature should be between 60 and 70°F.

For pre-cleaning, use piranha to clean new wafers or wafers stored in boxes for a while after previous cleaning. For wafers just done with a DRIE etch, use O_2 plasma ashing to remove residual polymer before piranha cleaning. A barrel-type asher is preferred because it can load wafers vertically so neither side of the wafers is in contact with anything. For wafers in MEMS devices to be assembled by Si fusion bonding, it is very important to protect the bonding surfaces, on both sides of all the wafers, from beginning to end of the processes.

(2) Priming of the wafer by HMDS
This step is to promote the adhesion of photoresist to the substrate using Hexamethyldisilazane (HMDS). There are specific HMDS recipes for each type of resist. After the HMDS treatment, which takes 20–30 min, the wafers should be cooled to room temperature before coating resist.

(3) Photoresist coating:
This step applies photoresist uniformly on the wafer by spinning the wafer at about 1000–5000 rpm on a vacuum chuck after dispense and/or spread of the resist on the wafer. The resist thickness, ranging from 1–2 μm for the OCG825 standard resist to 8–12 μm for AZ4620 and AZ9260 thick photoresist, decreases as the coater speed increases. These resist are positive resist, that is, the exposed region in the resist film will be washed away in developer.

Resist edge beads, which could be thicker than the resist film itself and cause inaccuracy in exposure by contact aligner and pinhole failure during DRIE, are removed by wiping the edge with a cotton swap in the last few seconds of the spinning. The edge beads on the flat of the wafers are wiped away when the coater stops. This step also avoids contamination of the wafers by resist debris generated when the edge beads are crashed on the photo masks.

The regular coater chuck used to mount the wafer on the coated is designed for single side photo work. It is usually smaller than the wafer, and resist dots always splash on the edge of backside of the wafers where it is not covered. The splashing of photoresist, if on the region to be patterned, could cause insufficient exposure where the dots with extra thickness are. Even for wafers to be patterned on only one side, the resist splashing on the back is undesired because it causes leaking of the cooling helium gas in DRIE chamber, which is sealed off by a rubber lip seal in contact with the back side of the wafer. To prevent resist splashing on the back side of wafers while spinning, we designed a coater chuck that has the same size as the wafers. The small inconvenience with this large chuck is that the wafers have to be grabbed on the edge manually when loading on the coater.

(4) Prebake
Prebake removes the solvent in the photoresist. It is done in a convection oven at 95°C for 30 min for thin resist and 60 min for thick resist. The wafers are loaded in Teflon carriers, normally vertically for better venting. It is important not to over bake the resist because this could result in longer expose time and cracking of the resist. On the other hand, under baking could cause resist failure in the form of "pox," circular pieces of resist popping off after 10–15 min DRIE etch as seen in Fig. 4.5 where both the popped and delaminated but not yet popped poxes are visible. This problem is more serious when the humidity and temperature is high. This is due to the out-gassing of the remaining solvent in the photoresist film in vacuum. It should be pointed out that contaminations on the wafer could also cause similar problems.

(5) Exposure
The prebaked resist is exposed using a contact aligner with wave length of 365 nm and light intensity of 10 mW/cm^2. The exposure time is typically 2 s for single coat of thin resist and 20 s for single coat of thick resist. The photo

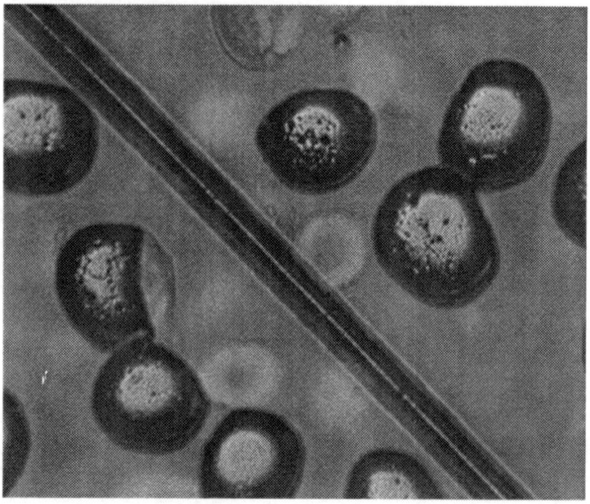

Fig. 4.5 pox failure in resist due to high humidity. The diagonal trench is a journal bearing. The rough circles with dark rims are etched Si exposed by poxes. The light dots near the journal bearing are poxes not popped yet

masks are sodalime glass plate with chrome patterns etched according to the device designs. The mask is put in contact with the coated wafer and then the UV light is turned on to print the mask pattern onto the wafer in a 1:1 ratio.

For wafers already patterned, the alignment between the patterns must be within the design specification. The alignment accuracy is especially important for the wafers that form the rotors of the devices because large misalignment means large rotor imbalance. The aligner we used can align a mask directly to the pattern on the top side of a wafer, or to the pattern on the bottom side of the wafer with an accuracy of approximately 2 μm. The front-to-front alignment and front-to-back alignment is made possible by global alignment marks etched onto each wafer on both sides at the beginning of the process on each side. In our masks, the global alignment marks are two cross hairs of 3 μm line width and 70–90 mm apart on the horizontal center line of the patterns. All subsequent patterns have same alignment mark pattern and are aligned to the global alignment marks. The global alignment marks should be shallow etched into Si, typically 0.5–1.0 μm, after oxide etch, since they are to be used in final bonding when oxide is stripped. Failure to etch the global alignment marks into Si would make wafer to wafer bonding alignment impossible since the alignment marks etched only in the oxide film would be washed away in the oxide stripping step before silicon to silicon fusion bonding.

The first time a wafer is patterned, it is critical to align the center of the mask pattern to the center of the wafer, so that when any two wafers are bonded their edges are flushed. Neglect of this pattern centering to the wafer could result in bonded wafer stack with large shifting, and difficulties in later processes. This can be done by setting initial positions of the mask and wafer precisely in the aligner, or more preferably using a global alignment mask that can locate and align to the edge, including the flat of the wafers.

The alignment accuracy depends on, in addition to machine resolution, the wafer bow and the resist type. Large wafer bow could cause defocus of one or both alignment marks. Thick resist over the shallow etched alignment marks could make the alignment marks difficult to focus on.

Some of our wafers need two coats of thick resist for long-time DRIE etch because the resist is also etched in DRIE, with a typical Si to photoresist etch selectivity in the order of 100: 1. This selectivity is a function of various parameters of the DRIE machine, including chamber pressure, powers, cycle times, and flow rates of etch and passivation cycles. The thickness of the double-coat resist film is typically 20 μm, about twice the thickness of a single-coat resist film. The exposure time, however, is much longer than double of the expose time for single-coat resist films. Usually the wafers are exposed using interval mode, in which a shutter opens and closes a few times to prevent the resist from burning by long-time continuous exposure to UV light. A typical interval mode exposure recipe is 15 s exposure and 15 s waiting for 5 cycles.

(6) Development

After exposure, the wafers are put in a developer solution which dissolves the UV light exposed regions in the resist and form the patterns in the resist. The developing time is about 1–2 min. We noticed that the edges of the features in a pattern are cleared first, often leaving well-defined trenches along the edges, probably due to optical effect at the interface of exposed and unexposed regions. This effect is more prominent in thick photoresist. We observed that the developing of narrow features, typically in the order of a few to 20 μm in the case of thick resist, are completed much earlier than larger features, and the edges of large features are cleared first in developer because of this edge effect. It is safe to stop the developing and start to rinse in DI water once the large open features are clear. After rinse the wafers are spin dried.

(7) Postbake

Postbake is the last photolithography step. It helps the adhesion of patterned resist to the substrate. The postbake is also done in a convection oven under the conditions of 120°C for 30 min for thin resist and 90°C for 30 min for thick resist. This step is important for wafers to be etched in BOE. It is also necessary for wafer that requires smooth and shallow resist profiles. Without postbake severe under cut could happen in BOE. On the other hand over baking could cause photoresist cracking. After postbake we should wait for some time for the resist to cool down and relax before BOE to avoid resist failure due to stress.

When postbaking thick resist with narrow features such as 10 μm journal bearings, it is important to avoid loading two wafers horizontally with resist sides closely facing each other, such as in neighboring slots of a wafer carrier. Under such conditions the solvent vapor from one wafer can reach the other wafer, melt the resist and change the trench width or even fuse the patterns thereon. Figure 4.6(a) shows the journal bearing patterns exposed with same mask but post-baked alone (left, 9 μm opening) and with other wafers nearby (right, 4 μm opening). Figure 4.6(b) shows a set of partially filled trenches due to solvent melting. The solvent melting is more serious when wafers are

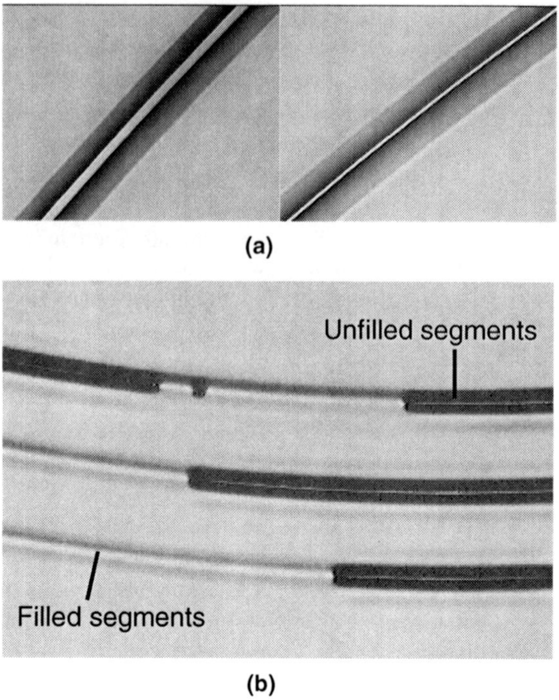

Fig. 4.6 Photo resist filling of patterned trenches during post bake due to solvent melting of the resist in small gaps between coated wafers loaded. (a) journal bearing trench width difference, left: without solvent melting, and right: with solvent melting. The exposed Si in the left picture is about 9 um wide. (b) partly filled pattern

horizontally loaded in the baking oven and the coated surfaces are close to each other. To avoid the damage by solvent melting, the wafers should be loaded vertically and far apart from each other, and if possible with patterns facing away from each other.

The postbake step has a big impact on the resist pattern profiles and therefore the dimensions of the DRIE defined components. After exposure and development the resist pattern has steep profiles, typically more than 60°. Postbaking makes the resist on the edge flow and forms a smooth shallow slope that could be as shallow as about 30–40°. Figure 4.7 shows the resist profiles before and after postbaking. On isolated small features where resist has limited space to flow the profile is steeper than on the edge of large features after postbake. The resist profile affects not only the dimensions but also the profiles of the DRIE features and will be discussed in more detail later.

4.3.3 Special Photo Processes

Below are two special photo processes that we developed to meet the needs of microengine devices.

(1) Double coats of resist
Sometimes the required DRIE etch time is too long for a single coat of resist to last to the end, such as in the case of narrow and deep journal bearing. We have

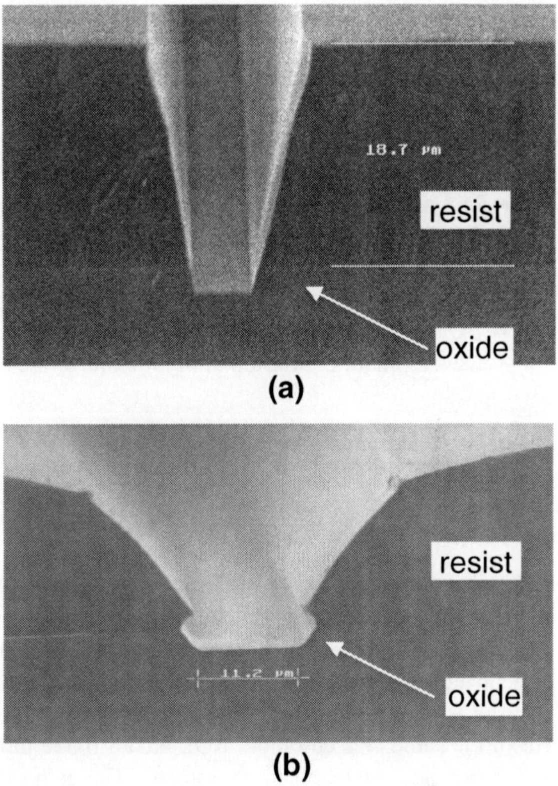

Fig. 4.7 thick resist profile before and after post bake

to make resist thick either by lowering the spin coating speed, or coat the wafers twice. Since there is a lower limit on the coater speed before we encounter problems of thickness uniformity, solvent out gassing and edge beads, we coat the resist at normal speed twice with a short baking in between. The spin parameter is the same for both coating. After first coating the wafer is pre-baked at 90°C for 15 min and cooled down to room temperature and coated again. The baking conditions after second coating is at 90°C for 75–90 min. The coater speed is variable according to resist thickness specification. Typically the thickness of the double-coated resist is about 20 μm, twice of a single coat. The total exposure time is 60–75 s, much more than double of the 20 s for single coat. The exposure is done with interval mode: 15 s exposures with 10–15 s cooling in between. The development time is about 2–3 min.

(2) Two-side photo process
When a wafer needs to be etched with the same process on both sides, it saves time to coat and pattern both sides in one photo process instead of two. The key factors are the prebaking and exposure steps. The following is a brief two-side photo process flow for a blank wafer.

- Preparation: clean the wafer, and wipe the coated chuck with solvents
- HMDS
- Coat one side with selected photoresist, remove edge beads with a clean room swap
- Prebake 1: 90°C, 5–7 min for thin resist, 10–15 min for thick resist
- Coat on the other side after the wafer is cooled to room temperature
- Prebake 2: 90°C, 30 min for thin resist, 60 min for thick resist
- Exposure 1: on one side, using bottom side alignment mode
- Development 1: put the wafer in developer with for 3–5 s, then rinse and dry the wafer
- Exposure 2: on the other side, under normal exposure conditions
- Development 2: develop until all features are clear. Rinse and spin dry

4.3.4 Wet Etching Processes

4.3.4.1 BOE Etch of Silicon Oxide

The wet processes used in the microengine device fabrication are oxide etch by BOE, a HF-based chemical solution, and nitride etch by hot phosphorous acid. There is no wet process for Si used for our devices mainly because the requirements of deep etch and precision control of dimensions are better met by DRIE processes.

BOE is a 7:1 mixture of NH_3F and hydrofluoric acid, HF. Compared to straight HF of the same concentration, BOE etches oxide more gently and more uniformly and causes less undercut to photoresist. The etch rate of silicon oxide by BOE is about 800 A/min. at room temperature. For thin oxide film less than 2 μm thick, photoresist can be used as the masking material for BOE. Keep in mind that to avoid resist failures such as lifting off, chipping, and cracking, postbake of the resist should not be skipped, and after postbake the resist should be set aside for a few hours of relaxation before BOE etch. This relaxation is especially important for thick resist and long oxide etch. For thicker oxide, since resist would suffer severe undercut due to the long-time etch, it is better to replace photoresist with a thin nitride film. The nitride has good adhesion to the oxide and the BOE undercut to depth ratio is about 1:1 for high-quality oxide. If there is a DRIE Si Etch after the BOE oxide etch, the overhung resist and nitride films should be stripped first. Otherwise flakes of the nitride film might fall off randomly and mask the Si, causing non-uniform etching of Si.

Often the BOE etch is done with the help of ultrasound agitation to drive out gas bubbles generated during the BOE etch and ensure a uniform etch. This is especially important to wafers with small features and thick resist. The ultrasonic agitation is done in an ultrasound bath filled with water. The degassing time is limited to 5–10 s every minute to prevent cracking and chipping of the resist from excessive exposure to ultrasound.

Unlike HF, BOE tends to roughen the Si surface slightly. It is therefore not the preferred chemical for oxide stripping before bonding. In the case, it has to be used for stripping oxide, care should be taken to limit the over etch time to no longer than a few minutes.

4.3.4.2 Nitride Wet Etch by Hot Phosphorous Acid

The wet etch process for nitride film uses phosphorous acid. The etch rate is about 2000 Å per hour when the temperature is set at 160°C. This process is mainly used to strip off the nitride film instead of selective etching with resist mask since none of the resist we use can survive in such high temperature. The phosphorous acid does not etch Si, silicon oxide, and even oxidized nitride. The LPCVD nitride we use is Si rich and when subject to thermal oxidation such as in local oxidation, a thin layer on the top of the nitride film would be oxidized. In this case, the wafer has to be dipped in BOE for about a minute to remove the oxidized layer before nitride stripping by hot phosphorous acid can proceed. After etching, the wafers are cleaned in DI water and spin dried.

4.3.5 Dry Etch Process: Reactive Ion Etch

RIE systems have been used in our process mainly to selectively etch silicon nitride film. Although most of the nitride etch can be done with our DRIE systems, the RIE system has the advantage of uniform etch across the whole wafer while the clamping fingers in DRIE systems leave unetched marks on the film. It is useful when we want to strip nitride completely on one side of the wafer for bonding purpose.

RIE process can also be used to etch oxide and silicon. However, in our processes oxide were mostly etched by a BOE wet process mainly for the control of sloped profile. Si etches, even shallow etches, were done by DRIE for precise dimension control.

4.3.6 Deep Reactive Ion Etch (DRIE)

DRIE basics

DRIE is the most important process for the microengine devices. We use DRIE machines to do all of our Si etches. Some of the microengine components have very tight tolerances with regards to lateral dimensions and aspect ratios that have never been demonstrated. Therefore, some of the basic issues of the DRIE process that are critical to our devices such as etch dependence on various parameters, uniformity, and profile control were systematically studied over the course of the microengine development [1–8].

Our DRIE machines are inductively coupled plasma (ICP) systems from Surface Technology Systems (STS) (see the photo in Fig. 4.8). It adopts the Bosch process to precisely etch high aspect ratio features into Si wafers. An ICP system has vacuum chamber that includes a shower head for feeding

Fig. 4.8 A photograph of the DRIE system by STS heavily used in microengine fabrication

the etching gas (SF_6), and passivation gas (C_4F_8), a RF coil that generates high-density plasma, and a capacitively coupled platen that provide ion acceleration. In our systems, the wafer to be etched is clamped on the platen by a set of ceramic fingers. The clamping force is provided by the weight of a shielding tube to which the ceramic fingers are attached. The wafer is cooled by helium on the back. The sealing of helium is achieved by a rubber lip seal on the platen.

The Bosch process consists of cycles of passivation by C_4F_8 and etching by SF_6, as schematically shown in Fig. 4.9. In the passivation cycle a thin polymer layer is isotropically deposited on the wafer, covering the surface of the wafer, the bottom and the side walls of the features being etched. In the etching cycle, the polymer film on the wafer surface and the bottom of the Si trenches is removed by the bombardment of positive ions from SF_6 plasma accelerated by the platen voltage. Since this ion bombardment is directed vertically to the wafer, the polymer on the side wall of the Si trench is mostly intact. Only the bottom of the trench is exposed to the etching species and etched down a little, 1–2 μm typically, in each cycle. It is the cycling of the passivation and etching steps that makes anisotropic etch of Si possible.

DRIE parameters

The various aspects of DRIE etch, such as etch rate, selectivity, and etch uniformity are determined by many parameters. Those that usually need to be adjusted in recipes are flow rates of SF_6 and C_4F_8, chamber pressure, cycle times of passivation step and etching step, coil powers, and platen powers.

Etch selectivity

The etch rate of Si in our systems is about 1–3 μm/min. and dependent on many of the DRIE parameters. The masking materials, the thick resists we used, AZ4620 and AZ9260, and thermal and CVD deposited oxide film also has a finite etch rate in DRIE, about 0.02 μm/min for resist and 0.01 μm/min for oxide, partly due to the ion bombardment and partly due to etching by the

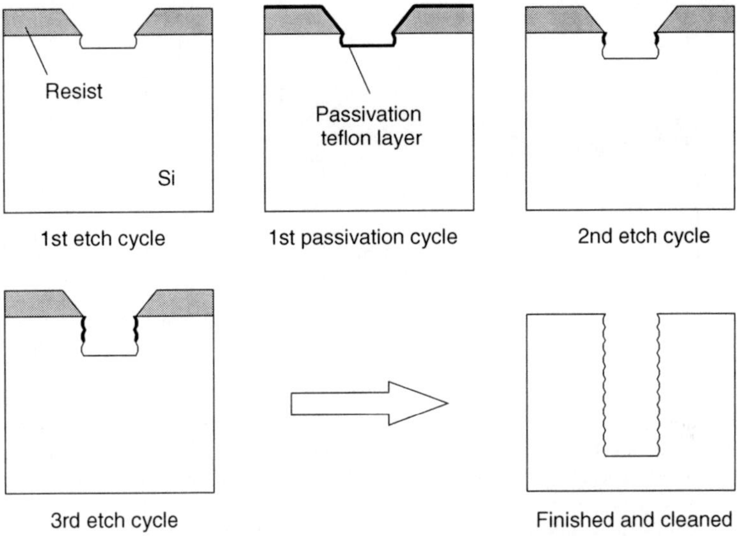

Fig. 4.9 Bosch process for anisotropic etch of Si. In the etch cycle Si is etch isotropically for a short time. In the passivation cycle a thin layer of fluoric polymer film is deposited isotropically. In the etching cycle the passivation film on the bottom of the Si trench is blasted off by ions accelerated at normal incidence and Si etch is resumed at the bottom while the side wall is protected by the passivation layer. Repeated etch and passivation cycles results in high aspect ratio etch of Si

active species in the SF_6 plasma. The selectivity of Si to resist is typically 80 to 1. The selectivity of Si to oxide is typically about 150 to 1. The etch selectivity is sensitive to chamber pressure and platen powers because they have more impact on the etch rates of resist and oxide. For example, Si to oxide selectivity can reach 250 to 1 with increase in pressure.

One consequence of the finite etch rate of resist and oxide is that the DRIE feature sizes increase noticeably when they are used as masking materials and have a shallow slope due to postbaking for resist, see Fig. 4.7 or isotropic BOE etch for oxide. As schematically shown in Fig. 4.10, when the resist or oxide with a slope is thinned down by DRIE, even in the case of completely anisotropic etch, the bottom lines would recede from the original position at a rate of

$$R_{//} = R \operatorname{ctg} \alpha$$

where R is the etch rate and α is the slope angle of masking resist or oxide. This effect sets a limit of the minimum feature size to $2R_{//}t$, where t is the DRIE etch time. Because of this, usually we have to offset the mask for the features that need long DRIE etch by about $2R_{//}t$ to achieve the desired feature size by the end of etch. 2 $R_{//}$t can also be taken as the lower limit of feature size.

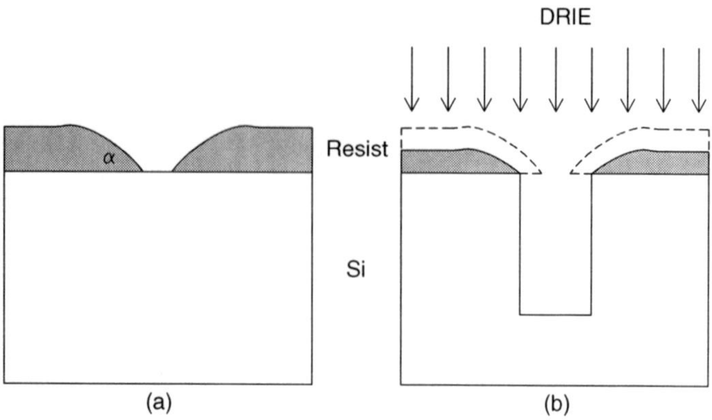

Fig. 4.10 sloped resist etch in DRIE causes trench widening, (a) after post bake, and (b) after DRIE. As the resist get thinner by DRIE, its bottom edge recedes away from each other, exposing the Si underneath. This effect set a lower limit on the trench width for a given depth

Since postbaking is an integral part of many processes such as those involves BOE etch of oxide, resist slope is unavoidable and therefore when precise feature size is required, it is necessary to offset the mask dimensions based on the resist baking conditions, feature size, DRIE recipe, and DRIE etch time. We will also see later that sloped resist or oxide profiles are desired for DRIE profile of narrow features like journal bearings.

Etch rate variations

DRIE etch rate varies with time and feature size. The etching slows down as the depth increases. Large features are usually etched faster than small features. This is the so-called loading effect. Both effects have something to do with the depletion of the etching species as the etch goes deeper. In the mask design one should try to avoid including both large and small features if depth uniformity is required. If both large and small holes are to be etched in a wafer, halo mask can be used to achieve uniform etch. As shown in Fig. 4.11, on the mask there are large and small donut shaped patterns with the same width. The outer diameter of the donuts defines the feature sizes.

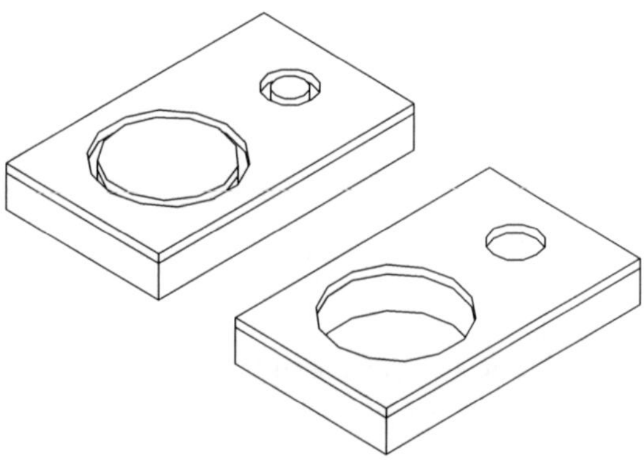

Fig. 4.11 Halo mask process. Halo patterns of different size but same width are patterned in the resist. The etch proceed uniformly because of the same narrow width. When the substrate is etched through the center pieces fell out and yield two holes of significant different size

DRIE will be uniform since the exposed Si areas have same width. When the wafer is etched through, the center pieces in the halos fall off, forming large and small holes in the wafer. The freed halo pieces should be collected and disposed of after the etching. The finished wafers should be checked for any halo pieces stuck inside their trenches or even worse on the surface. The former case could cause blockage and affect the function of the device. The later could cause bonding failure.

When large and small features have to be etched into a wafer from the same side to about the same depth but not through, the etching of the large features have to be delayed, say by an oxide film. This process will be discussed in detail later.

Etch uniformity

Because of the none-uniform distribution of plasma and etching species in the DRIE machine, Si etch rate typically is not perfectly uniform across the wafer, varying by a few percent. This is called global variation. Global variation is more of a problem in large loading patterns such as the early blade patterns. This is another reason to prefer halo etch because the widths of the donut shape can be designed small to reduce etch non-uniformity.

DRIE etch uniformity is also affected by the feature size variation, i.e., the loading effect as mentioned above, and the density of the patterns. The Si etch rate is lower in regions with crowded patterns than open area. Adjusting the DRIE parameters could to some extent improve the uniformity but sometimes compromise has to be made in device design.

Wafer mounting

In our DRIE system, the wafer is directly cooled by helium on the back. The helium gas is sealed from the plasma chamber by a rubber lip seal between the wafer and the platen. Large leak into the plasma chamber, which happens when the back side of the wafer is not clean and smooth or the wafer has through holes, could trigger an interlock and stop the etch. Therefore, it is obvious when a wafer is to be etched through, some type of back-side coverage is needed. A common practice is mounting the wafer to be etched on a handle wafer, either Si or quartz, using photoresist as binding material. The key to the success of this mounting method is good heat conduction through the resist film in between the device wafer and handle wafer. Both wafers should have good contact to the resist film. The resist should have venting channels patterned for the solvent in the resist to escape. Such a resist pattern looks like a target with venting channels as shown in Fig. 4.12. The mounting resist has to be baked just right: not too wet or too dry. Wet mounting could cause the solvent come out in the DRIE chamber where the pressure is low and the temperature is high, and push the device wafer away from the handle wafer. The result is larger gap, deteriorating thermal conduction between the wafers, heating up of the wafer, higher etch rate, burning of photoresist, and often loss of the device wafer. Over baking would make the resist fragile and a small impact during handling or from the clamping in the DRIE chamber could crack the resist and cause thermal conduction problem.

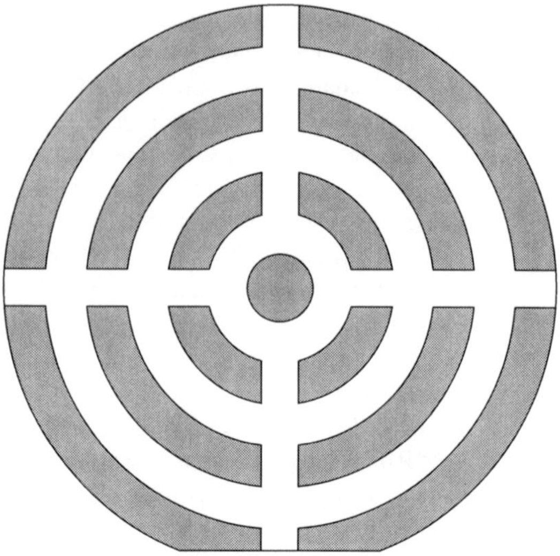

Fig. 4.12 Photo resist pattern in wafer target mounting. The dark areas are resist. There are 4 venting channels leading to wafer edges

Mounting failure is common with this common practice. Here is a process flow of the resist mounting process that proved relatively reliable.

- Choose a clean handle wafer. Wipe clean the backside of the device wafer with solvent.
- Set the coater at a speed of about 2000 rpm and a longer spin time. Put the handle wafer on coater, start the coater, and apply resist. Once the wafer is covered with resist, stop resist dispenser and wait for the resist to spread for 12–15 s.
- With the coater continuing spinning, squirt acetone onto the wafer, carefully from outside of the wafer and move in until a coin size of resist remains in the center. Stop squirting and wait for 12–15 s with the coater spinning.
- Start resist dispenser and move it from outside of the wafer inward till the resist reaches about 10–15 mm from the center spot. Stop the resist dispenser and wait for 12–15 s with the coater still spinning.
- Squirt acetone, carefully from outside of the wafer and move in until the newly coated resist becomes a ring of about 10 mm wide. Stop squirting and wait for 12–15 s with the coater still spinning.
- Repeat steps 4 and 5 till the wafer edge is covered.
- Remove edge bead by wipe the wafer edge with a clean room swap or fab-wipe while the wafer is spinning. Stop the coater and wipe the edge bead on the wafer flat. Make sure there is no edge beads left. Form four venting chennels using acetone soaked Q-tips to open the closed photoresist rings.
- Find a solid surface, put a clean fab-wipe on it and then put the handle wafer on the fab-wipe, resist side up. Put the device wafer on top of the

handle wafer and align them. Press the device wafer on places without patterns evenly throughout the wafer with a plastic tool. Apply enough force to initiate contact between the device wafers to the resist on the handle wafer, but be careful not to damage the wafers. The contact will be at the points pressed if seen through a quartz wafer. If the pattern on the device wafer is too crowded to press, put the device wafer face down on a clean fab-wipe and press on back of the handle wafers. Make sure to check the device pattern after pressing.

- Bake the wafer stack in an oven at 90°C for 20 min. Now the resist contact should be connected and is composed of uniformly distributed fine bubbles. Smooth resist surface means there is no contact between the wafers and should be avoided. The wafer is now ready for DRIE etching.

This mounting method is also used for wafers that would turn fragile after a half way DRIE etch because they tend to break in the DRIE chamber under the large clamping force.

Another method for sealing the wafer to be etched through is applying a clean room compatible tape on the back of the wafer to be etched. It is important to coat a layer of photoresist on the back of the wafer so the tape does not touch the wafer surface directly. Usually the tape can be peeled easily after baking at 90°C for a few minutes. If the wafer is fragile for the peeling off, put the wafer in acetone and the tape will detach quickly. It is not suitable, however, for wafer too fragile to sustain the force of the clamping fingers in the DRIE chamber.

Wafer edge protection and backside protection

For MEMS devices to be assembled using fusion bonding, it is important to protect the whole wafer throughout fabrication processes, including the backside and edges. Particles on the wafers, even on the edge where no device pattern is present, could result in large unbonded regions. Bonding defects due to particles strongly affect the yield of our multilayer devices. There are two major sources of particle on the wafers: (1) particles floating in the air and (2) particles from processing especially from DRIE etch. To minimize wafers exposure to the particles in the air, care should be taken to transfer the wafers in clean boxes from the RCA station to the bonder, and minimize the time the wafers have to be exposed during bonding alignment. During a long DRIE process, the wafer edge often suffers photoresist failure in the form of pox as seen in Fig. 4.5: small rounded pieces of resist popped up in the DRIE chamber due to out gassing of excessive solvent in resist edge beads on the wafer edges, and chipping from wafer handling. Subsequent DRIE etch of the exposed Si generates loose Si particles.

Close to the bonding surfaces, these particles could easily land on the wafer surfaces and cause bonding defects. Therefore, it is important to protect the edge of the wafers in our multilayer devices from DRIE damage. In addition to well-controlled photo process, sometimes we cover the edge of the wafers with a tape when doing deep Si etches. This is necessary because the edges of the wafers are not always covered by resist and the thermal oxide is usually not thick enough to last the long DRIE process. Again precaution is taken to apply tape always on photoresist rather than on the wafer surfaces directly to minimize the potential contamination of the wafer surfaces by the tape residue.

The back side of the wafer has the edge outside the helium sealing ring exposed to DRIE plasma and could be attacked if not protected properly because the SF6 etching radicals are isotropic and do attack the passivation polymer layer. For short etches, the oxide film is enough a protection layer for Si underneath, but for long etches we usually coat a ring of resist on the edge as protection.

Charging effect

The masking materials we used for DRIE are either photoresist or silicon oxide. They are insulating and tend to charge in the plasma environment by accumulating electrons, especially on the edges and pointed ends of the patterns. The negative charging on edges of the resist or oxide patterns bends the normal trajectory of the positive ions toward the side walls of the etched Si trenches. This could be a contributing factor of the vertical striped features often seen in the lower part of the side wall of the Si trenches, as shown in Fig. 4.13: the redirected positive ions attack the protective passivation layers on the side wall. The consequence is erosion of the Si at the side walls and change in the profiles of deep and narrow features such as journal bearings.

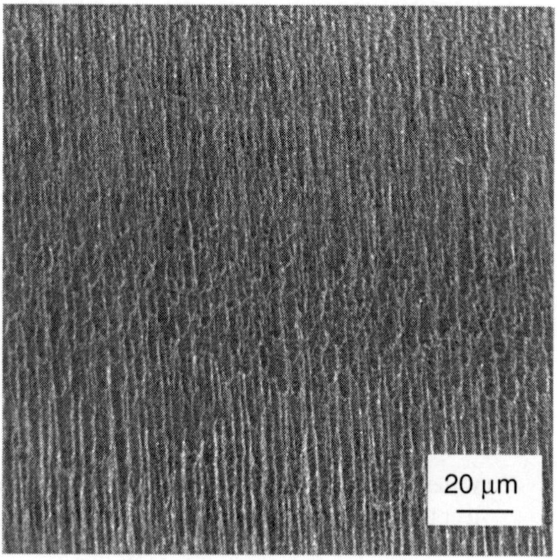

Fig. 4.13 SEM picture of vertical side wall roughness from DRIE

Profile control

As pointed out earlier, DRIE profile control is a critical issue for microengine devices because of the tight tolerances, especially those of journal bearings. It is in principle an issue of supply and demand of the etching species in the DRIE environment. Generally speaking, the profile of DRIE etched features tends to be negative tapered when the etching species supply is abundant such as in the cases of large feature size and high pressure, and positive tapered when the supply of the etching species is insufficient as in the case of small feature size and low pressure. It is normally enough to use recipes with fixed parameters to achieve the designed profiles in the narrow trenches of up to 350 μm deep. As for super deep (>1000 μm) narrow trenches such as blade plus journal bearing etch in SSE, it is necessary to use a recipe with parameter ramping to achieve high aspect ratio profiles. A parameter ramping DRIE recipe gradually changes one or more parameter over a set time automatically to cope with the slowing down of the etch rate over time.

In addition to DRIE recipe, the resist also plays an important role in profile control because the accelerated ions aimed to remove passivation layer could be deflected from the resist slope or redirected by the charging on edge of the resist pattern. More details will be given later in the discussion of journal bearing etch challenges.

4.3.7 Fusion Bonding

Wafer level silicon fusion bonding is used to assemble most of the microengine devices, except the generator which currently uses eutectic bonding on chip level as part of the assembly procedure due to the temperature limit set upon the device by the magnetic materials integrated in it. With respect to the fabrication success, wafer fusion bonding is the most critical technology because it is easily affected by wafer conditions such as surface smoothness, wafer bow, wafer thickness, contaminations, particles, stress, patterns in the wafer, and in the case of the wafer stack bonding, the process history of the stacks to be bonded. Because the bonding is usually done after the wafers are finished or heavily processed already, any defects or failure in bonding would cause heavy loss of resource and time. Bonding of up to 10 wafers with small and complicated patterns poses a big fabrication challenge that we have to deal with. Figure 4.13 shows as examples some IR images of bonded microengine wafers: (a) two-layer bonding in a turbo charger showing bonded compressor blades, (b) three-layer bonding in an SSE showing the bonded turbine blades and thermal shafts (dark dot in the center of the rotors), (c) six-layer bonding of a bearing rig showing overlapped features the rotor area, and (d) ten-layer bonding of a complete SSE.

The detailed microengine bonding process will be described in more details in the SSE fabrication section. Below is a general process flow for two wafer fusion bonding

Wafer preparation:

- Selection of wafer: the wafer must be highly polished on both sides; the wafer bow is preferably less than 10 μm for 6 in. wafers; the wafer TTV is preferably less than 2 μm and in the case of rotor layers less than 1 μm for 6 in. wafers.
- Protection of the bonding surfaces: the photolithography, wet and dry etching and cleaning process, if directly operated on the wafers, could render the Si surfaces unbondable. Our solution is growing a thin thermal oxide film on all the new wafers to protect the bonding surfaces, and strip this protective oxide film only right before the bonding.

Cleaning of wafers

- Since wafer bonding requires high level of wafer surface cleanness as mentioned earlier, wafer cleaning before bonding is of great importance for bonding success. Almost all of our wafers finished with a DRIE process that leaves a polymer coating on the surface and the side walls of the etched features. As mentioned in the photolithography section, the polymer coating can only be removed by oxygen plasma ashing, so the wafer cleaning process after DRIE etch follows the steps of ashing for 1 h in a barrel chamber, piranha cleaning for 10 min, rinse in DI water for 5–10 min depending on the pattern complexity in the wafers, and spin dry. The cleaning should be arranged in succession with the later steps of oxide stripping, RCA, and bonding. Storage in plastic boxes after cleaning for a long time is not recommended because out gassing from the wafer box could deposit onto the bonding surfaces a thin layer of plastic material that turns into bubbles during subsequent high temperature annealing.

Stripping of protective oxide film

- The bonding surfaces should be covered throughout the process until they are ready for bonding RCA cleaning step. The stripping of the protective oxide film should be done in BOE or HF solution. A dry process is not suitable here due to the risk of bonding surface damage. Only the surfaces to be bonded should be striped of the protective oxide film. For intermediate bonding steps where the wafers or wafer stacks are to be bonded only on one side, the backside, if to be bonded later, should have the oxide on. For wafers with only oxide films, this can be done by coating the backside with resist, carefully cleaning any resist residue from the bonding surface by squirting solvent on the wafer while it is spinning, using BOE to strip the oxide film on the bonding surface, and using piranha to strip the resist before RCA for bonding. It should be noted that before coating resist on the back side, the priming step should be skipped since we observed that the Si surfaces had spots that are not hydrophobic after BOE although the oxide is removed. Care should be taken to avoid other contaminations from baking oven too. Over etching in BOE should be avoided because long-time BOE etch tends to roughen the Si surface. By taking the wafer out at the end of the BOE stripping to inspect the "sheeting," the transition from hydrophilic oxide to hydrophobic Si, the over etch in BOE can be easily limited to within 1–2 min, much shorter than the time needed to roughen the Si surface. Once the oxide is stripped, the resist needs to be stripped with piranha before moving to next step.

For wafers with protective oxide film and stress compensation nitride film on top of the oxide, the nitride film should be etched with dry process first and then the oxide can be stripped with HF solution. There is no need to coat resist on the back side since the nitride film there protects the oxide film from stripping in HF while maintaining the flatness of the wafer. This is clearly a cleaner process than the resist covering method, with extra process steps though.

RCA cleaning

- All wafers to be bonded have to go through the same RCA cleaning as described in the thermal oxidation process. We usually add 1–2 pairs of new blank wafers in the batch to use as dummies in the wafer alignment and pressing steps later.

Bonding alignment and pressing

- The alignment of wafers to be bonded is done in an aligner similar to the exposure aligner used for lithography. It has two working modes: silicon direct bonding mode and fusion bonding mode. With the silicon direct bonding mode two wafers are aligned, put into contact, and pressed in the aligner. The bonded pair can be treated as a single wafer and bonded to another wafer. This way multiple wafers can be bonded in one batch, and the contacted wafer stack can be further pressed in a bonder chamber to promote the bonding area and strength. The contact in this mode is operated in atmosphere and therefore is only suitable to wafers that do not form sealed cavity once in contact because any sealed cavity filled with air will deform the wafers due to the pressure buildup during annealing at over 1000°C. The fusion bonding mode does not have this problem. In the fusion bonding mode three thin metal sheet flags are first put in between and around the edges of two wafers to be bonded, and then the wafers are aligned, clamped in a chuck and transferred to a bonder chamber. The bonder chamber is first pumped down to a preset vacuum level, usually about 10^{-3} mT, then a ceramic wafer bow pin is pressed onto the center region of the wafers to establish contact, the flags are then pulled out and the wafers are pressed together at the predetermined force and hold for certain time.

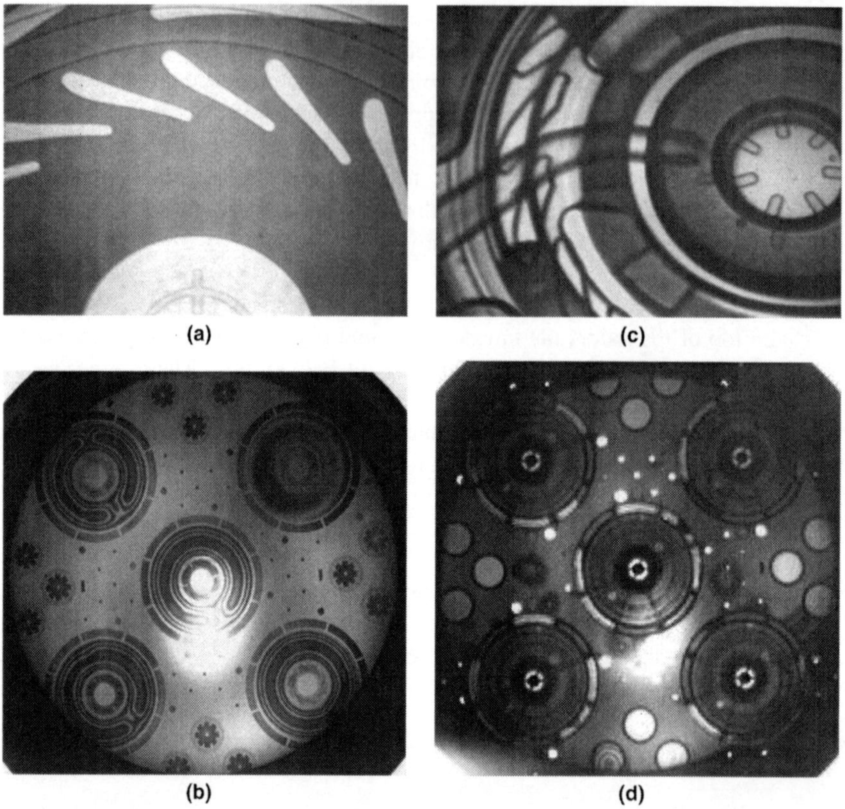

Fig. 4.14 IR images of bonded microengine wafers: (a) 2 layer bonding in a turbo charger showing bonded turbine blades, (b) 3 layer bonding in a SSE showing the bonded turbine blades and thermal shafts (bright dots in the center of the rotors), (c) 6 layer bonding of a bearing rig showing overlapped features the rotor area, and (d) 10 layer bonding of a complete SSE

Once the wafers are pressed the bonding quality is checked with infrared camera. Bonding quality inspection using IR imaging is the reason we choose the IR transparent N type wafers. Figure 4.14 shows some examples of the IR picture of bonded microengine devices.

Annealing

- The last step in bonding process is the annealing of the bonded stack in a diffusion oven to realize the fusion bonding of the Si wafers. The typical annealing conditions are 1000°C, 1–2 h, in flowing N_2. Because of the need of multilayer bonding in microengine devices, the annealing temperature in some of our intermediate bonding steps is as low as 800°C to reduce the risk of Si creeping, or plastic deformation under stress.

4.3.8 Anodic Bonding

We use anodic bonding to bond Si and borosilicate glass mainly for structure reinforcement and packaging. We choose Pyrex 7740 glass for its close coefficient of thermal expansion (CTE) match with Si at the bonding temperature. In the anodic bonding process the Si and glass are aligned, clamped and then heated to around 300°C at which the sodium ions became mobile. A high negative voltage is applied across the Si and glass stack. The electric field drives the sodium ions away from the Si to glass interface. The depleted glass surface is highly reactive and can form a strong solid chemical bond with Si.

Anodic bonding is more tolerant to surface conditions because of the strong electrostatic force involved. Therefore, the preparation of the wafers is not as rigorous as in Si fusion bonding. Usually a piranha cleaning is enough for anodic bonding.

The anodic bonding can be done on wafer level or on chip level depending on the device needs. For turbo charger and SSE, the package is a tri-layer structure of Si/glass/Si as schematically shown in Fig. 4.15.

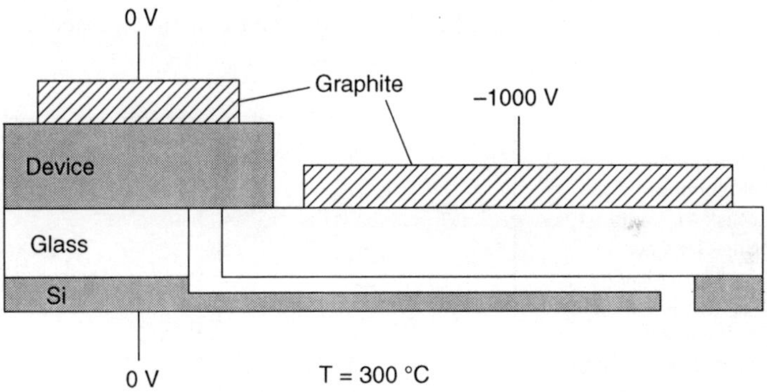

Fig. 4.15 Anodic bonding scheme for packaging of turbo charger and SSE. The two bondings between the device and the glass and between the glass and the Si are done at the same time. The graphite is used for better electric contacts

4.4 Critical Microengine Fabrication Processes

In addition to the three B challenges mentioned above, we also experienced other difficulties in microengine fabrication such as non-uniform etch of plenum with large loading spread that causes wafer failure, nozzle etch, rotor retaining on to the wafer before final bonding, and side wall protection in various situations. In this section, we will discuss the causes of these microengine specific issues and present our solutions. In solving individual fabrication problems, one has to keep in mind that the processes for a device are interconnected and a small change in one step could a cause complicated chain reaction and affect many other steps. Compromises are often made to ensure the completion of the whole process.

4.4.1 Plenum and Nozzle Processes

The plenum and nozzles are usually on the opposite side of the wafers next to the rotors, with the nozzle facing the rotor. Figure 4.16 shows the masks of the plenum layer on the turbine side of a turbocharger. The original numbers of each mask are used. The plenum mask, 17, includes large loading plumbing channels. In mask 21 are nozzles and other feeding holes connected to the plenum pattern. The nozzles are a ring of small holes in the center that feed air to the thrust bearing. Figure 4.17 shows the mask drawing and an SEM picture of the nozzles of turbo charger. The haloed features outside the nozzles are compressor inlet. The diameters of the nozzles are typically 10 μm in the wafer which means 7–8 μm in the mask considering the mask to wafer offset for DRIE due to resist etch. The depth of the nozzles is typically 100–150 μm. Mask 12 is for the tip clearance on nozzle side and for the plenum delay on the plenum side. Mask 15 is for the blade tip clearance.

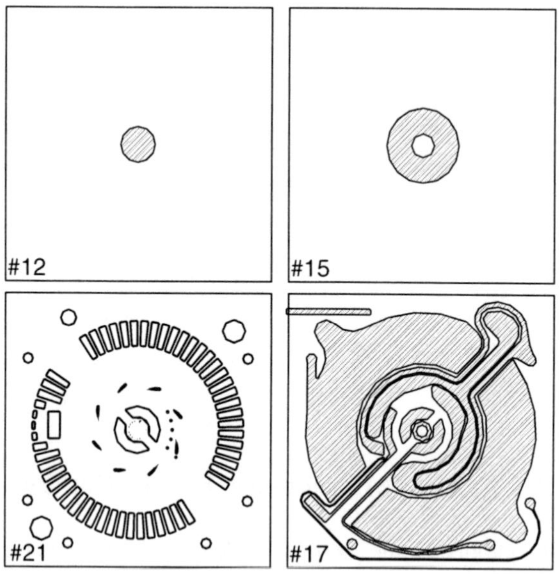

Fig. 4.16 Turbo charger masks for front plenum wafer, with original numbers in the mask set. Mask #12 is the thrust bearing clearance, mask #15 is the tip clearance, mask # 17 is the plenum, and mask #21 is the nozzle pattern. The dark areas are to be etched. The original mask numbers are used

There are two challenges associated with nozzle etch. One is that the DRIE time is usually long and it is hard to determine the depth because of the small size of the nozzles. The other is that the ring of nozzle holes distortion due to non-uniform resist reflow in postbake. As seen in Fig. 4.16, the nozzles are only about 30 μm to the edge of the resist pattern defined by the two halo shapes, and we know the reflow

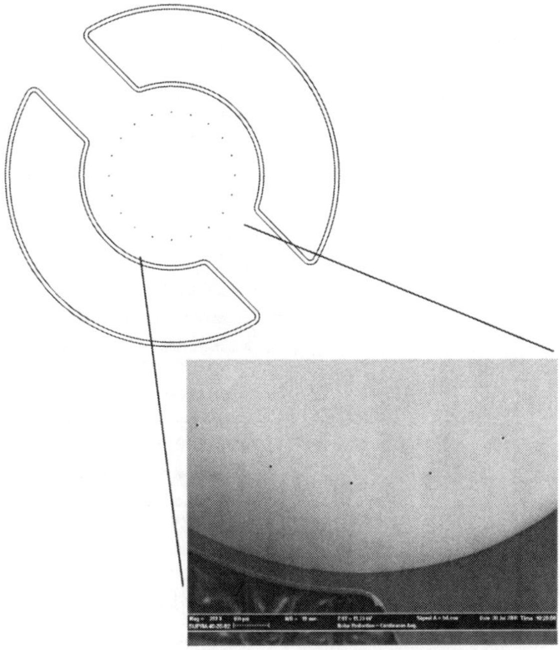

Fig. 4.17 Close up of mask 21 showing the nozzles and the haloed exhaust holes. The inset is a SEM picture of the nozzles etched in a turbo charger wafer

in postbake 10 μm thick resist extends more than 15 μm from the pattern edge. The proximity between the nozzles and the two halo patterns leads to distortion of the resist near the nozzles by postbaking. We have observed triangle nozzle holes after postbaking. The solution to this problem is to etch the nozzle first without post-baking, but with certain over etch so that when the plenum etch is finished the nozzles are easily visible. The side walls in the nozzles can be protected from plenum etch by a thin thermal oxide or a DRIE polymer deposition.

The major issue associated with the plenum is non-uniform etch across the wafer and between features of different sizes within a die. The plenum should be etched into a 450 μm thick wafer with 100–150 μm left, the depth of the nozzle. However, the plenum pattern in mask 17 could cause 30–50 μm etch depth variation, usually over etched if the nozzle length is to be achieved, across the wafer and within a die. This makes the large features very thin and fragile and the wafer easy to break from handling. In few cases of poor mounting on handle wafer, the large plenums in dies near the wafer edge were etched through before the nozzles in center dies were clear. To our relief, usually only the nozzle dimensions are critical and other features in the same wafer have larger tolerances, therefore, we can open up and etch the nozzles by means of nested mask and stop the plenum etching whenever the nozzle dimensions are reached.

A complete plenum nozzle process flow is shown in Fig. 4.18. It is the aft plenum wafer process for a turbo charger device, but the major steps are shared by both the forward and aft plenum wafers in most of the microengine devices. Photoresist is

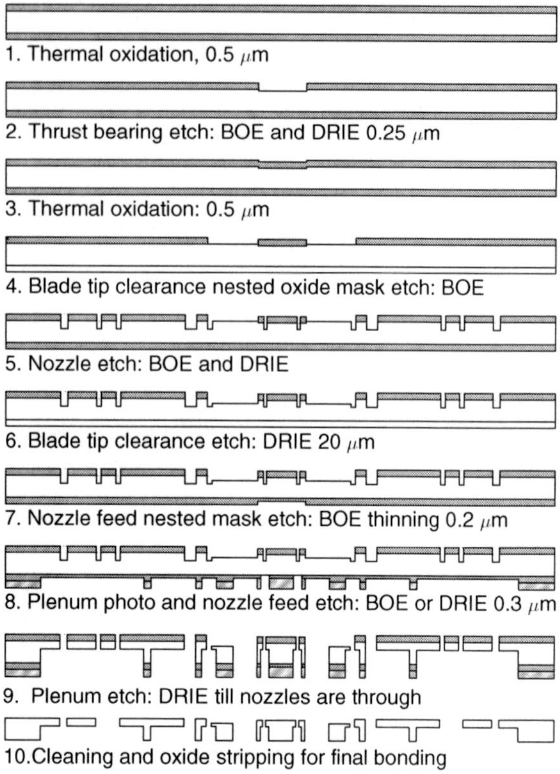

Fig. 4.18 Process flow of a plenum and nozzle wafer with features of large loading difference on plenum side

not shown in most of the steps for clarity, but keep in mind that except for the nested mask etch all of our etch steps start with photolithography process. Brief caption is given to each step in the figure. Below are detailed descriptions and explanations of each step.

Step 1: thermal oxidation

This oxidation is the same for all the wafers. The purpose is to protect the bonding surface from process damage. It starts with new blank wafers and follows the standard thermal oxidation procedures.

Step 2: thrust bearing etch

This step defines the thrust bearing clearance using mask 12. It is a circle that encloses all the nozzles in nozzle mask 21 in Fig. 4.17. The thrust bearing clearance is only 2.5 μm and takes less than a minute to etch in DRIE, so thin resist can be used in this step. After standard photo process, the oxide is etched by BOE with ultrasound agitation for uniformity. Then the thrust bearing clearance is etched into Si using DRIE. Dektek profiler is usually used to monitor the etch depth.

The wafer global alignment marks are etched in this step, the top-side alignment marks are from mask #12 and bottom side from a mask with only alignment marks. All of our masks have some global alignment marks and any of them can be used here on the bottom side as long as the device pattern is covered, either on the mask during exposure with a shadow cover or on the wafer by resist painting before postbake. Two-side photo processing as described in the basic process section is necessary in this step.

Step 3: thermal oxidation

The second thermal oxidation is to protect the thrust bearing clearance from the nested mask etching of tip clearance later. Since thermal oxide grows by consuming Si at the wafer surface, any defects in the starting wafer surface would potentially compromise the bonding surface quality. The first oxide has been subjected to a photo and DRIE process, thus its surface is not pristine anymore. To have the least process impact on the surface quality, the first thermal oxide is stripped before the second oxidation. This second oxide film also serves as a nested mask on the bottom plenum side that requires a thickness of 0.5 μm. The top side of the wafer needs a thinner oxide for easy nozzle etching. This can be achieved, though not shown in the flow drawing, by covering the bottom side of the wafer with resist and etching with BOE untill about 0.1–0.2 μm oxide is left on the top side.

Step 4: blade tip clearance nested mask formation

The tip clearance is 20 μm deep by design. Since it is difficult to have good resist coverage over 20 deep steps, it is not feasible to perform photolithography to etch both tip clearance and nozzle. The solution is the nested mask: define the tip clearance pattern in the oxide with mask 15, pattern and etch the nozzles with mask 21, strip the resist and clean the wafer and then etch the tip clearance already defined in the oxide. The nozzles will be etched a little deeper but this does not affect the process flow.

In this step, standard thin resist photo process is used to perform BOE etch of the oxide. Again ultrasound agitation is necessary for uniform etch of the oxide film. Wet etch of oxide is preferred here because it leaves a clean and undamaged Si surface.

Step 5: Nozzle etch

This step etches the nozzles and other through wafer holes using mask 21 and thick photoresist. One thing differs from the normal photo process is that the postbake step is skipped in order to avoid distortion of the nozzle shape. For the same reason, the wafer should not be put on a handle wafer, a process that requires baking and has negative impact on the etch uniformity. This is why we etch the nozzle before plenum: by the end of nozzle etch the wafer has no through holes and is still robust, therefore there is no need to mount the wafer on a handle wafer at the stage.

The thin oxide remained on the wafer will be etched in the DRIE process. Wet etch of oxide is not preferred here because thick resist without postbaking tends to fail in BOE with ultrasound. Without ultrasound agitation the less than 10 μm holes are hard to be uniformly etched. Since we have to etch through the oxide with DRIE, thinning down the oxide to the minimum on this side, done in step 3, is beneficial to the nozzle etch in terms of uniformity and dimension control. Because of their small size, the nozzle depth is impossible to measure with the optical microscopes we have. The etch time is determined by experience with a few dummies. Some over etch is accepted and actually preferred to guarantee the dimension of the nozzles.

Step 6: Tip clearance etch

As mentioned above, this step uses the nested oxide mask defined in step 3 for a 20 μm etch into the Si wafer. What needs to be pointed out about the nested mask etch is that the polymer residue from passivation cycles in DRIE must be completely removed first, or they would mask the area to be etched and destroy all the work done thus far. The most effective way to remove the polymer residue based on our experience is oxygen plasma ashing. The cleaning sequence after nozzle DRIE etch is: 1 h ashing in a barrel asher, followed by a piranha cleaning. The 20 μm etch is done in the DRIE machine with the nested oxide mask. A high Si to oxide selectivity recipe should be chosen. By loading the wafer into DRIE chamber without a handle wafer, a condition giving better cooling and less etch uniformity, the 0.2 μm thick oxide is enough for the 20 μm Si etch.

Step 7: nozzle feed nested mask formation

The purpose of this step is to thin down the bottom side oxide film on nozzle feed to create a oxide thickness difference and therefore a DRIE etch delay later in the large plenum feature relative to the smaller nozzle feed, so that the wafer would not be too fragile by the time the nozzle region is through. The same mask for thrust bearing clearance mask #12 is used here. Standard thin resist photo and BOE is adequate for this step. The oxide thickness is monitored to ensure the oxide film in the nozzle feed is about 0.2 μm. How precise the oxide thickness control should be depends on the oxide etch method in the next step.

Step 8: plenum photo and nozzle feed etch

This step defines the plenum on the bottom side using mask #17 and thick resist. The resist is drawn in this step to show the oxide thickness difference. The oxide can be etched by BOE or DRIE. The remaining oxide in the large plenum region should be about 0.2 μm thick when the nozzle feed region is cleared of oxide.

Step 9: plenum etch

Next the wafer is etched by DRIE. The wafer should be mounted on a handle wafer since it will be a through etch. To avoid the nozzles from being enlarged, before mounting on a handle wafer, the etched top side should be protected by an isotropic deposition of a passivation polymer film using the DRIE machine, the same deposition as in the passivation cycle. Thermal oxidation was also used for the same goal but it requires a nitride deposition right after the second oxidation and extra steps to remove the nitride. The oxidation of wafers covered with nitride film also leads to the formation of the so-called bird's beak on the oxide pattern which could be detrimental to wafer bonding. The polymer film protection method is much more convenient and in line with the regular post DRIE cleaning protocol. The 0.2 μm oxide should give about 40 μm delay in the plenum region by the time the nozzle holes are visible in the nozzle feed.

The protection layer on the nozzle side could mask the nozzles and leave a thin wall of Si around each hole when nozzles and other holes are negatively tapered and the plenum side is over etched. The Si residue can be removed by a short isotropic etch in the DRIE machine.

Step 10: post etching cleaning

When the DRIE is done, the wafer should be dismounted in acetone, cleaned in piranha, and ashed. Before final bonding the wafer should be cleaned, and stripped off oxide following regular pre-bonding cleaning protocol. Step 10 shows the finished layer with halo pieces removed and oxide stripped. Keep in mind that the first protective oxide should not be stripped from the wafers till they are ready for bonding.

4.4.2 Rotor Releasing Methods

The rotor is the most important component in a microengine device. It must be released from the wafer before the devices can be functional. How the rotors are released strongly affects the fabrication process flow, the overall yield, and the device performance. Generally, the releasing process should have the least impact on the wafer cleanness and process flow. It should not cause chipping or change of critical dimensions associated with the rotors. Substantial effort was made to develop reliable rotor releasing mechanism which has evolved from the mechanical fuse to oxide bonding pads and to free rotor method as our device becomes more and more complicated.

(1) Mechanical fuse method [18]

The mechanical fuse method was developed in the early stage of the program and was used in some bearing rig devices. The so-called mechanical fuses are a pair of Si bridges across the journal bearing on the blade side as schematically shown in Fig. 4.19. The rotor is held by the mechanical fuse, a pair of narrow

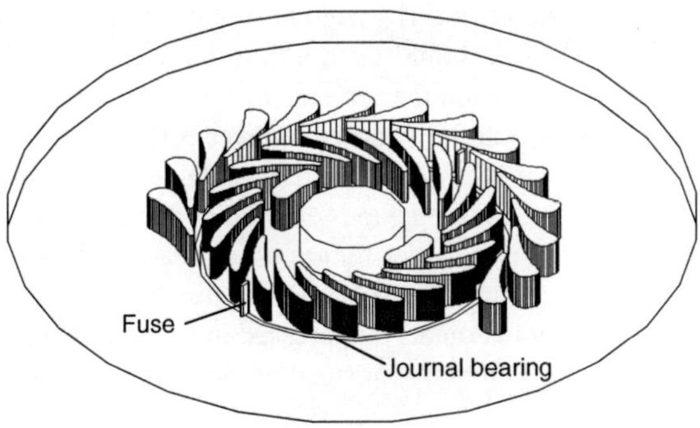

Fig. 4.19 3D drawing of mechanical fuses in a BR device. The two fuses are small bridges crossing the journal bearing that breaks open under small external force. The journal bearing and the blade and fuse height are not to scale for clarity

bridges across the journal bearing on the blade side during the whole process. They are set free before testing by breaking the fuses with a pin through holes etched in the outer layers.

The fabrication process with the mechanical fuse releasing method is straight forward and bonding friendly: all the finished wafers are clean and in one piece going into final bonding process and there are no extra loose parts. This method works fine in early bearing rig devices with single layer rotors, but it has the following drawbacks:

- The fuse breaking step generated particles that might be trapped in places such as a journal bearing or thrust bearing.
- There is no guarantee that the broken ends on the rotor are uniform and smooth. Large protrusions into the journal bearing could cause rotor crash during high-speed operation.
- The extra holes through the outer layers might unnecessarily affect the rotation of the rotors.
- In more advanced devices with multilayer rotors such as turbo charger and SSEs the mechanical fuses would make the process more complicated. For example, if we add mechanical fuses in each layer, it would be difficult and more unpredictable to break them and more debris would be trapped inside the device. If we have mechanical fuses in only one layer, all the other rotor layers would have to be bonded to the layer with the fuse first before etching through the rotor parts such as journal bearing, thus increasing the process complexity greatly. It is also a concern whether the fragile fuses can hold onto the heavier rotors in the more extended processes due to added rotor layers.

(2) Bonded oxide pad rotor releasing method
The above-mentioned difficulties in the mechanical fuse method leads to the next rotor releasing scheme, the bonded oxide pad method, in which the rotors are bonded to one of the two plenum/nozzle wafers next to it through a patterned oxide film. The rotors are held to the wafers throughout the process and released after dicing by dipping in straight hydrofluoric acid (HF). Figure 4.20 shows the rotor bonding and releasing procedures in a turbo charger device:

(i) Finish the photo and etching in layer 1 (L1), layer 2 (L2), layer 4 (L4) and layer 5 (L5) and have them ready for pre-bonding procedures. All the layers should have a thin thermal oxide film grown on them as the first step. Etch the top side of L3 to define the oxide bonding pads. Then etch the journal bearing in layer 3 (L3) and the running gap in layer 4 (L4), as well as other patterns such as seal clearance on the bonding surfaces. The running gap is also a circular donut trench similar to a journal bearing but it is wider than journal bearing. Bond the L3 and L4 together while keeping the oxide on the outer surfaces of the stack. The wafers should be put in contact and pressed in vacuum. Anneal the stack in flowing nitrogen at 600–800°C instead of the standard 1000°C to minimize Si deformation

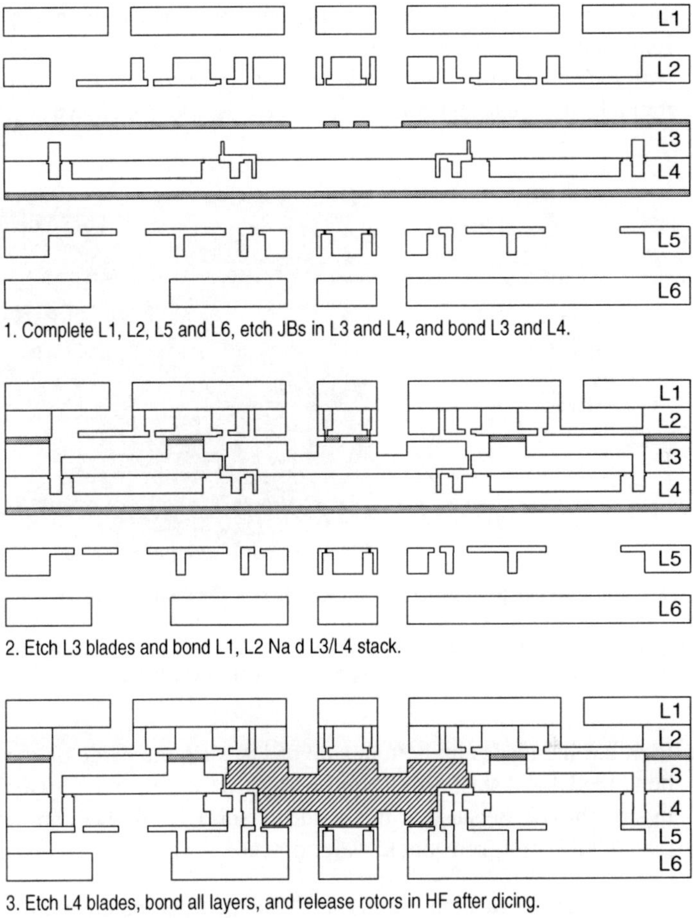

Fig. 4.20 Turbo charger assembly flow with oxide pads rotor releasing approach

due to ambient pressure on the sealed cavity formed at the journal bearing and running gap.

(ii) Etch the compressor blades on top side of L3 using DRIE. Note the journal bearing and blades are on the opposite sides of a single wafer, an earlier technology for blade formation when this releasing method was developed. Long over etch on the blade side should be avoided since it could result in blade height and journal bearing length deviation and change of the journal bearing profile. Clean wafers L1 and L2, and stack L3/L4 and bond them together. Use silicon direct bonding mode to contact the wafers since now the whole L1–L4 stack is an open structure and the annealing temperature can be raised to 1000°C. Since the opened cavity at the L3 journal bearing and the L4 running gap has a complicated shape, the cleaning of the wafer before bonding must be done thoroughly by repeated long-time rinses. After this bonding the top of the rotor is attached to the wafer L2 through an oxide pad. The pattern of the oxide pads and a photograph of a detached rotor with the oxide pad are shown in Fig. 4.21.

Fig. 4.21 (a) Oxide bonding pads pattern in a turbo charger. The oxide pads are the black lines, 500 mm long and 50 um wide, arranged in a circle at the center. The pattern of the compressor blades and speed bump, the toothed pattern in the center, is also shown. The perimeter of the rotor is define by the journal bearing pattern. (b) is a photograph of a detached rotor

(iii) Etch turbine blades on bottom side of L4, strip the oxide on the L4 side, bond the L1–L4 stack to L5 and L6, dice the wafer, and release the rotors in straight HF. Because of the bonded stack L1–L4, there are few things that are different from single wafer process:

- The regular vacuum wands are designed for picking up individual wafers. They may not be able to handle the heavy wafer stack reliably.
- The regular wafer carriers have slots too narrow for such thick stack. We enlarged some of the carrier slots by removing a spacer in a few places in a carrier so the thick wafer stack can fit in.
- The vacuum in the coater chuck does not hold the wafer stack reliably because of the weight and sometimes large bow in the stack. Check the coater vacuum seal before starting the machine, and in the case vacuum leak due to larger stack bow, use a few pieces of tape to fix the wafer stack on the coater chuck and then coat resist.
- The oxide stripping of L4 side should be done with BOE instead of HF for less impact on the oxide bonding pads on L3.
- All the cleaning processes for the stack should be finished with a thorough rinse to make sure there is no chemical trapped in its complicated internal structure.
- Dicing has to be done using multi-cut mode because normal single cut mode cannot go through the six-layer, 3 mm thick stack.

- The rotor releasing is done die by die in HF for 10–20 min. HF rather than BOE is used because it takes extremely long time for BOE to etch a 50 μm wide oxide pads laterally.

The oxide pad method solved the problems of particulates from breaking Si fuses in the device. The rotors are completed by DRIE etch and the HF releasing of the oxide pad leaves no solid particles in the device. However, the HF release unavoidably weakens other structures not only at the interface between L2 and L3 but also all the other Si-to-Si bonded interface. This is because there is always a native oxide layer on our wafers that are formed in RCA solutions and in air after oxide stripping, and HF can penetrate and attack the native oxide, leaving a fine gap. To determine the native oxide etch rate by straight HF, we carried out an experiment in which a Si wafer prepared by oxide stripping in HF, RCA cleaning, and nitride deposition were patterned and etched in HF. The etching of the native oxide was clearly visible through the nitride. The etch rate of native oxide, i.e., the propagating speed of the undercut front, is about 0.25 μm/min.

The major problem of this method is that the oxide pads do not hold the rotors reliably. Rotors often fall off in the pre-bonding oxide stripping in BOE and wafer cleaning processes. The rotor in the photo in Fig. 4.21 is an example of premature releasing: the detached rotor still has the oxide pads. This is because after being exposed to photo and etching processes the quality of the oxide surface is not ideal for bonding and the poorly bonded oxide pads could not sustain in BOE stripping of surface oxide before final bonding. One fix to this problem is to protect the bonding pad oxide with a layer of nitride and keep it on until the time of pre-bonding cleaning.

The deposition of nitride also provides a chance for us to protect the journal bearing profile from potential over etch on the blade side. Part of the turbo charger process flow with the nitride protection is shown in Fig. 4.22. After the journal bearing etch in the compressor wafer (step 1), with the nitride film still on, the wafer goes through a thermal oxidation (step 2). Since the nitride acts as a barrier to prevent the oxidation of the wafer surface, we are able to grow on the journal bearing side walls an oxide film thicker than the surface oxide. This is the so-called local oxidation of silicon (LOCOS). The nitride can be removed by wet process and a timed HF or BOE etch could strip off the surface oxide for bonding to the turbine wafer but only thin down the oxide on the side wall due to their thickness difference (step 3). The turbine wafer is prepared in the same way as the compressor wafer with LOCOS oxide. There is a seal clearance in the turbine wafer that overlaps the journal bearing in the compressor wafer and the running gap in the turbine wafer, so we do not have to be concerned about the bird's beak, extra oxide growth at the end of the nitride patterns, that could otherwise cause bonding problems. The remaining oxide film on the side walls protects the journal bearing from the blade etch (step 4). This oxide film will be removed in oxide stripping in the next bonding process.

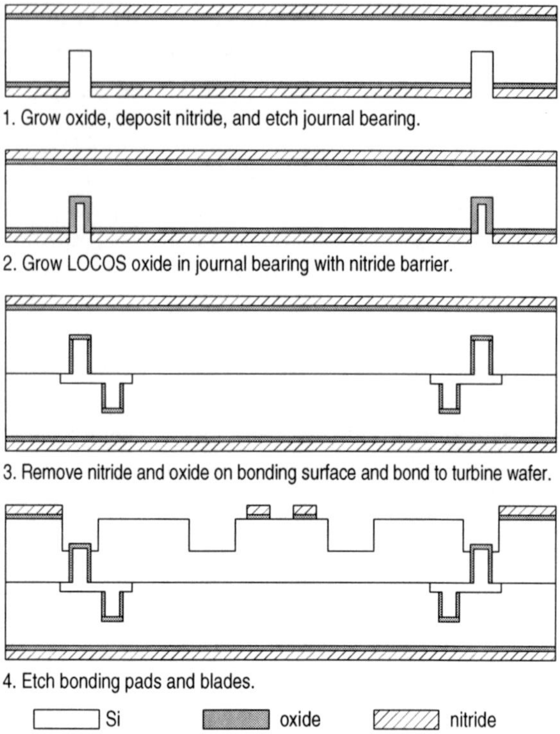

Fig. 4.22 Rotor oxide releasing process with nitride for oxide bonding surface protection and LOCOS oxide for journal bearing profile protection

When the journal bearing profile is negatively tapered, the LOCOS oxide would shield the blade etch and leave a thin Si wall at the base of the journal bearing, as seen in the exaggerated schematic view in Fig. 4.23. The height of wall depends on the overlapping on the etch depths of the journal bearing and blades. Such Si walls around the opening of the journal bearings on the blade side could have an adverse impact on the rotor performance. They can be easily removed by a short isotropic etch after blade etch.

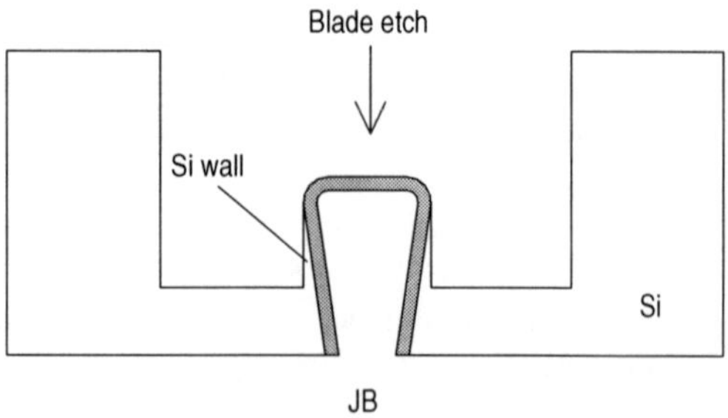

Fig. 4.23 Formation of thin Si wall by blade etch at the base of negative tapered journal bearing due to LOCOS oxide shielding

(3) Free rotor method

The third rotor releasing method was conceived from the experiences of the previous approaches. The main idea is to set the rotors free when the rotor wafer stack process is complete. The freed rotors are cleaned and put aside until the final bonding assembly when they inserted back into the wafer stack. This approach is therefore named free rotor method. This method is much simpler and cleaner because it eliminated the intermediate steps of retaining the rotor, patterning, etching, and cleaning to finish the rotors in bonded stacks. None of these steps are trivial due to the complicated 3D structure of the devices.

The free rotor method also brings in a major improvement to the process of the journal bearing. With previous rotor releasing methods, the rotors are fixed to their original location, and this means that usually the small journal bearing width must be defined in the mask. As a results the journal bearing specifications are restricted: the width cannot be too small because of the offset in the mask needed to count for the widening of the features in DRIE due to resist loss; the length cannot be too large because the journal bearing then will take a long time to etch since the etch rate quickly decreases with etching depth; and any change in the journal bearing specifications means rework of the process development since the DRIE process is very sensitive to the specifications especially the width. With free rotor method, a rotor does not have to go back to the original spot. Therefore, we can build rotors in locations away from the device spots and put them back into the device in the final bonding. This concept and its advantages are illustrated in Fig. 4.24. In (a) the rotors are formed in a rotor donor by through DRIE etch. A donut shape trench of the same width but different diameter is also etched in the device at the same time. The width of this trench is chosen based on optimized DRIE process for vertical profiles. The DRIE process is not constrained by the journal bearing width and can be standardized. The journal bearing width is determined by the diameter difference between the rotor donor and the device when the rotor is placed in the device in the final bonding step, see (b).

The main benefit of rotor allocation flexibility is the separation of DRIE process and various journal bearing specifications: we can fix the trench width in the mask and therefore standardized the DRIE process and form journal bearings of any width by matching the diameters of rotor and holes in the device. It is now a simple thing to have devices with different journal bearing width in a same build for in-depth study of the micro journal bearings.

The journal bearing process can be greatly simplified by fixing the widths of the DRIE in the mask for various journal bearing specifications. The trench width defined in the mask should preferably meet the needs for the journal bearing profile control, etch uniformity, and etch time. Usually the DRIE trench width is larger than that of the journal bearings for easy process control and shorter etch time. The restrictions on the journal bearing width and length are basically eliminated. Long and narrow journal bearing can be easily assembled, for example, by etching with a mask of relatively larger trench width and dropping a rotor to a hole of the right diameter.

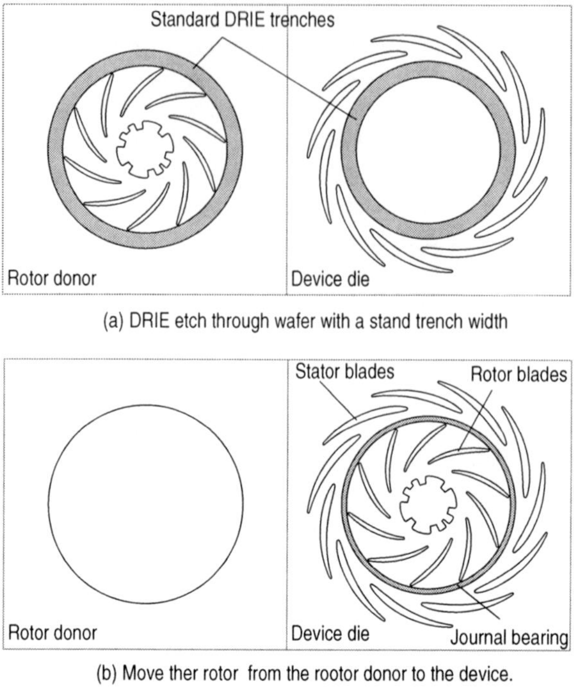

Fig. 4.24 Free rotor concept used to construct journal bearing with a standard DRIE process. (a) The rotor is formed in a donor at a different location from the device. The through etch trench width is the same in the donor device and the device but their diameters are different. (b) The rotor is placed in the device in the final bonding. The journal bearing width is determined by the diameter difference between the rotor donor and the device

Although Fig. 4.24 might imply that the donor's role is to provide rotors only, the donor itself can be a device too, if small variation in the rotor diameter is acceptable. Both approaches have been applied successfully in our devices.

4.4.3 Journal Bearing Process Development

JB definition and dimensions:

Journal bearing is one of the components that strongly affect the rotation speed of the microengine device. A journal bearing is a circular ring etched into Si and separating the rotor from the stator side of the wafer. Figure 4.25(a) shows the top view of the journal bearing together with the rotor blades, the stator blades, and the speed measurement bump in the center. The width of the journal bearing is magnified for clarity. Figure 4.25(b) is a 3D schematic of the same pattern with exaggerated vertical scale for clarity. A few of the stator blades are not shown for better view of the journal bearing. The journal bearings were etched on the opposite

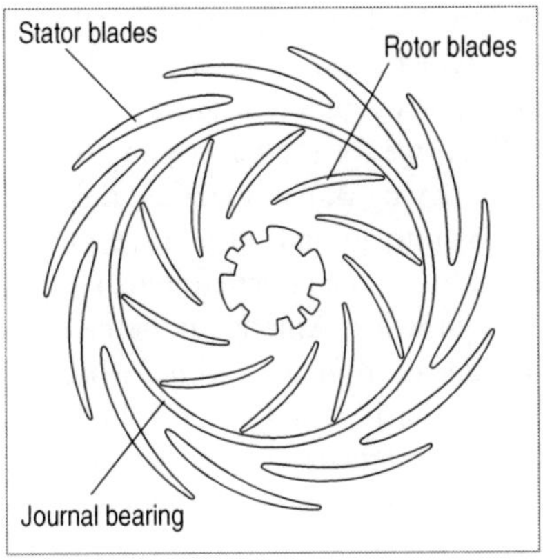

(a)

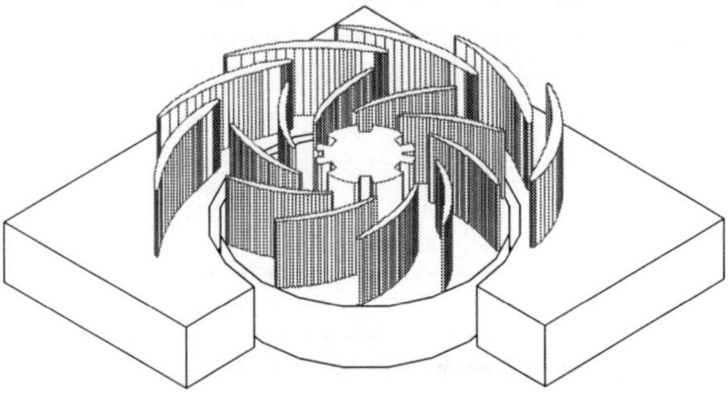

(b)

Fig. 4.25 (a) Top view of a turbo charger compressor showing the relative positions of the journal bearing, stator blades and rotor blades. The device size is 23 mm by 23 mm. The journal bearing width is exaggerates for clarity. (b) 3D schematic of the blades and journal bearing positions based on (a)

sides to the blades on same wafer in early designs but in later devices they are formed on a separate wafer with its length determined by the wafer thickness in order to lower the rotor imbalance.

The design specifications of journal bearings vary from device to device. The width of the journal bearing is 15–16 μm for the bearing rig and the turbocharger and about 30 μm for SSE. The length of the journal bearing is 330 μm for the bearing rig and turbo charger and 450 μm for the SSE. What makes the journal bearing etch a challenge is its tight dimension tolerances. The width tolerance is ±1 μm for the bearing rig and turbo charger, and ±2 μm for SSE. The length tolerance is ±5 m for all devices. Achieving the required narrow, deep, and straight profiles repeatedly proved to be very difficult with existing DRIE technology. Besides, such narrow and deep trenches take a long time to etch and their depth progress is hard to monitor with, for example, microscope, because the bottom of the trenches usually do not reflect enough light, especially when there are defects inside.

As described earlier, the Bosch process for DRIE uses neutral radicals to etch Si and deposits a fluoric polymer passivation layer in the pattern being etched for side wall protection. High aspect ratio Si etching is achieved by selectively removing the fluoric polymer deposition only on the bottom of the pattern by directional ion bombardment and cycling the etching and passivation steps.

The DRIE recipe for journal bearing is listed in Table 4.1. To ensure the journal bearing in the device wafers meet the tight tolerances, we usually run 2–3 dummy wafers first, dice them and inspect the cross section profiles. This is because the machine conditions change over time and the recipes need to be adjusted each time we run a critical wafer. Among all the parameters, we often use pressure and cycle times to fine tune the dimensions once a base recipe is fixed because the journal bearing profiles are sensitive to these parameters and the etching mechanisms involved are simple and straightforward. The chamber pressure can be controlled using the fixed exit valve opening mode or the fixed pressure mode of the machine. The fixed pressure mode is preferred over the fixed exit valve opening mode because we found that with fixed exit opening mode the chamber pressure has a surge in the first 10–15 min of the etch due to polymer accumulation in the chamber. This pressure surge changes from run-to-run depending on the polymer accumulation and a thorough burning with oxygen plasma is needed to have repeatable journal bearing profiles. For very deep journal bearings, etch pressure and etch time ramping in the DRIE recipe is necessary.

Table 4.1 Standard journal bearing DRIE recipe

	Etching cycle	Passivation cycle
SF6 flow (sccm)	105	0
C4F8 flow (sccm)	0	50
Cycle time (sec.)	15	11
Pressure (mT)	28	28
Coil power (W)	800	600
Platen power (W)	120	60

One major difficulty we encountered in the journal bearing etch is profile repeatability. Although very good journal bearing profiles can be obtained with our baseline DRIE recipe, defects such as blowup tops, steps, spikes, bellies, positive taper, and negative taper appear from time to time. This is because the DRIE recipe is not the only factor contributing to the journal bearing profiles. The masking material, usually thick resist and sometimes thick silicon oxide used in our processes, and their behavior in the DRIE environment also have a major impact on the journal bearing profiles. Failure to pay attention to the details of the masking resist and oxide often results in deviation from ideal journal bearing profiles. Some common defective journal bearing profiles, top blow-up, belly, and spikes are shown in Fig. 4.26.

The masking materials used in etching journal bearings are thick photoresist and thermally grown silicon oxide film that is needed the protection of the bonding Si surfaces. Since it typically takes more than 4 h to etch the narrow and deep journal bearings, sometimes two layers of the photoresist of about 20-22 μm were coated

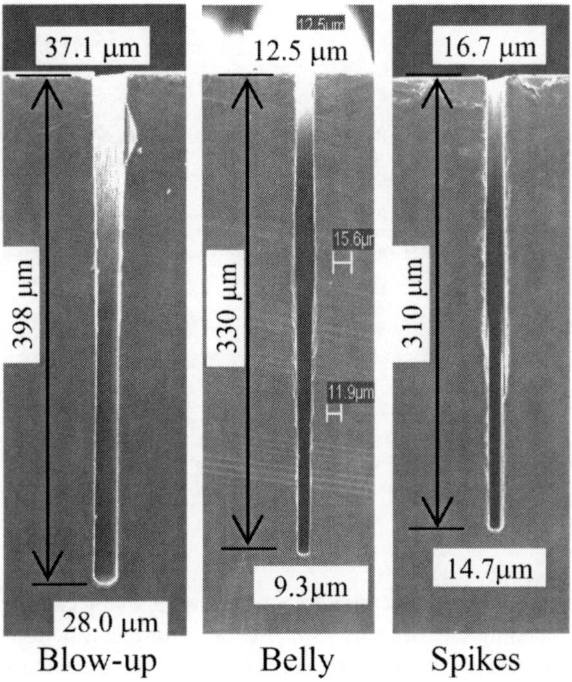

Fig. 4.26 Common defects in journal bearing profiles

on the wafers. The photolithography process includes integral exposure of 5 × 15 s, developing for 2–3 min, and postbake at 95°C for 30 min. The oxide film is etched in BOE before DRIE of Si. The photoresist pattern edges have 40–60° slopes that depend mainly on the baking conditions. The oxide, 0.2–0.5 μm thick, typically has a slope of 45° or less depending on the adhesion of the resist to the oxide film.

In DRIE process neither resist nor oxide last forever. In the etching cycle, the positive ions in the plasma, accelerated toward the wafer by the platen voltage, anisotropically mill away not only the passivation coating, but also the exposed resist or oxide once the passivation layer is gone.

The etch rate of resist is typically about 2–3 μm per hour, and it is very sensitive to pressure. As shown in Fig. 4.27, the etch rate decreased significantly with increasing SF6 pressure. Silicon oxide is etched in DRIE at about half the rate of resist. In our journal bearing recipe, the pressure is set at 28 mT for the resist to last long, as explained next, for better profile control.

The resist profile usually has a finite slope from postbaking as shown the inset photo of Fig. 4.27. As the resist is thinned down the bottom edges of the resist recede away from each other, therefore the journal bearing width keeps increasing. This is the main reason for photomask offset, and there is a lower limit for the smallest journal bearing width for a given length. Meanwhile, with increasing depth, the amount of etching species needed to remove Si at the bottom and on the side of a

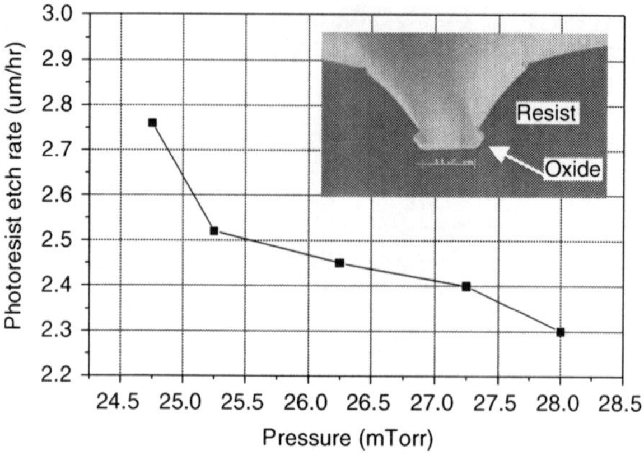

Fig. 4.27 Photo resist etch rate dependence on DRIE chamber pressure based on the journal bearing recipe. The inset is a SEM picture of the resist/oxide bi-layer profile after BOE etch of oxide. The resist slope is the result of baking before BOE. The opening is about 9 μm wide at the bottom of oxide

straight journal bearing trench increases. In the case of very shallow resist slope, the receding rate of the resist is too high for the system to provide enough etch species to the narrow trench already at a depth to achieve a vertical profile, and a blowup at the top as shown in Fig. 4.26 appears. In the case of thick oxide film beneath the shallow resist, a step could show up in the journal bearing profile due to the different receding rates of the resist and oxide.

In Fig. 4.28 are four SEM pictures showing the time dependence of the profiles of journal bearings patterned from same 8 μm wide mask and etched with same DRIE recipe. The oxide is 1 μm thick and the resist is about 22 μm thick for all these sam-

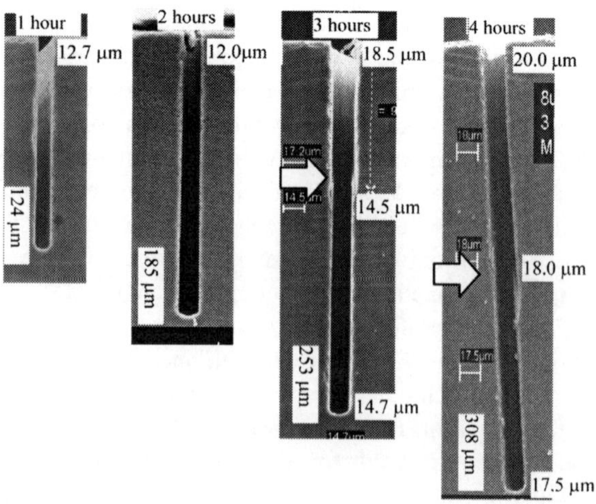

Fig. 4.28 Journal bearing profile evolution with DRIE time taken from 4 samples prepared with same recipes. The photo mask width was 8 mm. The resist and oxide are 22 mm and 1 mm thick respectively. Steps shown by the arrows start to show after 3 hours and moved downward. The tilting of the rightmost picture is due to SEM setting

ples. After 1 and 2 h the journal bearings remain straight. This is probably because of the low aspect ratio of the trench and the slower mask receding speed that is now still mainly determined by the BOE etched oxide slope. After 3 h, however, a step appears indicating a sudden change in mask receding speed when 1 μm thick and roughly 45° slope in oxide is completely etched away and resist starts dominating the mask receding. As the etch continues, the steps move further away from the top. This observation leads to the idea of using thicker oxide to delay the appearance of steps or sudden width change in journal bearing till after the desired journal bearing length is reached. Figure 4.29 shows the oxide thickness dependence of the journal bearing profiles. All three samples here were prepared in the same way except for their oxide thickness, 0.5, 1.0, and 2.0 μm from left to right. The DRIE etch was done for 4 h. Steps in the 0.5 μm sample are lower than in the 1.0 μm sample, and a straight journal bearing is obtained in the 2.0 μm sample.

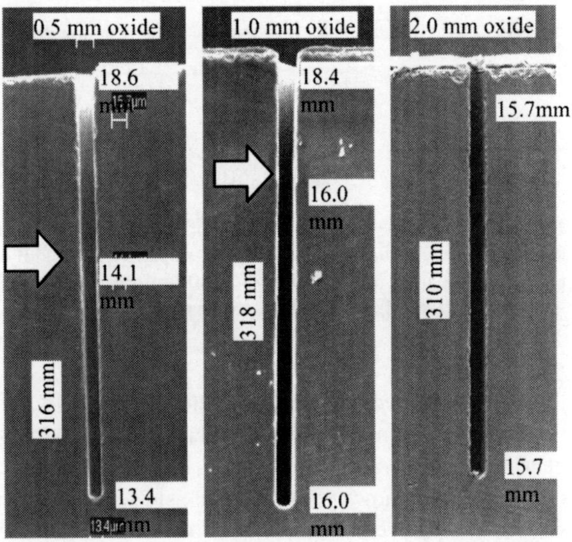

Fig. 4.29 Influence of oxide thickness on journal bearings in 3 samples with different oxide thickness. Steps labeled by the arrows were visible in 0.5 um sample, moved up in the 1.0 um sample and disappeared in the 2.0 um sample. Journal bearing depth is labeled on the left and width on the right of the trenches

The increased oxide thickness in the bilayer DRIE mask of resist and oxide helps to form straight journal bearing profiles when the resist profile is not optimized for journal bearing due to other process constraints. However, the cost is also obvious: the BOE etch becomes much longer and the risk of resist failure grows, the wafer bow increases significantly causing bonding problems, and in some cases whole wafer coverage by a thick oxide layer is simply not applicable to some exiting process flows.

To avoid wafer bow or full coverage of the wafer by thick thermal oxide, we can use a Local Oxidation of Silicon (LOCOS) oxide pattern for the journal bearing in our device. As an example, Fig. 4.30 shows the journal bearing process flow in a turbo charger. The critical requirement of this wafer is a journal bearing with one side open to a 30 μm clearance, and the journal bearing profile should be kept intact when forming the clearance. The highlights of the process are as follows. A nitride

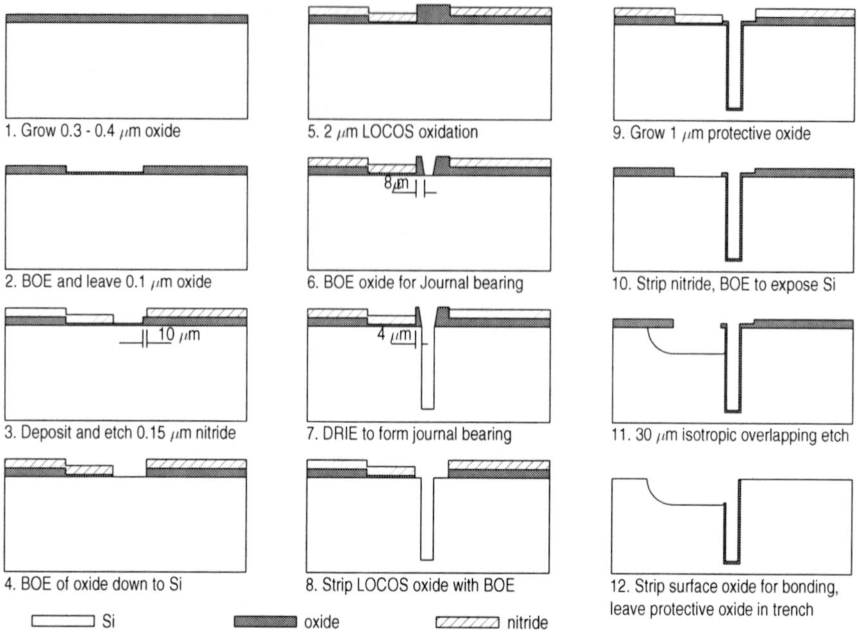

Fig. 4.30 Journal bearing process flow of a turbine layer in a turbo charger device. The journal bearing was etched using LOCOS oxide (steps 5-7) formed by a nitride barrier (step 3) and protected by oxidation (steps 8 and 9). A 30 um trench joined one side of the journal bearing by isotropic etch (step 11) through a nested mask

film was deposited and patterned, step 3, as the LOCOS oxidation barrier. A thick thermal oxide film was grown in step 5, followed by BOE and DRIE of the journal bearing in step 6 and step 7, respectively. To protect the journal bearing from the blade etch (not shown) from the back side, a layer of thermal oxide thicker than the surface oxide beneath the nitride film was grown in the trench, as seen in step 9. To form the 30 μm recess without damaging journal bearing profile, we used a nested mask pattern in step 2 and protected it by nitride through step 9, and isotropically etched away a small Si ring adjacent to the journal bearing in step 10, which is now under the protection of oxide.

Another commonly seen defect in journal bearing profile is the spikes appearing often at the bottom of the relatively straight trenches but also in the middle of the bellied trenches. The former is more likely to find its way into the devices because usually the latter is considered unacceptable for devices wafers once observed in a dummy. Figure 4.31(a) shows the SEM picture of the cross section of a turbo charger compressor monitor journal bearing. At the bottom there is a spike over 30 μm long. Figure 4.31 (b) shows a spike over 100 μm long attached to a bearing rig rotor. The blade is visible on top of the picture and the spike is upside down. Closer investigation tells us that the spikes are needle shape Si structure pointing up to the opening of the journal bearings. The base diameter ranges from 2 to about

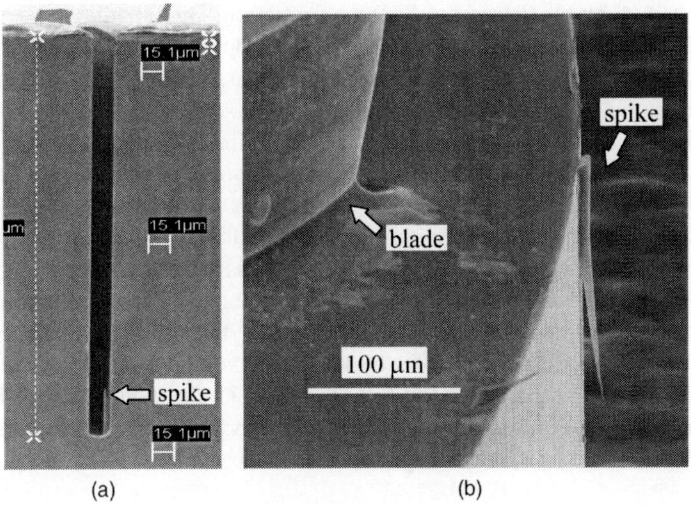

Fig. 4.31 SEM pictures of spikes. (a) cross section view of a journal bearing with a 30 um long spike at the bottom. (b) a spike over 100 um long attached to a bearing rig rotor. Note the blade is on top and the spike is upside down

6 µm and the length can be 150 µm long. A spike as shown in 4.31(b) could be disastrous to the device spinning at high speed.

We noticed that spikes usually do not appear in journal bearings with negative tapers. This observation leads us to believe that the spikes are caused by the micro masking effect by the debris from the resist. It is a known fact that the resist is etched in the DRIE plasma and its surface sometimes becomes rough and full of pits after the etch. It is likely that some small debris pieces are knocked off the resist surface and fall in the journal bearing trenches. As illustrated in Fig. 4.32 the positively taped trench tends to collect the debris on the side wall and result in the

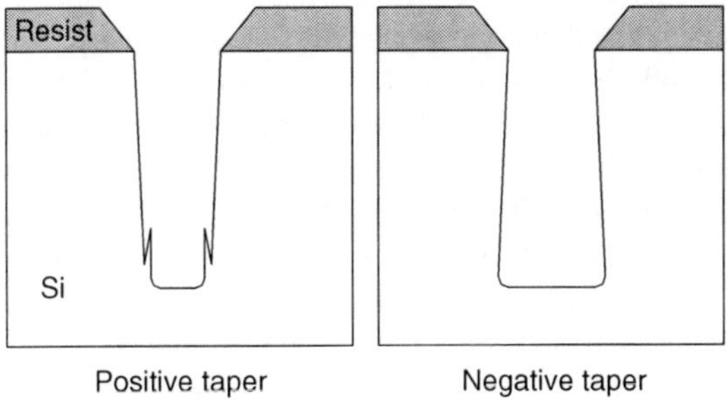

Fig. 4.32 The mechanism of spike formation. The spike probably originates from debris from the resist due to the plasma etch. The debris tends to land on the side wall of the positive tapered trenches and cause irregular local masking that result in spikes. The negative taped trenches do not collect debris on the side wall as easily and therefore usually have no spikes

micro masking by the debris and Si spike formation. On the other hand, the resist debris pieces do not easily cling on the negatively tapered trenches and form spikes. It is therefore necessary to modify the journal bearing process to require slightly negative tapered profile in order to eliminate the spikes.

From the study above it is tempting to think that increasing the resist slope angle can reduce its receding speed and therefore preventing the blow-ups and steps. The reality, however, is that slopes steeper than 45° often accompany bellies, the widening or the journal bearing in the middle. The reason for this defective widening must be some kind of damage to the protective passivation layer and then attack on the Si on the side walls. The only culprit is the positive ions that are used to mill away the passivation layer on the bottom of the trenches. The positive ions are generated by the coil, accelerated to the wafer by the platen voltage, and supposedly hit the wafer at normal incident angle. The fact that the Si in the middle of the side wall is attacked means that the trajectory of the some of positive ions is redirected to the side wall. We will try to give an explanation of the causes using Fig. 4.33.

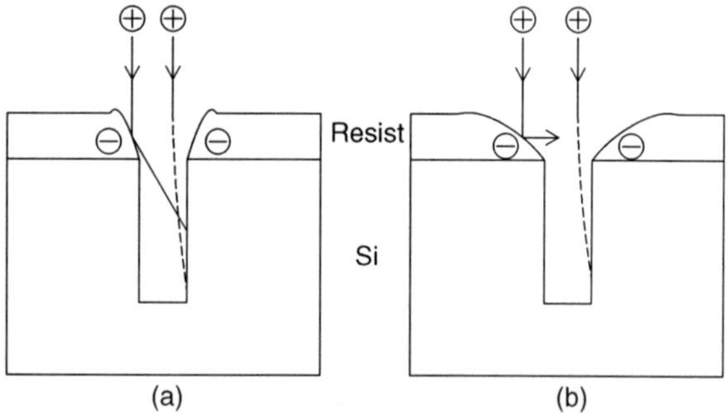

Fig. 4.33 The influence of resist profile, accelerated positive ions and charging on resist on the profile of journal bearing trenches

Before the DRIE etch, the wafer has its resist and oxide (not shown) already patterned and baked. Those positive ions that hit on the resist slope are deflected, and if the resist slope is steeper than 45° the reflected ions [29] would hit the side walls of the journal bearing trench, as in the case of Fig. 4.33(a). Another contributing factor is the negative charging on the tip of the insulating resist. The electric field from the accumulated charges at the edge of the resist pattern tends to attract the positive ions, both the normal incident ones and the deflected ones, and direct them toward the side walls. The result of the attraction on the deflected ions is that they hit the opposite side wall lower than expected for pure deflection. The distribution of the normal incident ions bent by the resist tip charging should be more or less continuous assuming homogeneous plasma. The ions deflected from the resist slope, however, is more concentrated on the side walls because of the small variation in the slope angle and close proximity of the ion trajectory to the charges on the resist. It is therefore likely that the deflected ions make the main contribution to the formation of journal bearing bellies. If this is true, to solve the belly problem, the first solution is naturally to reduce the resist slope angle to 45° or slightly below, so that the deflected positive ions could not enter the journal bearing trenches, as seen in Fig. 4.33(b). However, as mentioned earlier, the resist slope should not be too shallow or else the journal bearing will have a top blow-up. The optimized baking conditions for 45° slope in the AZ4620 and AZ9260 resist we used are: 95°C for 17–20 min. Our investigation shows that once the resist profile is controlled below 45°, the journal bearing profiles do not have bellies. Figure 4.34 shows SEM pictures of three straight journal bearings of different widths etched, demonstrating the success of the improved resist profile control. Figure 4.35 shows the repeatability in journal bearing etch we achieved: SEM pictures of consistent journal bearing profiles with small variations from device to device in wafer, and from wafer to wafer.

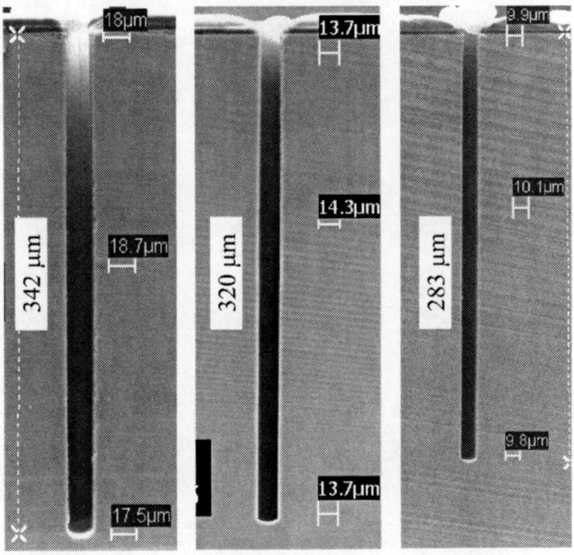

Fig. 4.34 SEM picture of 3 high quality journal bearing profiles with well controlled resist profile. The widths of the journal bearings from left to right are 18 um, 14. um, and 10 um with spread within the tolerance. On top of the trenches the resist slopes are visible

With the resist recipe fixed, we are able to fine tune the DRIE recipe for best journal bearing profiles. For example, we found out that by increasing pressure or increasing slightly the mask opening the profiles can change from positively tapered, wider on top and narrower on bottom, to negatively tapered, narrower on top and wider on bottom. Increasing the platen power would make the journal bearing straighter but wider. For journal bearings about 15–16 μm wide and over 300 μm long, the DRIE etch time is usually over 4 h. If one coat of thick resist, about 10 μm thick, is to last that long, the DRIE chamber pressure should be high enough and the wafer should be well cooled during the DRIE.

The complexity of resist profile control forced us to look for more reliable approaches, and we found that thicker deposited oxide etched by BOE can replace photoresist as the masking material and gives better dimension control We believe this is due to better adhesion of the oxide to the silicon surface, thinner film and therefore smaller ion reflecting slope areas, smaller etch rate and more uniform slope angle. This approach is especially suitable to longer journal bearings that otherwise have to use double coat thick resist. The price to pay for the thick oxide mask is increased process steps and uncertainty as in any new process.

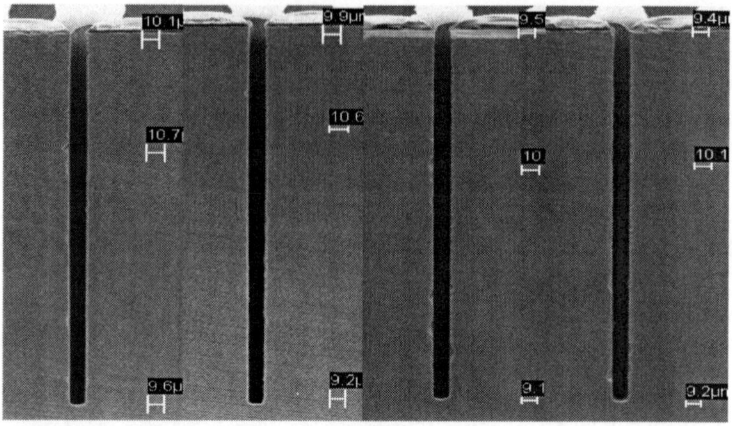

Wafer A

Wafer B

Fig. 4.35 Journal bearing profile control quality demonstration: the SEM pictures of journal bearings in two wafers each are shown. The length is about 300 mm. The width spread from 9.1 μm to 10.7 μm

We use CVD deposited oxide for this purpose, with the thickness varying from 2 to 5 μm. The protective thermal oxide film should not be removed before the CVD oxide deposition due to concern of damage to the bonding surface by the CVD process. The CVD oxide etches the same way in BOE as thermal oxide. The patterning of thick oxide should be done with BOE for sloped profile. Because it takes a long time to etch thick oxide with BOE, we cannot use resist as the masking material. In our process, we deposit a thin layer of LPCVD silicon nitride. LPCVD nitride is chosen for its strong adhesion to the oxide film, low stress, and pin hole free coverage. The nitride etch is done by DRIE or RIE using standard resist as mask. BOE etch of oxide should be done in an ultrasound bath to drive out bubbles in the pattern that otherwise could cause non-uniform etch. Before the DRIE etch of the journal bearing, the nitride film should be stripped to avoid masking of the journal bearing trench by the overhung nitride film above the oxide pattern. Attention must be paid to the final stripping in that the oxide film after a long-time journal bearing etch is not uniform. The variation could be 3–4 μm because the ceramic clamping fingers completely shield the spots they touch. Using BOE for final stripping would cause long-time exposure of the bonding surfaces to BOE and make it rough. So the final stripping should be done in HF whenever possible. The key steps of etch the journal bearing with oxide as a masking layer are summarized as follows:

(1) Grow thermal oxide 0.2–0.5 μm
(2) Deposit CVD oxide 2–5 μm depending on device needs
(3) Deposit LPCVD nitride right after CVD oxide deposition
(4) Pattern the wafer with standard lithography
(5) Etch nitride using DRIE or RIE
(6) Etch oxide in BOE with sporadic ultrasound agitation
(7) Strip resist with piranha
(8) Strip nitride in hot phosphorous acid
(9) Etch journal bearing with DRIE
(10) Strip oxide in HF before final bonding.

The DRIE recipes with fixed parameters above is sufficient for journal bearings of up to 350 μm in length, such as in the bearing rig and turbo charger devices. Some of the parameters, often the pressure and cycle times, are adjusted for fine tuning of the journal bearing profiles.

In the SSE process, there is a need to etch journal bearing and blade with a single mask. The total length of the trench is 1110 μm, 450 μm for the journal bearing and 660 μm of the turbine blade. In addition to all these, the DRIE etch has to be stopped in the middle for the 660 μm nested mask etch for the blades. This job is clearly beyond the ability of photoresist and thick oxide is the only choice. Fortunately, we can apply the free rotor method here to construct the journal bearing, so the deep straight etch can be done with 30–70 μm mask although the journal bearing width is 30 μm.

With the SSE specifications in mind, we studied deep trench etch using thick CVD oxide as mask for DRIE. The oxide was deposited and patterned according to the process just described. First we run some dummy wafers using fixed parameter recipes as path finders. Figure 4.36 is the evolution of the trench widths at the top, middle, and bottom for two samples. The recipes used is based on the recipe in Table 4.1, and they differ only in the etch cycle time, 15 s for sample in (a) and 13 s for samples in (b). The trench profiles were obtained by taking the sample out every hour and cover a couple of identical devices with resist and tape. The devices were diced after the etch and measured using a microscope with 0.5 μm precision.

The sample etched at 15 s etch cycle time, in Fig. 4.36(a), has negative taper throughout the etch. The sample etch at 13 s etch cycle time, in Fig. 4.36(b), has smaller width spread till 600 μm when a positive taper starts to form. The taper grows to 20 μm by the depth of 900 μm. Later tries at intermediate etch cycle time

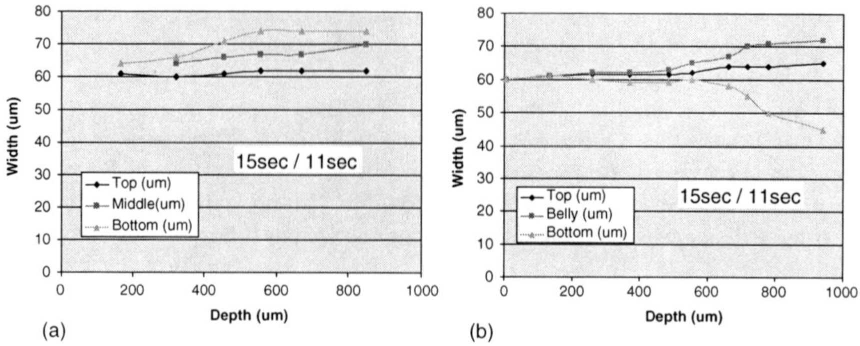

Fig. 4.36 Deep trench width vs. etch depth. Thick CVD oxide was used as DRIE mask. Fixed parameter journal bearing recipe was used for both samples. For sample (a) the etching cycle time is 15 sec. and for sample (b) the etching cycle time is 13 seconds

did not resolve the profile spread issue. These results show that for these trenches to be straight not only the DRIE masks have to be made of thick oxide film for long-time DRIE etch, but also the DRIE recipe have to change to parameter ramping mode for better journal bearing profile control.

With the experience on fixed recipe test, the parameter we choose to ramp is the etching cycle. All the other parameters are the same as in the standard journal bearing DRIE recipe listed above. The etching cycle time was set at 13 s for the first 4 h and then was ramped up 0.2 s every hour thereafter till the tenth hour. Figure 4.37 shows the optical pictures of the trench profiles in a single wafer after 4, 5, 6, 7, 8, and 9 h. The horizontal line in some of the pictures is from multi-cut dicing of the thick wafer. The trench width at the top, middle, and bottom were measured for each sample and plotted in Fig. 4.38. The journal bearing is defined by the lower 450 μm of the deepest trench, and its spread is less than 5 μm, meeting the design specification. At the end of the etch there is still more than 1 μm oxide left. This recipe is applied to SSE device fabrication.

4.4.4 Blade Process

As mentioned at the beginning of this section, the key to high-rotation speed of the microengine devices is low imbalance in the rotor mass. The factors that contribute to the multilayer rotor imbalance are, as schematically shown in the 3D drawing of a turbo charger rotor in Fig. 4.39.

(1) Blade height uniformity Δz
(2) Pattern misalignment from photo process between blade and journal bearing $\Delta b\text{–}j$
(3) Bonding misalignment between the compressor axis and the turbine axis Δa.

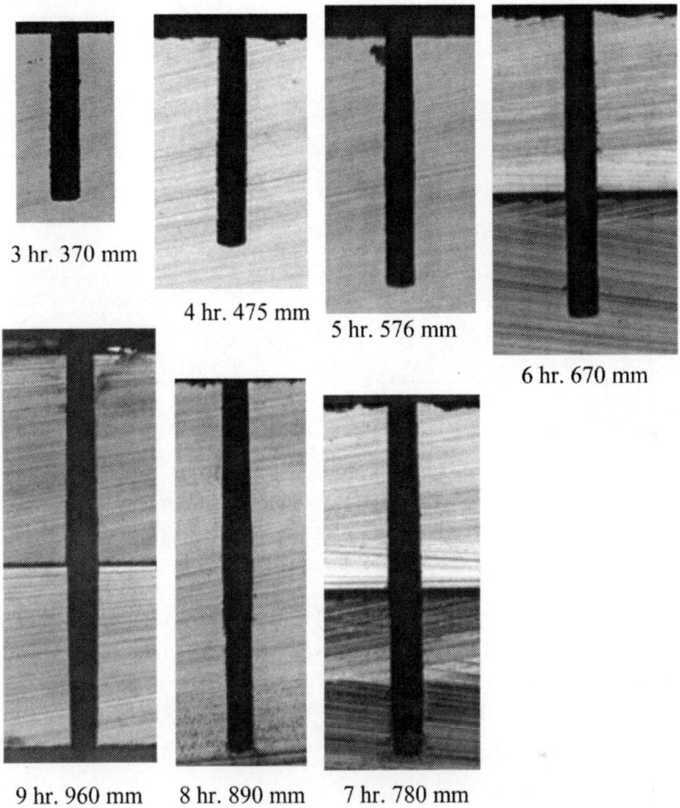

Fig. 4.37 Photo graph of the deep trench profiles at different time. This wafer was etched using deposited thick CVD oxide as DRIE mask and etch cycle time ramping recipe, 0.2 second every hour from 13 second and 4 hours. The horizontal dark lines in some of the photos are from the multi-cut dicing of the thick wafer. The width spread is acceptable to SSE device

The bonding misalignment Δa is contributed by the bonder aligner. The pattern misalignment Δb–j is the result of a photo aligner precision limit. The major contribution to the rotor imbalance is from the blade etch non-uniformity Δz, and it is the only one factor that could get worse with rotor dimension change: the blade etch non-uniformity grows with increasing blade height as desired for high power efficiency. We will first discuss the blade etch uniformity issues and present our solution of bonded blades to this problem, and then extend the same idea to address the issue of blade to journal bearing pattern misalignment issue.

In early devices, the blades are etched by DRIE on one side of a wafer and the journal bearing on the other side. Around the blades there is a large open area to be etched away. Such large loading DRIE etch unavoidably leads to large non-uniformity in blade heights and therefore in the thickness of the more massive journal bearing disks due to global and local etch variations. Figure 4.40 shows (a) the blades and journal bearing of the compressor and turbine of a turbo charger etched

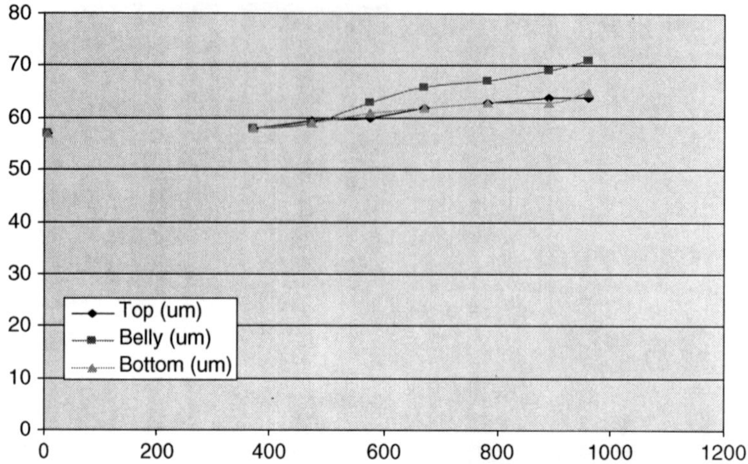

Fig. 4.38 plot of the trench width vs. depth for the sample shown in Figure 4.3. J13. the top and bottom are close throughout the etching. The center has a belly of up to 6 μm

half way into the wafers by DRIE, and (b) blade height measurement results of a compressor. To minimize the blade height variation [5], in addition to DRIE recipe tuning, the number of devices in early turbo chargers was reduced to four in a 4 in. wafer, as shown in Fig. 4.41. Also, during the DRIE process the wafers are taken

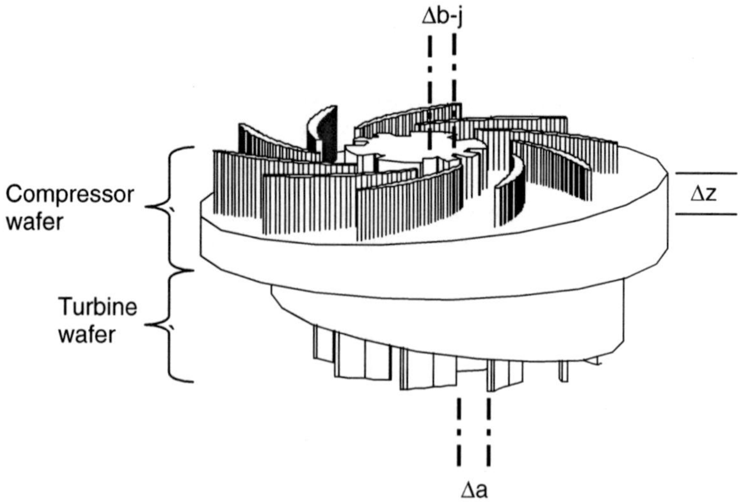

Fig. 4.39 Schematic 3D drawing of a turbo charger rotor. The compressor and the turbine both have their blades etched halfway into the wafer using large loading masks. The rotor imbalance are contributed to by the blade height uniformity Dz, the pattern misalignment between the blades and journal bearings Db-j, and bonding misalignment between the compressor and turbine axes Da

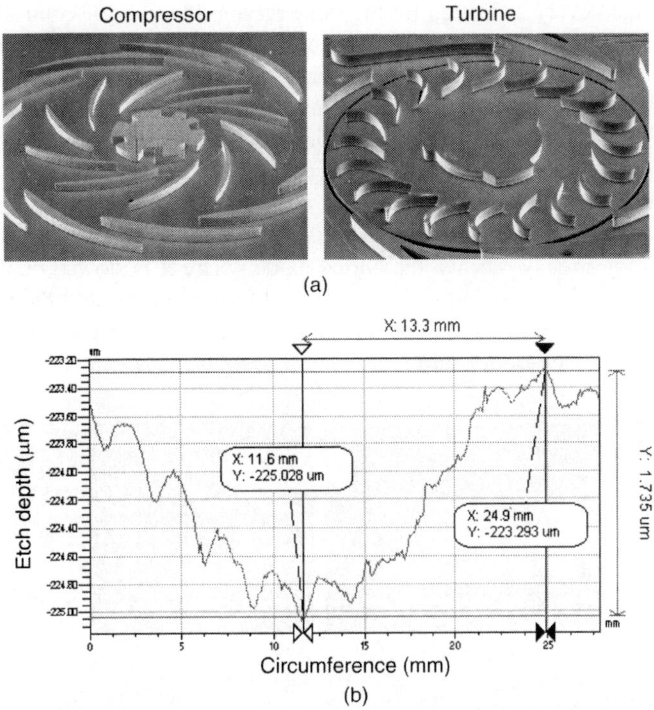

Fig. 4.40 (a) SEM pictures of the compressor blades and turbine blades in a turbocharger device. (b) variation around a compressor perimeter under best DRIE conditions. The variation in etch depth is 1.7 μm, and the resulting blade height is in the range of 223 μm to 225 μm

Fig. 4.41 4 devices in a turbo charger wafer for uniform blade etch. The pattern in the devices are the compressor blades

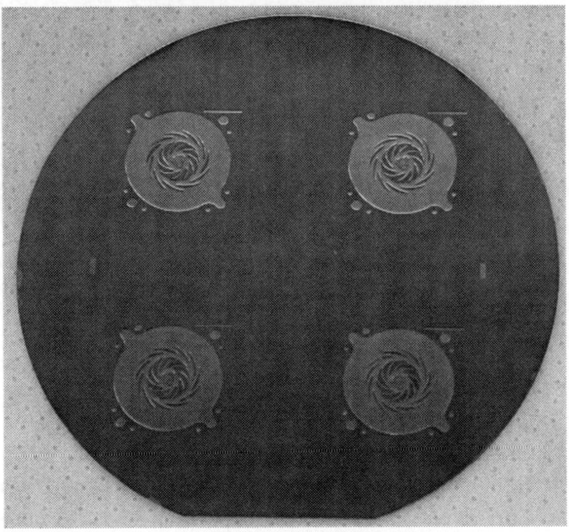

out and rotated every 7–15 min for the same reason. In this well-controlled sample the blade height difference across the rotor is about 2–3 μm. The blade height is about 220 μm and this is as deep as the blade etch can go without overstepping the design specification on blade etch uniformity although taller blades are desired for more power output. A clearer visualization of non-uniformity of the large loading blade etch is shown in the optical photographs of all the four dies in an SOI wafer in Fig. 4.42. The blades are etched into a device layer 100 μm thick. The relative positions of the four devices in the individual photos are the same as in the wafer. The etch front already reached the buried oxide where it is slowed down by more than 150 times. The remaining Si is in the lighter regions around the blades. The irregular shapes of remaining Si on the darker buried oxide give an enhanced view of the blade etch non-uniformity.

Fig. 4.42 Blade etch uniformity test on a SOI wafer with the same pattern as in Figure 4.41. The lighter regions around the blades are the remaining Si and the darker regions are the buried oxide that acts as a etch stop

This blade construction method has following drawbacks, (1) the blade height is defined by DRIE etch which intrinsically has 3–5% non-uniformity. (2) The uniformity control leads to insufficient use of the wafer space: only four 23 mm × 23 mm devices in a wafer. (3) There is not much room to increase the blade height in order to enhance the performance of the device. Deeper etch of the blades means more height variation and therefore more rotor imbalance. And (4) the large loading etch over long time tends to weaken the blade structure due to negative tapering by

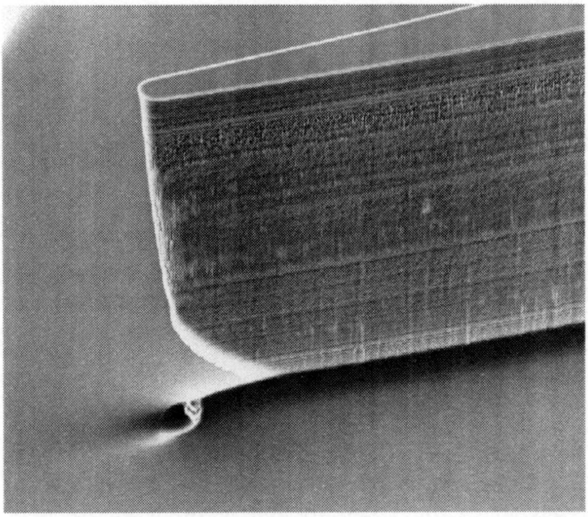

Fig. 4.43 A SEM picture of the tip of a turbo charger compressor blade. There is severe undercut at the tip, a typical phenomenon in large loading DRIE etch. The blade height is about 220 mm

charging in DRIE. An example of this can be seen in the SEM picture of a blade tip in Fig. 4.43.

It is clear that the etch non-uniformity in current DRIE technology will not meet the need for low rotor imbalance in microengine devices. We have to find a way to go around the limitation of the DRIE to achieve tall blades with drastically reduced height variation for high-power output. Our solution is using two wafers instead of one wafer as in the existing design to construct the compressor or turbines, as illustrated in the 3D drawing of a turbo charger compressor in Fig. 4.44. With two wafers, we can define the blade height with the commercially well-controlled wafer thickness by bonding the blades to another wafer. This bonded blades concept and its advantages compared to the single wafer approach are described in the cross section drawings in Fig. 4.45. As we already know, the single wafer approach forms journal bearing and blades on opposite sides of the same wafer, and their heights are determined by the depth of DRIE etching. With the two-wafer approach the blades and journal bearing are in separate wafers that are bonded together rather than DRIE. The blade height and journal bearing length are defined by the wafer thickness. TTV of commercial available DSP Si wafers, irrelevant of the wafer thickness, can be as

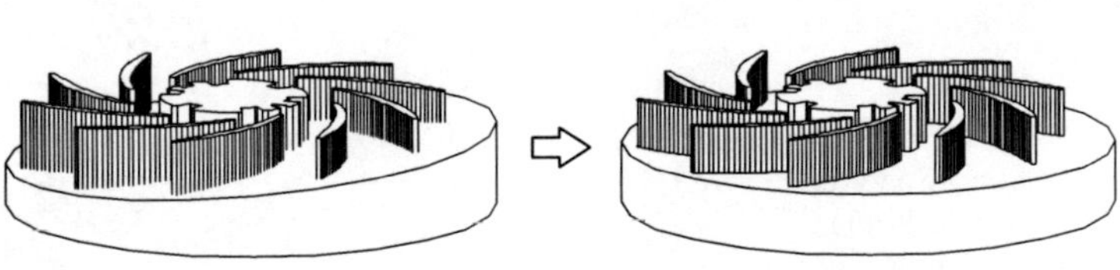

Fig. 4.44 From single wafer to two wafers, the concept of constructing blades by wafer bonding instead of DRIE

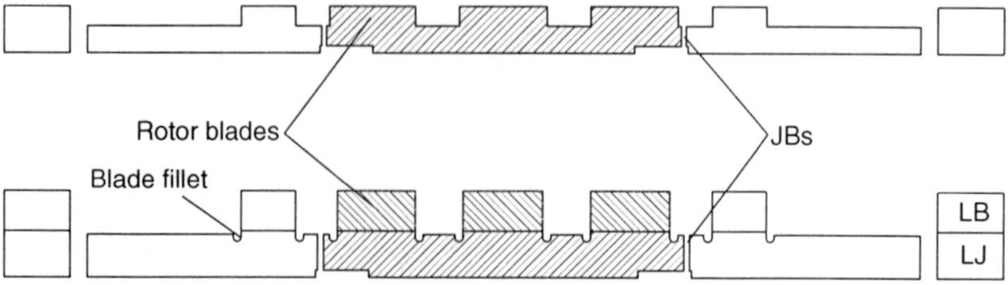

Fig. 4.45 Cross sections of the compressor wafers formed by single wafer approach on top and the and the two wafer approach, or bonded wafer approach at bottom. The blade height and journal bearing length are defined by DRI in the single wafer approach, but by wafer thickness in the bonded blades approach

small as 0.5 μm across a 6 in. diameter, and the thickness variation is gradual, only about one period across the wafer. This means that the blade height variation within each device could be more than an order smaller than that etched by DRIE, and the blade height could be much larger.

Forming the blades and journal bearing in separate wafers also provides a chance to halo etch the blades, a technique necessary to define blades height with wafer thickness. The halo etch needs to go slightly into the journal bearing wafer to form fillets at the blade base as will be discussed next in the process flow. The etch non-uniformity in the halo is negligible compared to single wafer approach because its loading is more than an order smaller and the etch depth. This can be seen in the mask patterns comparison in Fig. 4.46.

Another benefit from the improved etch uniformity is that now we can fit as many devices as the wafer can hold. Figure 4.47 shows the wafer level haloed blade mask drawing of the turbo charger compressor. There are 19 dies tightly packed in a 6 in. wafer. There are 12 extra rotors outside the devices that allow for rotor selection and for journal bearing width variation study owing to the adoption of free rotor method.

To further explain the bonded blade method, a process flow including only the key steps is shown in Fig. 4.48. This process can be applied to both compressor and turbine. The masks associated with turbo charger compressor are listed in Fig. 4.49

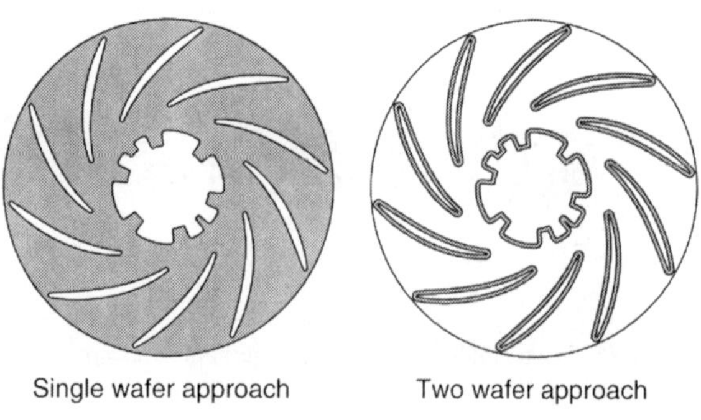

Fig. 4.46 The masks for rotor blade etch with single wafer approach and two wafer approach showing the loading (dark areas) difference that is about 19:1. The two wafer approach uses halo mask. The relative width of the halo is exaggerated for clarity

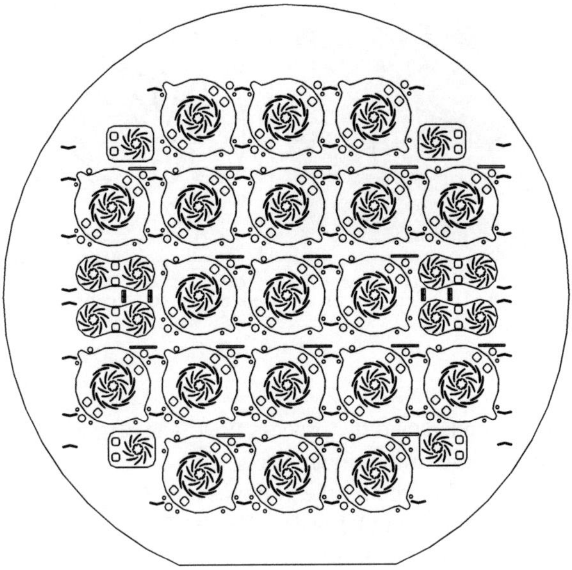

Fig. 4.47 The wafer level masks for haloed rotor blade etch of turbo charger compressor. The negligible etch none uniformity of the halo pattern allows for compact filling of the 17 dies in a 6 in wafer. There are 12 extra rotors outside the devices for quality selection of the rotors and for journal bearing variation study, thanks to the free rotor method

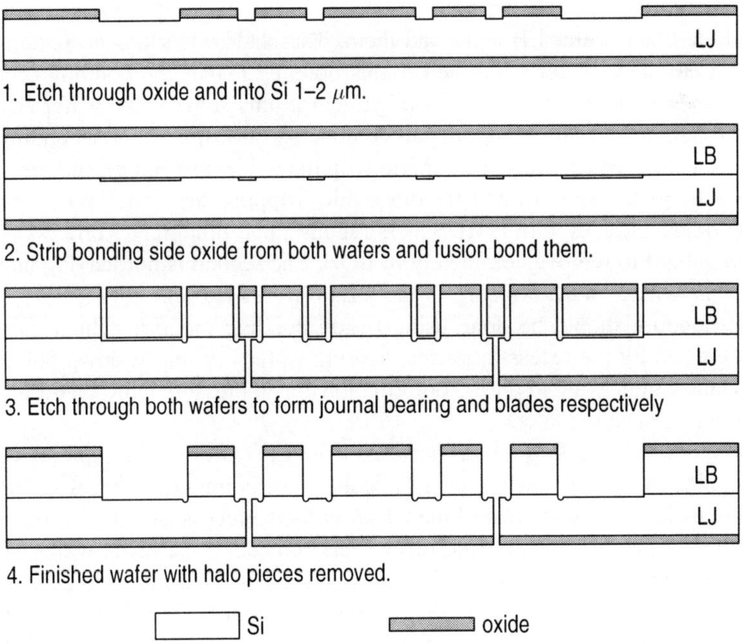

Fig. 4.48 Process flow for making the bonded blades. This process can be applied to compressor or turbine of the microengine devices

Fig. 4.49 Turbo charger masks used to form the compressor with bonded blades

to help understand the process flow. The original mask numbers used in the whole turbo charger mask set are used.

Step 1: Prepare all the wafers involved with thin thermal oxidation and pattern the journal bearing LJ wafer with BOE and 1–2 μm DRIE using blade seats mask #33. The blade seats are shaped like blades but slightly wider than the actual blades. The reason of this offset is to make sure the blades are wholly bonded and there is no risk of misalignments. The large recess area around the blade seats is for halo etch of blades in LB. The back side of the wafer should also be covered with resist to keep the oxide there from being removed.

Step 2: Strip the oxide on bonding surfaces of both the journal bearing wafer LJ and the blade wafer LB and bond them. The blade wafer has no pattern on it. Keeping the blade wafer blank before this bonding makes the bonding procedure simpler by eliminating the need for alignment in this step, so the wafer edges are perfectly aligned. Oxide stripping can be done by covering the non-bonding outer surfaces with resist and do a BOE dip to strip the oxide on bonding surfaces. There are two things to keep in mind for this oxide stripping step. The first is not treating the oxide surfaces with HMDS as in regular photolithography process because HMDS is hard to remove completely in BOE. The second is not leaving any resist residue anywhere on the bonding surfaces for obvious reason.

The bonding should be done using fusion bonding mode in which the sealed cavity created by the recess in journal bearing wafer is pumped down before contacting the wafers to avoid plastic deformation due to pressure build up during high-temperature annealing later.

The oxide thickness on outer surfaces should be as thin as the process allows to minimize the wafer bow induced by oxide. This is important because the blade seats are isolated from the rest of the LJ wafer by the recess, and thick oxide would create large bow that prevents the blade seats contacting the blade wafer LB even with vacuum in the recess cavity. If for any reason, such as when thicker oxide is needed for a specific process, and the oxide is still too thick before bonding, we can strip the resist after bonding surface oxide stripping, and do a time etch in BOE or more preferably diluted HF to thin down the outer surface oxide. It is obvious that the remaining oxide thickness is pattern dependent. According to our experience, to ensure good contact of the blades seats to the blade wafer LB, the oxide thickness should be controlled to about 1000 A for rotors about 8 mm in diameters.

To further secure the bonding contact of the blade seats, we can deposit a layer of nitride on top of the thermal oxide to compensate the stress of oxide film. We found that 0.2 μm of LPCVD nitride matches the stress of the 0.1 μm thermal oxide very well. The added process step is removing nitride selectively on bonding surfaces with a dry etch process. There is no need to cover the other side in oxide stripping because nitride is etched much more slowly in BOE or HF.

Step 3: Etch the blades and journal bearing using the journal blade halo mask #6 and the journal bearing mask #8 in Fig. 4.49, respectively. The width of the blade and journal bearing trenches are defined in mask according the etch depth and device design. They are about 30 μm for bearing rig and turbo charger and about 70 μm for SSE. Note since we are using the free rotor method to construct the final device, the rotors do not go back to the same spot they are etched from, so the trench width of journal bearing are wider than it should be for easy DRIE process control. The blade halo trench width is set to be similar to that of the journal bearing so they share the same standard DRIE recipe.

For devices with only one set of blades such as the bearing rig and combustor, the rotors are formed in this step. From the blade halo mask in Fig. 4.49 we can see that the rotor blades and the stator blades are surrounded by a large halo piece. This halo piece serves two purposes: one is to etch the blades more uniformly with narrow trenches, and the other is to protect journal bearing wafer from DRIE etching. By the time the blade etching is finished the halo piece should fall off. The blade etch should not stop just at the interface between these two wafers. It should go 20–30 μm further into the journal bearing wafer to form a rounded fillet away from the bonding interface as schematically shown in Fig. 4.45. Figure 4.50 is an SEM picture of some bonded turbine blades with the halo pieces removed. It clearly shows that the narrow halo trenches around the blades are etched into the journal bearing wafer. The fillet is formed naturally at the DRIE etch front of such narrow trenches. The halo etch

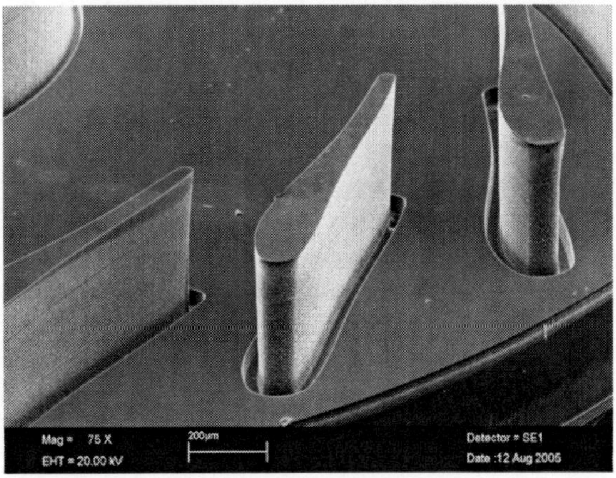

Fig. 4.50 SEM picture of finished bonded turbine blades showing narrow halo trenches around the blades that are etched into the journal bearing wafer

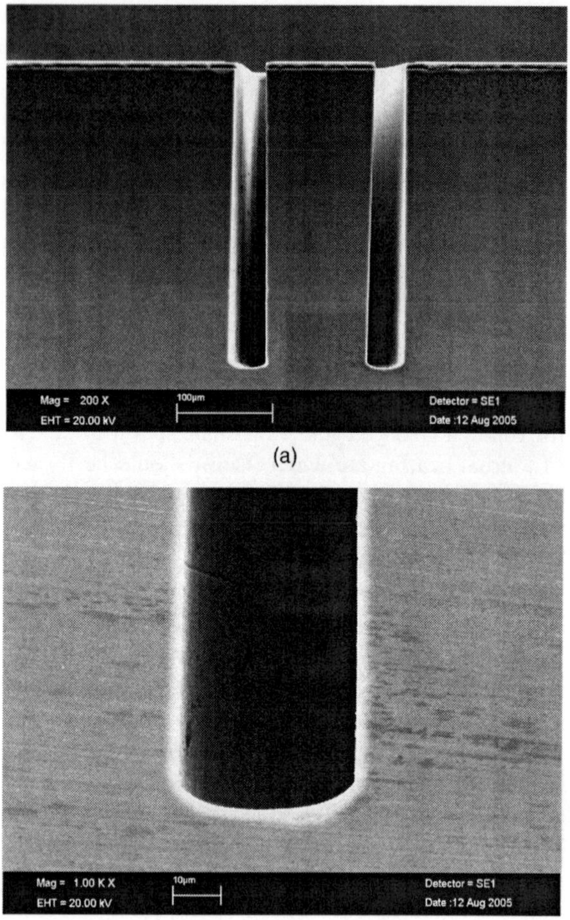

(b)

Fig. 4.51 (a) SEM picture of the cross section of a blade showing the halo trenches. The halo width is about 30 mm. (b) the SEM picture of the trench bottom close-up showing the rounded fillet naturally formed in the narrow halo trenches by DRIE

cross section of blade and a close-up of the fillet are shown in Fig. 4.51. The blade halo width is designed to be slightly larger than the outer halo width for the same piece so that the blade can be etched deeper into the journal bearing wafer while the halo piece is still holding due to loading effect.

Step 4: Remove the halo piece and clean the wafer. Figure 4.52 shows a finished dummy turbocharger compressor chip with the rotor and the halo piece taken out (a), and an SEM picture showing the rotor and stator blades, the journal bearing and the blade fillets. The cloudy region in the chip is (a) is the result of slight surface roughening of the journal bearing wafer by the end blade etch. The eight holes in the halo piece are added for easy handling of the halo piece. An important note is that the final oxide stripping should be done in BOE with over etch controlled to

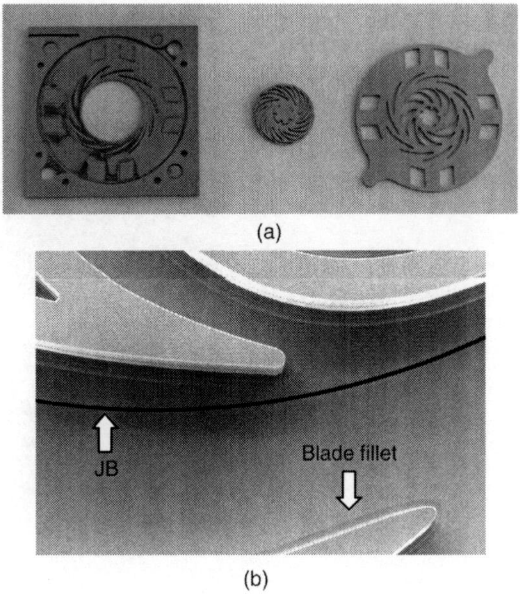

Fig. 4.52 (a) A finished dummy turbo charger compressor chip with its rotor, and halo piece. (b) A close up SEM picture showing the blades, journal bearing and the blade fillets

minimum. HF should be avoided because there is always a layer of native oxide in between the two layers, and HF could penetrate below the narrow blades and weaken them. In fact it only takes a few hours to lift the whole blades. Keep in mind this process flow only demonstrates the formation of bonded blades. It can be readily applied to the bearing rig device. For devices with more than one set of blades such as turbo charger and SSE the rotor process has more steps, a complete process flow of SSE which has six layers in the rotor will be presented in next section.

A legitimate question about the bonded blades is the quality of the bonding. To test the bonding strength, we tried to push on the blades on dummy wafers till they break, and found that all the blades tested break in the Si instead of at the bonding interface, indicating high-bonding strength. Figure 4.53 shows (a) the SEM picture of a dummy compressor die showing the side view of the bonded blades breaking in the Si, (b) the top view of the broken blades, and (c) the infrared photo of the bonded blades. Our effort on bonding quality control study is the basis for the successful application of the bonded blade concept in the microengine devices. Nearly perfect yield on the blade bonding is routinely achieved. Figure 4.54 shows the SEM picture of a turbo charger compressor after blade wafer to journal bearing wafer bonding. The patterns are those of the shallow blade seats etched in the journal bearing wafer. All the 620 blades from 31 rotors are bonded. The seat cavity areas around the blades have brightness variation because of their height variation under the ambient pressure.

The bonded bladed process described above basically eliminated the DRIE etch non-uniformity. However, to further reduce the rotor imbalance, we have to address the misalignment between the blades and the journal bearing patterns that

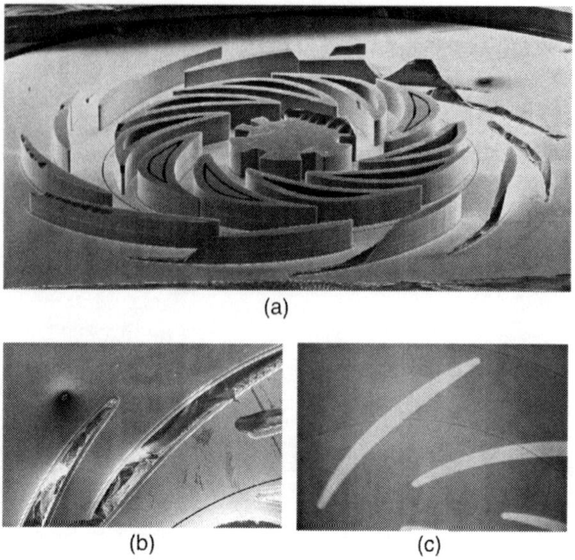

Fig. 4.53 (a) SEM picture of a dummy compressor die showing the side view of the bonded blades breaking in the Si in strength test. (b) The top view of the broken blades. (c) the infrared photo of the bonded blades

is unavoidable if the blades and journal bearing are defined in different masks. The only way to eliminate this misalignment is defining the blades and journal bearing in a single mask and use nested mask technique to achieve the etch of blades and journal bearing. Moreover, the sealed cavity formed in step 2 of Fig. 4.48 by the recess in LJ is not suitable to be opened from the journal bearing side and then be subject to more wet cleaning and fusion bonding step, because the shape of this cavity is large in area and very thin in height, it is very hard to clean and dry once subjected to wet process. For device with more than one set of blades and journal bearings

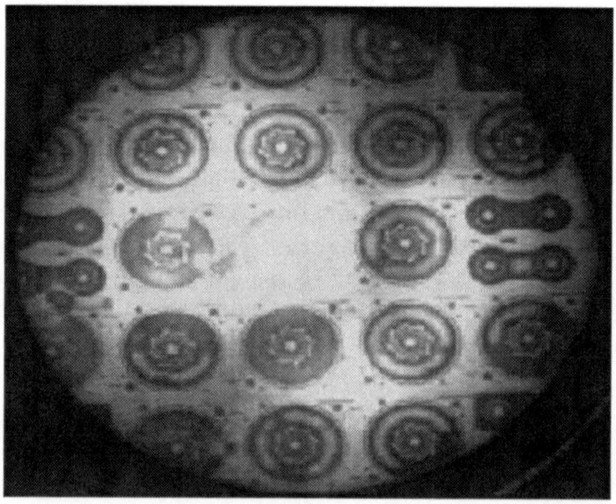

Fig. 4.54 SEM picture of a turbo charger compressor after blade wafer to journal bearing wafer bonding. The patterns are those of the shallow blade seats etched in the journal bearing wafer. All the 620 blades from 31 rotors are bonded. The seat cavity areas around the blades have brightness variation because of their height variation under the ambient pressure

such as turbo charger and SSE, it is desired to have a way of etching the blades and journal bearing from same side if the bonded blade concept is to be applied.

The solution to this problem is defining the blades and journal bearing patterns with a single mask. The goal of this combined blade etching method is to completely eliminate the misalignment between the blade and journal bearing that used to be defined by separate masks. This means we need to use nested mask and etch narrow trenches through both the blade and journal bearing wafer using the same mask. For SEE this combined trench and the nested blade trenches are over 950 and 500 μm, respectively, for the compressor, and 1110 and 660 μm for the turbine. Photoresist apparently is not the right choice, so we have to use deposited oxide film as the masking material. The process flow in Fig. 4.55 shows key concepts of this method.

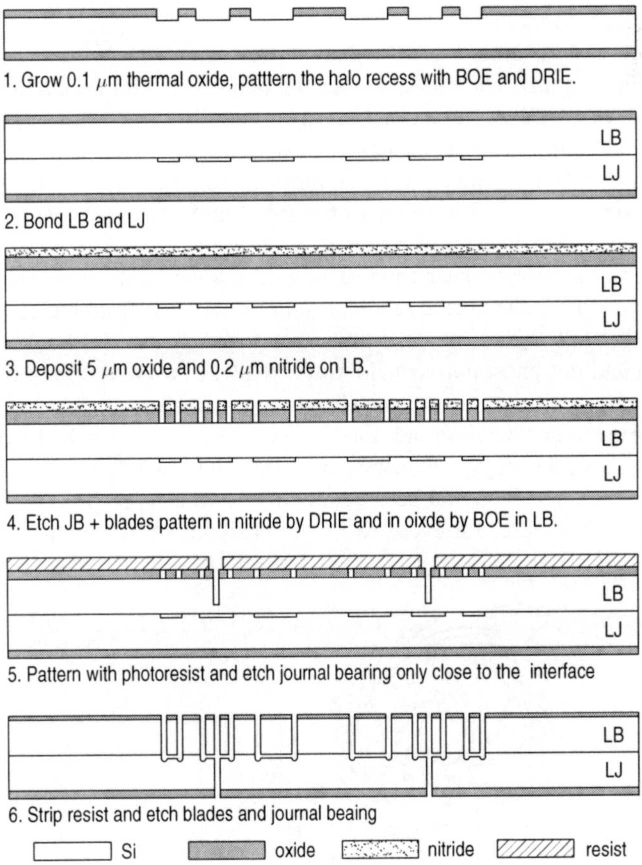

Fig. 4.55 Process flow of the combined blade method

Steps 1 and 2 in Fig. 4.55 are the same as those in Fig. 4.48. The blade halo recess is etched into the journal bearing wafer JB and then the blade wafer LB is bonded to the journal bearing wafer by fusion bonding. The difference is in that this new method requires much high-bonding quality on the wafer level. This is because the bonded stack will be subjected to wet processes such as hot phosphorous acid, piranha, and RCA, and high-temperature processes such as fusion bonding later, it is critical to maintain a clean bonding in step 2. Even the edges of the wafers should be defect free to avoid liquid being trapped in between the wafers.

Step 3 is where the process flow difference starts. A thick layer of oxide is deposited by CVD on top of the LB over the existing thin oxide film, and an LPCVD layer is deposited immediately after. The thickness of the CVD oxide film is determined by the total thickness and the etch time of the stack. The nitride is for masking the oxide later in BOE etch and its thickness is not critical as long as it provides good coverage, and is typically about 1000–2000 A. Note that for clarity reason the thickness of the wafers and the films are not to the scale as in all of the cross section schematic drawings.

Step 4 defines the blades and the journal bearing patterns with one mask that is shown in Fig. 4.56(a). The blades are haloed and the width of the halos is slightly smaller than the width of the journal bearing width. Because nitride residue means a major problem to the oxide etching in next step it is important to make sure the nitride etch is complete. Standard photo process is applied on LB for the blade and journal bearing patterns. The etching of nitride film is done by DRIE or RIE. The oxide is then etched in BOE for sloped oxide profile needed for journal bearing profile control. It is ok to keep the resist on during the oxide etch. The resist may actually help to keep the nitride film under cut by the BOE from breaking up. However, after the BOE etch of oxide, both the resist and the nitride should be stripped, by piranha and hot phosphorous acid, respectively, to avoid the masking of the Si etch by overhung nitride films or even break-away nitride film pieces later.

Because the large thickness and isotropic nature of the BOE etch of oxide, it is not practical to define the alignment marks with 3 μm wide cross hairs. So the alignment marks on top of the LB should be etched before nitride deposition, for example, in step 1 or 2. In this case, for every wafer the positions of alignment marks should

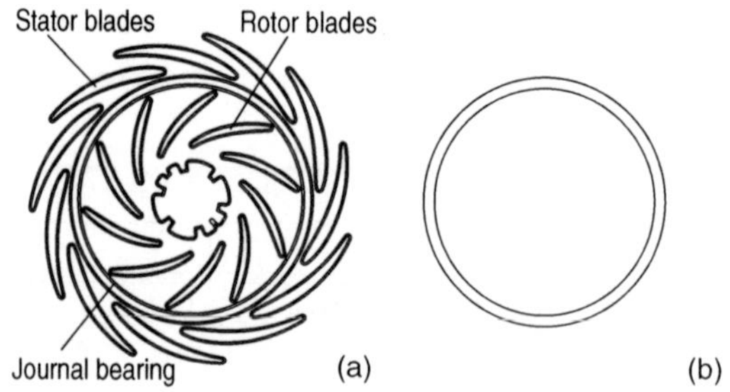

Fig. 4.56 The masks for the combined blade method. In (a) is the mask that include both the blade and journal bearing to be patterned on a thick oxide to be used as nested mask. The blades are haloed and the halo width is slightly smaller than the journal bearing width to compensate for the etch delay due to their depths difference. (b) is the mask that cover the blades but opens the journal bearing

be consistent so that when we align two wafers according to the existing alignment marks there is minimum shifting between the wafers.

Step 5 uses thick resist to cover both sides of the wafer stack and open only the journal bearing with mask (b) in Fig. 4.56. The opening is wider than the journal bearing to ensure the complete exposure of the journal bearing pattern in the oxide film under normal machine related misalignment. Then the journal bearing is etched in the DRIE system to a few tens of microns from the bottom of LB. Effort should be made to prevent premature breaking through the sealed cavity in between the wafers by over etching or cracking because the open cavity could trap liquid inside and cause disaster for devices that have to go through high-temperature processes.

Step 6 finishes up the blade and journal bearing by nested mask etching. First the resist and the polymer residue from DRIE is stripped by standard post DRIE cleaning procedure as described in the basis processes. It is critical to have a thorough polymer removal by oxygen plasma ashing for the nested mask etch. Then the wafer is etched in DRIE to finish the blade starting from the top of LB, and the journal bearing starting from near the bottom of LB. The blade halo width is designed to be smaller than the journal bearing width in order to compensate the DRIE etch rate decrease with increasing depths. Careful monitoring during the etching is necessary for depth control.

By the end of this process the blade height variation and misalignment between blade and journal bearing, two plaguing contributions to the rotor imbalance are virtually eliminated. The realization of zero misalignment between the journal bearings and blades by this combined blade and journal method is not possible without the development of the long journal bearing process that adopted thick nested oxide masks. This method was successfully used in the SSE fabrication.

4.4.5 Bonding Processes

Microengine device are constructed by fusion bonding of 5–10 wafer. As mentioned earlier, multilayer bonding is one of the major challenges to microengine device fabrication because of its stringent requirements on the wafer quality, its impact on the device yields, and its overriding restriction on the process flow design. Although it is ideal to have all the wafers finished and bonded together in one run, the complexity of the microengine devices, especially the multilayer free standing rotors, unavoidably lead to various multiple bonding steps and complex processing in between. For example, the 10-layer SSE devices undergo six separate bonding steps and the photo and etch processes grow more complicated after each bonding step. Such processing of bonded stack works against the normal bonding protocol and careful wafer cleaning and process flow design is needed. There are also issues unique to the step-by-step multilayer bonding such as back surface protection and stress compensation in the wafer. They must be addressed to achieve successive assembly of the devices. We have described general bonding processes in the basic process section, next we will discuss the special techniques associated with the multilayer device bonding. A complete bonding sequence of the SSE will be described later in section 4.5.

(1) Wafer cleanness
Based on our experience, keeping the wafer and the bonded wafer stack clean is the most important task for microengine device process. Less than perfect bonding quality in the individual bonding steps would result in significant low yield in the final device. For example a 90% yield in each of the nine bonding steps between the 10 wafers in SSE gives a final yield of only about 38.7% without considering defect overlapping. Unlike in other single layer devices, large defects outside the device areas of our wafers also affect the wafer bonding quality and therefore have a big impact on the multilayer device yield. The bonding defects open to the wafer edges could be disastrous because these defects are prone to trap liquid from wet processes. If not properly cleaned and dried they could bring contaminations to other bonding surfaces, or even burst the wafer stack when being annealed. The protocol we have adopted in maintaining wafer bonding surface cleanness is summarized as follows:

- Protect the bonding surfaces with a thermal oxide film from the very beginning. The oxide film is to be stripped just before the bonding cleaning procedures.
- After DRIE process, the wafers must be thoroughly ashed in oxygen plasma to remove the passivation polymer residue, followed by piranha, HF or BOE stripping of oxide, and RCA.
- All the cleaning steps should proceed continuously without pause. If for any reason the cleaning has to be stopped in the middle and wafers stored more than a few hours in a wafer box, it is prudent to start the cleaning all over again. This is because the plastic wafer carriers and storage boxes might outgas and contaminate the bonding surfaces.
- Every bonding tool, especially those that will be in contact with the wafers should be wiped clean right before the wafers are transferred from the RCA cleaning facility. If possible choose a light traffic time in the cleanroom. When transferring wafers from carrying box to the aligner, try not to expose the wafer to the environment for long to minimize the particle accumulation on the wafers. Try not to touch the bonding surfaces with anything during the bonding process unless it is absolutely necessary.

Following our bonding cleaning procedures, we are able to achieve high yield in the multilayer device bonding. Figure 4.57 shows the IR pictures of a pair of turbo charger wafers (a), a six-layer rotor wafer of SSE. The later has been through three bonding steps and DRIE processes in between. As for final device yield, for example, a six-layer bearing rig build yielded 46 good devices out of 48 in. the mask design. A 10-layer SSE build yielded 4 good devices out of 5, and 13 good six-layer rotors out of 16.

2) Stress compensation
Except for the final bonding, all the bonding steps include stripping all the films on the bonding sides to expose the Si bonding surface while keeping the protecting oxide films on the opposite side of the wafers. Although the one side stripping of oxide can be easily achieved as mentioned earlier, the remaining

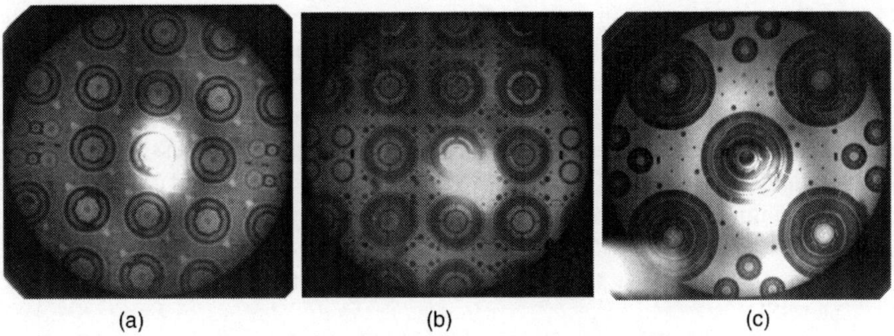

Fig. 4.57 IR pictures of bonded microengine wafers, (a) turbo charger turbine blade to journal bearing wafer bonding, (b) 4 wafer bonding for shaft test, and (c) 6 wafer bonding of SSE rotor that includes bonding of small shafts and compressor and turbine blades

oxide film on the back side would cause excessive wafer bow to prevent proper contact of the wafers, especially in the region with small isolated bonding areas such as blades and thermal shafts. This is because the propagation of the bonding front is disconnected at the recess area isolating the blades and thermal shafts, making the contact to them unpredictable. Besides, the bonding force from the small area of isolated blades and thermal shafts may not match the separating force by the oxide induced stress even with the help of ambient pressure in the case of vacuum sealed cavities. A quick solution is to reduce stress by thinning down the oxide. We found that 0.1 μm oxide is acceptable for devices with small recesses areas around the blades such as the bearing rig which has a 4 mm rotor and 9.4 mm recess span. Oxide thicker than 0.1 μm often results in unbonded blades. It is not safe to rely on thin oxide to bond the turbocharger (TC) and SSE with recess span of 20 and 35 mm, respectively. The differences in the recess area sizes of the bearing rig (BR), turbo charger (TC), and SSE can be seen in their blade seats mask design in Fig. 4.58, in which the recessed areas are hatched. The rotors of the SSE are in different locations from the stator blades. They are to be put back to the space inside the stator blades in the final bonding step.

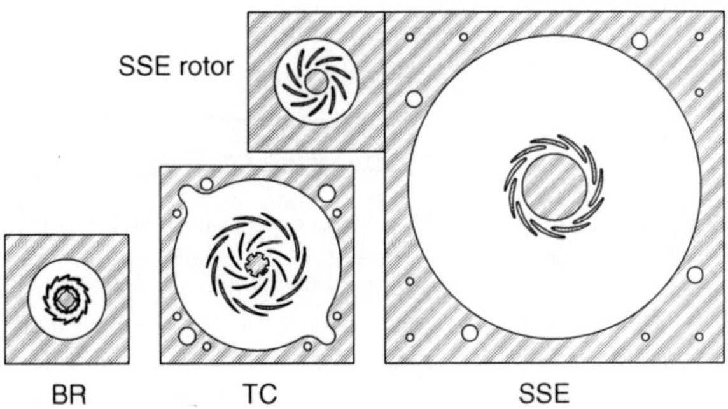

Fig. 4.58 The blade seat mask drawings for bearing rig, turbo charger, and SSE. The size of the device as well the seat cavity area grows from BR to TC and SSE, making the bonding more challenging

The blade seats patterns are chosen as an example since this problem is the most prominent there. In Fig. 4.58, the white regions are the blade seat recesses etched into the journal bearing wafer and the dark region are to be bonded. These drawings are copied from the real blade seats recess mask designs of the devices. The die sizes are 15, 23 and 41 mm, and the recess spans are 9.4, 20, and 35 mm for BR, TC, and SSE, respectively. The ratios of bonding area including the blades and speed bumps in the center to the recess area are 12%, 6.7%, 7.2%, and 22.1% for BR, TC, SSE, and SSE rotor respectively, indicating to some extent the increased contacting and bonding difficulties for TC and SSE.

From BR to TC and to SSE the patterning, etching, and cleaning processes grow more complicated and bonding more risky because of the increased size and reduced number of devices in a wafer as discussed above. We have to take a new approach to deal with the oxide induced stress and ensure high-bonding yield for the intermediate bonding steps. Our solution is compensation of the compressive oxide film with tensile nitride film. Figure 4.59 shows the results of wafer bow evolution test with the deposition and removal of oxide and nitride films. The measurement was done with a Tencor stress analyzer. Two scans, one parallel to the wafer flat, the x-bow, and one perpendicular to the flat, the y-bow, were made over the top of the wafer. Positive bow number means the wafer top side is convex: its center is higher than the edge. Negative bow number means the wafer top side is concave: the center is lower than the edge. Below the chart is a cartoon drawing showing the process flow. The starting wafer is a 6 in., 450 μm thick DSP Si wafer with 1000 A thermal oxide grown on both sides in out lab at 1000°C, step 1. Once the test starts the wafers sides are tracked. The

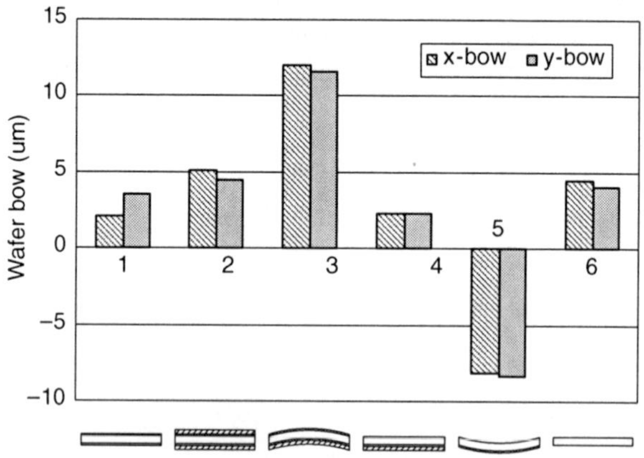

Fig. 4.59 Thermal oxide film stress compensation by LPCVD nitride film on a 450um thick 6 in Si DSP wafer. The chart gives the wafer bow changes with following process flow, as schematically shown in the drawing below the chrt: 1. 0.1 um thermal oxidation, 2. 0.2 um LPCVD nitride deposition, 3. removal of top side nitride by RIE, 4. removal of top side oxide by BOE, 5. removal of bottom side nitride by RIE, and 6. removal of bottom side oxide by BOE. The drawing below the chart schematically shows the wafer bow in the each steps

wafer bows up about 2–4 μm after oxidation. Then a 2000 A thick nitride film was deposited on both sides by LPCVD, and the upward wafer bow increases to nearly 5 μm, step 2. In step 3, the top-side nitride was removed by RIE etch, and the wafer bow jumped to 12 μm. This wafer bow jump is the contribution of the now unbalanced top oxide film. the top side oxide film is then removed by BOE, step 4, and the wafer bows returned to the 2 μm level, same as after the oxidation. This low level of wafer bow when one side of the wafer is stripped to Si is what we want for Si wafer bonding with backside protection. The bottom side nitride was removed in step 5 by RIE again, and only the bottom side oxide film remained on the wafer. The wafer bow is about −8 μm, indicating that it is concave because of the stress in the bottom oxide film. Finally in step 6 the bottom oxide was stripped and the wafer is bare Si. The wafer bow returned to the same level as after the deposition. These results show that the 0.2 μm thick LPCVD nitride film provides a good stress balance to the 0.1 μm oxide.

The wafer bow measurement gives support to the concept of wafer stress compensation that is useful to multilayer Si wafer bonding where protection is needed for the outer surfaces. This concept is shown in the schematic drawing in Fig. 4.60 where the lateral scale is the same as the SSE journal bearing to blade wafer bonding but the vertical scale is exaggerated. The lower journal bearing wafer is etched for shallow blade seats and a center circle, see Fig. 4.58 for reference.

Without nitride to compensate the stress in oxide, as in (a), there is a pulling force F_s on the wafers due to the stress in the oxide films. The pressing force F_{ap} is from the ambient pressure since the blade seat cavity is supposed to be pumped down before contacting the two wafers. The van de Wals force of bonding does not act without contact. In the case of thick oxide and/or large recess span the pulling force could overcome the pressing force and separate the wafers. One could argue that during the high-temperature annealing the stress in

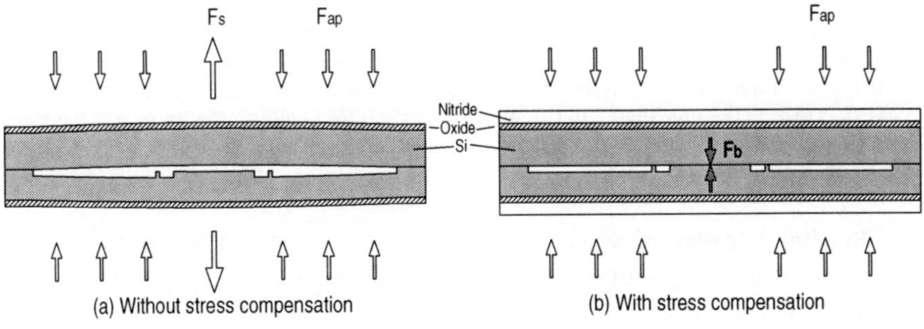

Fig. 4.60 Schematic drawing showing the advantage of stress compensation by nitride in bonding of wafers that forms sealed cavity that is in vacuum. The bonding sides of both wafers are stripped to Si surfaces. Without nitride compensation (a) the outer surfaces are stressed by the thermal oxide film kept on for surface protection for later bonding, and the separating force due to this stress F_s could overcome the pulling force F_{ap} by the ambient pressure and prevent bonding. With nitride compensation and the stress induced force negligible, the net force is contributed mainly by F_{ap} and the bonding van de wals force F_b, so good bonding is achieved

oxide might be relaxed but the ambient pressure stays the same, and therefore the blades could be bonded. The uncertainty is, however, in the sealed cavity there might be minute amount of process residues that might cause damage to the bonding surfaces before they have a chance to contact. With a nitride film to cancel out the stress from the oxide as shown in (b), the pulling force does not appear, the wafers are in contact at room temperature, and the bonding van de Wals force F_b kicks in. The pressing forces F_{ap}+ F_b dominates and the wafers are bonded. Figure 4.61 shows the IR photos of the SSE wafers with nitride stress compensation: (a) turbine journal bearing wafer to shaft wafer, and (b) compressor blade wafer to the compressor journal bearing and shaft stack. In each wafer there are 5 dies with 35 mm circular pattern and 4 cluster of 8 mm rotors (only 3 of the 4 rotors in each cluster are visible due to camera view field limit). The bright dot in each rotor areas in (a) is the bonded 0.9 mm shafts which is already etched through in the shaft wafer at this stage. In (b) the compressor rotor blades in the rotors and the stator blades in the dies are clearly visible. The fringes in the dies indicate the wafers are pressed and bent at the sealed cavities and they in general get more irregular with increasing number of wafers.

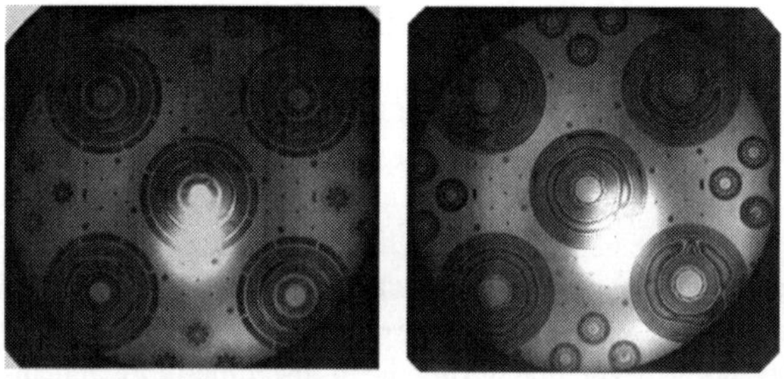

Fig. 4.61 IR pictures of bonded SSE wafers using nitride compensation: (a) turbine journal bearing wafer to shaft wafer, and (b) compressor blade wafer to the compressor journal bearing and shaft stack. In each wafer there are 5 dies with 35 mm circular pattern and 4 cluster of 8 mm rotors (only 3 of the 4 rotors in each cluster are visible due to camera limit). The bright dot in each rotor areas in (a) is the 0.9 mm shafts. In (b) the compressor rotor blades in the rotors and the stator blades in the dies are clearly visible. The fringes in the dies indicates the wafers are pressed and bent at the sealed cavities

3) Bonding of patterned wafers
The growing irregularity of the fringes in the sealed cavities is probably in part due to the distortion of the bonded stack, and/or particles on the wafer surfaces. Another potential contributor could be the way the bonding front propagates on the patterned surfaces. From the SSE IR pictures we can see that the center of the wafers is an 8 mm bonded circle surrounded by a large cavity. With the fusion bonding mode, the wafer contact starts at the center of the wafer by a wafer bow pin pushing on the center point but the contact front is interrupted at the cavity and could continue at any point or points across the cavity when the flags separating the wafers on the edges are pulled out and a piston is activated

Fig. 4.62 shows the IR pictures of a SSE L3-L5 stack, compressor blade, journal bearing and shaft wafers, before (a), (b) and after annealing (c). The top rotor in the lower cluster is black in (a) because the two wafers were not in contact. It snapped to contact under a light push with a tweezers tip and become bright in (b) and stay bonded after annealing (c). The fringes in the device cavities in (a) and (b) are smoothened after annealing at 900°C (c) indicating plastic deformation. The deformation is visible to naked eyes after annealing

to press on the whole wafer area. The consequences are that each cavity, including the rotor ones, could be clamped differently and not so smoothly given the possible warp of the wafer stack and random particle distribution. Figure 4.62 shows the IR pictures of an SSE L3–L5 stack, compressor blade, journal bearing and shaft wafers, before (a), (b) and after annealing (c). Right after pressing in bonder we see in (a) that the fringes in the dies are not smooth, with the lower left die showing a large defect area probably due to large particles in the cavity, and the top rotor in the lower cluster is black indicating a separation and the wafer is buckled at this rotor. Touched by a clean tweezers tip with a small force on the blade wafer, the black rotor snaps to white (b) indicating contact at the pattern. After annealing at 900°C for 2 h, the rotor stays bonded and the fringes in the dies are much smoother, apparently as a result of stress relaxation due to Si plastic deformation under ambient pressure and high temperature during annealing. The smoothing of fringes also indicates that there is no significant pressure buildup from out gassing in the sealed cavity during annealing, giving proof of a good wafer cleaning work.

It can be concluded from the pattern change in Fig. 4.62(c) that the wafer bending pattern changed after annealing, so the surfaces of the sealed vacuum cavities were permanently deformed. This part of the wafer should not be used as bonding area again, and this point should be a guide line in patterned wafer bonding. However, the deformation does not necessarily mean that there is no stress in the cavities. The halo pieces defined by those cavities, when etched through, usually jump off the wafer, and sometimes crack the base of the bonded blades and shafts or even the whole wafer if the pattern widths are not carefully designed. It is desired to have the critical features such as the shafts and blades done with the through etch before the edge of the halo pieces are through by varying the widths. Around the halo edge the trench width should also have width variation so that the halo pieces are released gradually to minimize the stress damage.

(4) Bonding of sealed cavities

In the multilayer assembly of microengine devices, we encounter sealed cavities in all the bonding steps except the final one. From a fabrication point of view, it is necessary to seal the patterns already etched inside the wafer stacks in the intermediate bonding steps because (1) the post bonding lithography process requires a flat wafer surface. Any patterns on the surface with steps on same level as the resist thickness will cause exposure and developing difficulties, and result in loss of dimension control or even damage to the wafers. (2) The post bonding steps usually involve wet processes. If the cavities are opened by etched holes in the wafers, they could trap chemicals and are difficult to clean and dry, causing catastrophe to the wafer stack, for example, during annealing in next bonding step. And (3) the sealed cavities, pumped down as required by the bonding annealing protocol, are naturally pressed against each other by the ambient pressure. This is critical to the success of the bonding in the isolated blade seats and shaft areas.

The need of sealed cavities in the intermediate bonding steps puts some constrains on the process flow of the devices. The process sequences should meet following guidelines. (1) The patterns on the bonding interfaces should be final since it is very hard to do any more processing to features buried in between the bonded wafers. (2) Any patterns on the bonding surfaces of wafers or wafer stacks should be completely open and wide enough for easy cleaning and drying prior to bonding. Narrow and long side way channels should be strictly avoided unless they are formed in the final bonding. (3) Have the long and deep features etched last. (4) Protect the buried features with tight dimension tolerance from DRIE etch. And (5) in the mask design avoid bonding of the features that might be subject to high-temperature Si plastic deformation. A good example of the application of these guidelines is demonstrated in the SSE process flow in next section.

The depth of the sealed cavity for blade seats is about 1–2 μm. It is chosen to be small enough to have negligible contribution to rotor imbalance, but large enough to alleviate potential pressure build-up in the recess cavity due to the evaporation of any process residue.

The bonding annealing temperature should not be set too high when there are sealed cavities in the wafer stack to avoid serious plastic deformation in the recess region due to ambient pressure. We usually anneal the wafer stack with sealed cavities at 800–850°C. In the final bonding, when all the sealed cavities are open, the annealing is done at 1000–1050°C for high-bonding strength.

4.5 SSE Fabrication

4.5.1 Introduction

SSE is designed to burn fuel and generate enough power to keep running without external power source. Compare to a turbo charger an SSE has an additional ther-

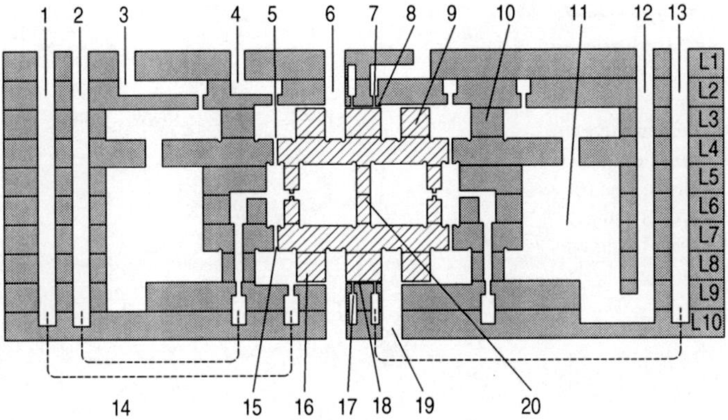

Fig. 4.63 Cross section of SSE. It includes a thermal shafts in layer 5 and layer 6, bringing the total to 10 layers layers and has 10 layers. Its fabrication adopted the new processes of bonded blades and shafts, free rotor releasing, thick oxide as DRIE mask, and combined journal bearing and blade etch with a single mask

mal isolation shaft formed by two wafers between the compressor and the turbine making the total number of wafer layers to 10. Figure 4.63 is a schematic cross section drawing of the SSE device with the critical components of the device labeled in numbers. Some of the hidden internal connections are shown by dashed lines. The thickness of the wafers are 450 μm for L1, L2, L9, and L10, 500 μm for journal bearing in L4 and L7, 330 μm for shaft layers in L5 and L6, 500 μm for compressor blades in L3, and 660 μm for turbine blades in L8. The total thickness of device is over 4.5 mm. There are total of 27 masks used. Three of them are shared by two wafers. It has a much larger combustor and as a result the die size is 41 mm by 41 mm and there are only 5 dies in a 6 in. wafer. The arrangement of the die and rotors are shown in the overlapped mask drawing in Fig. 4.64. The rotors are in 4 clusters, each with 4 rotors, out side the dies due to the free rotor approach. Below is a list of the components shown in Fig. 4.63.

(1) feed to turbine
(2) JB inlets. There are four separated such inlets for anisotropic feed to the JBs
(3) fuel inlet
(4) feed to compressor
(5) compressor journal bearing
(6) air inlet to compressor
(7) forward thrust bearing inlet, plenum, and nozzles.
(8) forward thrust bearing
(9) compressor rotor blades
(10) compressor stator blades
(11) combustion chamber
(12) combustion pressure tap
(13) aft thrust bearing inlet

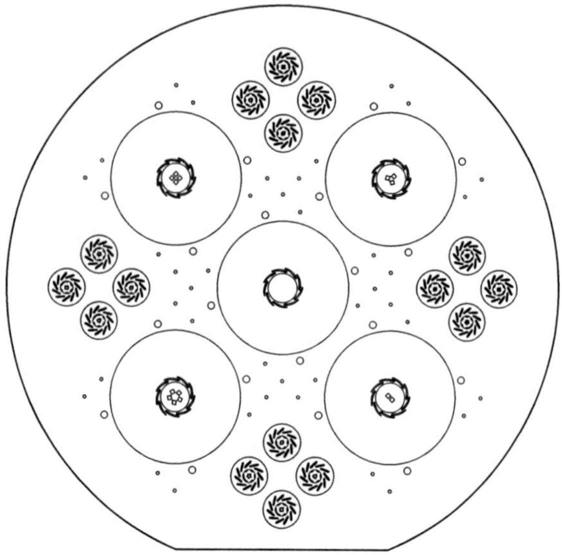

Fig. 4.64 SSE wafer level masks, compressor blade seats and rotor ID, showing the device arrangement on 6 inch wafer. There are five 41 mm x 41 mm dies and 16 rotors in 4 clusters. The rotors are outside the dies and are to be chosen and put in the dies in the final bonding

(14) blade compensation rings
(15) turbine journal bearing
(16) turbine rotor blades
(17) turbine thrust bearing plenum and nozzles
(18) aft thrust bearing
(19) turbine exhaust
(20) thermal shaft

Among the long list, we are going to discuss only those that pose major fabrication challenges. The first thing to notice is the number of wafers has increased to 10 due to an addition of the shaft. The shafts are formed in two layers, L5 and L6 so that the joints of the shaft to the compressor and to the turbine are at the bottom of a DRIE etch. This arrangement allows for the formation of controlled fillets at the joints to minimize the high stress concentration there when the rotor spins. With the addition of the two shaft layers and adoption of bonded blades for both the compressor and the turbine, the rotor is now composed of six wafers. How to process and bond these layers, 5 times for six wafers, and construct the rotors with minimum imbalance are the major challenges in SSE fabrication.

4.5.2 Rotor Process: L3–L8

Figure 4.65 is a process flow for the SSE rotor. The rotor includes layer 3 (L3) to layer 8 (L8). All the new process technologies discussed in previous section are adopted in this device: the delay mask for plenum etch uniformity, the bonded blades method, the long journal bearing etch with deposited oxide as DRIE mask, the free

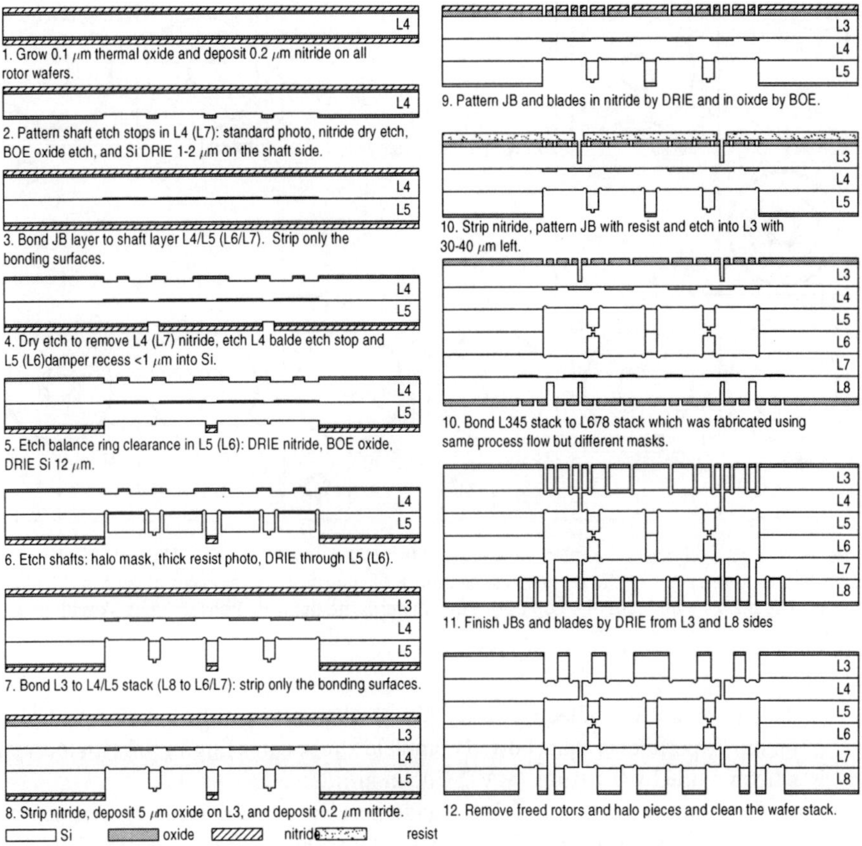

Fig. 4.65 SSE rotor process flow. The SSE rotor include 6 wafers: L3 to L8

rotor releasing method, the combined journal bearing and blade patterning with single mask and nested mask etch, and multilayer bonding. The die level masks for the rotor layers are shown in Fig. 4.66 arranged in columns for L3–L8 from left to right. The smaller drawings attached to lower right of the die patterns are those of the rotors in the rotor cluster in the same mask. The masks with four circular copies of same patterns are the rotor clusters in mask with no die features. The original numbering of mask is kept in Fig. 4.66, where A denotes top side and B bottom side of the wafers. Next we will discuss the process steps in more details.

(1) Wafer preparation

The wafers in the rotors, layer 3 (L3) to layer 8 (L8) as in Fig. 4.63, are all DSP Si wafers with low TTV and low wafer bow. The first step for these wafers, as a microengine standard for bonding surface protection, is thermal oxidation. The oxide thickness for all L3–L8 wafers is 0.1 μm. Right after oxidation 0.2 μm LPCVD nitride was deposited to compensate the stress of

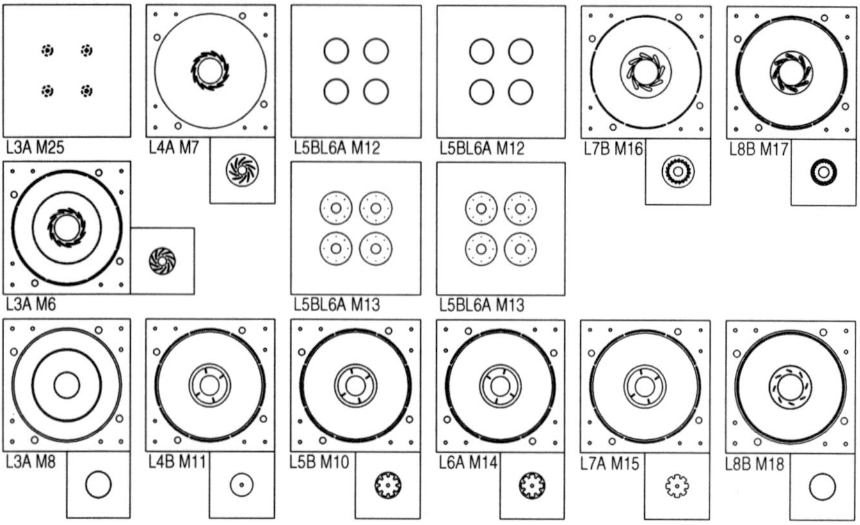

Fig. 4.66 Die level masks for rotor wafers L3-L8 from left to right columns. The smaller drawings attached to lower right of the die patterns are those of the rotors in the rotor cluster in the same mask. The masks with 4 circular copies of same patterns are the rotor clusters in mask with no die features

0.1 μm oxide. The selection of the film thickness of nitride and oxide is based on our experiments that showed complete stress canceling in 0.2 μm thermal silicon oxide by 0.1 μm LPCVD silicon nitride.

The compressor layers L3–L5 undergo basically the same process as the turbine layers L8, L7 and L6, except for an additional rotor ID shallow etch on top of L3. In the drawing, however, only L3–L5 are shown in more details. Here in step 1 only L4 is shown but L7 wafers are processed in the same batch as L4 using different masks.

(2) Shaft etch stop patterning on L4 and L7

Since the shaft length takes the whole wafer thickness of L5 and L6, as the bonded blades in L3 and L8, a shallow recess, shaft etch stop, is etched into L4 (L7) on the side facing the shaft layer L5 (L6) to prevent the unwanted parts the shaft layers to be bonded onto L4 (L7). The corresponding mask is L4B M11 for L4 and L7A M15 for L7. The etch stop should be shallow enough to avoid any imbalance due to etch non-uniformity; however it should not be too shallow in order to absorb pressure buildup due to possible minute out gassing in the cavity formed after bonding to another wafer. Typically the depth we chose is 1–2 μm.

The first step in this part is a dry etch of nitride on the bottom side of L4 (we will neglect the equivalent turbine layers from now on) to expose the oxide. Then the wafer is coated on both sides with standard thin resist and baked. Global alignment marks are patterned on the top side and the shaft etch stop is patterned on the bottom side. The top-side nitride alignment marks are etched by DRIE or RIE until the oxide is exposed. Then the oxide films on both sides are etched by BOE wet process. It takes about 2–3 min to clear the 0.1 μm oxide in BOE. Finally both sides are etched by DRIE, 0.5 μm on the top side and 1–2 μm on bottom side. Here DRIE is chosen to etch nitride and Si purely for convenience. Other RIE tools should work fine for these shallow etch. All the masks have global alignment marks included; therefore, in this step the wafer should have alignment marks permanently etched into Si on both sides, a necessity for wafer-to-wafer alignment during wafer bonding. In later steps when other masks are used, the global alignment marks will be etched into the resist for alignment accuracy check but they should be covered before any etching. For BOE etch the exposed and developed alignment marks should be covered by a bit of resist. For DRIE, since the thick pileup of resist sometime causes resist failure during long-time DRIE etch, a piece of tape which is compatible with the plasma environment should be added on the painted resist. Direct contact of the wafer surface by the tape should be avoided due to the risk of surface contamination that might affect bonding quality.

The reason to remove nitride on the L4 bottom side first is for the convenience of stripping the bottom side oxide before bonding to L5. As mentioned before, it is important to protect the bonding surfaces with thermal oxide and not to strip it until right before fusion bonding. In the case of L4, we want to keep the nitride on the oxide on the top side so that the wafer is flat when bonding to L5. Maintaining wafer flatness is critical to the bonding of small isolated features such as the shaft which is 0.9 mm in diameter. Removing nitride from one side of a wafer could not be easily done by wet process that involves hot phosphorous acid. Because photoresist cannot survive the hot process, covering the other side means unnecessary complication of the process flow. Dry process has no problem to etch nitride on one side only but the potential problem is that if the Si is already exposed it will be further etched by the dry nitride etch recipe.

When the etching of Si is done, the wafer is cleaned by ashing and piranha. Ashing by O_2 plasma is necessary to remove polymer depositions from DRIE chamber as mentioned earlier.

(3) Bonding of L4 to L5

The first step here is stripping the bonding surface while keep the oxide and nitride films on the other side. For L5 the top-side nitride should be removed by RIE or DRIE. Then both wafers are put in straight HF for a few minutes to make sure the oxide is completely removed. It is preferred to use HF to strip oxide before bonding since HF does not attack the Si surface. BOE etches Si, albeit very slowly. When BOE has to be used for oxide stripping as in the case when the other side of a wafer is also oxide and is protected by photoresist, every effort should be made to make the over etch time minimum.

Next the wafers are cleaned in RCA and spin dried. A few clean dummy wafers are usually included in the RCA cleaning. As a standard procedure, these dummy wafers are to be bonded before the devices wafers in order to clean the tools the device wafers are going to touch and check the machine conditions and bonding recipes. Prior to the test bonding using dummy wafers, the wafer aligner and bonder should be thoroughly cleaned. The dummy wafer first procedure is important because the device wafers are now stripped and any mistake could permanently damage them. To reduce the probability of particle collection on the wafer, it is a good practice to choose a time of less traffic and limit access to the bonder area.

Since L4 has alignment marks on both sides, L5 can be a blank wafer to save alignment time and to have minimum misalignment between the two wafers. Because the shaft etch stop recess will form a sealed cavity once L4 and L5 are in contact, fusion bonding mode is used for this bonding. In fusion bonding mode, the two aligned wafers are put in contact in vacuum and pressed so the cavity will not show Si plastic deformation due to high pressure during high-temperature annealing.

The pressing conditions are 2500 N (we use 6 in. wafers) for over 5 h. 2500 N on 6 in. wafers is equivalent to about 1.4 atm. The long-time pressing is to make sure the wafers with complicated patterns and small isolated bonding areas are in contact. Normally the ambient pressure pressed the wafers together if the cavity is sealed in vacuum, assuming a prefect contacting across the wafer. However, sometimes the wafer buckles at the cavity due to uneven bonding front propagation, causing the isolated shafts or blades in the cavity lose contact to the other wafer. The edge of the wafer should also be carefully inspected since unbonded areas on the edge could trap liquid in later wet processes and cause wafer contamination and bonding failure.

Once the wafers are pressed and inspected under the IR camera, they are annealed in a tube furnace. The annealing condition is 800°C for 3–4 h. The annealing temperature is lower than the usual 1000°C to avoid undesired deformation around the sealed cavity at higher temperature. Remember for the 10-layer stack to be completed we have to conduct fusion bonding nine times. Figure 4.67 shows the IR pictures of the bonded L4/L5 stack (a) and bonded the L6/L7 stack (b).

(4) Compressor blade etch stop and shaft damper recess patterning

The bonded L4/L5 stack has nitride film on both sides. The next bonding step is L3 to L4/L5, so the nitride on the L4 top side has to be removed first by dry etch while the L5 bottom side nitride stays for stack flatness control. Either RIE or DRIE can be used for the nitride removal. RIE is preferred because our

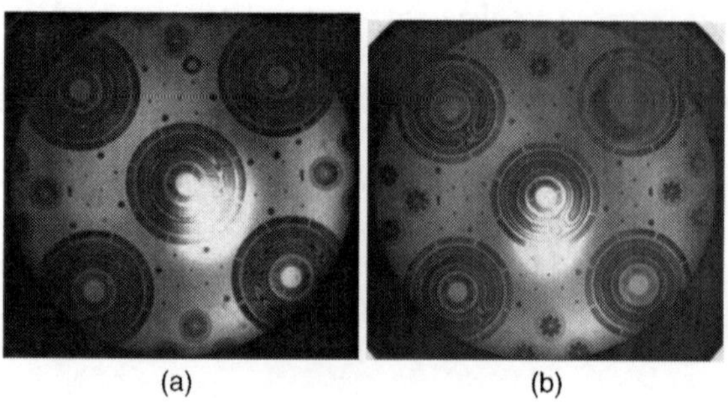

Fig. 4.67 IR pictures of bonded L4/L5 stack and bonded L6/L7 stack. A SSE wafer includes 5 device and 16 rotors that are formed in 4 clusters. Only 3 rotors in each cluster are shown in the pictures because of the IR camera view field limit

DRIE machine clamps the wafer with ceramic fingers during etch for cooling by He on the back, and the wafer has to be rotated multiple times to clear the clamp finger marks.

Once the nitride on top of L4 is removed, the L4/L5 stack is coated with standard resist on both sides. Since the L5 is still blank at this stage, the alignment marks on top of L4 are used for the patterning of compressor etch stop on top of L4 and the shaft damper recess on bottom of L5. The masks used for the blade etch stop are L4A M17 for L4 (and L7B M16 for L7, not shown in the process flow). The masks used for the shaft damper recess are L5BL6A M12 for L4 (and also for L6). After the photolithography process, the L5 side nitride is etched by DRIE. Since the clamp finger marks from DRIE machine are not an issue here, DRIE is preferred for better photoresist etch selectivity. The exposed oxide on both side of the wafer stack is then etched by BOE. Since there are small features on these patterns, the BOE etch is performed with ultrasonic agitation to avoid incomplete and non-uniform oxide etch in those small features due to trapped gas. The standard thin photoresist is robust enough to survive normal ultrasonic agitation. However, thick resist often suffers serious undercut or cracking during BOE etch with normal ultrasonic agitation, if its photo process is not carefully controlled. It is therefore a good idea to avoid BOE etch of thick resist patterns in process designs, especially when small features are involved.

After BOE etch of the oxide, the exposed Si is etched by DRIE one side at a time. The depth of the Si etch is 1–2 μm for the L4 side blade etch stop and less than 1 μm on the L5 side shaft damper recess. From now on the L5 side of the stack has a set of alignment marks for future pattern-to-pattern alignment and wafer-to-wafer alignment. The cleaning of the wafer is done again by oxygen plasma ashing and piranha, followed by spin dry.

(5) Blade balance ring clearance etch
The blade balance ring is a 12 μm etch into the Si on the shaft side. No work needs to be done on the L4 side in this step. The mask used, L5BL6A M13, is shared by L5 and L6. The photolithography process on L5 is the same as in last step: standard thin resist, nitride etch by DRIE, oxide etch by BOE with ultrasonic agitation, and DRIE etch. The cleaning procedure of ashing and piranha is also the same. The only difference is that the depth etched into Si is 12 μm instead of 1–2 μm.

(6) Thermal shaft etch
The thermal shaft is formed in this step by a through L5 DRIE etch using a halo mask, L5B M10 (L6A M14 for turbine side), same as the bonded blades are formed as described earlier. The shaft is the only structure connecting the hot turbine to the compressor. It has a small cross section, 0.9 mm in diameter, and long length, 0.74 mm, to minimize the heat loss from the hot turbine to the compressor. From steps 1–6 one can see that the shaft is bonded to L4 and come into shape when the halo etch is completed in step 6. The details of all the shaft mask layouts on the rotor are shown in Fig. 4.68, in which (a) and (b) are the damper recess and balance ring clearance shared by L5

and L6, (c) is the L5 shaft mask, and (d) is the L6 shaft mask. The dampers are eight circular posts, 8 μm in diameter and 12 μm tall each, uniformly distributed in a ring 7.5 mm in diameter. Their top view can be seen in B. The dampers in L5 and L6 sit on the blade compensation ring and are aligned to each other. The function of the compensation rings in L5 and L6 is preventing excessive bending of the three-layer compressor assembly and the three-layer turbine assembly during high-speed rotation at high temperature. They bend to each other due to weight distribution of the blades during high-speed rotation. The dampers are etched shallow to stop the bending from beginning but not contacting to avoid direct heat conduction between the compressor and the turbine. Because turbine blades tend to bend more on to the compressor side due to their larger thickness and higher operating temperature, extra inward protrusions are included in the compensation ring in L6 as seen in Fig. 4.68(d).

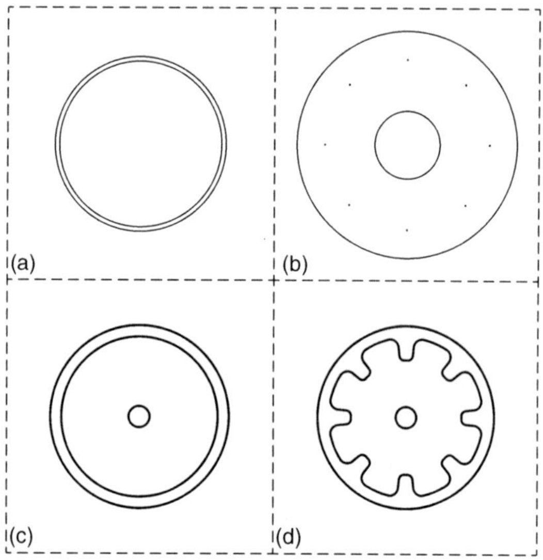

Fig. 4.68 Masks used for the SSE shaft etch

The reason that the dampers are small in cross section and long is again to minimize the heat loss from the turbine to the compressor. The halo piece is formed by 50 μm trenches around the shafts and the compensation rings as seen on the masks (c) and (d) in Fig. 4.68. The donut shape Si pieces, called halo pieces, between the shaft and compensation ring are not bonded to anything. They are separated from the journal bearing layers, L4 and L7 by the shaft etch stop that were etched in step 2, so they should fall off, when the halos are etched through. The width of the halo trenches was chosen to maintain the etch uniformity across the wafers, so that the Si halo pieces are released from the wafers at about the same time. If the halo releasing spreads over a long time, the journal bearing disks first exposed would suffer from over etching that causes none uniform disk thickness and therefore rotor imbalance. The dimensions and the fillets of the thermal shafts exposed first could also be negatively affected.

Because the through wafer DRIE etch of the shaft start on the 12 μm deep compensation ring clearance pattern, a double coat of the thick resist is necessary to completely cover the steps in the pattern, especially those at the 8 μm dampers, even though one coat is enough for the through wafer etch. A single coat of the thick resist, typically about 10 μm thick, could expose the edge of the dampers to the long-time DRIE, usually about 3.5 h. As a result the dampers become smaller or are etched away completely.

As seen from the rotor process flow in Fig. 4.4.3 and the mask design in Fig. 4.68, the nitride and oxide films are still on where the shafts are ready for etching, so the same process of dry etch of nitride, BOE of oxide with ultrasonic agitation and DRIE as in previous steps are applied here to. Since the oxide is thin and the BOE only takes about 2–3 min, resist failure in ultrasound is not a big problem here. In retrospect, however, in the masks the halo ring around the shaft should be wider than the halos around the compensation rings in order to etch through the shafts before the compensation rings for good fillts at the shaft bases. We took a makeshift approach in this build by put a small piece of tape across each compensation ring and starting the DRIE etch for about 15 min before removing the tape. This way the completion of compensation rings is delayed till the shafts are finished. The cleaning procedure is the same as in other step, except for longer ashing because of thicker polymer deposition in the wafer from the long DRIE etch.

(7) Bonding of blade wafer to journal bearing/shaft stack

This bonding between L3 and L4 is still done in vacuum because of the blade etch stop on top of L4 that could form a sealed cavities when the two wafers are contacted. Before the bonding, the L3 top side is patterned with the rotor ID and speed marks using mask L3A M25 in Fig. 4.66. An example of enlarged rotor ID and the speed mark from the mask design is shown in the shaded areas in Fig. 4.69. Also shown in Fig. 4.47 are the compressor blades to mark the

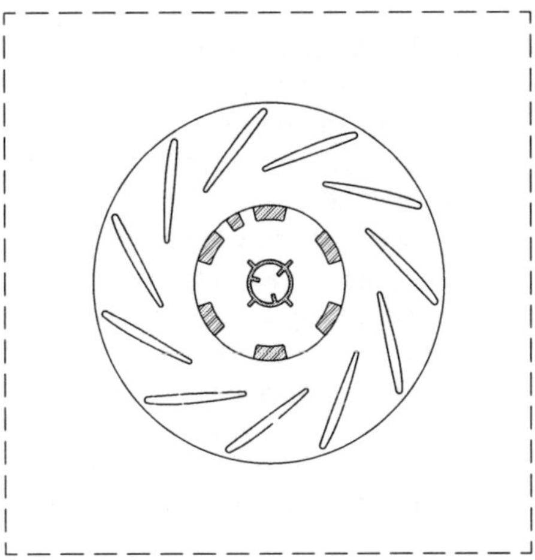

Fig. 4.69 Rotor ID and speed bump mask design, in shaded hatch, together with the compressor blades

relative position of the rotor ID and speed marks. The wafer global alignment marks on this mask are etched into the L3 Si for future pattern-to-pattern and wafer-to-wafer alignment. The global alignment marks in all other masks on this side of L3 are not going to be etched. They are used only for alignment and are covered before etching, usually by resist painting. The same practice applies to all the wafers: The global alignment marks on either side of a wafer are etched only once in the first shallow etch. For wafers without shallow etch, we can either add a photo and shallow etch step for the global alignment marks, or stop the deep etch after 1 min and cover the global alignment marks and then resume the DRIE etch.

In our test set up, we direct a laser beam on the edge of the rotor. The reflected light from the etched surface has lower intensity than from the unetched surface, and the intensity difference of the reflected light is registered for the rotor rotation speed measurement. To have larger reflected light intensity difference, we usually choose an etch recipe that gives relatively rougher etch surface than the unetched Si surface. In the SSE case, however, this is not very critical since in later steps an oxide layer will be deposited on top side of L3 and will partly remain on the speed marks. The oxide film deposited on the etched surface is much rougher than on the unetched surface.

The process for etching the rotor ID and speed marks is the same as in previous steps: standard thin resist lithography, dry etch of nitride, BOE of oxide, and DRIE of Si for a short time. An alternative to the BOE etch of oxide is dry etch since the 0.1 μm oxide does not take too long to etch in DRIE, about 6–8 min, and the continued dry etch makes the final Si surface rougher because the roughness accumulates as the dry etch proceeds. This etch is preferably done before the L3 to L4/L5 bonding so the process of the L345 stack is simpler. The condition is that this first pattern on L3, as well as all the first patterns etched into each wafer are well centered so that when the wafers are aligned according to these patterns for bonding, there is no wafer shifting. The wafer shifting, resulted from inconsistent location of the first patterns, could cause serious problems in later boding and DRIE etching. An alternative to the patterning before bonding approach is bond a blank L3 wafer to the L4/L5 stack and do the L3 rotor ID and speed mark patterning afterward. Since this approach does not adjust the wafer location during bonding there is usually no wafer shifting. The drawback of this method is additional processing of the bonded stack and hence increased risk of damaging the more heavily invested wafer stack.

Since the shaft halos are already released, the bonding between L3 and L4/L5 stack is actually between the L3 and L4 at the rotors, so the blade seats are pressed by ambient pressure from both sides. If the shafts are not etched, there would be a sealed cavity defined by the blade seats on top of L4 and another cavity defined by the shaft etch stops on the bottom side of L4, so the blade seats are pressed only from the L3 side once the stack is pumped to vacuum. Finishing halo etch in L4/L5 stack before bonding to L3 makes the L3 to L4/L5 bonding much easier.

The pre-bonding preparation of L3 is the same as in step 3 for L4 and L5. The nitride film on bottom of L3, still blank, is removed by dry etch. The oxide stripping is done in straight HF. The L4 bonding surface has only oxide now but its stripping is different because of the need of protecting the shaft bonding interface with L5. As we know although the L4 to L5 bonding is normally called Si fusion bonding, in reality the wafers had a native oxide layer before bonding and annealing, and the bonded interface usually has a nanometer thick oxide layer. If this interface is exposed to HF it will be undercut, and the shaft or blades bonded similarly will be severely weakened or even lift off. We take an approach that completely eliminates the chance of HF attacking the bonding interface. First the L5 shaft side is coated with resist and then covered with a plastic tape that is CMOS compatible so there is no concern of contamination to the clean room machine. Then the L4 side oxide is stripped by HF vapor: the L4/L5 stack is loaded in a wafer carrier and put in a beaker with straight HF with L4 facing down but a few millimeter above the HF liquid surface so only the HF vapor can reach the L4 oxide. The oxide is removed by the HF vapor in a few minutes. The wafer is rotated a few times to fully expose the edge of the wafer that are blocked by the wafer carrier to make sure the whole wafer is stripped of oxide. The purpose of the tape is to seal the L5 side of the L4/L5 stack from HF vapor. The resist is to prevent wafer contamination from direct contact of the tape to the wafer surface.

Once the wafer stack L4/L5 is stripped of oxide on the L4 side and rinsed, the tape is peeled off after heating up to 90°C for about 5 min in an oven. The wafer stack is then cleaned by ashing and piranha. Next the L3 and L4/L5 stack go through pre-bonding RCA cleaning together with some dummy wafers. As in step 3, the wafer aligner and bonder are cleaned and a dummy pair is bonded to clean the tools and check the machine conditions before the device wafer bonding. Fusion bonding mode is chosen again because of the sealed cavity formed by the blade etch stop in L4. The wafers are pressed at 2500 N in vacuum for over 5 h, inspected with an IR camera, and annealed at 900°C for 3 h. Figure 4.70 shows the IR pictures of the bonded L345 stack and L678 stack.

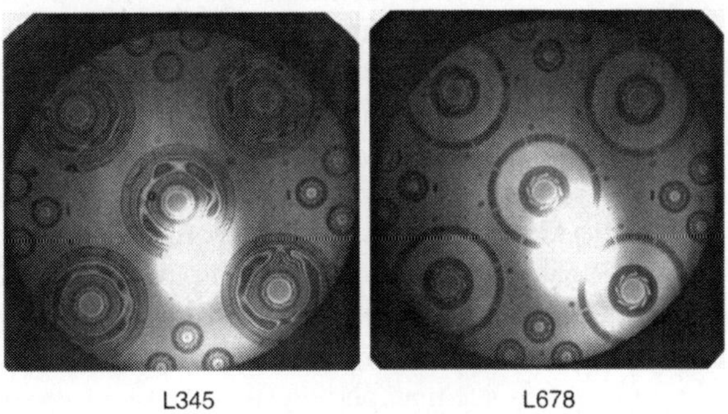

Fig. 4.70 IR pictures of bonding L345 stack and L678 stack. The L678 dies has no fringes as in L345 because the there is no cavities between the L7 and L8 in the same area of the dies

(8) Thick masking oxide deposition on blade side of L345 and L678 stack

The bonded blade method basically eliminates the contribution to rotor imbalance by non-uniform blade heights typically seen in processes that defines blades by DRIE etching blades halfway into a wafer and this allows us to increase the blade height according to the device performance requirement without compromising the rotor balance. However, to further reduce the rotor imbalance, we have to address the misalignment between the blades and the journal bearing patterns that is unavoidable if the blades and journal bearing are defined in different masks. The only way to eliminate this misalignment is defining the blades and journal bearing in a single mask and use nested mask technique to achieve the etch of blades and journal bearing as discussed in previous section. Since the blades are in L3 and the journal bearing is in L4 and the etch has to start from top of L3, even with the help of free rotor method that allows for wider trench for journal bearing formation, a durable masking material is needed to etch through the 950 μm L3/L4 stack and 1100 μm L7/L8 stack and also to provide a shadow mask for tall blade etch, 500 μm for the compressor, and 660 μm for the turbine. Double coating of photoresist can survive the long-time etching job but is not suitable for the nested mask. Our solution to this problem is deposited oxide thick film by CVD for both the journal bearing and blade patterns.

In this step, we only discuss the oxide deposition processes. The nested mask etch process will be left to the next step. The first action after annealing of the L345 stack is stripping of the nitride film by wet process, i.e., hot phosphorous acid etch. The nitride has to be stripped because the oxide has to be deposited on oxide so the BOE etch in next step will not be stopped in the middle. The temperature of the phosphorous acid is set at 160°C and the nitride etch rate is about 2000 per hour. After the stripping of nitride, the wafer stack should be completely dried by pumping in vacuum for a long time and/or baking at about 90°C and/or repeating the pumping and baking cycles for a few times. This is a precaution measure to remove any moisture in the common unbonded regions near the edge of the wafer stack. Otherwise the moisture could quickly vaporize in the CVD oxide deposition chamber which is set at 600°C, damaging the wafer stack, and contaminating the deposition system.

Unlike the LPCVD nitride that covers both side of a wafer, the CVD oxide is deposited on one side of the wafer only. The thickness of the deposited oxide is 5 μm. The requirement on the quality of deposited oxide is that it behaves similarly in BOE as the thermal oxide: forming nice 45° slope. Test runs were done with dummy wafers that went through same processes as the device wafers before the device stack was deposited. Some more dummy wafers are also added to the device batch for future use in fine tuning of the processes.

Right after oxide deposition the wafers in the batch are transferred to an LPCVD machine to have a layer of 0.2 μm nitride deposited. This nitride is to be used as the mask for BOE etching of the 5 μm thick oxide later. Nitride is a better choice than photoresist because of its superior adhesion on the oxide film. Now the bottom side of L5 is covered with nitride too but it does not have the thick oxide.

The original rotor process design had the oxide deposition after all six rotor layers are bonded. It was abandoned because the CVD machine we have cannot accommodate the 2.8 mm thick and heavy wafer stack L3-8. A benefit of this approach, however, is that current process flow moves the oxide deposition and some other process a few steps ahead of the final rotor bonding, therefore reduces the overall process risk.

(9) Pattern blades and journal bearing into the oxide with a single mask
The concept of patterning the blades and journal bearing in a single mask is born from the pursuit of low imbalance in the rotors. Traditionally the blades and journal bearing were patterned using different masks, and the machine error always leads to a misalignment, typically 2–3 μm, between the two patterns. Putting the blades and journal bearing in the same masks eliminates the misalignment problem. The consequence of this new concept is more complicated process which is never applied to any devices before. First the journal bearing etch is much longer than before since it has to be started from top of L3 and ended at the bottom of L4. This problem is greatly relieved by the free rotor method since the journal bearing DRIE trench can be much wider than the designed width. It is 70 μm wide on the mask and 30 μm in the assembled device. Secondly, the two-step etch of journal bearing and blades requires nested mask that lasts for 9–10 h and can sustain elevated temperature when some features are thermally isolated. For example, at the end of the last blade etch of the six layer rotor stack, the pattern being etched is 2.8 mm away from the He cooling of the DRIE chamber, connected by the shaft whose designed function is thermal isolation. Oxide is clearly a better choice over resist to survive potential local heating due to feature isolation.

In order to control journal bearing and blade profile, the oxide profile should be sloped at about 45°. This means the etching of the thick oxide is better to be done by BOE. The BOE etch rate of the deposited oxide is about 800 A/min. It takes over an hour to clear the 5 μm oxide. Resist usually could not survive in BOE this long without large undercut, which causes widening of the journal bearing profile. This is the reason why nitride is deposited as the masking material for BOE.

The sequence of the processes involved in this step is as follows

1. photolithography with thin resist. The mask, L3A M6 (L7B M16 for the turbine side) includes haloed blades and journal bearing. The halo width is about 70 μm.
2. nitride etch by RIE or DRIE.
3. oxide etch by BOE with ultrasonic agitation. The etch time is about 60 min. Check to make sure the oxide is completely clear. If not, etch a little longer.
4. clean the wafer with piranha.

(10) This step is the so-called nested mask etch, in which a thick resist film is patterned over the existing oxide pattern using mask L3A M8 (L8B M18).

Only part of the oxide pattern, the journal bearing and other features that go through both L3 and L4, is exposed and etched to a depth close to the bottom of the L3. The trench over the journal bearing in this mask is 40 μm wider than the journal bearing so there is no need for journal bearing realignment in this step. The consequence of the wider trench is that the oxide film exposed here will go through the DRIE etch of both L3 and L4. This is why we need a thick oxide film. The oxide film is deposited by CVD, as described in step 8. Although thermal oxide has better quality control, its thickness is limited to about 1.5 μm with our facilities. To ensure clean Si etch in DRIE, the nitride film has to be stripped first because it is now overhanging the oxide trench due to BOE undercut in step (9). The oxide pattern after nitride stripping is 5 μm deep with a shallow slope, and the Si etch is about 450 μm deep for L3 and 610 μm for L8 (with about 50 μm left), so a single coat of thick resist is enough.

The deep Si etch is done in this step instead of after L3-8 bonding for the following reasons: (1) to reduce the process work to the six-layer stack as much as possible, and (2) to make the wafer stack more flexible and easier to bond.

The sequence of this step is as follows:

1. nitride stripping by hot phosphorous acid, 1 h for 0.2 μm nitride film.
2. thick resist lithography using a mask that exposes with journal bearing.
3. DRIE etch of Si into L3 about 400 μm.
4. Wafer cleaning by 1 h ashing and piranha.

(11) Fusion bonding of compressor layers L345 to turbine layers L678

This is the last bonding for the six-layer rotor. The L678 stack, not shown till now for clarity reason, follows exactly the same processes as the L345 stack. They differ only in L3 and L8 wafer thickness and mask layouts. This bonding is also a fusion bonding, i.e., the wafers stacks are aligned in air but put in contact and pressed in vacuum, because of the sealed cavity in L5 and L6. While the pressing force remains at 2500 N, the pressing time is longer to ensure the only contact region, the shafts, 0.9 mm in diameter, has good contact. The nested mask etching in step (10) and the vacuum in the shaft cavity must overcome the stress in the 5 μm oxide on top of the blade layers L3 and L8. The fact that the oxide film on the rotors is etched through in step (9) in Fig. 4.65 using the combined blade and journal bearing mask L3A M6 in Fig. 4.66 also help to reduce the local bow near the shaft. This approach is demonstrated in the first SSE build where thirteen of the sixteen rotors were bonded successfully. Three rotors separated at the L5/L6 shaft interface because of particles. Figure 4.71 shows an IR picture of a bonded L3-8 stack after annealing.

There is a 0.1 μm thermal oxide film on bottom of L5 and top of L6, not shown in Fig. 4.2, before bonding. This oxide film has to be stripped without affecting the previous shaft bonding interface between L4 and L5 and between L6 and L7. This can be done by covering the shaft bases with resist and then strip the oxide film using BOE or diluted HF. It is not necessary to cover the

Fig. 4.71 IR image of the bonded SSE rotor stack L3-8

blade sides, top of L3 and bottom of L8, because the oxide films there are over 3 μm thick. A list of the oxide stripping procedures for the bonded stack is as follows:

1. cover the base of the shafts in L5 and L6 with photoresist and bake for 60 min at 120°C
2. strip the 0.1 μm oxide on L5 and L6 shaft side by BOE or 5% diluted HF.
3. strip the resist with piranha
4. RCA
5. fusion bonding of the stacks
6. anneal at 1000°C in N2 for 2 h.

(12) Finish the blade and journal bearing by DRIE

After bonding and annealing the rotor stack is ready for final etch. The patterned thick oxide film, with the journal bearing already half way etched, will be the mask for both the blade and journal bearing. Only the global alignment marks need to be covered by resist painting and taping, which is a simple process itself but has a big impact on the final bonding process.

The SSE journal bearing etch is very deep, 850 μm for the compressor side and 1110 μm for the turbine side, and takes two steps, the first in the blade wafer before the 3 + 3 bonding, or the compressor to turbine bonding as described in step 10, and the second in this step using nested mask.

The L3 side was etched first until the L3 blades and journal bearing in L4 are through and all the halo pieces are released. It is important to have all the features are finished at about the same time, with the blades about 20–30 μm earlier to form smooth fillets at the blade bases. This can be done by selecting the right trench width for each feature and the right DRIE recipe whose major criteria is straight journal bearing profile. Early release of the blade halo results in large loading etch of the blade base surface on L4 and therefore additional non-uniformity, i.e., rotor imbalance. More etch after the journal bearing is through might cause unwanted erosion to the already finished L5 and L6 structures. In our mask design the blade trench is a little narrower than the journal bearing trench. The blade halo trench width is about the same as that of the blade trench, but with a few narrow places so the blade halo piece is free at about the same time as the journal bearing is etched through.

When a halo piece is released, it usually jumps a little and shift position, often land on an unfinished trench. Therefore, it is necessary to inspect the wafer stack frequently during the final minutes, stop the etching and remove any freed halo pieces before resuming the etching. Usually this over etch with the halo pieces removed only happens in the last few minutes of the etch and therefore has a small impact on journal bearing profile and blade height uniformity.

When the L3 side DRIE is done, a thin layer resist is applied on it and a tape is applied over the resist after baking. This tape is to seal the L3 features from the cooling He gas when the L8 side is etched through and connects to the L3 side trenches. Before the L8 etch starts, however, the tape is poked with a sharp tool at each cavity to avoid ballooning of the tape in the DRIE chamber because the chamber pressure is much lower than the He pressure. These punched holes in the tape are sealed with small pieces of tape once the L8 side etch is through in any place of a cavity to avoid excessive He leaking into the DRIE chamber. High helium leakage usually stops the etching and the machine will give a warning message. The etch process of L8 side is the same as L3 side.

This step is an otherwise normal two-side DRIE etch except for the large stack thickness and weight. There is not photo work needed for this step, but the global alignment marks should be covered by resist since the etch is to go through the blade and journal bearing layers, L3/L4 on the compressor side and L7/L8 on the turbine side. This will leave a 5 μm oxide in the covered region after the DRIE etch, while other part of the blade wafers will have only about 2 μm oxide and less than 1 μm around the journal bearing due to the nested mask etch in step (8). The uneven oxide thickness on the stack raises the same issue of stripping the bonding surfaces while protecting the exposed stator blade bases in the final 10-layer device bonding, only this time the oxide is 20–50 times thicker. Using BOE only will over etch the device areas in the L3-8 stack surfaces for a long time, rendering them unbondable. Our solution is covering the stator blade bases with photoresist and baking at 120°C for over 1 h, stripping oxide first with BOE till the thinner region is clear of oxide, judging by the first sign of "sheeting," and stripping in straight HF for 1–2 min till both side of the stack is free of oxide. The hard baked resist survives the short HF dipping.

We only mentioned the stator blade protection because the rotors are already released from the stack before the final bonding. Our intention is not to subject the rotors to any wet process once they are released because it has a large shaft cavity between L5 and L6. This cavity has an opening of only 24 μm, between the blade compensation rings. If we treat the rotors in the same way as the L3-8 stack, we would not be able to protect the shaft bonding interfaces and straight HF would be trapped in the cavity and cause damage to the shaft.

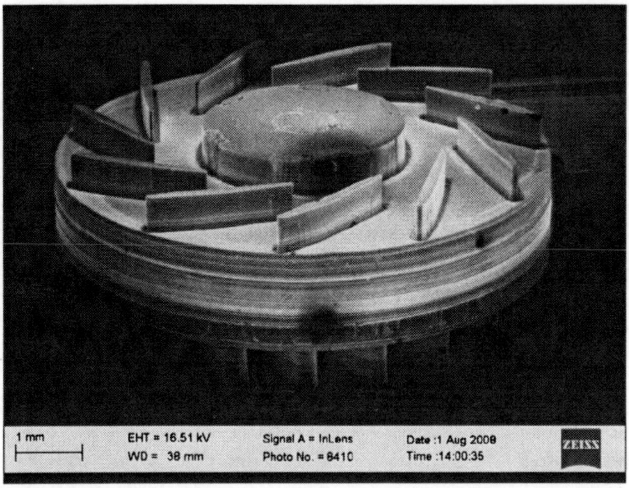

Fig. 4.72 SEM picture of a released SSE rotor with compressor blades on top

The rotors will be separately cleaned by ashing after releasing. They will be dropped in the L3-8 stack with oxide still on. The overall rotor height increase is compensated by more thrust bearing clearance etch in L2, after of course precise measurement of the oxide thickness on the rotors. To have parallel progress of all the wafer and save time, the L2 process can be designed in such a way that the thrust bearing etch is the last step using an oxide mask, and is carried out according to the rotor oxide thickness measurement.

Figure 4.72 is an SEM picture of a released six-layer rotor with compressor blades on top. The rotor ID and speed marks are on the center cylinder. The fact that the rotor holds in one piece means good bonding of the blades and especially the shaft. The turbine side of a rotor with blades on top and a turbine halo piece can be seen in the SEM pictures in Fig. 4.73. Figure 4.74(a) shows

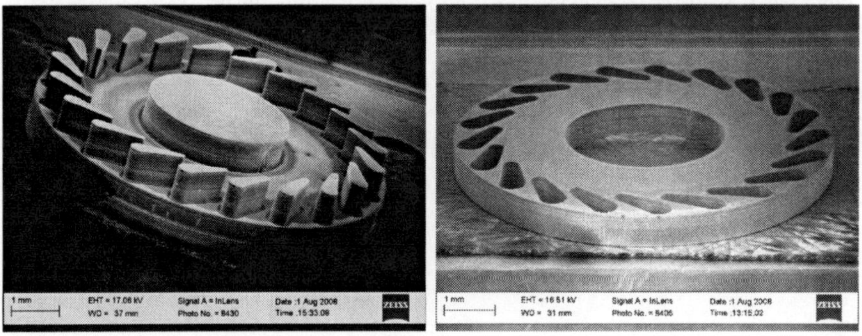

(a) (β)

Fig. 4.73 SEM pictures of SSE turbine blades (a) and its halo piece (b)

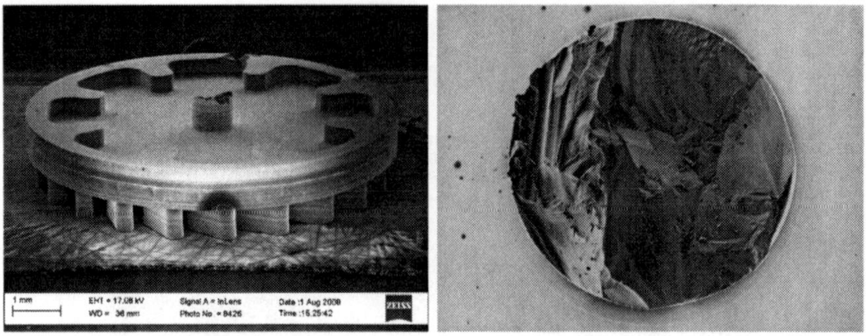

(a) (b)

Fig. 4.74 SEM picture of a broken turbine shaft seen from the turbine side, (a) the full side view and (b) close-up top view, showing the shaft broke in the Si rather than athe bonding interface

the SEM picture of the shaft side of a turbine. The shaft is broken rugged and is above the L6 surface, meaning the shaft broke in the Si on the compressor side and the bonding of the 900 μm shaft is strong. The detail of the breakage can be seen in the enlarged top view of the shaft 900 μm in diameter on the compressor side is shown in Fig. 4.74(b).

4.5.3 Fabrication of L1 and L10

The process for the top and bottom wafers, L1 and L10, both 450 μm thick, are straight forward and the dimension tolerance is not as tight as the rotor wafers. They use three masks each, one for channels halfway into the wafer, one for through holes, and one for dicing lines and device ID. Their process starts with the standard thermal oxidation. The dicing lines are etched first on the top side of L1 and the bottom side of L10. This is because the total thickness of the device is too large for a one side cut with the dicing machine we have and two side dicing is necessary. The halfway channels and through holes are patterned with two-side photo process and etched with DRIE in sequence. Figure 4.75 lists the masks for L1 and L10

4.5.4 Fabrication of L2 and L9

The plenum/nozzle wafers L2 and L9 are processed using the same process flow for TC as described in previous section. The masks for L2 and L9 are shown in Fig. 4.76.

4.5.5 Final Assembly

In the final bonding assembly all the wafers, L1, L2, L3-8, L9, and L10 are to be bonded in one batch. This is a straight forward process but needs to be done with extreme care for cleanness because a great deal of fabrication effort has already been

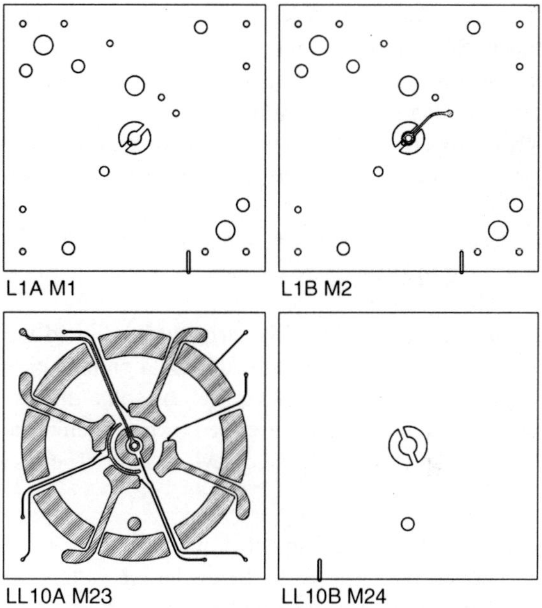

Fig. 4.75 Die level mask drawings of SSE L1 and L10 that shares same process flow

invested in these ten wafers, and they are stripped of the protective oxide and can not be reworked easily once contaminated by particles. Moreover the thick stack L3-8 is much more demanding on surface cleanness than single wafers because it is much more rigid and very hard to bend and bond around particles. Moreover, the heavily patterned rotor stack might have warping that make contact to other wafers difficult.

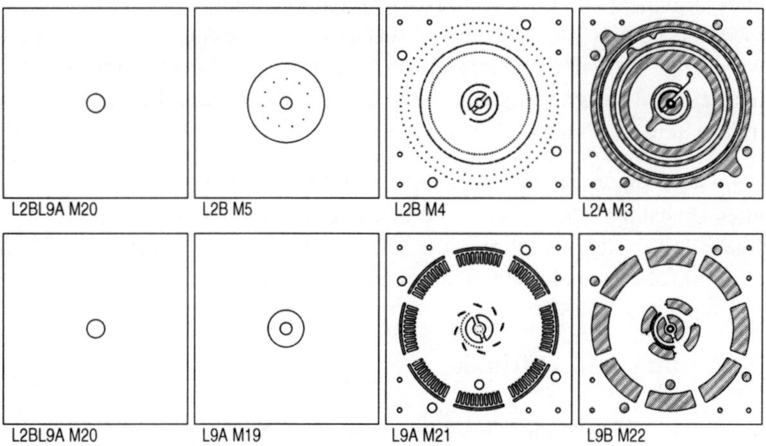

Fig. 4.76 Die level mask drawings of SSE L2 and L9 that shares same process flow

There are four-wafer contacting steps for this set of four wafers and one six layer stack. Silicon direct bonding mode was used for contacting except in the last step where fusion bonding was used for assisted contacting and pressing of two thick stacks. Silicon direct bonding mode is preferred because it does not need to transfer wafers to a bonder for pressing where the outer bonding surfaces are exposed to the pressing tools. The wafers are all through and open so silicon direct bonding mode is allowed.

Like in other bonding steps, the key to the final bonding is still wafer cleaning. First the wafers are cleaned by our standard cleaning procedures of ashing and piranha. Then the oxide is stripped by straight HF for L1, L2, L9, and L10. The rotor wafer stack should be stripped by diluted HF as described above for bonded feature protection, and with additional piranha cleaning. Standard RCA cleaning is done to all device wafers, as well as some dummy wafers at the same time. Here again the dummy wafers are used to test machine conditions and clean the tools that are to be in contact with the device wafer. The final bonding sequence is shown in the drawing in Fig. 4.77.

(1) Bond L1 and L2 together with silicon direct bonding mode and then set the contacted stack aside. Load the L1 wafer first so its top side, not a bonding surface, touches the bonder aligner chuck.
(2) Bond L9 and L10 again using silicon direct bonding mode. Load L10 first so the non-bonding bottom surface of L10 is in contact with the aligner tool.
(3) Continue with the silicon direct bonding mode and load the L9/L10 stack first and then L3-8 stack and bond them. Make necessary machine adjustment according the stack thickness.
(4) Insert the rotors into the device carefully. Try not to touch the bonding surfaces with the rotors to avoid particle generation.
(5) Bond L1/L2 stack to L3-10 stack with rotors inserted. Load the L1/L2 stack first. Use fusion bonding mode, so the whole 10-layer assembly can be clamped after alignment and there is no risk of separation failure of the thick stacks due to minor defects on the bonding surfaces. The pressing in the bonder afterward helps the wafer contact and bond together. Anneal the wafer stack after IR inspection. The annealing conditions are 1000°C for 2 h. Dicing is done using multi-cut mode of the dicing machine on both sides.

Figure 4.78 shows the IR images of a finished 10-layer SSE wafer stack before and after annealing. The device yield is 4 out of 5 and the rotor yield is 13 out of 16. Figure 4.79 is the picture of a finished die and two rotors, one with compressor blades on top and the other with the turbine blades on top.

4.6 Summary and Outlook

In this chapter, we have introduced the basic MEMS fabrication processes and the processes specific to microengine devices, discussed the fabrication challenges and our solutions, and presented the complete process flow of the SSE.

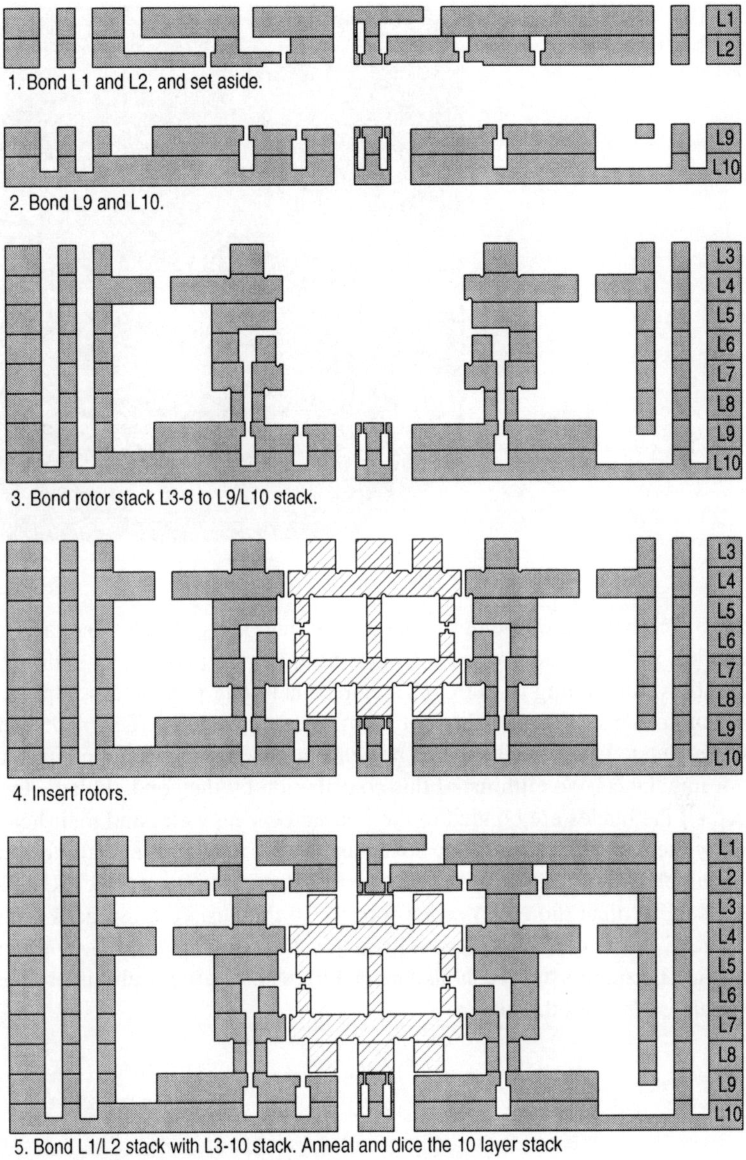

Fig. 4.77 SSE final bonding sequence

The multilayer structure and tight tolerances of many of the specifications of the microengine devices are the main causes of the microengine fabrication challenges. The most representative and difficult ones among them are the 3B: bearing, blade, and bonding challenges. They are all originated from the need of high-speed rotation operation of the rotors. The bearings, especially the journal bearings are

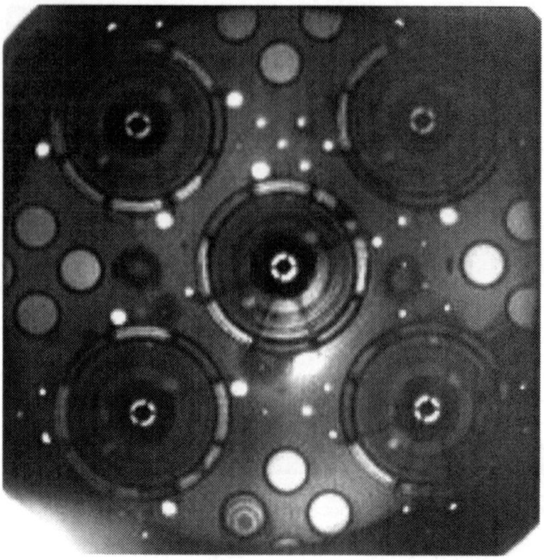

Fig. 4.78 IR image of a complete SSE wafer stack after final annealing

difficult to etch because of the tight tolerance of their narrow width and large depth. Our solution is careful control of the masking material, photoresist and/or silicon oxide profiles, fine tuning of the DRIE recipes including parameter ramping, addition of monitoring structure, and adoption of free rotor method. The blade had been a challenge due to DRIE machine etch non-uniformity that is a major contribution to the rotor imbalance. We eliminated this contribution by the bonded blade approach in which all the blades are bonded to the journal bearing wafer and their height are defined by the wafer thickness. The advantage of this approach is that the commercially available wafer thickness variation, irrelevant of the wafer thickness, is over an order smaller than the etch non-uniformity so the desired tall blades with low-height spread can be readily achieved. In conjunction with the free rotor method and nested oxide mask, this approach also helped to realize zero misalignment between the journal bearing and the blades.

Fig. 4.79 Photo of a diced SSE device and two rotors showing the compressor blades and turbine blades respectively

Bonding quality is of utmost importance to microengine device yield. Multiple bonding steps of the multilayer structures, extra etching and cleaning between bonding steps, limited number of the large size devices in a build either increase the probability of device failure or contribute to the bonding surface contamination. With the number of bonding steps determined by the device structure, we address the wafer cleanness issue by protecting the bonding surfaces with thermal oxide film that is grown at the beginning of the wafer processes and stripped just before wafer bonding, and by handling the wafers with extreme care to avoid particle accumulation. In process flow design, minimizing and simplifying the intra bonding steps is a top priority. The thermal oxide induced stress is compensated by nitride films.

In addition to the 3B challenges, we also solved the etch non-uniformity in plenum patterns with mixed large and small features by oxide film delay of the DRIE etch in the large features. The nozzle geometry was ensured by eliminating postbaking of the photoresist. Several side wall protection approaches using oxide film, passivation film, and nitride films were successfully adopted in the wafer that need two-side etching. The releasing of rotors evolved from mechanical fuse method, oxide bonding pad, and settled down on the free rotor method. The free rotor method not only simplified the process flows and increased the device yield, but also revolutionized the journal bearing and blade processes. It allows for standard journal bearing etch at a universal processing width for all journal bearing design by the rotor donor and receiver scheme.

In the fabrication effort, we have been able to go around most of the machine limitations to achieve high precision as required by the microengine devices and demonstrated in the process of 10-layer SSE. Many of the innovations involve chain reaction like modifications of many of the processes. One exception is the alignment of the compressor side to turbine side that is determined by the bonder aligner and can be controlled within 2–4 μm, enough for current turbo charger and SSE designs. Further improvement we believe can be obtained by modification of some parts in the bonder and/or improvement of the process flow.

Adding a shroud layer on top of the compressor blades is considered to be beneficiary to higher power efficiency. This is achievable from the basis of SSE process, with certain extra rotor imbalance due to the unavoidable shroud to compressor blade misalignment and added bonding difficulties.

By design, the power efficiency of the devices can be increased with higher turbine temperature. Since the current design temperature is already close to the creeping point of Si, further increase in temperature requires integration of stronger materials such as SiC in the device. Some work has been done on the SiC material and its bonding to Si [30–32]. More investigations on the impact of SiC integration on the microengine processes if needed.

References

1. Ayon, A.A.; Braff, R.A.; Bayt, R.; Sawin, H.H.; Schmidt, M.A., Influence of coil power on the etching characteristics in a high density plasma etcher, *Journal of the Electrochemical Society*, v 146, n 7, July, 1999, pp 2730–2736

2. Ayon, A.A.; Braff, R.; Lin, C.C.; Sawin, H.H.; Schmidt, M.A., Characterization of a time multiplexed inductively coupled plasma etcher, *Journal of the Electrochemical Society*, v 146, n 1, January, 1999, pp 339–349
3. Ayon, A.A.; Nagle, S.; Frechette, L.; Epstein, A.; Schmidt, M.A., Tailoring etch directionality in a deep reactive ion etching tool, *Journal of Vacuum Science and Technology B: Microelectronics and Nanometer Structures*, v 18, n 3, May, 2000, pp 1412–1416
4. Ayon, A.A.; Bayt, R.L.; Breuer, K.S., Deep reactive ion etching: A promising technology for micro-and nanosatellites, *Smart Materials and Structures*, v 10, n 6, December, 2001, pp 1135–1144
5. Ayon, A.A.; Zhang, X.; Khanna, R. Anisotropic silicon trenches 300–500 μm deep employing time multiplexed deep etching (TMDE), *Sensors and Actuators, A: Physical*, v 90, n 3, July 15, 2001, pp 381–385
6. Spearing, S.M.; Chen, K.S., Micro-gas turbine engine materials and structures, *Ceramic Engineering and Science Proceedings*, v 18, n 4 B, 1997, pp 11–18
7. Chen, K.-S.; Ayón, A.A.; Zhang, X.; Spearing, S.M., Effect of process parameters on the surface morphology and mechanical performance of silicon structures after deep reactive ion etching (DRIE), *Journal of Microelectromechanical Systems*, v 11, n 3, June, 2002, pp 264–275
8. Ayon, A.A.; Ishihara, K.; Braff, R.A.; Sawin, H.H.; Schmidt, M.A., Application of the footing effect in the micromachining of self-aligned, free-standing, complimentary metal–oxide–semiconductor compatible structures, Journal of Vacuum Science and Technology A v 174, 1999, pp 2274–2279
9. Zhang, X.; Turner, K.; Choi, D.; Miller, B.; Nagle, S.; Spearing, S.M., Low temperature silicon wafer bonding for MEMS applications, *Proceedings of the IEEE Micro Electro Mechanical Systems (MEMS)*, 2002, pp 411–414
10. Miki, N.; Zhang, X.; Khanna, R.; Ayon, A.A.; Ward, D.; Spearing, S.M., Multi-stack silicon-direct wafer bonding for 3D MEMS manufacturing, *Sensors and Actuators, A: Physical*, v 103, n 1–2, January 15, 2003, pp 194–201; Mirza, A.R.; Ayon, A.A., Silicon wafer bonding for MEMS manufacturing, *Solid State Technology*, v 42, n 1–8, August, 1999, pp 73–78
11. Turner, K.T.; Spearing, S.M., Role of wafer bow and etch patterns in direct wafer bonding, *Proceedings – Electrochemical Society*, v 19, *Semiconductor Wafer Bonding VII: Science, Technology, and Applications – Proceedings of the International Symposium*, 2003, pp 166–174
12. Turner, K.T.; Thouless, M.D.; Spearing, S.M., Mechanics of wafer bonding: Effect of clamping *Journal of Applied Physics*, v 95, n 1, January 1, 2004, pp 349–355
13. Turner, K.T.; Spearing, S.M., Mechanics of direct wafer bonding, *American Society of Mechanical Engineers, Micro-Electromechanical Systems Division Publication (MEMS)*, v 5, *Micro-Electro-Mechanical Systems (MEMS) – 2003*, 2003, pp 163–168
14. Epstein, A.H.; Senturia, S.D., Macro power from micro machinery, *Science*, v 276, n 5316, May 23, 1997, p 1211
15. Epstein, A.H.; Senturia, S.D.; Anathasuresh, G.; Ayon, A.; Breuer, K.; Chen, K.-S.; Ehrich, F.; Gauba, G.; Ghodssi, R.; Groshenry, C.; Jacobson, S.; Lang, J.; Mehra, C.-C.; Mur Miranda, J.; Nagle, S.; Orr, D.; Piekos, E.; Schmidt, M.; Shirley, G.; Spearing, S.; Tan, C.; Tzeng, Y.-S.; Waitz, I. , Power MEMS and microengines, *Tranducers 97. 1997 International Conference on Solid-State Sensors and Actuators. Digest of Technical Papers (Cat. No.97TH8267)*, v 2, 1997, pp 753–756
16. Epstein, A.H.,Millimeter-scale, micro-electro-mechanical systems gas turbine engines, *Transactions of the ASME. Journal of Engineering for Gas Turbines and Power*, v 126, n 2, April 2004, pp 205–226
17. Lin, C.-C.; Ghodssi, R.; Ayon, A.A.; Chen, D.-Z. ; Jacobson, S.; Breuer, K.; Epstein, A.H.; Schmidt, M.A., Fabrication and characterization of a micro turbine/bearing rig, *Technical Digest. IEEE International MEMS 99 Conference. Twelfth IEEE International Conference on Micro Electro Mechanical Systems (Cat. No.99CH36291)*, 1999, pp 529–533

18. Teo, C.J.; Liu, L.X.; Li, H.Q.; Ho, L.C.; Jacobson, S.A.; Ehrich, F.F.; Epstein, A.H.; Spakovszky, Z.S., High-speed operation of a gas-bearing supported mems air turbine, *Proceedings of STLE/ASME International Joint Tribology Conference, IJTC 2006*, v 2006, *Proceedings of STLE/ASME International Joint Tribology Conference, IJTC 2006*, 2006, p 12
19. Frechette, L.G.; ;Jacobson, S.A.; Breuer, K.S.; Ehrich, F.F.; Ghodssi, R.; Khanna, R.; Wong, C.W.; Zhang, X.; Schmidt, M.A.; Epstein, A.H., High-speed microfabricated silicon turbomachinery and fluid film bearings, *Journal of Microelectromechanical Systems*, v 14, n 1, Feb. 2005, pp 141–152; Frechette, L.G.; Jacobson, S.A.; Breuer, K.S.; Ehrich, F.F.; Ghodssi, R.; Khanna, R.; Wong, C.W.; Zhang, X.; Schmidt, M.A.; Epstein, A.H., Demonstration of a microfabricated high-speed turbine supported on gas bearings, *Technical Digest. Solid-State Sensor and Actuator Workshop (TRF Cat. No.00TRF-0001)*, 2000, pp 43–47
20. Wong, C.W.; Zhang, X.; Jacobson, S.A.; Epstein, A.H., A self-acting thrust bearing for high speed micro-rotors, *Technical Digest. MEMS 2002 IEEE International Conference. Fifteenth IEEE International Conference on Micro Electro Mechanical Systems (Cat. No.02CH37266)*, 2002, pp 276–279
21. Mehra, A. (Dept. of Aeronaut. & Astronaut., MIT, Cambridge, MA, USA); Zhang, X.; Ayon, A.A.; Waitz, I.A.; Schmidt, M.A.; Spadaccini, C.M., A six-wafer combustion system for a silicon micro gas turbine engine, *Journal of Microelectromechanical Systems*, v 9, n 4, Dec. 2000, pp 517–527
22. Spadaccini, C.M.; Zhang, X.; Cadou, C.P.; Miki, N.; Waitz, I.A., Development of a catalytic silicon micro-combustor for hydrocarbonfueled power MEMS, Proceedings of the 15th IEEE International Conference on Micro Electro Mechanical Systems MEMS 2002, Las Vegas, NE, January 20–24, 2003
23. Spadaccini, C.M.; Peck, J.; Waitz, I.A., Catalytic combustion systems for microscale gas turbine engines, *Journal of Engineering for Gas Turbines and Power*, v 129, n 1, January, 2007, pp 49–60
24. Spadaccini, C.M.; Mehra, A.; Lee, J.; Zhang, X.; Lukachko, S.; Waitz, I.A. ,High power density silicon combustion systems for micro gas turbine engines, *Journal of Engineering for Gas Turbines and Power*, v 125, n 3, July, 2003, pp 709–719
25. Li, H.; Savoulides, N.; Ho, L.; Jacobson, S.; Khanna, R.; Teo, C.-J.; Wang, L.; Ward, D.; Epstein, A.; Schmidt, M., Technical Digest, Solid-State Sensor, Actuator and Microsystems Workshop, Hilton Head Island, SC, June 6–10, 2004, pp 258–261
26. Fabrication and testing of a high-speed microscale turbocharger Savoulides, N. (Loomis Sayles & Co., Boston, MA, USA); Jacobson, S.A.; Li, H.; Ho, L.; Khanna, R.; Teo, C.-J.; Protz, J.M.; Wang, L.; Ward, D.; Schmidt, M.A.; Epstein, A.H., Source: *Journal of Microelectromechanical Systems*, v 17, n 5, October. 2008, pp 1270–1282
27. Savoulides, N., "Development of a MEMS turbocharger and gas turbine e
28. ngine," Ph.D. dissertation, MIT, Cambridge, MA, Feb. 2004
29. Steyn, J.L.; Kendig, S.H.; Khanna, R.; Lyszczarz, T.M.; Umans, S.D.; Yoon, J.U.; Livermore, C.; Lang, J.H., "Generating electric power with a MEMS electroquasistatic induction turbine-generator," in *Proceedings of the 18th IEEE Int. Conf. MEMS*. Miami Beach, FL, January 2005, pp 614–617
30. Yen, B.C., "A fully integrated multi-Watt permanent magnet turbine generator", Ph.D. dissertation, MIT, Cambridge, MA, September, 2008
31. Li, H.Q.; Ono, R.H.; Rudman, D.A.; Vale, L.R.; Liou, S.H., Multilayer processing of High-TC films and stacked bicrystal Josephson junctions, *Applied Superconductivity*, v 6 n 10–12, 1998, pp 711–717
32. Lohner, K.A.; Chen, K.-S.; Ayon, A.A.; Spearing, S.M., Microfabricated silicon carbide microengine structures, *Materials Research Society Symposium – Proceedings*, v 546, 1999, pp 85–90

33. Choi, D.; Shinavski, R.J.; Steffier, W.S.; Hoyt, S.; Spearing, S.M., Process development of silicon-silicon carbide hybrid micro-engine structures, *Materials Research Society Symposium – Proceedings*, v 687, 2002, pp 197–202
34. Moon, H.-S.; Choi, D.; Spearing, S.M., Development of Si-SiC hybrid structures for elevated temperature micro-turbomachinery, *Journal of Microelectromechanical Systems*, v 13, n 4, August, 2004, pp 676–687

Chapter 5
Fabrication of Microscale Rotating Magnetic Machines

David P. Arnold and Mark G. Allen

5.1 Introduction

In this chapter, microfabrication techniques are discussed to enable miniaturized magnetic machines for use with silicon microturbomachinery. These rotational machines could be used as either motors (for motoring or fluidic pumping) or as generators for electrical power generation. Challenges and methods for microfabricating microscale magnetic machines are described, and examples of previously implemented devices are discussed.

Prior to considering turbomachinery-specific micromagnetics, it is instructive to consider the historical perspective that led to the current state-of-the-art in microfabrication of magnetic microactuators. Beginning in the 1980s, initial investigations into actuators focused almost exclusively on electrostatically driven devices, in spite of the fact that in the macro-domain, almost all electromechanical actuators are based on magnetic principles [48]. This approach was taken for a number of interrelated reasons. First, the fabrication of capacitors, conductors, and other electrostatic actuation subcomponents is a straightforward extrapolation from standard planar integrated circuit fabrication technologies. Second, the materials set required for electrostatic actuation was a familiar one to integrated circuit fabrication researchers, especially when compared with the non-standard metal alloys or oxides required for magnetic actuation. Third, fabrication approaches to the three-dimensional structures required to produce flux linkage in the magnetic domain were not widely used by MEMS researchers at the time. Finally, the power consumption characteristics of electrostatics were considered more favorable than magnetics, which often carry penalties in coil losses due to the need for relatively large currents when compared with electrostatic devices.

Scaling laws describing the functionality of magnetic and electrostatic actuators lent further weight to the apparent superiority of electrostatic approaches in the small scale. Since stored magnetic energy scales as the volume, or cube of a system's

D.P. Arnold (✉)
Department of Electrical and Computer Engineering, University of Florida, 229 Benton Hall, Gainesville, FL 32611-6200, USA

J.H. Lang (ed.), *Multi-Wafer Rotating MEMS Machines*, MEMS Reference Shelf,
DOI 10.1007/978-0-387-77747-4_5,

characteristic length, while stored electrostatic energy scales as the area, or square of a system's characteristic length, electrostatics would be favored as characteristic lengths decreased [48]. This argument was further bolstered by the Paschen phenomenon, in which the maximum breakdown-limited electrostatic energy density stored in charged conductors separated in air would increase as the spacing between the conductors decreased.

Electrostatics did have certain issues, such as small gaps leading to relatively lower strokes in some actuator geometries, field emission limiting the maximum voltage that could be applied between electrodes in spite of the Paschen phenomenon, and dynamics issues such as damping and viscous losses caused by operating in these narrow-gap regimes. These issues, coupled with rapid advances in fabrication technology as the MEMS field developed, spurred renewed interest in magnetics [10].

Initial approaches to permanent magnet (PM) actuation utilized planar coil technologies couples with assembled PMs, relying on fringing fields to produce the required three-dimensionality of the system [49], and accepting the resultant loss of efficiency compared to a flux-guided system. To move beyond conductor and magnet approaches required the incorporation of magnetic core materials. Although fabrication technologies for the realization of 3D structures as well as the use of magnetic materials were unusual in the traditional integrated circuit facility, lithographic approaches to the fabrication of magnetic systems had been utilized in the past for thin film magnetic tape heads [40]. Scaling up these approaches to the many tens or hundreds of microns of thickness were achieved by advances in X-ray lithography [20] as well as the development of thick optical photoresists [21, 30]. These approaches enabled the fabrication of variable reluctance rotary magnetic micromotors as some of the early MEMS-based magnetic actuators possessing integrated three-dimensional windings [1]. These prior efforts set the stage for the realization of complex electrical machines.

The current challenge, and the focus of this chapter, is the microsystem integration of electrical machines with silicon-based turbomachinery. In general, two approaches are possible. In the first, the machine can be monolithically integrated within the turbomachinery. This offers tighter integration, but presents significant fabrication challenges at the microscale, especially for integrating the magnetic components within the silicon-based superstructure. An alternative approach is to microfabricate the electrical machinery separately from the turbomachinery and then mechanically couple the two systems together. This simplifies the respective fabrication, but presents other challenges for mechanical coupling and likely results in a larger overall size/mass. Both of these approaches have been explored for the fabrication of magnetic generators for silicon-based microturbine engines, and these efforts will be discussed in detail.

This chapter is divided into several sections as follows. Section 5.2 introduces the basic principles and structures used for electromechanical magnetic interactions. Section 5.3 discusses the primary fundamental and technological challenges for implementing magnetic machines at the microscale. Sections 5.3, 5.4, and 5.5 describe microfabrication approaches for the key components of a magnetic system, namely coils, soft magnets, and hard magnets, respectively. Section 5.7 highlights some specific issues relating to integration of these microfabricated magnetic components within a multi-wafer silicon stack. Section 5.8 reviews several magnetic machines developed for power generation via a microturbine engine.

5.2 Basic Principles/Structures

In electrical motor/generator systems, linkage between the mechanical energy domain and electromagnetic energy domain is governed by the electromechanical forces acting on static and dynamic (moving) charges. The force acting on a discrete carrier of charge q moving with velocity $\vec{v}$ in an electrical field $\vec{\mathbf{E}}$ and magnetic flux density $\vec{\mathbf{B}}$ is given by the Lorentz equation,

$$\vec{\mathbf{F}} = q(\vec{\mathbf{E}} + \vec{v} \times \vec{\mathbf{B}}). \tag{5.1}$$

Electrostatic (or capacitive) devices rely on the forces created by electric fields. In contrast, magnetic machines rely on the electrodynamic effects between magnetic fields and moving electrical charges (currents). For current, I, in a filamentary wire, the magnetic portion of Eq. 5.1 can be expressed as

$$\vec{\mathbf{F}} = I \oint_c \mathbf{d}\vec{l} \times \vec{\mathbf{B}}. \tag{5.2}$$

Also, the electromotive voltage, V, induced on a conductor moving with velocity in a static magnetic field is given by

$$\mathbf{V} = \oint_c (\vec{v} \times \vec{\mathbf{B}}).\mathbf{d}\vec{l}. \tag{5.3}$$

As can be seen by Eqs. (5.2) and (5.3), large magnetic fields and/or large currents are required in order to maximize electromechanical forces and voltages. The goal of the motor/generator design is thus to form structures that maximize these interactions.

5.3 Challenges for Microscale Magnetic Machines

The use of MEMS microfabrication strategies for constructing rotating magnetic machines imposes many design and fabrication constraints not typically encountered at the macroscale. The biggest fundamental difference is that microscale machines are generally restricted to planar geometries due to the constraints of surface/bulk micromachining techniques, whereas macroscale machines typically use three-dimensional cylindrical geometries. Since the machine design and fabrication are intimately related, this presents many interesting scale-related design tradeoffs.

5.3.1 Machine Design Issues

As introduced above, the design and optimization of a planar, microscale magnetic machine presents many unique and interesting challenges. Specific modeling strategies are discussed in Chapter 8, but included here are some general trends, observations, and tradeoffs related to the machine design.

For example, one disadvantage of a planar machine geometry is that inner and outer end turn connections can take up considerable area and typically account for a large proportion of the overall winding resistance. The space required for the inner end turns of a planar machine limits the minimum radius of the magnetically active area, while the space required for the outer end turns can double the surface area of the stator. In contrast, in a macroscale machine, end turns make up a small percentage of the overall size and resistance of the winding.

Microfabrication also limits the thickness and aspect ratio of the windings, resulting in highly constrained wire cross sections and winding patterns. Additionally, relatively low copper "fill factors" are common in microfabricated windings as compared to traditional wound-wire windings. These effects combine to increase the net winding resistance and thus limit the machine performance. For example, macroscale machines typically have many coil turns/pole ($\sim$20 turns/pole). However, due to space limitations and fabrication complexity, microscale machines will generally have only a few turns/pole (perhaps 2–4 turns/pole). Increasing the number of turns/pole would significantly increase the resistance and reduce power output due to minimum feature size constraints.

Microfabrication issues also affect the magnet design. Macroscale PM machines use discrete PMs, which are assembled together to form the machine. In contrast, microscale machines, such as the ones presented here, may use a mechanically contiguous multi-poled PM annulus. The annulus contains "transition regions" where the rotor magnetization smoothly transitions between north and south poles. The size of these "transition regions" can be on the order of a magnetic pole width, resulting in non-ideal magnetization patterns, and thus, reduced voltages induced in the windings.

Microfabrication issues can also affect high-level machine design decisions. For example, the difficulty of using microfabrication to bury windings in slots and closing them over with hats, as is typical in the stators of macroscale magnetic machines, makes surface-wound stators preferable at the microscale. This may limit performance, but one advantage of using surface-wound stators is that their larger air gaps and small armature reactance make them are less prone to demagnetizing the rotor PM, thus improving reliability.

Despite these design limitations, MEMS fabrication also offers some advantages for microscale magnetic machines. The stator windings can be photolithographically defined, which permits optimized winding patterns with variable-width conductors. In addition, high-resolution, fine-pitch features can help reduce certain losses; skin effects can be reduced by using windings with small cross-sections and eddy current losses in both the windings and magnetic core can be reduced by using laminations of micron scale thicknesses.

The ratio of surface area to volume also increases as the length scale is reduced. Heat conduction is proportional to surface area while heat production is proportional to volume. Microfabrication can be used to create flat conductors which have a larger surface area to volume ratio than round conductors. By placing these flat conductors in intimate contact to their substrate, cooling via thermal conduction greatly improves. This allows for much higher current densities in the windings than possible in macroscale machines. Current densities of up to 10^9 A/m^2 have been achieved in microscale magnetic machines while their macroscale counterparts are generally limited to 10^7 A/m^2.

Lastly, any proposed device should be able to operate over a wide temperature span. The challenge for magnetic devices is high temperatures, whether from ambient environmental conditions or from the system operation, e.g., a heat engine in a microengine generator system. All ferromagnetic materials lose their magnetization at a material-specific temperature, the Curie temperature. This temperature is most critical for hard magnetic materials, which in a device would provide a source of magnetic flux. For example, the Curie temperature of NdFeB, a high-performance PM, is only ~300°C, and the maximum operating temperature before significant performance loss is only ~150°C. Thus, judicious material selections must be made when considering hot operating environments.

5.3.2 Fabrication Issues

In addition to the machine design issues, there are significant technological hurdles and inherent limitations for fabrication of magnetic machines at the microscale. Magnetic transduction schemes demand uniquely different construction in comparison to electrostatic, piezoelectric, electrothermal, or other transduction schemes more often employed in MEMS.

First, for best performance, high currents and high magnetic fields are desired, as described in Section 5.2. This motivates the need for relatively voluminous structures: large conductor cross-sections to avoid electromigration; voluminous hard magnetic (PM) structures to provide large magnetic fields; and thick soft magnetic structures to avoid magnetic saturation. The dimensions of the desired conductor and magnetic components could range from 1 to 100+ μm, much larger than the dimensions commonly found in modern integrated circuits or many MEMS devices.

Second, because of the dipole nature of a magnet, three-dimensional structures are required to direct and guide the looping magnetic fields. This contrasts with the more planar 2D architectures that are suitable for most other MEMS devices. Three-dimensionality is also desirable to achieve the multi-turn windings for electromagnets, sense coils, inductors, or other coil-like structures. Spiral, toroidal, and other complex shapes are all common on the macroscale, but because of the planar microfabrication approaches, these structures are challenging to achieve on the microscale. The use of multiple metal layers for the construction of windings greatly increases the fabrication complexity.

Third, materials and temperature compatibility must be addressed when designing a multi-step process flow. For example, incorporation of magnetic components (copper windings or magnetic materials) into the process flow may require subsequent microfabrication steps to occur at low temperature. Long-term reliability issues such as adhesion, diffusion, oxidation, etc. must also be considered.

5.4 Coils

Electrical coils are the first key component to a magnetic machine. These coil windings serve as the linking interface between the electrical, magnetic, and mechanical energy domains. Current may be passed through a coil in order to create a magnetic field, or current may flow through a coil as a result of a magnetically induced electromotive force. In either case, the purpose of a coil is to simply serve as a shaped conductor of electrical current.

Naturally, where power consumption is a concern, coils should have low electrical resistivity to minimize resistive losses and Joule heating. Thus, resistivity is the primary material property concern. Secondary material properties include electromigration resistance, thermal stability, and chemical stability. Other integration issues such as thermal expansion and dielectric isolation must also be considered.

Silver has the lowest room temperature resistivity, 1.6 $\mu\Omega\cdot$cm, but on the macroscale is costly compared to copper, which has a resistivity of 1.7 $\mu\Omega\cdot$cm. Note, for small microfabricated devices, the material costs are usually not of great concern because of the small quantities required. With a significantly higher resistivity of 2.65 $\mu\Omega\cdot$cm, aluminum is the dominant metal classically employed for microfabricated integrated circuits, primarily for its chemical stability—silver and copper readily diffuse in silicon-based compounds. Recently, however, copper has become increasingly popular for its lower resistivity and lower susceptibility to electromigration, and suitable barrier metallurgies have been developed to prevent diffusion into silicon. Gold is also often considered, with a resistivity of 2.2 $\mu\Omega\cdot$cm and little susceptibility to oxidation, but again, cost and diffusivity are strong deterrents.

Since Joule heating or other thermal sources may cause temperature increases during operation, the temperature dependence of these materials should also be considered in the machine design. Figure 5.1 shows some resistivity data for common metals extracted from Dyos and Farrell [19]. As can be seen, the resistivities of most of the metals double at a temperature of only $\sim$600 K. This can significantly degrade the power production of a magnetic generator, where for example, low coil resistance is critical for minimizing conductive losses.

5.4.1 Basic Fabrication

In addition to material properties, the microfabrication approach must also be considered when considering a coil integration strategy. Sputtering and electrodeposition are the most common methods for microfabrication of conductive metal coils.

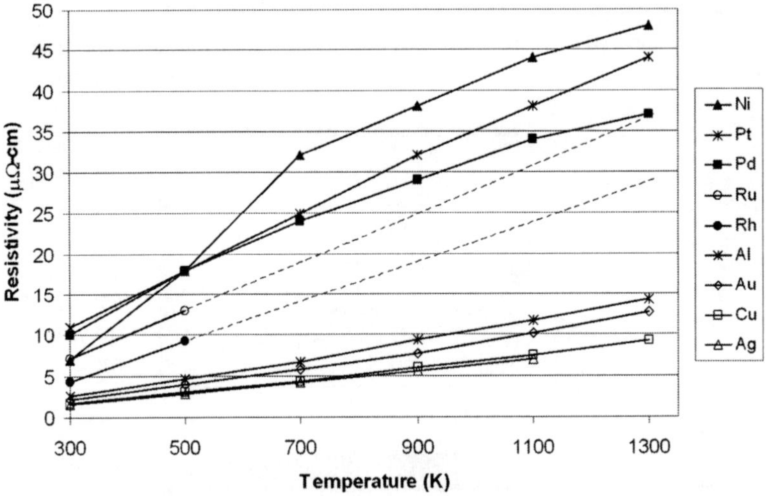

Fig. 5.1 Temperature dependence of resistivity for common metals. Data compiled from Dyos and Farrell [19]

Sputtering allows the deposition of nearly any material (pure elements or complex alloys), but is usually practically limited to films less than ~1 μm in thickness due to relatively slow deposition rates. Because relatively large currents are often desired (see Section 5.2), thicker films may be required, so electrodeposition is often more attractive.

Electrodeposition offers a fast, low-cost method for depositing a variety of materials [42]. It is well suited for depositing pure metallic elements, such Au, Ag, Al, Cu, Cr, etc. Also, the electroplated region is precisely defined by a photolithographically-defined photomask. For these reasons, electrodeposition is very useful for thick coil fabrication. An overview of the approach is shown in Fig. 5.2 and is described as follows.

The process is assumed to begin with a semiconducting Si wafer or other similar substrate. The substrate must be non-conductive or have a dielectric layer to isolate the coils from the substrate. Since the electroplating process requires an electrically conductive surface on which the electrodeposit grows, a thin metallic "seed" layer is required, most often Cu or Au. These conductive seed layer metals exhibit poor adhesion to Si and SiO_2, so an adhesion layer, usually Cr or Ti, is typically employed. A typical process would consist of 10 nm of Cr or Ti for adhesion, followed by a 100 nm Cu seed layer, deposited by sputtering or evaporation. After the metal layers, a mask is defined using photosensitive polymers such as photoresist, polyimide, or SU-8 epoxy. These non-conductive polymers mask off regions of the substrate and serve as a mold for the growing electrodeposit. The electroplated metal fills the polymeric molds, taking on the sidewall profile as it grows. For dimensional control, the electroplating is usually stopped before it reaches the top of the mold. Otherwise, the electrodeposit will "mushroom out" of the mold, growing

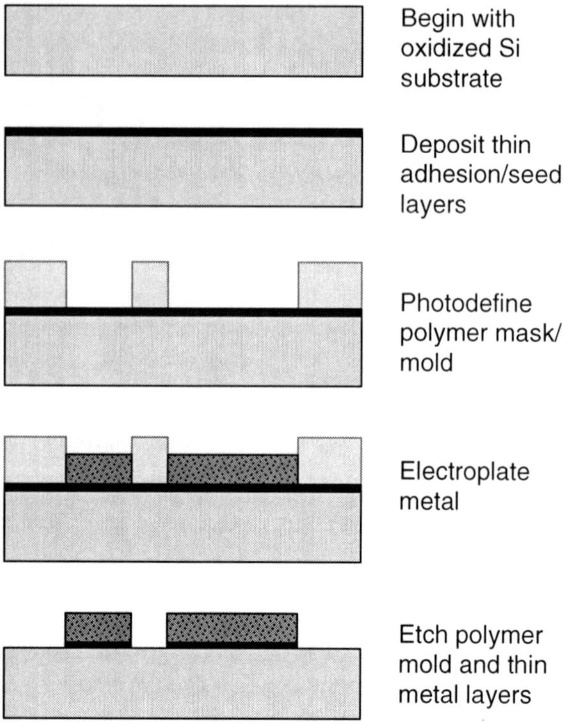

Fig. 5.2 Basic electroplating process to form single layer conductive windings

in all directions. After electroplating, the polymer molds are stripped using a solvent and/or oxygen plasma etch. At this point, the electroplated structures are still all electrically connected to one another via the thin metal seed/adhesion layers. These thin metals in the fields surrounding the plated structures are subsequently wet etched. This wet etch may also etch the electroplated structure, but assuming the electrodeposit is much thicker than the seed layers, the etch removes a negligible amount of coil material. Figure 5.3 shows several examples of electroplated Cu structures.

Fig. 5.3 Example images of ~30 μm thick single-layer electroplated Cu structures

5.4.2 Multilevel Coils

The previous section described a method for photolithographically defining and electroplating metal coil elements on a single layer. Many times, however, multiple metal levels are necessary to enable more complex winding patterns. Consider for example a simple multi-turn spiral coil winding. A "cross-over" or bridge is needed in order to make an electrical connection to the end of the winding in the center of the spiral.

Multilevel coils can be achieved by extending the previously described method using multiple masking and electroplating steps. An inter-layer dielectric material must be selected to electrically insulate between the metal layers to avoid unexpected short circuits. Considering that the top surface of the first metal layer may be fairly rough (typical of many electroplated metals), the dielectric must be able to conformally and completely coat the top and side surfaces of the lower metal layer. Even the smallest pin-hole defect in the inter-layer dielectric can cause the winding to fail. Commonly used inter-layer dielectric materials include PECVD or sputtered oxide/nitride, spin-on glass, or a chemically stable spin-coated polymer such as polyimide or SU-8. The intended temperature of operation must also be considered, since polymer dielectrics may not suitable above a few hundred degrees Celsius.

An example process for achieving a two-layer coil is shown in Fig. 5.4. The process begins by depositing the first metal layer as described previously. Then the inter-layer dielectric is deposited and waysare opened to provide interconnects between the two metal layers. For many films, this step requires the patterning of a photoresist mask and a subsequent etch. In contrast, the use of a photosensitive polymer dielectric offers the opportunity for a simple, one-step photodefinition.

After the inter-layer dielectric is formed, a new adhesion/seed layer is required, since a conductive electrical surface is needed for electroplating. In order to provide sufficient step-coverage over the topography created by the first metal layer, sputtering is the preferred method. It should be noted that adhesion of the metal to a polymer inter-layer dielectric can be quite low. If necessary, adhesion promoters or an oxygen plasma can be used to improve adhesion of the metal [23, 32].

After the seed layer deposition, a polymer mold is photodefined to create the pattern for the second metal layer, and the metal is subsequently electrodeposited to the desired thickness. Afterwards, the polymer mold and thin metal layers are etched away. If necessary a passivation layer can be deposited on top of the second metal layer for protection from the environment. Also with the right selection of materials, the interlayer dielectric can be selectively etched away to create "free-standing" coil segments that are isolated by air.

5.5 Soft Magnets

Magnets are the second key component to a magnetic machine. While the subjects of magnetism and magnetic materials are well beyond the scope of this document, a brief summary of the fundamentals is presented here to provide a basis for

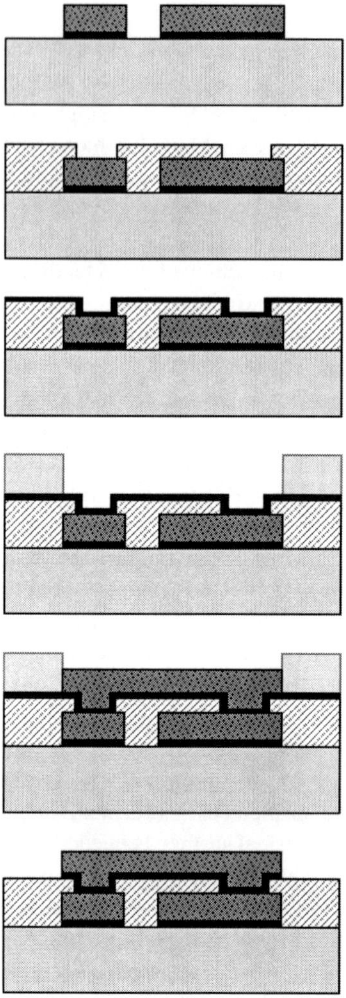

Fig. 5.4 Electroplating process to form multi-layer conductive windings

discussion for machine design. The reader is directed to O'Handley [33] or Cullity and Graham [15] for more thorough treatments.

Under the application of a magnetic field, all materials exhibit a magnetic response. The macroscopic magnetic behavior is characterized by a measurement of the magnetic flux density versus the magnetic field, the B–H curve (or similarly by the magnetization versus applied field, the M–H curve). This B–H relationship quantifies the constitutive material relationship, $B = \mu_0(H + M)$. Some materials exhibit a large magnetic response to an applied field, and it is these materials, known as ferromagnets, which are of most interest for engineering applications. In addition to their large response, ferromagnets exhibit a hysteretic behavior in their B–H loop.

Figure 5.5a shows the typical B–H relationship for a ferromagnetic material along with two-key parameters. The remanence, B_r, is the magnetic flux density produced by the magnet when the magnetic field is zero; the larger the remanence, the "stronger" the magnet. The coercivity, H_c, is the magnetic field required in opposition of the magnetization to reduce the magnetic flux to zero. The larger the coercivity, the harder it is to overcome the material magnetization. In practice, the coercivity sets a field limit for stray magnetic fields that may spuriously demagnetize or remagnetize a magnet.

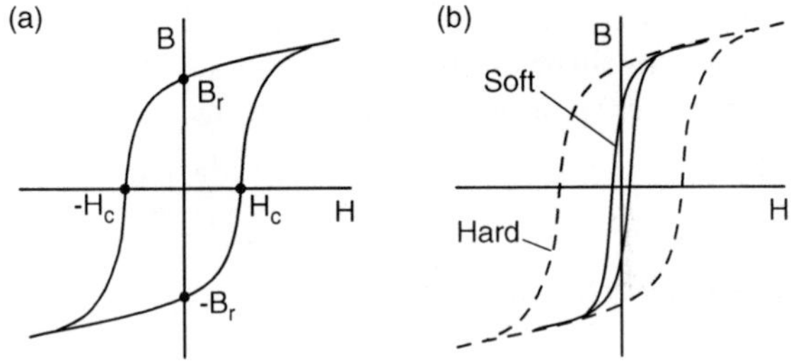

Fig. 5.5 (**a**) Typical B–H hysteresis loop for a ferromagnet, indicating coercivity, H_c, and remanence, B_r; (**b**) Comparison of soft and hard ferromagnets

Ferromagnets are further subdivided as either soft or hard depending on their coercivity. Generally, soft magnets have coercivities <1 kA/m, and hard magnets >10 kA/m. In effect, soft magnets are easily magnetized, whereas hard magnets are difficult to magnetize, but retain their magnetization. Figure 5.5b shows an example comparison between a hard magnet and a soft magnet. A soft magnet has a narrow, tall B–H curve, while a hard magnet has a broader, fuller B–H curve.

The topic of this section is soft magnets; hard magnets are covered in Section 5.6. Soft magnets are used to direct and guide magnetic fields in a magnetic circuit. These structures, collectively known as the "core" of a magnetic motor/generator, are especially important for maximizing the fields from hard ferromagnets by "closing" the magnetic loop. The magnetic flux density of a hard magnet can be increased above its "open-circuit" value by creating a low-permeability path for the magnetic flux and concentrating the field in a well-defined air gap.

Since the usual function of a soft magnetic core is to guide a time-varying magnetic flux, two key functional traits are desired. The first is the ability for the core to carry magnetic flux, known as the permeance (or more commonly characterized by the inverse known as reluctance), which depends on geometry and relative permeability. The second is the resulting power losses involved in the time-varying magnetic fields, necessary for device operation. This power loss depends on geometry and a host of other material properties, along with the amplitude and frequency of the magnetic fields, which are not usually uniformly distributed within the core. In most devices, this quickly requires complex analysis.

The key performance metrics for a soft magnetic material can be encompassed in its saturation flux density B_s, relative permeability μ_r, coercivity H_c, electrical resistivity ρ, magnetostriction λ, and corrosion resistance. High-saturation flux density is desired to avoid magnetic saturation with minimal core cross-sectional area (film thicknesses). Reasonably high permeability is desired to reduce the magnetic reluctance of the core, but too high of permeability decreases the magnetic skin depth leading to larger eddy current losses. Low coercivity and high resistivity are desired to minimize hysteresis and eddy current losses, respectively. Low magnetostriction is desired to avoid variations with stress. Good corrosion resistance is desired for long-term stability, especially if exposed to air.

In general, different material classes offer strengths and weaknesses in one or more of these areas. For example, soft ferrites offer high resistivity but low saturation and metal alloys vice versa. Thus, ferrites are often preferred at high frequency where power losses can dominate the design, but this often necessitates large cores. For low frequencies, metal alloys are preferred (or usually silicon steels for large scale power application) to minimize the total core volume.

5.5.1 Fabrication

While bulk fabrication of soft magnetic alloys relies heavily on casting and conventional metallurgical techniques, microfabrication is usually achieved via physical vapor deposition (e.g., evaporation, sputtering) or electrodeposition. Vapor deposition is suitable for thin films, but electroplating of alloys of Ni, Fe, and Co has been proven to be more effective for achieving the relatively voluminous cores desired for magnetic devices. In contrast to vapor deposition methods, electrodeposition also allows precise, selective deposition via photolithographic patterning. Most often, a polymer is used as a plating mask (molds), including PMMA, polyimide, SU-8 epoxy, and photoresist. Moreover, the development of high-performance soft magnetic materials has been crucial for the advancement of read/write heads for the magnetic recording industry [13], so there has been strong industrial and commercial backing for material investigations. This combination of simplicity, cost-effectiveness, performance, and abundance of research literature makes electrochemical deposition nearly ubiquitous for the fabrication of soft magnetic structures in magnetic MEMS.

The most widespread soft magnetic bath is the $Ni_{80}Fe_{20}$ permalloy alloy, which offers good performance with reliable fabrication from a simple bath. As summarized in Table 5.1, an array of more complex NiFe, CoNi, and CoFe alloys offer increasing saturation flux density, usually at the expense of coercivity, magnetostriction, stress, and/or corrosion resistance. These baths are usually electroplated from a sulfate or chloride electrolyte baths, with various additives and complexing agents to fine-tune the properties. The films are also often deposited in a static magnetic field in order to induce a preferred magnetic direction (easy axis), known as magnetic anisotropy. Some of the more complex compositions also require high-temperature post-deposition annealing to induce a specific microstructure or phase transition.

Exhaustive details of fabrication processes and their application can be found in various archival sources. A good starting point is the contributed proceedings from the International Symposium on Magnetic Materials, Processes, and Devices

Table 5.1 Summary of typical properties for soft magnetic electroplated alloys (summarized from Cooper et al. [13])

Material	Saturation flux density B_s (T)	Easy-axis coercivity H_c (Oe)	Resistivity ρ ($\mu\Omega\cdot$cm)	Magnetostriction λ (ppm)	Film stress (MPa)
$Ni_{80}Fe_{20}$	1.0	0.2	20	< −3	100
$Ni_{45}Fe_{55}$	1.7	0.5	40	+20	160
CoFeCu	1.8–2.0	< 1		± 3	
CoNiFe	2.0–2.2	< 2	30	+3.5	115
$Ni_{20}Fe_{80}$	2.2	2.5	35	+25	240
CoFe	2.4–2.5	5–10		+45	845

(Index Available: http://research.ihost.com/symposiummmpd/index.html). It should be noted that, in general, the magnetic properties of thin- and thick-films can differ significantly from bulk samples of similar composition. The details of this are beyond the scope of this chapter, but the most striking impact is that these films tend to have significantly lower relative permeabilities (usually in the 100 s) as compared to bulk alloys (usually in the 1000 s).

5.5.2 *Laminations*

One key design consideration for devices that utilize soft magnetic cores is the minimization of core losses. Core losses result from hysteresis effects and eddy current effects [29]. Hysteresis losses can be minimized by careful control of the ferromagnetic alloy composition and are typically considered minor losses. Eddy current losses are minimized by (1) using high-resistivity magnetic materials (e.g., ferrites) to increase the skin depth or (2) by laminating the magnetic material into dielectrically isolated thin sheets in order to prevent large scale eddy currents from flowing. Laminates of conductive metallic alloys such as NiFe are often preferred over high-resistivity ferrites because they can support larger saturation flux densities and thus higher power densities. In macroscale magnetic devices, low-loss laminated cores are typically achieved by mechanically laminating alternating layers of core material and insulating material to form a multi-layer stack. This approach is not feasible for microscale laminations and thus presents a manufacturing challenge.

The lamination thickness should be reduced to the order of the magnetic skin depth (typically 1–100 μm range) while maintaining large total core thicknesses (10 μm–1 mm) to prevent saturation. These requirements dictate large numbers of thin, high-aspect-ratio laminations, which are difficult to achieve using conventional microfabrication techniques.

Several approaches for microfabricating laminations have been investigated: alternating sputtering of thin-film magnetic and dielectric layers [44]; one-step electroplating of lamination structures in high-aspect-ratio micromolds [52]; repetitive deposition of polymer insulator, seed layer, and electroplated magnetic layers [34]; mechanical lamination of polymer-coated magnetic foils [35]; alternating electrodeposition of magnetic and sacrificial metal layers [36]; and even direct electroplating on to the sidewalls of etched silicon trenches [2]. Although these approaches have all demonstrated reduction of eddy currents, processability and scaling remain unaddressed issues. Moreover, the suitability of any of these techniques depends on the required core geometry and lamination thickness.

5.6 Hard Magnets

Hard magnets—ferromagnets with coercivities >10 kA/m—are particularly important for microscale magnetic machines. Since fields produced from an electromagnet scale unfavorably with diminishing size while fields from a hard magnet are scale-invariant [14], PMs are often the only viable way to achieve useful magnetic fields in a microscale device. Hard magnets provide a "permanent" and "free" source of magnetic flux. They are often coupled with soft magnetic structures to form magnetic "circuits" that guide and concentrate the magnetic fields in specific regions of the device. While hard magnets have been known for thousands of years, it was only in the late 20th century that fundamental understanding of magnetism and materials led to advancements in their magnetic properties. Today, high-performance magnets are nearly ubiquitous in everyday life.

Hard magnets are characterized primarily by the portion of the B–H curve in the second-quadrant, known as the demagnetization curve. During use, the magnet operates somewhere on this curve—the magnetic flux is in a direction opposite to the magnetic field. As such, the most relevant properties for a hard magnet are the coercivity, H_c, the remanence, B_r, and maximum energy product, BH_{max}. The maximum energy product is the value where the product $|B \cdot H|$ is maximized. It is the point on the demagnetization curve where the magnet supplies the most magnetic energy. For maximum performance, the size and shape of the magnetic circuit should be designed such that the magnet is operating in this "sweet spot" [11].

As described in Section 5.5, methods for integration of magnetically soft ferromagnetic materials, such as electroplated NiFe or CoFe alloys, are fairly common. In contrast, methods for integrating hard magnets are much more challenging. First, like soft magnets, there are several different classes of hard magnetic materials, each having strengths and weaknesses, as summarized in Table 5.2. Temperature effects, chemical stability, and available fabrication methods all play a large role in the material selections. Second, the material properties of most hard magnets are highly sensitive to the microstructure, not just chemical composition. This requires careful control over the processing conditions to achieve optimal properties. Lastly, most hard magnetic materials require a magnetization step to achieve the desired properties and magnetic orientation. This complicates the integration process, particularly for wafer-scale batch-fabrication.

Table 5.2 Summary of typical bulk properties of hard magnetic materials (summarized from MMPA Standard 0100-00 [31], O'Handley [33], and Furlani [22])

Material	Energy product (kJ/m^3)	Curie temp. (°C)	Max. operating temp. (°C)	Corrosion resistance	Bulk manufacturing methods
Ferrite	6–32	450–470	400	Excellent	Powders
Alnico	11–75	720–900	450–550	Very good	Cast, powders
CoPt/FePt	70–95	475–825	–	Very good	Cast
$SmCo_5$	130–180	685–750	250–300	Good	Powders
Sm_2Co_{17}	190–240	750–970	350	Good	Powders
$Nd_2Fe_{14}B$	190–400	312	125–150	Poor	Powders

5.6.1 Fabrication

Macroscale magnets are most commonly created by powder metallurgical processes, and to a lesser degree, casting [22]. Typically, powders of the raw magnetic materials are pressed and sintered into molds to create a desired shape. Subsequent heat treatments are often required to induce precipitation, diffusion, or microstructural changes in order to achieve optimum magnetic properties. Magnetic fields can also optionally be applied during the compression step to achieve preferred magnetic orientations (anisotropy). Because most of the magnetic materials are brittle, the magnetic powders can be augmented with organic binders to create "bonded" magnets. These bonded magnets are more mechanically robust and can be formed into more complex shapes using extrusion or injection molding processes. However, these bonded magnets typically exhibit lower magnetic properties.

The methods for macroscale hard magnet fabrication—casting and powder processing—are not very common in microfabrication. For MEMS and other applications such as magnetic recording media, magnetically hard thin-films are usually achieved via physical vapor deposition, chemical vapor deposition, or electrochemical deposition. Additionally, non-traditional fabrication strategies incorporating magnetic powders have been investigated for microfabricating thicker ($>10\,\mu m$) hard magnets for microsystem applications. The materials of interest can be broadly categorized as low-energy density ferrites (e.g., $BaFe_{12}O_{19}$, $SrFe_{12}O_{19}$), medium-energy density metal alloys (e.g., CoNi, FePt, CoPt alloys), and high-energy density rare-earth alloys (e.g., $SmCo_5$, Sm_2Co_{17}, $Nd_2Fe_{14}B$). The metal alloys and rare-earth alloys offer the potential for the highest energy densities, whereas the ferrites offer much better chemical stability.

Table 5.3 summarizes some of the materials and microfabrication approaches investigated. The design of a particular magnetic machine/actuator dictates certain material properties, physical dimensions, chemical stability, temperature of operation, etc. These performance requirements direct the selection of both the material and fabrication method, and there are inherent dependencies between material and

Table 5.3 Summary of commonly employed microfabrication methods for hard magnetic materials

H	BH_{max} theoretical (kJ/m^3)	Electro-deposition	Sputtering	Pulsed laser deposition	Powder processing
$BaFe_{12}O_{19}$	50				X
Co_3Pt	378	X	X		
$CoPt - L1_0$	200	X	X		
$FePt - L1_0$	407	X	X		
$SmCo_5$	220		X	X	X
Sm_2Co_{17}	333		X	X	X
$Nd_2Fe_{14}B$	514		X	X	X

method. For example, the metal alloys can most easily be deposited by either electrodeposition or sputtering. Electroplating offers a relatively low-cost, simple, and high-deposition rate approach for achieving multi-micron films, as well as the ability to use photoresist masks to pre-define the deposition pattern. In contrast, sputtering is usually slower, more costly, and requires post-deposition etching or polishing steps to define the pattern. As another example, ferrites and the rare-earth magnets cannot be electrodeposited from aqueous solutions, so sputtering, pulse laser deposition, or powder processing methods are the only option. Compared to sputtering, pulsed laser deposition generally offers higher deposition rates, but usually is restricted to deposition areas of only a few square centimeters. There are many tradeoffs to consider in the selection of a material and fabrication approach.

In general, deposited films of FePt alloys and rare-earth alloys require in situ or post-deposition annealing at high temperatures (>400°C) to achieve their high magnetic properties. Thus, the temperature compatibility of other materials in a device must be considered. In contrast, the electroplated Co-rich CoPt alloys (approximately 80:20 Co:Pt atomic ratio) show high coercivities and high anisotropy in the as-deposited state. Also, with the exception of the CoNi alloys, which have relatively low magnetic performance, most of the sputtered or electroplated films are limited to a few microns in thickness. This is due to intrinsic stresses that develop during deposition and/or thermal mismatch stresses that arise during annealing steps. Also, the magnetic performance of a film tends to degrade with thickness due to variations in the microstructure, e.g., grain size and shape.

As a result of the film-thickness limitations of conventional deposition methods, high-performance bulk magnetic powders are an attractive alternative for achieving micromagnets with dimensions in the 10 μm–1 mm range. Naturally, the size of the magnetic particle places constraints on the minimum magnet size; the particles must be smaller than the desired feature size to ensure a quasi-homogenous distribution of particles. Commercially available ferrite powders are available below 1 μm, SmCo down to 5–10 μm, and NdFeB down to ~50 μm. The minimum particle size is dependent on a combination of the microstructural grain size, the magnetic domain size, and thermal stability [11].

Different strategies have been investigated for incorporating magnetic powders into microfabricated devices. In one approach, coined "magnetic composite electroplating" (MCE), non-soluble magnetic particles are incorporated into conventional magnetic alloy electroplating baths to create metal-particle composites [26, 27]. The magnetic particles are mechanically and electrochemically captured into the growing electrodeposit, with the goal of minimizing stresses in thick electroplated films. Other approaches emulate macroscale bonded magnet powder methods using polymers/resins/inks loaded with magnetic particles. These composites can be spin-coated, screen-printed, or "squeegeed" into cavities to form magnets with micron to millimeter scale features. Bonded micromagnets have been achieved using ferrite, SmCo, and NdFeB powders in epoxy, SU-8, PDMS, PMMA, and resin binders. As in bulk-manufactured bonded magnets, the final magnetic properties are usually lower than those of the original magnetic powder. The coercivity generally remains constant, but the remanence is proportional to the particle fill factor, typically only 60–80%, thus reducing the net energy density.

Tables 5.4 and 5.5 summarize some of the previous results. As in bulk materials, the best magnetic performance can be achieved using rare-earth alloys. Dempsey et al. [17] recently reported 5 μm-thick sputtered NdFeB magnets that approach the theoretical performance limit for that material. However, the annealing temperature requires for these films may preclude their widespread use. Where high temperatures are not permissible, CoPt alloys offer good magnetic performance and are highly integrable via electrodeposition methods. Where suitable, micropowder methods are also emerging as viable alternatives.

Many of the aforementioned magnetic films were obtained under ideal conditions and without regard for compatibility or integration into existing MEMS processing. Examples of incompatibilities include using high-pH electroplating solutions at elevated temperatures, which will dissolve standard photoresist masks; requiring non-standard substrate materials or specific crystalline orientations, which limits versatility; requiring in situ or post-deposition heat treatments, which exceed temperature limits of other common MEMS materials; and lacking means for photolithographic patterning. Clearly, if these magnetic films are to be integrated into more complex micromachined MEMS transducers, these integration issues must be resolved. Moreover, the process compatibility and long-term stability of these films with regard to pre- and post-deposition thermal cycling, chemical exposure, mechanical stress, etc. are largely unknown. Thus, while strides have been made over the last decade to develop suitable materials, there are still many opportunities for continued investigation.

5.6.2 Magnetization

In addition to the deposition challenges, hard magnetic films must be magnetized after deposition in order to provide a permanent magnetization. The issue of magnetization is particularly challenging for batch-fabricated microscale devices. In macroscale devices, discrete magnets can be first magnetized and then assembled

Table 5.4 Examples of conventionally deposited hard magnetic films

References	Material	Fabrication method	Integration notes	Thickness (μm)	H_c (kA/m)	B_r (T)	BH_{max} (kJ/m^3)
Guan and Nelson [25]	CoNiP	Electro-deposited	None	1–52	55–105	0.06–0.1	1.3–1.8
Yufeng et al. [53]	CoNiMnP	Electro-deposited	0.2 T magnetic field	25	40–210	0.06–0.2	0.6–10
Thongmee et al. [45]	FePt - $L1_0$	Electro-deposited	400–800°C anneal	0.1–0.8	320–1440	–	–
Wang and Arnold [50]	CoPt	Electro-deposited	(110) Si substrate	10	330	1.0	69
Budde and Gatzen [9]	SmCo	Sputtered	560° anneal, Glass/alumina substrate	3–50	1200	0.7–0.75	75–90
Dempsey et al. [17]	NdFeB	Sputtered	500°C deposition, 750°C anneal	5	1280	1.4	400

Table 5.5 Examples of microfabricated hard magnets using magnetic powders

Ref	Magnetic powder	"Carrier" material	Fabrication method	Thickness (μm)	H_c(kA/m)	B_r(T)	BH_{max}(kJ/m^3)
Guan and Nelson [27]	BaFeO	CoNiP	MCE	15–50	120–175	0.2–0.25	6–9
Lagorce and Allen [28]	SrFeO	Polyimide	Spin-casting	10–20	320	0.16–0.28	5–12
Cho and Ahn [12]	SrFeO	Epoxy resin	Squeegee	60–70	350–360	0.03	2–3
Dutoit et al. [18]	SmCo	SU-8	Spin-casting	15	140–550	0.25–0.34	22
Bowers et al. [8]	SmCo	none	Dry-packing	15–500	130–140	0.3–0.5	18–23
Pawlowski and Töpfer [37]	NdFeB	Resin	Screen-printing	10–50	300–900	0.2–0.4	–
Guan and Nelson [26]	NdFeB	NiP	MCE	5–20	50–190	0.24–0.35	2–3
Romero et al. [41]	NdFeB	PMMA	"Rain-dropping"	500	1090	0.25–0.7	–
Wang et al. [51]	NdFeB	Wax particles	Dry-packing	320	720–740	0.22–0.34	8–17

to form a magnetic structure. However, the order must be reversed for microscale implementation. The question is how to magnetize the device, ideally with spatially varying magnetization patterns in a batch-fabrication manner?

Magnetization of a hard magnetic material is achieved by applying an intense magnetic field (typically 3–5 times the value of the coercivity) in order to fully saturate the magnet. To achieve such high fields, particularly for modern high-performance magnets, a pulse-discharge magnetizer is most commonly used. This device uses a capacitor bank to store a large electrical charge that is subsequently discharged onto a conductive winding. Upon discharge, the high-current density current pulse creates a brief (< 1 ms), but strong magnetic field which can be used to magnetize a sample.

For example, a solenoidal coil can be used to create a unidirectional magnetic field. A wafer of hundreds of MEMS devices can be placed inside the coil and magnetized in one direction. Of course, other structures or integrated-circuits that may exist alongside the magnetic components must be capable of withstanding the high fields applied during the magnetization process. This approach is fairly straightforward to implement, but the restriction of uniform magnetization in a single direction limits the versatility of PMs within the magnetic machine design.

It would be very beneficial if spatially varying, three-dimensional field patterns could be achieved. A simple and useful starting point is the development of a series of alternating north and south poles, oriented out-of-plane. This pattern is useful for creating time-varying magnet fluxes for magnetic machines. Uniformly magnetized patterned magnets will create stray fields that exhibit this spatial variation (e.g., stripes of equally separated magnets magnetized out of plane), but the field strengths are limited, and the positive and negative fields are not equal and opposite. Higher magnetic fields are desired by patterning a magnetic layer with distinct magnetization patterns to achieve the desired spatially varying magnetic fields, keeping in mind that batch-fabrication should be maintained and fine-pitch magnetization patterns may be required.

Rhen et al. [38] demonstrated an approach to achieve mm-scale pole pitch patterns by patterning two different magnetic materials, one with a high coercivity and one with a lower coercivity. A large magnetic field was used to saturate both materials in one direction. Then, a weaker field was applied in the opposite direction to reverse the magnetic material with the lower coercivity. Ideally, this did not affect the magnetization of the higher coercivity material. While interesting and most likely scalable to the microscale, this approach inherently requires a material with inferior magnetic properties (lower coercivity) and also doubles the fabrication complexity by requiring the deposition of two different materials.

An alternative approach to achieving spatially varying magnetization requires shaping of the magnetizing fields to "direct" the fields to specific locations of the magnetic film. Shaping of the magnetic fields can be achieved by appropriately shaped coil windings [47, 39], but this approach does not scale well to the microscale (less than $\sim$1 mm pole pitch) because of intolerably high-current densities. The field produced by a current density-limited conductor scales proportional to the linear size [14].

An alternative method to concentrating the magnetic fields in specific regions is using high-permeability, ferromagnetic magnetizing heads of specific geometries to guide the magnetic fields. This approach has so far been scaled down to mm-scale pole patterns for magnetic rotors for micromachines. Gilles et al. [24] demonstrated a closed-field magnetizing head with an integrated coil winding, as shown in Fig. 5.6. In contrast, Zana et al. [54] used an open-field magnetizing head placed inside a pulsed field, as shown in Fig. 5.7. This latter method shapes and concentrates the fields across specific regions of magnetic materials and offers the possibility for wafer-level batch-microfabrication using micromachined magnetizing heads.

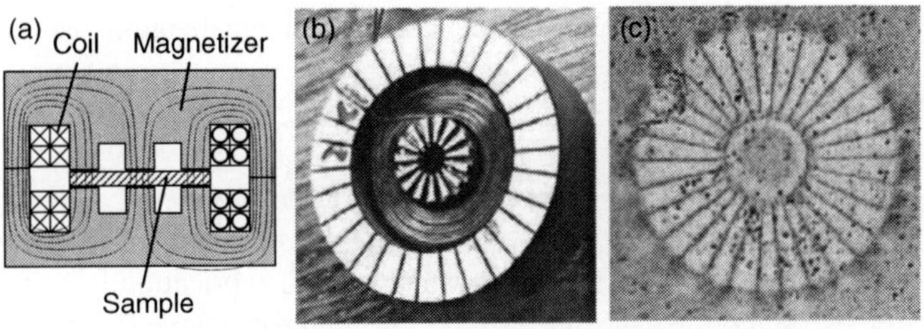

Fig. 5.6 (**a**) Schematic of closed-field magnetization process; (**b**) photograph of half of closed-field magnetizer (adapted from Gilles et al. [24] © 2002 Elsevier); (**c**) resulting 30-pole magnetization pattern on magnetic annulus imaged via "magnetic viewing paper" (adapted from Gilles et al. [24] © 2002 Elsevier)

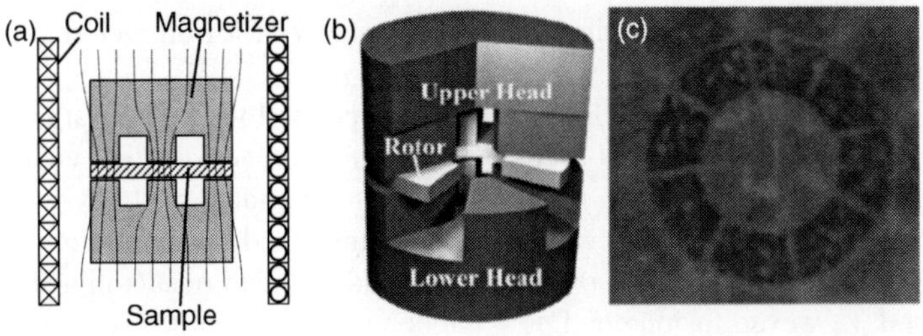

Fig. 5.7 (**a**) Schematic of open-field magnetization process; (**b**) cutaway drawing of eight-pole open-field magnetizer with magnetic rotor (reprinted from Arnold et al. [6] © 2006 IEEE); (**c**) resulting pole pattern magnetization pattern on magnetic annulus imaged via "magnetic viewing paper"

For either of these processes, the magnetizing head is designed in a shape similar to the desired magnetization pattern, and usually capitalizes on some symmetry. The heads are made of a high-permeability, high-saturation material such as $Fe_{49}Co_{49}V_2$ (Hiperco 50) to enable flux crowding while avoiding magnetic saturation.

The magnetization process requires two steps. First, the magnetic sample is magnetized in one direction to create, say, north poles. Then, the sample is shifted or rotated to another position, where it is magnetized in the opposite direction to create south poles. During each of these magnetization steps, only the areas in between the magnetizer "teeth" are magnetized. The areas in the gaps between the teeth experience some magnetic fields due to leakage and fringing, but these fields are insufficient for magnetization.

This method requires careful design of the magnetizing head and control of the magnetization fields. Too small of fields will not be insufficient to achieve full magnetic saturation. Conversely, too large of fields will cause excess flux leakage and fringing, resulting in regions that are unintentionally magnetized.

The procedures described here were used to magnetize bulk magnets with millimeter scale pole patterns. The magnetizing heads were conventionally machined. However, one could imagine scaling this procedure, particularly the open-field method, to batch fabricate samples with smaller dimensions by microfabricating the magnetizing heads. Achieving more complex magnetization patterns at the microscale is an open topic for ongoing investigation.

5.7 Integration in Multi-Wafer Silicon Process

Bulk-etched silicon MEMS differ significantly in their fabrication from micromagnetic machines. Silicon MEMS are usually bulk-micromachined using subtractive (etching) processes to "cut" silicon into the desired shapes. In contrast, magnetic machines are usually surface-micromachined using additive processes to build up layers of coils and magnetic structures. To achieve monolithic integration, these two technologies must be merged to enable magnetic machines embedded within bulk-micromachined silicon devices.

Bulk micromachining of silicon MEMS typically relies on wet etching, using KOH or other aqueous etchants, or dry etching, such as DRIE. In conjunction with deep silicon etching, wafer bonding is commonly used to achieve large mechanical structures, fluidic channels, sealed cavities, membranes, floating elements, etc. [43]. Fusion, or direct, bonding is preferred to other bonding methods (e.g., anodic, eutectic, polymer, glass frit, etc.) because it is simple and low cost, eliminates thermal mismatch issues, requires no intermediate layers, and results in a strong, uniform bond. However, fusion bonding of silicon requires chemically clean, polished surfaces and a post-bond anneal to strengthen the bond. Typically, an RCA clean is performed before bonding, and post-bond annealing is performed at temperatures from 300 to 1100°C [46].

A bond strength approaching that of the silicon itself can be achieved using a bonding anneal of 1100°C for 1 h under nitrogen. Obviously, however, many metals required for magnetic machines cannot withstand such extreme temperatures (e.g., the melting point of Cu is 1085°C). Thus, alternative bonding processes are needed.

For applications where extreme bond strength is not needed, lower-temperature bonding processes may be plausible. The bond strength directly correlates to the bonding time and temperature [46]. However, even if successful wafer bonding can be achieved at lower temperatures, there are numerous integration issues that must be addressed for wafer bonding with metallic structures, including oxidation, diffusion, thermal expansion coefficient mismatch, chemical stability, etc.

Process steps during the bonding process can alter the material properties, or even completely destroy the material functionality. For example, during high-temperature annealing, a metal like Cu may diffuse through a dielectric silicon dioxide layer creating an electrical short circuit to the silicon substrate. Similarly, high temperatures can cause permanent microstructural changes in magnetic materials that significantly affect their magnetic performance.

In addition to these thermal effects, fusion-bonded wafers typically undergo rigorous chemical cleaning steps, which are very harsh to most metal (and organic) structures. The hydrochloric acid/hydrogen peroxide-based bath (SC2) in the commonly used RCA pre-clean is not compatible with most metals; in fact, it will react violently when large quantities of metal are present! Thus, alternative chemical cleaning steps (wet chemistries and/or plasma etches) and bonding procedures (lower temperatures) are required.

Arnold et al. [3] explored the low-temperature fusion bonding of silicon wafers with embedded electroplated metals. They showed that strong bonds could be achieved using a wafer bonding machine at 500°C for 4 h with physical bonding pressure of 200 kPa in an evacuated ambient environment. For the pre-clean step, the SC2 (HCl-based) step in the normal RCA process was replaced by a reduced-concentration SC1 (NH_4OH-based) step to "activate" the surface silicon for bonding. The approach was validated via mechanical, electrical, and magnetic measurements. Figure 5.8 shows an example of Cu coils embedded in fusion-bonded silicon wafers. While this work indicated the suitability of electroplated metal structures within a complex etched silicon structure, further studies may be necessary to validate other embedded metals and/or improve the bond strength.

In summary, it would be ideal to monolithically integrate magnetic machines within multi-wafer fusion-bonded silicon structures. However, there are many significant fundamental and technological issues that must be resolved to achieve this integration. Thus, depending on the overall system performance goals and constraints, in some cases it may make more sense to fully integrate, while in others it may not. For microturbomachinery, this system partitioning of the mechanical and electrical components must be considered during the early stages of the system design.

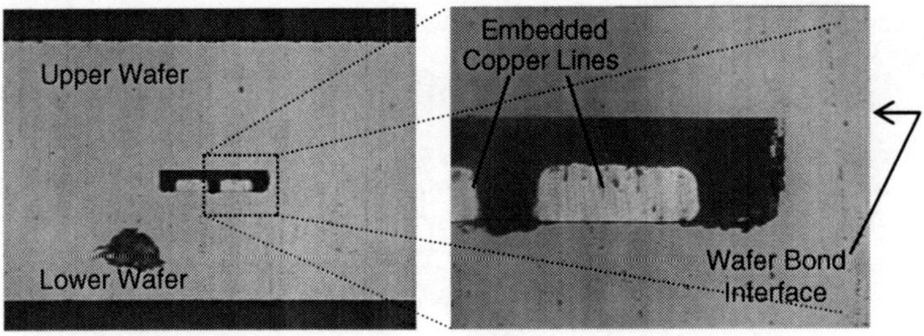

Fig. 5.8 Cross-sectional images of fusion-bonded (500°C, 4 h) silicon wafers with embedded Cu coils showing a seamless bond interface

5.8 Example Devices

Two coin-sized magnetic machines were designed and fabricated as candidate electrical generator technologies for integration with a microturbine engine to form a power generation system. Note, only the fabrication of the machines is included here. The design, modeling, and characterization results are included in Chapter 8. The first device, an induction machine, was monolithically integrated into a multi-wafer silicon process. The second device, a PM machine, was built using non-silicon process technologies.

5.8.1 Induction Machine

The first device was a planar induction machine [5], comprised of two primary components: a two-phase, eight-pole stator, and a composite annular rotor, as shown in Fig. 5.9. Electromechanical energy conversion occurred through the interaction of a traveling magnetic wave in the rotor–stator air gap and the eddy currents induced in the rotor by the traveling wave. In motoring mode, a traveling magnetic wave was established by applying two sinusoidal currents 90 degrees out of phase to the two stator phases. This resulted in a motoring torque as well as a pull-in force acting to close the air gap between the rotor and stator. If, on the other hand, a rotating mechanical power source (e.g., an engine or turbine) forced the rotor to rotate faster

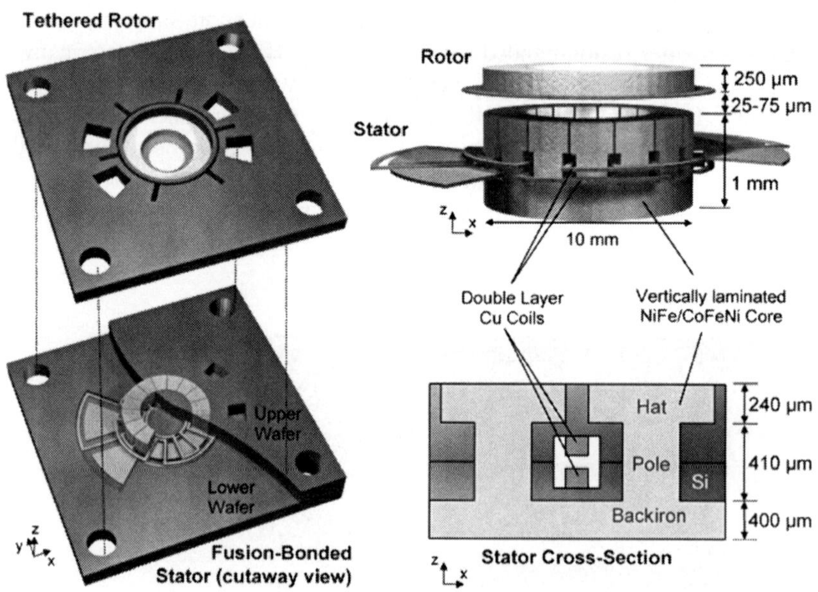

Fig. 5.9 Renderings of the magnetic induction machine (not drawn to scale) (reprinted from Arnold et al. [5] © 2006 IEEE)

than the traveling magnetic wave, net electrical power would be supplied, and the machine would function as a generator.

In the prototype design, the stator contained two planar Cu windings in two separate metal layers forming the electrical phases. These windings were wound in a three-dimensional, vertically laminated soft magnetic core, all supported by a silicon frame. The core laminations formed onion-like concentric rings, where the lamination thickness was designed to be approximately one skin depth ($\delta \approx 30$ μm at 1 Mrpm) to limit eddy current losses in the stator. The rotor was a 250-μm thick ferromagnetic annular ring (10-mm OD, 6-mm ID) with a 20-μm thick surface layer of Cu. The Cu extended beyond the magnetic region (12-mm OD, 4-mm ID) to enhance eddy current generation in the rotor for maximum electromechanical coupling.

Machines were fabricated using either $Ni_{80}Fe_{20}$ or $Co_{65}Fe_{18}Ni_{17}$ as the core material. The CoFeNi alloy had a permeability and resistivity similar to NiFe, but offered higher saturation flux density (higher power density) and higher Curie temperature (higher operating temperature) at the expense of a slight increase in coercivity (larger hysteresis loss) [24].

For testing purposes, the rotor was suspended above the stator using six, 35-μm wide, 475-μm tall, 2.5-mm flexible silicon tethers that permitted angular rotation and a controllable air gap while avoiding the difficulties of supporting a fully spinning rotor. The use of tethered rotors simulated a blocked-rotor test (slip frequency equals the electrical frequency and no mechanical losses), commonly used to test macroscale machines [21]. For a fully rotating machine, these tethers would be eliminated, and a release etch would used to free the rotor.

As depicted by Fig. 5.10, the stator was fabricated by embedding coils inside a pair of fusion-bonded silicon wafers and then building the magnetic core around the coils. The process began with two low-resistivity (~0.001 Ω·cm), n-type, 100-mm diameter, double-side-polished, silicon wafers. Alignment marks were first etched into the backsides of the wafers. Then, on the front side, 70-μm deep cavities were etched using deep reactive ion etching (DRIE). A 200-nm thermal SiO_2 layer was then grown to serve as a dielectric. A 400-nm Ta diffusion barrier (to prevent interdiffusion of the Cu and SiO_2) and 200-nm Cu electroplating seed layer were then DC sputtered covering the wafer surface and down into the cavities.

Next, a thick negative photoresist was patterned in the cavities to define a plating mold for the coils. A 35 μm-thick layer of Cu was then electroplated to form the coils. After plating, the photoresist was stripped, and the thin Ta and Cu metal layers and SiO_2 were sequentially wet-etched. The short etch times required to etch each of these thin films limited any undercut underneath the Cu coils and/or etching of the Cu itself. At this point, the coil was recessed in the cavity, and the top surface was pristine silicon, as shown in Fig. 5.11.

The two wafers were then aligned and fusion bonded, sealing the coils inside. The bonding was performed at 500°C for 4 h under vacuum ($<2\times10^{-6}$ torr) with 200-kPa clamping pressure, resulting in a strong, uniform bond without degrading the embedded copper [3]. From the outside, the bonded silicon pair appeared as fresh wafer, and processing continued to form the stator core. First, a 3-μm PECVD oxide

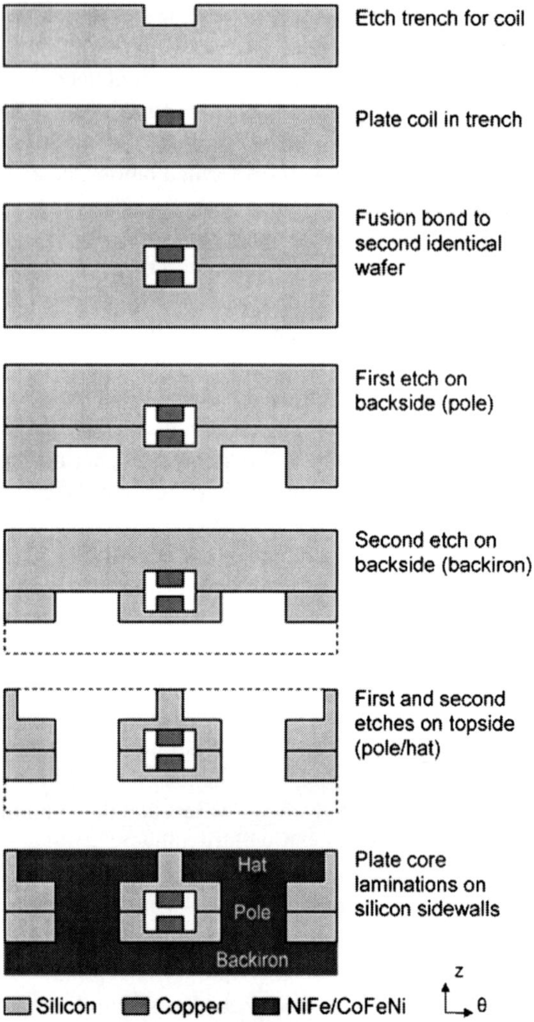

Fig. 5.10 Simplified fabrication process flow for wafer-bonded machine stator, depicting cross sections in θ-z plane: two coil conductors pass radially through the laminated core structure (laminations not shown) (reprinted from Arnold et al. [5] © 2006 IEEE)

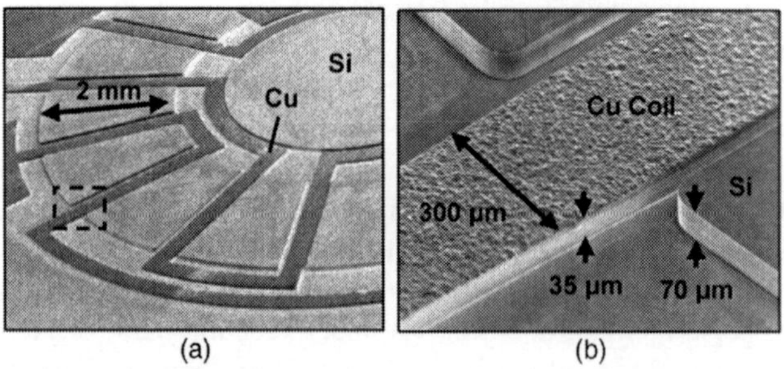

Fig. 5.11 (**a**) SEM of 35 µm thick, 300 µm wide, copper coil recessed in 70 µm deep silicon cavity before wafer bonding to a second identical wafer; (**b**) detail SEM (reprinted from Arnold et al. [5] © 2006 IEEE)

was deposited on the top and bottom surfaces of the bonded silicon stack. Using this oxide layer and photoresist as masks, nested deep silicon etches were performed from the bottom and top sides to form a silicon skeleton for the laminated core. Then ~30 μm of NiFe or CoFeNi was selectively electroplated onto the conductive silicon sidewalls to form the magnetic laminations, using a process described in [2]. The metals plated only on the etched silicon regions, with the SiO_2 layer on the top and bottom surfaces acting as a mask. Finally, contact openings were made from the top side by etching the silicon down to the buried coil contacts. Figure 5.12 shows images of the completed stator, with corresponding cross-sections shown in Fig. 5.13.

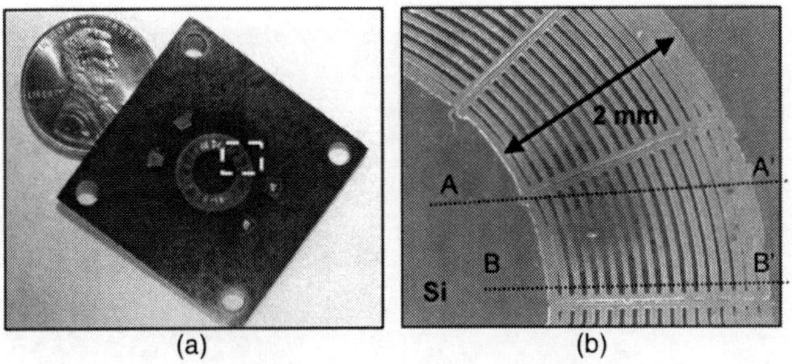

Fig. 5.12 (**a**) Fabricated wafer-bonded machine stator; (**b**) detail SEM showing NiFe vertical magnetic laminations (reprinted from Arnold et al. [5] © 2006 IEEE)

For rapid prototyping to demonstrate machine functionality, the metallic rotors and silicon tether structures were fabricated separately and hand assembled. Rotors were built on a silicon substrate by first electroplating a 250-μm thick NiFe or CoFeNi annular ring on top of a 20-μm thick Cu ring using SU-8 molds, as shown in Fig. 5.14. The SU-8 was then mechanically removed, and the silicon substrate was etched away using KOH. An electroplated rotor, still attached to the silicon substrate, is shown in Fig. 5.15. The silicon tether structures were formed separately by through-etching a wafer using DRIE. The rotors were subsequently glued into the tethered retaining ring. Clearly, for full integration, an alternative fabrication strategy would be required.

5.8.2 Permanent-Magnet Machine

A PM machine was also designed and fabricated [16, 6, 7]. As described previously, the integration of high-performance hard magnets into a micromachined silicon process flow is still a significant challenge. Thus, for initial demonstration of the suitability of the machine, a fabrication strategy was pursued that did not rely on monolithic integration in silicon. However, the basic operation and dimensions explored do not preclude future silicon-based implementation.

The devices were three-phase, eight-pole, axial-flux, synchronous machines, as shown in Fig. 5.16. Each comprised a rotor with an annular PM and soft magnetic back iron, and a stator with multi-turn surface windings on a soft magnetic substrate

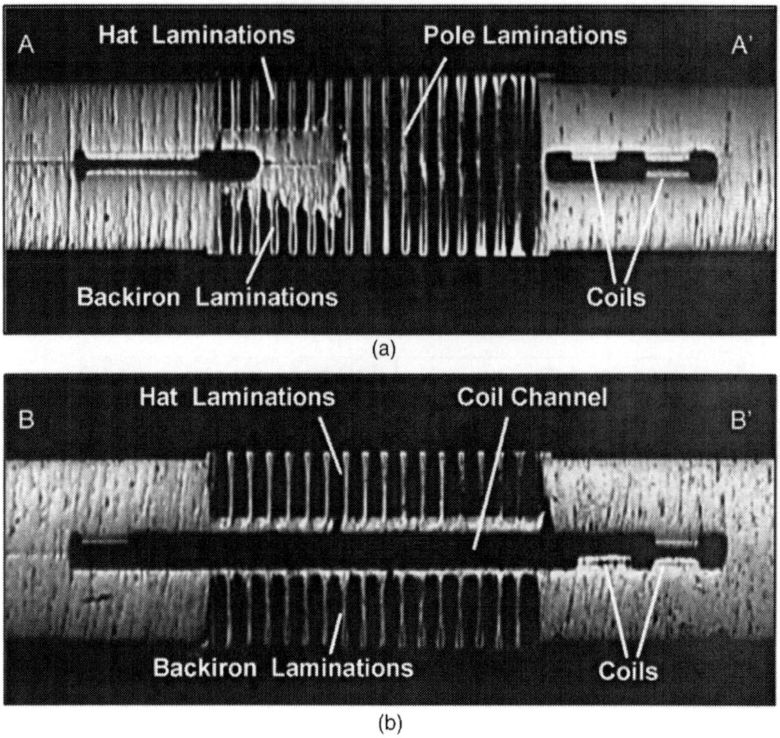

Fig. 5.13 Cross-sections along (**a**) A–A' and (**b**) B–B' for induction machine stators, as shown in Fig. 5.12b (reprinted from Arnold et al. [5] © 2006 IEEE)

(back iron). In operation, rotation of the multi-poled magnetic rotor created time-varying magnetic fields, which induced voltages on the stator windings, as governed by Faraday's Law [7]. When the windings were connected to an electrical load, current flowed and electrical power was supplied from the device. The PM provided a permanent magnetic field source, and the soft magnetic substrate and rotor back iron reduced the magnetic reluctance in order to maximize the magnetic flux for greater power density.

The stator used interleaved, electroplated Cu windings that were dielectrically isolated from a 1-mm thick $Ni_{80}Fe_{15}Mo_5$ (Moly Permalloy) substrate by a 5-μm polyimide layer. The rotor consisted of an eight-pole, 500 μm thick, annular SmCo PM, and a 500 μm thick $Fe_{49}Co_{49}V_2$ (Hiperco 50) back iron. The rotor magnet and back iron had an inner diameter of 5 mm and an outer diameter of 10 mm. The stator and rotor components were fabricated using a combination of microfabrication and precision machining/assembly, and subsequently tested using an off-the-shelf gas-powered turbine to spin the rotor over the stator.

The stators were fabricated using two layer electroplated Cu windings on $Ni_{80}Fe_{15}Mo_5$ (Moly Permalloy) soft magnetic substrates. The complete winding pattern is shown in Fig. 5.17. First, a ~5 μm thick layer of PI-2611 polyimide

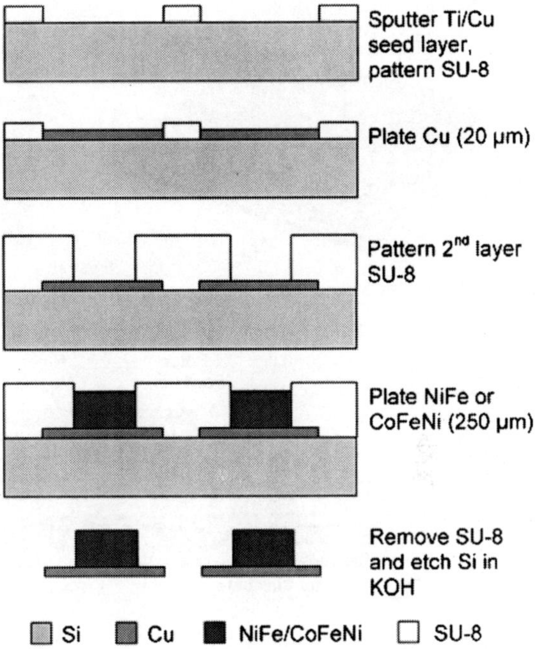

Fig. 5.14 Process flow for induction machine rotors (reprinted from Arnold et al. [5] © 2006 IEEE)

was deposited to isolate the coils from the conductive substrate. The windings were formed using a two-layer electroplating process (see Section 5.4.2) to achieve 60–80 μm-thick Cu layers with 40 μm-thick SU-8 epoxy as the interlayer dielectric. Figure 5.18 shows images of the windings after fabrication.

The rotor consisted of an eight-pole, 500 μm thick, annular SmCo PM, and a 500 μm thick FeCoV (Hiperco 50) back iron, as shown in Fig. 5.19. The rotor magnet and back iron had an inner diameter of 5 mm and an outer diameter of 10 mm. In contrast to the microfabricated coils, the rotor was completely conventionally machined. Commercially available, annular SmCo magnets were purchased in the correct dimensions, and the eight-pole magnetization pattern was achieved using a two-step pulse magnetization process (see Section 5.6.2). The rotor back iron and

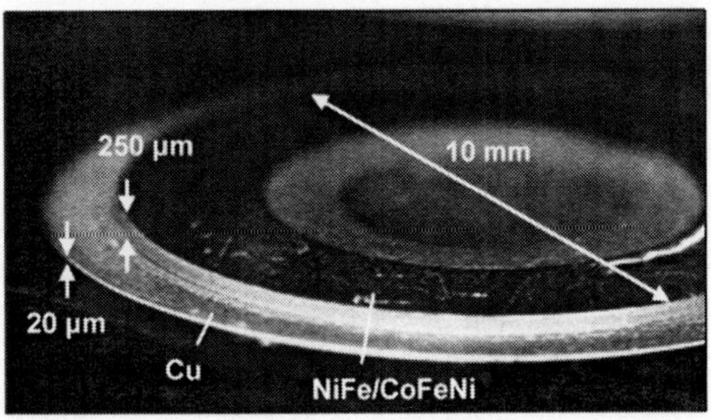

Fig. 5.15 Photograph of electroplated induction machine rotor on Si substrate before release (reprinted from Arnold et al. [5] © 2006 IEEE)

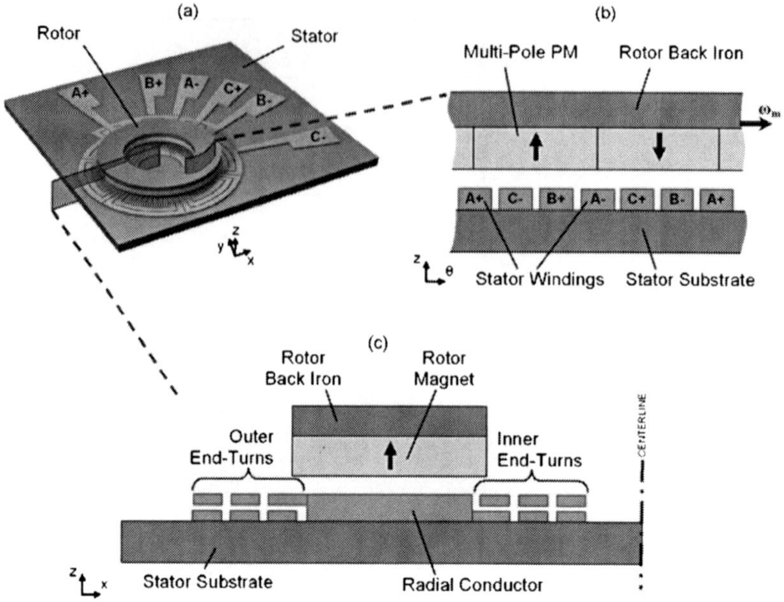

Fig. 5.16 Conceptual drawing of the PM generator: (**a**) perspective view, (**b**) angular cross section, and (**c**) radial cross section (adapted from Arnold et al. [6] © 2006 IEEE and Arnold et al. [7] © 2006 IOP)

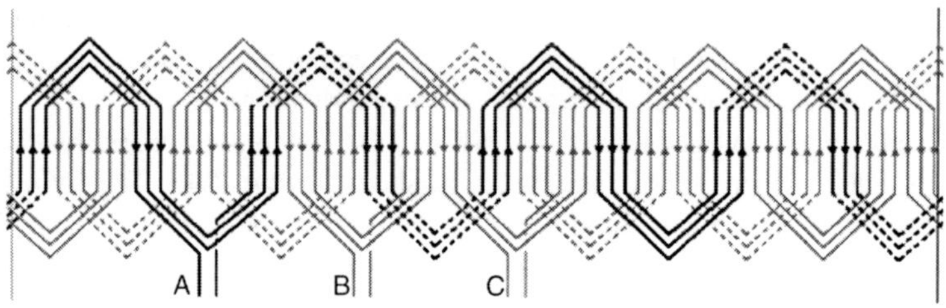

Fig. 5.17 Three-phase winding diagram for 3-turn/pole, eight-pole PM machine achieved in two metal layers. *Solid lines* represent metal layer 1, and *dashed lines*, layer 2. Phase A is darkened for reference (reprinted from Arnold et al. [7] © 2006 IOP)

mounting adaptor were conventionally milled from Hiperco 50 and Ti, respectively. The PM and rotor back iron were then glued into the mounting adaptor, which was fit onto a 1.6 mm shaft for testing.

5.9 Summary

Electrodeposition has proven to be an effective method for building the thick (tens or hundreds of micrometers) conductive and magnetic components to support the high currents and magnetic fluxes required for optimum electromechanical coupling

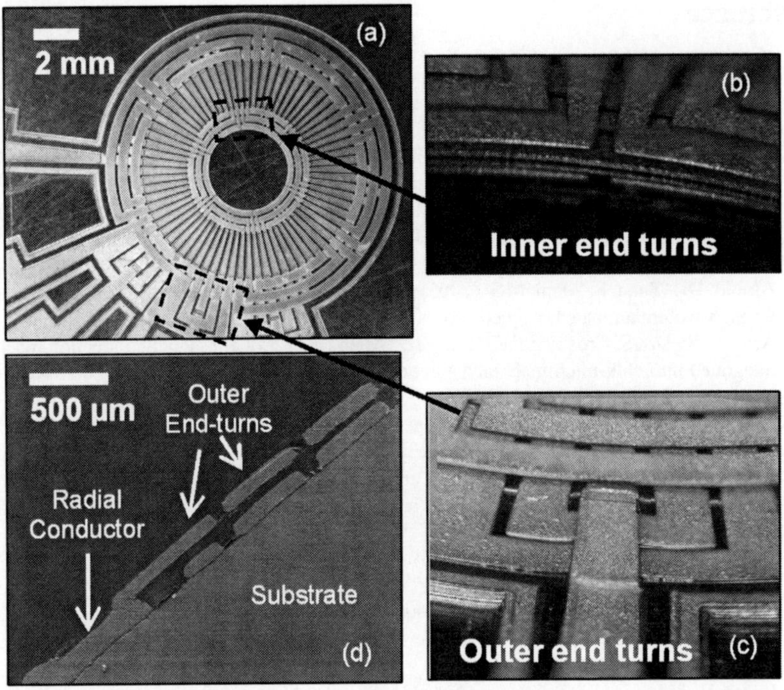

Fig. 5.18 (**a**) Fabricated 3-turn, eight-pole stator; (**b**) Close-up of inner end-turns; (**c**) Close-up of outer end turns and electrical connection to bond pad; (**d**) SEM cross-section image of the windings (compare with Fig. 5.16c) (reprinted from Arnold et al. [7] © 2006 IOP)

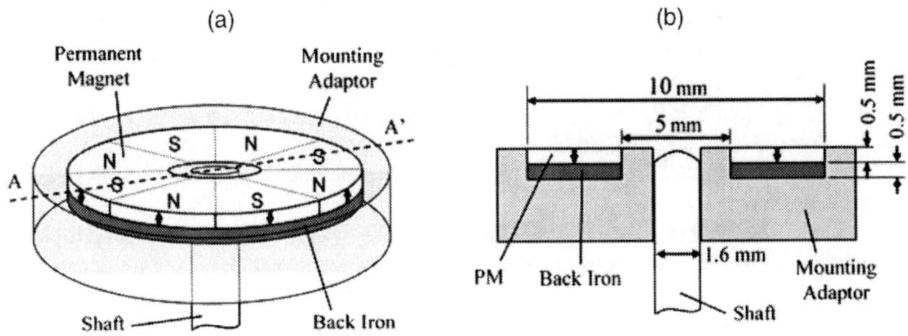

Fig. 5.19 (**a**) Perspective view and (**b**) cross-section of PM rotor assembly (adapted from Arnold et al. [6] © 2006 IEEE)

and maximum power density. Moreover, multi-wafer silicon-based planar magnetic machines have been shown to be suitable for integration with microturbomachinery for electrical power generation, using a variety of fabrication strategies. Clearly, more integration efforts are required to successfully implement a fully integrated device. However, much of the groundwork has been laid towards this goal.

References

1. Ahn CH, Kim YJ, Allen MG (1993) A planar variable reluctance magnetic micromotor with fully integrated stator and wrapped coils. Proc. IEEE MEMS Conf., Ft. Lauderdale, FL, Feb. 2003, 1–6
2. Arnold DP, Zana I, Cros F, Allen MG (2004a) Vertically laminated magnetic cores by electroplating Ni-Fe into micromachined Si. IEEE Trans. Magn. 40(4):3060–3062
3. Arnold DP, Cros F, Zana I, Veazie D, Allen MG (2004b) Electroplated metal microstructures embedded in fusion-bonded silicon: conductors and magnetic materials. J. Microelectromech. Syst. 13(5):791–798
4. Arnold DP, Zana I, Allen MG (2005) Analysis and optimization of vertically oriented, through-wafer, laminated magnetic cores in silicon. J. Micromech. Microeng. 15(5): 971–977
5. Arnold DP, Das S, Cros F, Zana I, Lang JH, Allen MG (2006a) Magnetic induction machines integrated into bulk-micromachined silicon. J. Microelectromech. Syst. 15(2):406–414
6. Arnold DP, Das S, Park J-W, Zana I, Lang JH, Allen MG (2006b) Microfabricated high-speed axial-flux multiwatt permanent-magnet generators – Part II: design, fabrication, and testing. J. Microelectromech. Syst. 15(5):1351–1363
7. Arnold DP, Herrault F, Zana I, Galle P, Park J-W, Das S, Lang JH, Allen MG (2006c) Design optimization of an 8-Watt, microscale, axial-flux, permanent-magnet generator. J. Micromech. Microeng. 16(9):S290–S296
8. Bowers BJ, Agashe JS, Arnold DP (2007) A method to form bonded micromagnets embedded in silicon. Tech. Dig. 14th Int. Conf. Solid-State Sensors, Actuators, and Microsystems (Transducers '07), Lyon, France, June 2007, vol.2, 1581–1584
9. Budde T, Gatzen HH (2006) Thin film SmCo magnets for use in electromagnetic microactuators. J. Appl. Phys. 99:08N304
10. Busch-Vishniac IJ (1992) The case for magnetically-driven micro-actuators Sens. & Actuat. A33:207–220
11. Campbell P (1994) Permanent magnet materials and their application. Cambridge University Press, Cambridge, UK, Ch. 2
12. Cho H, Ahn C (2003) Microscale resin-bonded permanent magnets for magnetic micro-electro-mechanical systems applications. J. Appl. Phys. 93(10):8674–8676
13. Cooper EI, Bonhôte C, Heidmann J, Hsu Y, Kern P, Lam JW, Ramasubramanian M, Robertson N, Romankiw LT, Xu H (2005) Recent developments in high-moment electroplated materials for recording heads. IBM J. Res. Develop. 49:103–126
14. Cugat O, Delamare J, Reyne G (2003) Magnetic micro-actuators and systems (MAGMAS). IEEE Trans. Magn. 39:3607–3612
15. Cullity BD, Graham CD (2009) Introduction to magnetic materials. IEEE Press/Wiley, Hoboken, New Jersey
16. Das S, Arnold DP, Zana I, Park J-W, Allen MG, Lang JH (2006) Microfabricated high-speed axial-flux multiwatt permanent-magnet generators – Part I: modeling. J. Microelectromech. Syst. 15(5):1330–1350
17. Dempsey NM, Walther A, May F, Givord D, Khlopkov K, Gutfleisch O (2007) High performance hard magnetic NdFeB thick films for integration into micro-electro-mechanical systems. App. Phys. Lett. 90:092509
18. Dutoit BM, Bese P-A, Blanchard H, Guérin L, Popovich RS (1999) High performance micromachined Sm_2Co_{17} polymer bonded magnets. Sens. Actuat. 77:178–182
19. Dyos GT, Farrell T, Eds. (1992) Electrical resistivity handbook. Peter Peregrinus Ltd., London
20. Ehrfeld W, Gotz F, Munchmeyer D, Schelb W, Schmidt D (1988) "LIGA process: Sensor construction techniques via x-ray lithography," Proc. IEEE Solid State Sensor and Actuator Workshop, 1–4.
21. Frazier AB, Allen MG (1992) High aspect ratio electroplated microstructures using a photosensitive polyimide process, Proc. IEEE MEMS Conf., Travemuende, Germany, 87–92

22. Furlani EP (2001) Permanent magnet and electromechanical devices. Academic Press, San Diego, CA, Ch 1
23. Ge J, Turunen MPK, Kivilahti JK (2003) Surface modification and characterization of photodefinable epoxy/copper systems. Thin Solid Films 440:198–207
24. Gilles P-A, Delamare J, Cugat O (2002) Rotor for a brushless micromotor. J. Magnet Magn. Mat. 242–245:1186–1189
25. Guan S, Nelson BJ (2005a) Pulse-reverse electrodeposited nanograinsized CoNiP thin films and microarrays for MEMS actuators. J. Electrochem. Soc. 152(4):C190–C195
26. Guan S, Nelson BJ (2005b) Fabrication of hard magnetic microarrays by electroless codeposition for MEMS actuators. Sens. & Actuat. A 118:307–312
27. Guan S, Nelson BJ (2006) Magnetic composite electroplating for depositing micromagnets. J. Microelectromech. Syst. 15(2):330–337
28. Lagorce L, Allen MG (1997) Magnetic and mechanical properties of micromachined strontium ferrite/polyimide composites. J. Microelectromech. Syst. 6(4):307–312
29. Lammeraner J, Stafl M (1966) Eddy current. Iliffe, London, ch.1–2.
30. Lorenz H, Despont M, Fahrni N, LaBianca N, Renaud P, Vettiger P (1997) SU-8: A low-cost negative resist for MEMS. J. Micromech. Microeng. 7:121–124
31. Magnetic Materials Producers Association, MMPA Standard No. 0100-00, Standard Specifications for Permanent Magnet Materials. Available www.intl-magnetics.org
32. Nordström M, Johansson A, Noguerón ES, Clausen B, Calleja M, Boisen A (2005) Investigation of the bond strength between the photo-sensitive polymer SU-8 and gold. Microelec. Eng. 78–79:152–157
33. O'Handley RC (2000) Modern magnetic materials. Wiley, New York
34. Park JY, Han SH, Allen MG (1999) Batch-fabricated microinductors with electroplated magnetically anisotropic and laminated alloy cores. IEEE Trans. Magn. 35: 4291–4300
35. Park JW, Park JY, Joung YH, Allen MG (2002) Fabrication of high current and low profile micromachined inductor with laminated Ni/Fe core. IEEE Trans. Comp. Packag. Technol. 25:106–111
36. Park JW, Allen MG (2003) Ultralow-profile micromachined power inductors with highly laminated Ni/Fe cores: Application to low-megahertz dc-dc converters. IEEE Trans. Magn. 39:3184–3186
37. Pawlowski B, Töpfer J (2004) Permanent magnetic NdFeB thick films. J. Mat. Sci. 39: 1321–1324
38. Rhen FMF, Backen E, Coey JMD (2005) Thick-film permanent magnets by membrane electrodeposition. J. Appl. Phys. 97:113908
39. Riley CD, Jewell GW, Howe D (1998) The design and analysis of axial field multipole impulse magnetizing fixtures. J. App. Phys. 83:7112–7114
40. Romankiw LT, Croll I, Hatzakis M (1970) Batch-fabricated thin film magnetic recording heads. IEEE Trans. Magn. 6(3):597–601
41. Romero JJ, Cuadrado R, Pina E, de Hoyos A, Pigazo F, Palomares FL, Hernando A, Sastre R, Gonzalez JM (2006) Anisotropic polymer bonded hard-magnetic films for microelectromechanical system applications. J. Appl. Phys. 99:08N303
42. Schlesinger M, Paunovic M, Eds. (2000) Modern electroplating, 4th Ed. Wiley, New York
43. Schmidt MA (1988) Wafer-to-wafer bonding for microstructure formation. Proc. IEEE 86: 1575–1585
44. Sullivan CR, Sanders SR (1995) Microfabrication process for high frequency power-conversion transformers. Proc. 26th Annu. Power Electronics Specialists Conf., June 1995, pp. 658–664
45. Thongmee S, Ding J, Lin JY, Blackwood DJ, Yi JB, Yin JH (2007) FePt films fabricated by electrodeposition. J. Appl. Phys. 101:09K519
46. Tong Q-Y, Gösele U (1999) Semiconductor wafer bonding: science and technology. Wiley, New York

47. Töpfer J, Christoph V (2004) Multi-pole magnetization of NdFeB sintered magnets and thick films for magnetic micro-actuators. Sens. Actuat. A 113:257–263
48. Trimmer W (1989) Microrobots and micromechanical systems. Sens. Actuat. 19:267–287
49. Wagner, B, Kreutzer M, Benecke W (1992) Linear and rotational magnetic micromotorsfabricated using silicon technology. Proc. IEEE MEMS Conf., Travemuende, Germany, 183–189
50. Wang N, Arnold DP (2008) Thick electroplated Co-rich Co–Pt micromagnet arrays for magnetic MEMS. IEEE Trans. Magn. 44:3969–3972
51. Wang N, Bowers BJ, Arnold DP (2008) Wax-bonded NdFeB micromagnets for microelectromechanical systems applications. J. Appl. Phys. 103:07E109
52. Xu M, Liakopoulos TM, Ahn CH, Han SH, Kim HJ (1998) A microfabricated transformer for high-frequency power or signal conversion. IEEE Trans. Magn. 34:1369–1371
53. Yufeng S, Wang H, Ding G, Cui F, Zhang W, Chen W (2005) Electroplated hard magnetic material and its application in microelectromechanical systems. IEEE Trans. Magn. 41(12):4380–4383
54. Zana I, Herrault F, Arnold DP, Allen MG (2005) Magnetic patterning of permanent-magnet rotors for microscale motor/generators. Proc. 5th Int. Workshop Micro Nanotechnology For Power Generation and Energy Conversion Apps. (PowerMEMS 2005), Tokyo, Japan, Nov. 2005, 116–119

Chapter 6
High-Speed Gas Bearings for Micro-Turbomachinery

Zoltán S. Spakovszky

6.1 Introduction

The mechanical design and architecture of high-speed rotating machinery, independent of size or scale, are strongly governed by the rotordynamic behavior of the spool and its bearing arrangement. Large-scale gas turbine engines yield multi-spool shaft constructions where the rolling contact bearings are close to the centerline of the engine supporting the shaft and disk assemblies as shown in Fig. 6.1 on the left.

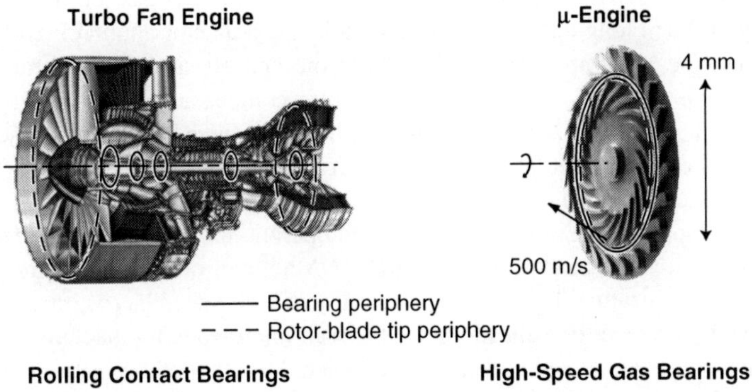

Fig. 6.1 Rotordynamic arrangement of radial bearings for large-scale gas turbine engines (*left*) and MEMS-based microengines (*right*). Note that the microengine blade-tip speed is equal to the bearing surface speed

This architecture is governed by the achievable surface speed of the bearings, commonly characterized by the *DN* number. The *DN* number is defined as the

Z.S. Spakovszky (✉)
Gas Turbine Laboratory, Massachusetts Institute of Technology, Bldg 31-265, 77 Massachusetts Avenue, Cambridge, MA 02139, USA

J.H. Lang (ed.), *Multi-Wafer Rotating MEMS Machines*, MEMS Reference Shelf,
DOI 10.1007/978-0-387-77747-4_6,

diameter of the shaft, D, in mm times the shaft speed, N, in rpm. Conventional ball bearings can reach DN numbers of up to 1–2 million mm-rpm. Since the turbomachinery is operating at much higher speeds than the bearing surface speed, the bearings must be located close to the shaft centerline as indicated by the solid red circles in Fig. 6.1 on the left. The architecture of MEMS-based micro-gas turbine engines is vastly different compared to large-scale engines. The powerMEMS devices are made out of multiple wafers stacked and bonded together to form the mechanical structure.

Dictated by DRIE fabrication constraints and the inherently flat rotor architecture, the bearings are located on the outer periphery of the rotor. Since the bearing surfaces are running at the blade-tip speed, the bearings must be designed to operate at DN numbers of up to about 10 million mm-rpm, which is one order of magnitude higher than in conventional rolling element bearings. To enable the required shaft speeds and DN numbers, fluid film bearings,[1] more specifically, plain cylindrical, axially fed gas journal bearings are employed. This yields a number of challenges and requires novel theories and engineering solutions.

This chapter focuses on high-speed gas bearings and rotordynamic considerations necessary to enable high-power density MEMS engines. Figure 6.2 depicts an MEMS air-turbine that served as a test bed for gas-bearing technology assessment and the demonstration of stable high-speed bearing operation [33]. The device consists of a six-wafer stack and the turbine is driven by an external pressurized gas source. The turbine disk is supported radially by journal bearings which are supplied with a pressurized gas from feed plena. The axial flow gas journal bearing is typically 300 μm long and has a nominal radial clearance of about 15 μm. Hydrostatic thrust bearings are employed on the front and aft side of the rotor disk. A thrust balance plenum on the aft side is used to compensate axial loads from turbomachinery pressures. A plain annular seal and an outer labyrinth seal separates the flow systems at different pressures and reduces rotordynamic coupling effects of the leakage flows. Spin tests were performed to demonstrate repeatable, stable gas-bearing operation. Controlled high-speed operation at a rotation rate of 1.7 million rpm, a rotor-tip speed of 370 m/s and a DN number of 7 million mm-rpm was achieved demonstrating the technical feasibility of high-speed gas bearings required to achieve high-power densities in MEMS-based micro-rotating machinery.

The chapter is organized as follows. Section 6.2 briefly outlines related work and Section 6.3 describes the challenges and design requirements for the stable operation of high-speed gas bearings. Section 6.4 introduces concepts related to the rotordynamics and bearings for MEMS turbomachinery, setting the stage for a detailed analysis of hydrostatic gas journal bearings in Section 6.5. The modeling of journal bearings is followed by an in-depth investigation of whirl instability in Section 6.6. Hydrostatic gas thrust bearings are discussed in Section 6.7, followed by an analysis of annular seals in Section 6.8. Experimental considerations and results are discussed in Section 6.9, and the chapter is concluded with design guidelines given in Section 6.10.

[1] Magnetic bearings are not feasible due to Curie point limitations and regular rolling element bearings cannot be manufactured by DRIE.

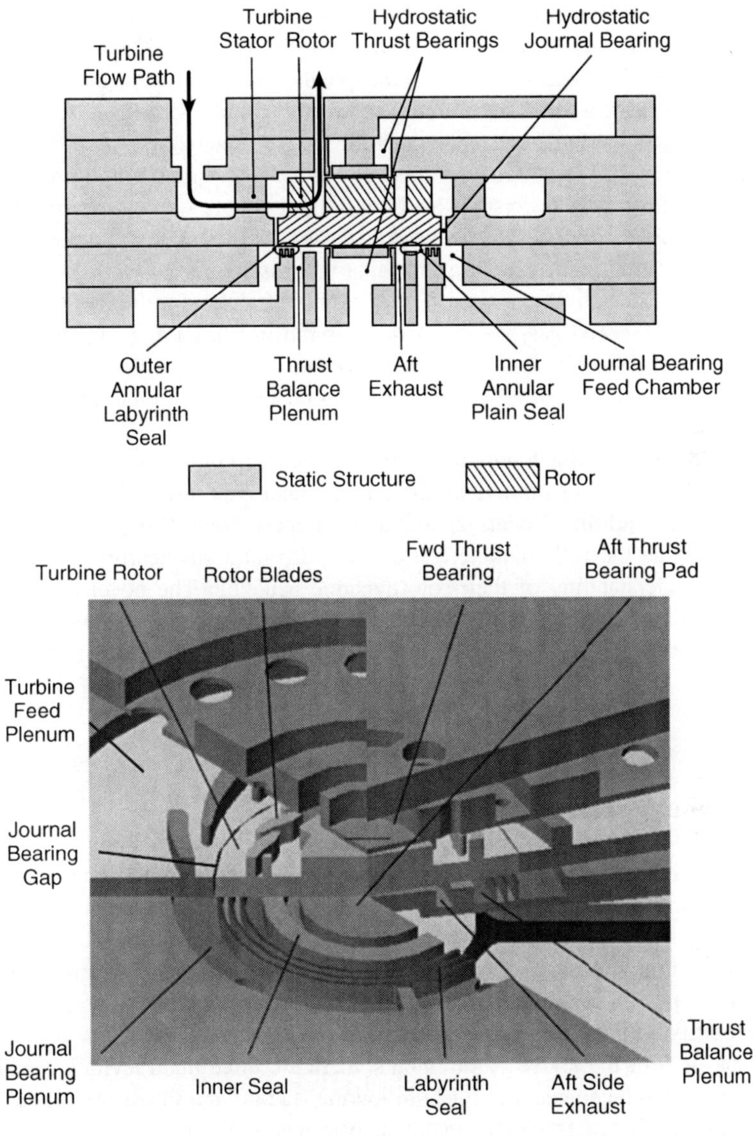

Fig. 6.2 Cross section and isometric view of gas-bearing supported MEMS air-turbine [33]

6.2 Related Work

A summary of large-scale hydrostatic and hydrodynamic gas bearings can be found for example in Lomakin [20], Larson and Richardson [15], Lund [21], and Childs [4]. High-speed bearings for MEMS-based micro-turbomachinery applications exhibit vastly different dynamic behavior compared to their large scale counterparts. The radial bearings used in the microengine project were originally designed as hydrodynamic journal bearings. A detailed discussion and analysis of hydrodynamic journal bearings can be found in Piekos et al. [25]. The simulation results showed that high-speed operating points are more stable than their lower speed counterparts. It was conjectured that the problem of hydrodynamic high-speed bearing operation became one of establishing the required loads to maintain reduced running clearances. Potential solutions such as side-pressurization, unbalance and non-circular geometry were suggested but did not enable stable high-speed operation. As a consequence, the journal bearing design evolved using a simple axial-flow hydrostatic configuration (see Fig. 6.2) that was discovered to be already inherent in the original design. An overview and summary of the development of the first generation high-speed gas bearings can be found in Ehrich and Jacobson [8] and Frechette et al. [9]. Although a small number of devices operated at high speeds, stable and repeatable operation near the target design speed could not be achieved at that time.

The technical approach adopted in this chapter combines first principles based modeling, high-fidelity numerical simulations, and experimental testing of high-speed journal- and thrust-bearings and annular seals. New theories are introduced which facilitate the full characterization of hydrostatic gas bearings and yield an improved understanding of their rotordynamic behavior. The novel theories both support and enable the design of such gas bearings. Repeatable, stable high-speed operation of micro rotors at the required blade-tip speeds is demonstrated experimentally.

6.3 Challenges and Unknowns

The choice of bearing technology is directly linked to the requirements imposed by the application. The following are key considerations that govern the selection of the rotor support system and bearings: (1) low friction, (2) long life, (3) stable operation at high speed, (4) critical (resonant) frequencies outside steady-state operating envelope, (5) sufficient load capacity, (6) high-temperature operation, and (7) compatibility with MEMS fabrication and manufacturing.

In the light of the above requirements, there are three main levitation and support concepts to be considered for both journal (radial) and thrust (axial) bearings: rolling element bearings, electro-magnetic bearings, and fluid film bearings. As discussed earlier, the logical choice for MEMS turbomachinery applications is fluid film bearings. More specifically, gas (i.e., air) bearings have no temperature limit, yield high-load capacity, are relatively simple to manufacture, and allow oil-free operation. Gas bearings are also found in large-scale application such 30–70 kW micro-GTs, APUs, turbochargers, gyroscopic instruments, and dentist drills.

Returning to Fig. 6.1, the gas journal bearings supporting the microturbomachinery are located at the outer periphery of the rotor. Given a typical rotor diameter of 4 mm and a blade-tip speed of about 500 m/s, they must be designed to operate at *DN* numbers of up to about 10 million mm-rpm, equivalent to a rotor shaft speed of

2.4 million rpm. Since these journal bearings are also very short compared to their relatively large bearing diameters, the bearing length to diameter ratio is typically 0.1. This inherently alters the driving flow mechanisms in the bearing journal and the rotordynamic stability of the rotor-bearing system, setting new challenges in the design, implementation, and operation of very high-speed gas bearings. Figure 6.3 depicts the achievable *DN* number as a function of bearing length-to-diameter ratio, *L/D*, for various bearing technologies. Note that the *L/D* of micro-gas bearings (crosses) is about one order of magnitude smaller and the *DN* number is about one order of magnitude larger than in conventional gas bearings.

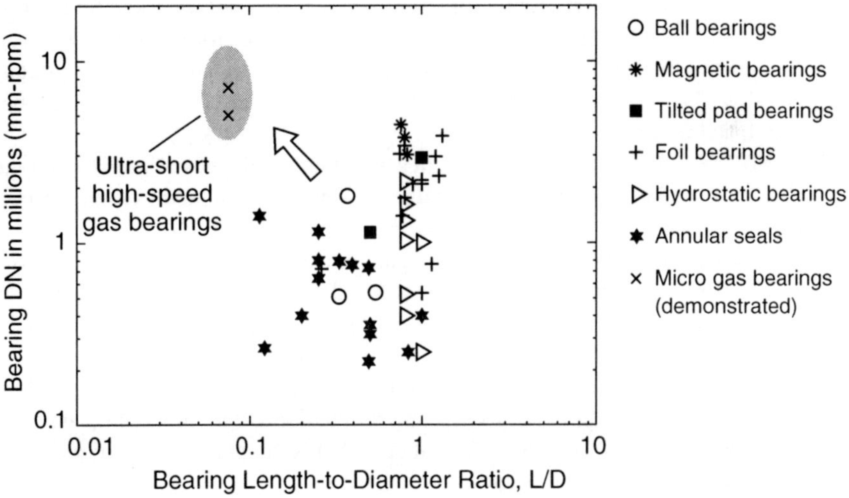

Fig. 6.3 Bearing DN number as a function of bearing L/D for various bearing technologies

The ultra-short hydrostatic gas journal bearings of the power MEMS devices are fed axially with air. They essentially act as short Lomakin seals due to the very low *L/D* (see Section 6.4 for more details). Conventional, large-scale annular seals can have *L/D*s as low as 0.1 (open stars in Fig. 6.3). Table 6.1 compares the geometric and fluid dynamic similarity parameters of micro-gas bearings with short annular seals to illustrate that the micro-gas-bearing designs are outside of the explored parameter space.

Table 6.1 Similarity parameters for micro-gas bearings and short annular seals

	Micro-gas bearings	Short annular seals
L/D	0.07	0.1 ... 1.0
C/R	0.007	0.005 ... 0.01
Re	~ 300	30,000 ... 300,000
M	> 0.5	0.1 ... 0.3
$\beta = \omega L/U$	~ 1	$<< 1$

The Reynolds number based on radial clearance and through flow velocity is in general of order 30,000, implying that the flow in conventional large scale annular seals is turbulent. The ultra-short hydrostatic gas journal bearings of the micro devices operate at Reynolds numbers of order 300, two orders of magnitude lower than conventional annular seals. The laminar flow regime sets new challenges in establishing sufficient damping and improving whirl instability. A significant amount of work has been done on short seals for turbulent flows (see Childs [4] for a summary). Unfortunately, the concepts that hold for large-scale annular seals do not apply to ultra-short hydrostatic gas journal bearings used in the micro-devices because of the laminar flow regime. In addition, the bearing flow-through time governed by the hydrostatic differential pressure is much shorter than the time scale of viscous effects. The diffusion of vorticity away from the bearing surfaces is therefore limited and the flow in the bearing land does not fully develop. These unsteady compressible flow effects impact the bearing dynamic behavior and the conditions for whirl instability.

The MEMS manufacturing process entails further challenges related to fabrication tolerances and uncertainties (for details, see Chapter 3). In addition, the inspection of the fabricated bearing geometry is impossible due to the fusion bonding of the wafer-stack such that non-destructive characterization methods need to be developed. Furthermore, the small scale of the MEMS devices makes the instrumentation for rotordynamics yet another challenge, limiting the number and types of diagnostic probes.

In summary, the stable high-speed operation of micro-gas bearings for MEMS-based turbomachinery applications sets many challenges in modeling, analysis, design, fabrication, and operation. The focus of this chapter is to introduce new approaches and theories to address these challenges.

6.4 Rotordynamic Considerations for MEMS Turbomachinery

In general, there are two types of fluid film bearings: hydrodynamic and hydrostatic bearings. The hydrodynamic, or self-acting, bearing generates stiffness from the fluid film trapped between the shaft surface and the journal. The motion of an off-centered shaft acts like a viscous pump and creates a constriction building up pressure in the fluid film. The induced pressure field together with the viscous stress field lead to a hydrodynamic restoring force. Since the resulting stiffness yields a component perpendicular to the shaft offset (so-called cross-coupled stiffness), whirl instability can be promoted.

Hydrostatic bearings are externally pressurized and the restoring force in the fluid film is generated through the pressure differential supplied by an external source. For zero shaft rotation, the resulting stiffness acts in the direction of the shaft displacement. Lomakin [20] first explained the large direct stiffness that can be developed by annular seals. Assuming a plain annular seal with an off-centered shaft and imposing an axial pressure difference across the seal, the pressure drop due to the inlet loss into the seal will be higher at the large clearance than at the small clearance due to the increased local flow rate. Thus, the axial pressure gradient inside the seal will be lower at the large clearance compared to the axial pressure gradient at the small clearance. The difference in the pressure gradients provides a restoring force which opposes the shaft displacement, yielding a potentially large direct stiffness. In summary, the so-called "Lomakin" effect stems from the combined effect of the inlet loss of the axial flow into the seal and the axial pressure gradient across the seal.

Figure 6.4 depicts the conceptual arrangement of the hydrostatic gas journal bearing radially supporting the rotor of the MEMS device shown in Fig. 6.2. The stiffness and damping coefficients of the bearings are indicated by springs and dashpots. The goal of the analysis is to appropriately describe and to model the stiffness and damping coefficients which can then be used in a rotordynamic analysis. To illustrate the dynamic behavior only the radial motion is considered here. Coupling phenomena between the axial and radial motion will be discussed later.

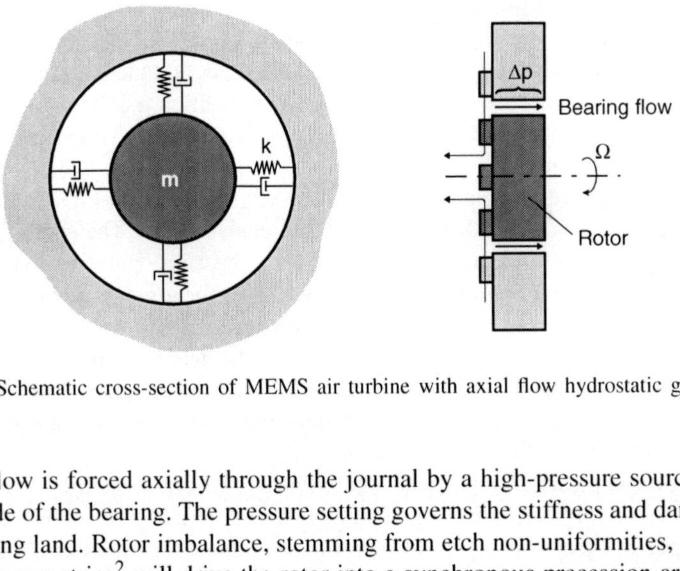

Fig. 6.4 Schematic cross-section of MEMS air turbine with axial flow hydrostatic gas journal bearings

The flow is forced axially through the journal by a high-pressure source on the lower side of the bearing. The pressure setting governs the stiffness and damping in the bearing land. Rotor imbalance, stemming from etch non-uniformities, and geometric asymmetries[2] will drive the rotor into a synchronous precession around the geometric center of the journal. In other words, an eccentric whirl at rotor frequency will be superimposed to the rotor spinning around its geometric center. This can be described using the equations of motion for a two-degree-of-freedom rotor model.[3] Assuming small departures from the precessing motion with dynamic eccentricity

[2]Geometric asymmetries of the rotors are mostly due to local and global variations in etch depth and out-of-roundness effects are negligible. For a more detailed discussion see Chapter 3.

[3]The analysis of the hydrostatic thrust bearing dynamics and the tilting degree of freedom are discussed later in Section 6.7

ε_0 the equations of motions can be linearized and yield an eigenvalue problem. The imaginary part of each of the eigenvalues is the mode frequency Ω, which in the absolute frame is equal to the natural frequency Ω_N. The real part of the resonance is the growth rate and determines whether the mode will grow in time (positive real part, unstable) or decay in time (negative real part, stable). Thus, the onset of whirl instability is determined by the rotor frequency $\Omega = \Omega_W$ at which the real part of the least stable mode becomes zero. An important parameter indicating the limit speed is the ratio of whirl instability speed to natural frequency denoted as the whirl ratio $\Re = \Omega_W/\Omega_N$. The modeling of the whirl instability limit is discussed in more detail in Section 6.6.

The response of the rotor-bearing system to a given imbalance a can be found by solving the steady state form of the linearized equations of motion (for details see [19]):

$$(ma\Omega^2)^2 = (F^{\mathrm{hs}}(\varepsilon_0) + m\varepsilon_0\Omega^2)^2 + (F^{\mathrm{hd}}(\varepsilon_0,\Omega) + C_y(\varepsilon_0)\Omega\varepsilon_0)^2. \tag{6.1}$$

Note that the hydrostatic and hydrodynamic forces and the damping coefficient are a function of eccentricity and introduce the non-linear behavior of the journal gas bearing. In the limit of $\Omega\rightarrow\infty$, the dynamic eccentricity becomes the imbalance $\varepsilon_0 = a$. The hydrodynamic and hydrostatic forces and the damping coefficient are obtained from the bearing model described in Section 6.5. The purpose of the analysis here is to illustrate the rotordynamic behavior of a typical device during transcritical operation.

For large departures, a numerical simulation of the full non-linear equations of motion was conducted together with a stability analysis [17]. The results show that the linearized stability treatment is valid for the bearing operating conditions of interest here. Figure 6.5 depicts the response curve for a range of rotor speeds Ω and three different levels of imbalance at a low-bearing pressure setting of $\Delta p =$ 0.25 psi. The skew of the response curve is due to the increased hydrostatic stiffness when the rotor approaches the wall. Close to the wall the damping is also increased due to the non-linear effects of the hydrodynamic and the damping forces which influence the width of the resonance peak.

Figure 6.6 depicts the rotor speed at whirl instability onset and the corresponding response curve for bearing pressure settings Δp of 1.0, 2.0, 3.0, and 6.0 psi and an imbalance $a = 1$ μm. Raising the bearing differential pressure increases the stiffness of the bearing and thus shifts the natural frequency to the right. However, the distance to contact when the rotor reaches peak amplitude (the bearing clearance in this calculation is 12 μm) decreases and the model results suggest that the natural frequency should be crossed at low Δp settings when transitioning from subcritical to supercritical speeds.[4] Note that the whirl number $\Re$ is about five and stays roughly constant when the bearing differential pressure is raised. This is reflected in a shift

[4] At subcritical speeds (below the natural frequency), the rotor is spinning about its geometric center. After crossing the natural frequency and transitioning to supercritical speeds, the rotor is spinning about its center of mass.

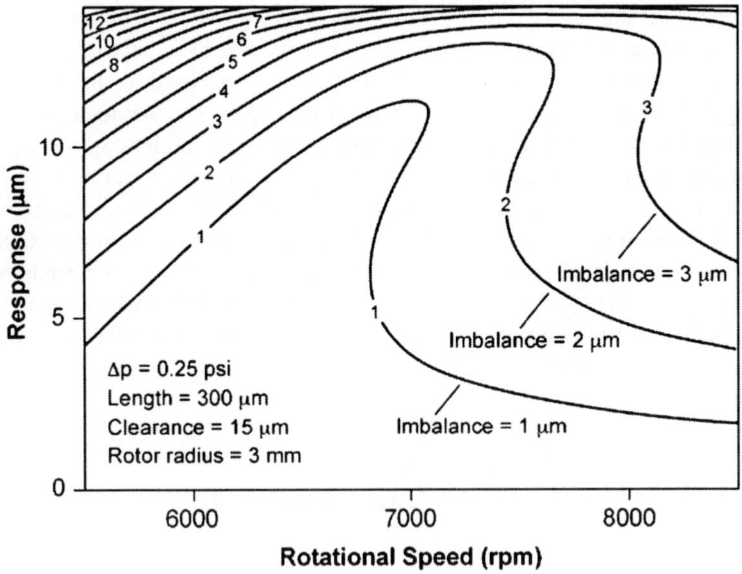

Fig. 6.5 Response curves for three different levels of imbalance $a = 1$, 2, and 3 μm at a low-bearing pressure setting of $\Delta p = 0.25$ psi [19]

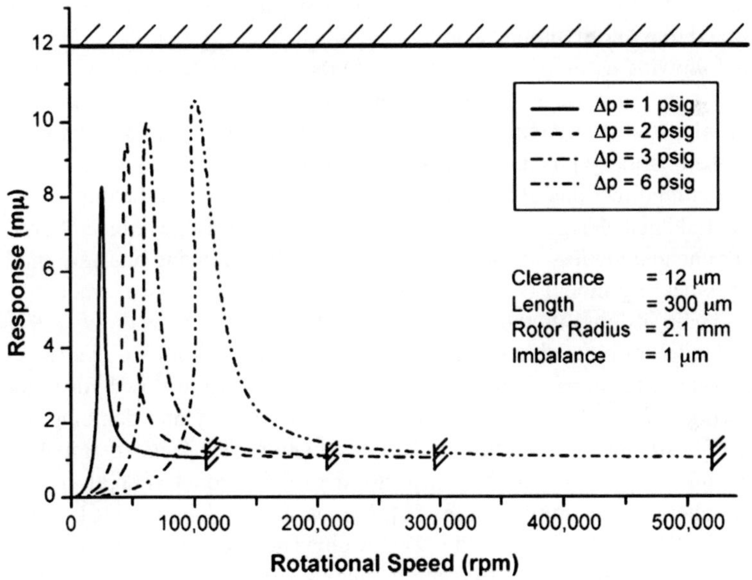

Fig. 6.6 Response curves and whirl instability limit for bearing differential pressure settings $\Delta p = 1.0$, 2.0, 3.0, and 6 psi and a relatively low imbalance of $a = 1$ μm [19]

of the whirl instability limit to the right, similar to the natural frequency. Using the model results, acceleration schedules and guidelines for the bearing pressure settings can be established. The established guidelines are used in the experimental testing of the micro-bearing device and are discussed later in Section 6.9.

It is useful to summarize the observations made in the preliminary analysis discussed above. The calculations show that the ultra-short micro journal gas bearings yield relatively low-damping ratios (and hence relatively small distances to wall contact). The damping ratio is further reduced at higher bearing pressure settings and the rotor imbalance has a strong influence on the bearing dynamic behavior. Furthermore, the stable operating range is limited by the whirl ratio which, to first order, is independent of the bearing pressure setting.[5] However, the whirl ratio depends strongly on the level of rotor imbalance and journal bearing clearance as will be shown later.

6.5 Analysis of Hydrostatic Gas Journal Bearings

First, the gas journal bearing forces are analyzed near transcritical speeds to investigate the bearing behavior. Assuming that the flow induced by the rotor motion can be separated from the flow induced by the static pressure drop across the journal, the fluid forces exerted on the rotor can be split into a direct-coupled hydrostatic force, a cross-coupled hydrodynamic force, and a damping force. The modeling of these three main force components constitutes the fluid dynamic model which can be coupled to a rotordynamic analysis.

The following assumptions are made: (1) the pressure drop across the journal is relatively small and the flow in the bearing gap is assumed incompressible, (2) the journal inlet loss coefficient is assumed independent of Reynolds number, (3) the flow in the bearing gap is modeled as fully developed, 2-D viscous flow in the axial and tangential directions. It will be shown in Section 6.6.2 that the third assumption does not hold near design speed because at those conditions the reduced frequency is of order one and the flow is inherently unsteady. The following analysis and hydrostatic gas-bearing model is valid at rotor speeds near the critical frequencies and useful to gain insight into the mechanisms at play.

The model introduced below is capable of dealing with all the elements of, (1) micro-devices, (2) dynamic response characteristics of axial flow hydrostatic journal bearings, (3) evaluation of stiffness, natural frequency and damping, (4) evaluation of instability boundaries, and (5) evaluation of effects of imbalance and bearing anisotropy. Some of these aspects have been studied in depth by a number of researchers (for example Smith [30], Larson and Richardson [15], Fuller [10], Childs [4], San Andres and Wilde [27], Arghir and Frene [2], Orr [23]). The key feature of this model is that it is able to analytically deal with all of these aspects together.

[5] The effect of unsteady flow on the whirl ratio will be discussed in Section 6.5.2.

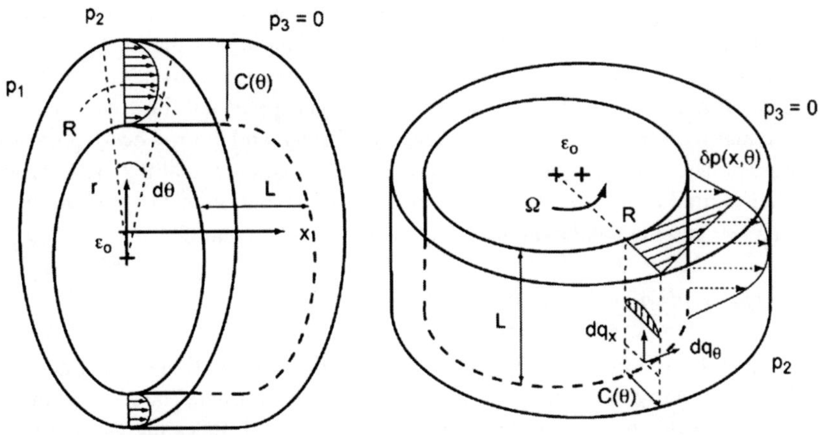

Fig. 6.7 Flow field model for hydrostatic force (*left*) and hydrodynamic force (*right*) [19]

6.5.1 Hydrostatic Force Model

Consider the simple journal bearing geometry depicted in Fig. 6.7. Assuming that the rotor is stationary and that the center of the rotor is offset from the center of the journal as depicted in Fig. 6.7 on the left, the static pressure drop across the gap sets up a Poiseuille type flow in the axial direction for each circumferential segment $C(\theta)\cdot Rd\theta$. The local journal clearance is denoted by $C(\theta) = C_0 - e\cos\theta = C_0\cdot(1 - \varepsilon\cos\theta)$, where e and R are the eccentricity and the mean journal radius respectively, and $\varepsilon = e/C_0$ is the eccentricity normalized by the bearing clearance C_0. The local axial flow rate is obtained by integrating the Poiseuille flow profile in the radial direction and yields

$$q_x(\theta) = -\frac{C(\theta)^3}{12\mu}\frac{\partial p}{\partial x} \tag{6.2}$$

From continuity and the assumption of purely axial flow the local flow rate $q_x(\theta)$ remains constant along the gap. Thus, the axial pressure gradient is constant as well and the pressure distribution is linear in the axial direction with $\partial p/\partial x = -p_2(\theta)/L$. Including the effects of inlet loss, the pressure balance across the bearing from stations 1 to 3 ($p_3 = 0$) can then be written as $\Delta p = p_2(\theta) + \omega_{\text{inlet}}/2 \cdot v_2(\theta)^2$. The inlet pressure loss coefficient ω_{inlet} is assumed 1.5, which is a value commonly tabulated in handbooks of hydraulics. This assumption was verified by a steady 3-D CFD calculation of the micro-bearing geometry [17]. The inlet velocity distribution $v_2(\theta)$ is found from continuity and becomes

$$v_2(\theta) = \frac{C(\theta)^2}{12\mu}\frac{p_2(\theta)}{L} \tag{6.3}$$

The axial velocity through the hydrostatic journal gas bearing was computed and the mean axial Mach number was found to be less than 0.3 for the range of bearing pressure settings of interest. This verifies the incompressible flow assumption made earlier. The pressure distribution in the bearing gap $p_2(\theta)$ can be found from the above equation as a function of geometry, eccentricity, and journal bearing pressure setting Δp. The direct-coupled hydrostatic force F^{hs} and the total flow rate through the axial journal bearing Q can be determined by integration of the pressure distribution $p_2(\theta)$ and the local axial flow rate $q_x(\theta)$ around the annulus.

6.5.2 Hydrodynamic Force Model

As stated in the assumptions, the flow induced by the rotor motion is separated from the flow induced by the static pressure drop across the journal. The rotating motion of the rotor drags the flow along the circumference, and under the assumption of incompressible, fully developed viscous flow, the combination of a Couette flow and a Poiseuille flow is locally set up. For this type of flow the local tangential flow rate is

$$q_\theta = \frac{\Omega R C(\theta)}{2} - \frac{C(\theta)^3}{12\mu}\frac{\partial \delta p}{R\partial\theta}. \tag{6.4}$$

The first term is due to the Couette flow and proportional to the local journal clearance. ΩR is the rotor surface speed. When the rotor moves closer to the wall, the fluid will be forced to flow also in the axial direction and thus builds up pressure as depicted in Fig. 6.7 on the right. Assuming again Poiseuille flow in the axial direction (see Eq. 6.2), the hydrodynamic pressure distribution in the journal can be obtained from continuity

$$\frac{\partial q_x}{\partial x} + \frac{1}{R}\frac{\partial q_\theta}{\partial\theta} = 0,$$

or

$$-\frac{C(\theta)^3}{12\mu}\frac{\partial^2 \delta p}{\partial x^2} + \frac{\Omega}{2}\frac{\partial C}{\partial\theta} - \frac{1}{R^2}\frac{\partial}{\partial\theta}\left(\frac{C(\theta)^3}{12\mu}\frac{\partial \delta p}{\partial\theta}\right) = 0. \tag{6.5}$$

Since L/D is much smaller than 1, the third term in the above equation is negligible compared to the other two terms. Assuming $\delta p(0,\theta) = \delta p(L,\theta) = 0$ as boundary conditions, the hydrodynamic pressure distribution then yields after integration

$$\delta p(x,\theta) = \frac{3\mu\Omega}{C(\theta)^3}\frac{\partial C}{\partial\theta}\cdot x\cdot(x-L). \tag{6.6}$$

Figure 6.8 shows the circumferential pressure distribution along the bearing centerline ($x = L/2$) for eccentricities $\varepsilon = 0.2$ and $\varepsilon = 0.4$ and a rotor speed of 2.4 million rpm. The hydrodynamic force can be found by integrating the pressure

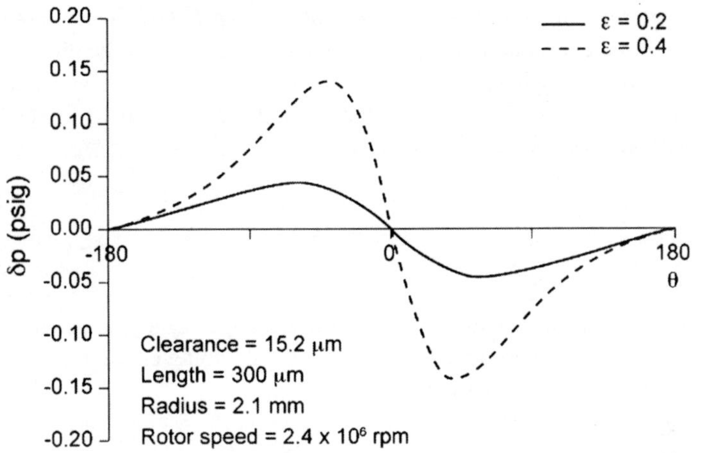

Fig. 6.8 Hydrodynamic pressure distribution for eccentricities $\varepsilon = 0.2$ and $\varepsilon = 0.4$ [19]

distribution $\delta p(x,\theta)$ and the viscous stress distribution $\tau(\theta) = \mu\Omega R/C(\theta)$ around the circumference yielding

$$F^{\mathrm{hd}} = \frac{\pi}{2}\mu\frac{RL^3}{C_0^2}\Omega\frac{\varepsilon}{\left(1-\varepsilon^2\right)^{1.5}} - 2\pi\mu\frac{R^2L}{C_0}\Omega\frac{1-\sqrt{1-\varepsilon^2}}{\varepsilon\sqrt{1-\varepsilon^2}} \tag{6.7}$$

The first term is due to the hydrodynamic pumping effect and the second term is due to the viscous drag effect of the rotor motion. The hydrodynamic force acts perpendicular to the rotor deflection and is thus a cross-coupled force that can potentially contribute to whirl instability.

6.5.3 Damping Force Model

The damping force can be calculated from the pressure field that is setup as the flow is squeezed in the gap due to the radial rotor motion toward the wall. From continuity, the local axial flow rate can be approximated as

$$q_x(\theta) = v_r \cdot \left(x - \frac{L}{2}\right) \tag{6.8}$$

where v_r is the rotor translational velocity in the radial direction. Assuming again Poiseuille type flow in the axial direction, the pressure distribution is obtained by combining the above equation with

$$-\frac{C(\theta)^3}{12\mu}\frac{\partial p}{\partial x} = v_r \cdot \left(x - \frac{L}{2}\right) \tag{6.9}$$

Applying appropriate boundary conditions $p(0,\theta) = p(L,\theta) = 0$, the damping force components in the direction of the translational motion and in the direction normal to it are determined by integration of $p(x,\theta)$ around the annulus. The damping force is a function of eccentricity ε, rotor velocity and bearing geometry, and in the limit of $\varepsilon \to 0$ the damping coefficient $\tilde{C}$ yields

$$\tilde{C} = \pi \mu R \frac{L^3}{C_0^3} \tag{6.10}$$

The derived models for the hydrostatic force, the hydrodynamic force, and the damping force can be introduced as stiffness and damping coefficients in the rotordynamic models discussed earlier.

6.5.4 Model Limitations

The analytical fluid dynamic modeling approach is limited to incompressible flow in the journal bearing gap. This implies that effects of compressibility at high bearing differential pressure settings and large eccentricities might not be captured. Furthermore, the model assumes that the bearing flow field can be decoupled into an axial and a tangential flow field, neglecting transverse boundary layer development and 3-D effects. These effects are investigated and addressed in detail in Section 6.6.2.

Since the micro-bearing turbine test rig is operated with cold shop air, the gas properties were assumed to be at standard ambient conditions. However, the model is not limited to ambient conditions and can also deal with hot operating conditions as they can occur in turbocharger and gas turbine applications.

The effects of compressibility on the hydrodynamic force are briefly investigated to assess the model limitations. Assuming a rotor surface speed of ~500 m/s at design conditions, the mean tangential Mach number is about 0.8. A comparison between an incompressible analytical short bearing solution and a pseudo-spectral, compressible, numerical method (SPECTRES) is reported in Piekos [24]. The results show that the analytical, incompressible short bearing solution is accurate up to an eccentricity of $\varepsilon = 0.8$ and a bearing number[6] $\Lambda = 10$. Given the micro-bearing geometry of $C_0 = 12\ \mu\text{m}$, $L = 300\ \mu\text{m}$, and $R = 2.1$ mm, and choosing a bearing number of $\Lambda = 1$, the rotor speed Ω is about 3 million rpm. Thus, the incompressibility assumption in the hydrodynamic force model is suggested to be valid up to the design speed of 2.4 million rpm.

---

[6] The bearing number Λ is used to non-dimensionalize the rotational speed Ω and is defined as $\Lambda = 6\,\mu\Omega/p_0 \cdot (R/C)^2$. For an ideal gas the bearing number may also be expressed in variables more familiar to the fluid dynamicist, that is $\Lambda = 6\gamma M^2/Re_C \cdot (R/C)$.

6.5.5 *Scaling Laws for Ultra-Short Hydrostatic Gas Journal Bearings*

Scaling laws for the stiffness and damping coefficients and the natural frequency of hydrostatic micro-gas journal bearings are established to guide the design and conception of engineering solutions critical for stable high-speed bearing operation. The derivation of the scaling laws is based on the above analysis and assumptions made for ultra-short bearing geometries. A simplified Lomakin bearing model is discussed first to shed more light on the underlying mechanisms for the hydrostatic stiffness of ultra-short gas journal bearings.

6.5.5.1 Hydrostatic Stiffness – Simplified Lomakin Bearing Model

Assuming that the rotor is stationary and that the center of the rotor is displaced from the center of the journal by a small offset e as shown in Fig. 6.9 on the left, the bearing can be approximated as two half-seals acting in parallel between the same bearing pressure drop Δp. With a nominal bearing clearance C, the upper half of the seal experiences on average a reduced clearance $C - e$ whereas the lower half of the seal experiences on average an enlarged clearance $C + e$.

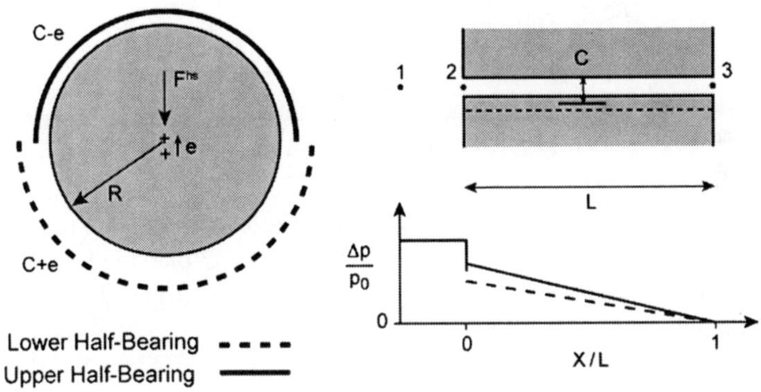

Fig. 6.9 Simple model of an ultra-short hydrostatic gas journal bearing [31]

Considering the bearing axial cross-section shown in Fig. 6.9 on the right and an inlet stagnation pressure loss coefficient $\omega_{\text{inlet}} = (p_{t1} - p_{t2})/(\frac{1}{2}\,\rho\, v_2^2)$, the externally imposed total pressure drop $\Delta p = p_{t1} - p_{t2} + p_{t2} - p_{t3}$ can be written as $\Delta p = \frac{1}{2}\,\rho\, v_2^2 \cdot \omega_{\text{inlet}} + p_2$, where $p_3$is assumed zero and $v_2 = v_3$from continuity. The imposed total pressure drop Δp is the same for both the upper and the lower half of the bearing. However, the flow rate through the upper half of the bearing will be lower compared to the lower half due to the increased resistance (reduced clearance). As a consequence, the stagnation pressure loss at the inlet will be reduced and an increased axial pressure gradient will prevail in the upper half of the bearing as shown in Fig. 6.9 on the right.

The pressure gradient in the bearing gap sets up a Poiseuille type flow with a radially averaged axial velocity of

$$v(x) = -\frac{C^2}{12\mu}\frac{\partial p}{\partial x} \tag{6.11}$$

From continuity and the assumption of purely axial flow, the local flow rate remains constant along the gap. Thus, the axial pressure gradient is constant as well and the pressure distribution is linear in the axial direction with $\partial p/\partial x = -p_2/L$. The pressure balance across the bearing between stations 1 and 3 can then be written as

$$\Delta p = p_2 + \frac{1}{2}\rho\omega_{inlet}\frac{C^4}{144\mu^2}\frac{p_2^2}{L^2} \tag{6.12}$$

The above equation can be used for both the lower and the upper half-bearings by substituting $C \pm e$ for C. Assuming that there is no reverse flow, the inlet pressures p_2 of the two half-bearings become

$$p_2^U = \frac{A_0}{(C-e)^4}\cdot\left[\sqrt{\frac{2\Delta p(C-e)^4}{A_0}+1}-1\right] \tag{6.13}$$

$$p_2^L = \frac{A_0}{(C+e)^4}\cdot\left[\sqrt{\frac{2\Delta p(C+e)^4}{A_0}+1}-1\right] \tag{6.14}$$

where $A_0 = 144\mu^2L^2/\rho\ \omega_{\text{inlet}}$. Assuming linear pressure profiles in the axial direction as sketched in Fig. 6.9 and using $p_3 = 0$, the direct-coupled hydrostatic restoring force becomes $F^{\text{hs}} = DL/2\cdot(p_2{}^U - p_2{}^L)$. The hydrostatic stiffness $K^{\text{hs}} = \partial F^{\text{hs}}/\partial e$ then yields for small shaft offsets $e/C << 1$

$$K^{\text{hs}} \approx \frac{4DLA_0}{C^5}\cdot f\left(\frac{2\Delta pC^4}{A_0}\right) \tag{6.15}$$

where

$$f(z) = \frac{z/2+1-\sqrt{z+1}}{\sqrt{z+1}} \tag{6.16}$$

For the range of z of interest $f(z)$ is approximately linear. Rearranging terms the hydrostatic stiffness is found to scale with the bearing parameters according to

$$\frac{K^{\text{hs}}}{p_o C} \propto \left(\frac{L}{D}\right)\left(\frac{C}{R}\right)^{-2}\left(\frac{\Delta p}{p_o}\right) \tag{6.17}$$

where the stiffness is non-dimensionalized by the ambient pressure p_0 and the bearing clearance C.

6.5.5.2 Natural Frequency

Assuming that the rotor can be modeled as a simple disk with mass $m = \rho_d \pi R^2 L$, the natural frequency of the rotor-bearing system is defined as $\Omega_N{}^2 = K^{\mathrm{hs}}/(\rho_d \pi R^2 L)$. Using Eq. (6.17) the natural frequency can be written as

$$\Omega_N^2 \propto \left(\frac{L}{D}\right)\left(\frac{C}{R}\right)^{-2} \frac{\Delta p\, C}{\rho_d\, R^2 L} \tag{6.18}$$

In order to non-dimensionalize the natural frequency in terms of a bearing number, both sides of the above equation are multiplied by $(\mu R^2/p_0\, C^2)^2$ such that

$$\frac{\mu^2 \Omega_N^2}{p_o^2} \frac{R^4}{C^4} \propto \left(\frac{L}{D}\right)\left(\frac{C}{R}\right)^{-4}\left(\frac{\Delta p}{p_o}\right)\left(\frac{\rho_o}{\rho_d}\right) \cdot \frac{\mu^2}{\rho_o^2 C^2 (\Omega R)^2} \frac{\rho_o (\Omega R)^2}{p_o} \frac{C}{L} \tag{6.19}$$

Collecting terms and using the definition of Reynolds number Re_C based on journal clearance C, Mach number M based on the velocity at the bearing surface ΩR, and bearing number Λ, the natural frequency in non-dimensional form becomes

$$\Lambda_N^2 \propto \left(\frac{L}{D}\right)\left(\frac{C}{R}\right)^{-4}\left(\frac{\Delta p}{p_o}\right)\left(\frac{\rho_o}{\rho_d}\right) \frac{\gamma M^2 C}{Re_C^2 L} \tag{6.20}$$

Expressing the bearing number in terms of M and Re_C, the scaling law for the natural frequency in non-dimensional form yields

$$\Lambda_N \propto \left(\frac{C}{R}\right)^{-1}\left(\frac{\Delta p}{p_o}\right)^{1/2}\left(\frac{\rho_o}{\rho_d}\right)^{1/2} \Lambda^{1/2} Re_C^{-1/2} \tag{6.21}$$

Note that the above result suggests that the natural frequency of an ultra-short hydrostatic gas bearing-rotor system does not scale with the bearing length to diameter ratio L/D.

6.5.5.3 Hydrodynamic Stiffness and Damping Ratio

Similarly, the hydrodynamic stiffness can be cast in non-dimensional form. From Eq. (6.7), it can be shown that in the limit of e • 0 the hydrodynamic stiffness $K^{\mathrm{hd}} = \partial F^{\mathrm{hd}} / \partial e$ becomes

$$K^{\mathrm{hd}} = \pi \frac{\mu L (\Omega R)}{C} \left(\frac{L^2}{2C^2} - \frac{R}{C}\right) \tag{6.22}$$

Introducing again the Mach number and Reynolds number, the hydrodynamic stiffness yields in non-dimensional form

$$\frac{K^{\mathrm{hd}}}{p_o C} = \pi \left(\frac{1}{2} \frac{L^3}{C^2 R} - \frac{L}{C} \right) \frac{\gamma M^2}{Re_c} \tag{6.23}$$

Using the bearing number, the hydrodynamic stiffness in non-dimensional form then becomes in terms of common bearing parameters

$$\frac{K^{\mathrm{hd}}}{p_o C} = \frac{2\pi}{3} \left[\left(\frac{L}{D} \right)^3 \left(\frac{C}{R} \right)^{-2} - \frac{1}{2} \left(\frac{L}{D} \right) \left(\frac{C}{R} \right)^{-1} \right] \Lambda \tag{6.24}$$

The total hydrodynamic stiffness can be split into a pure hydrodynamic pumping part k^p and a pure viscous drag part due to rotor rotation k^v such that $K^{\mathrm{hd}} = k^p - k^v$. Note that the two effects oppose one another: the hydrodynamic force due to pumping acts in the direction of rotor rotation whereas the hydrodynamic force due to viscous drag acts in the direction against rotor rotation. The two opposing forces have an impact on the whirl instability limit which will be discussed in detail in Section 6.6. The scaling laws for the two hydrodynamic stiffness terms then become

$$\frac{k^p}{p_o C} \propto \left(\frac{L}{D} \right)^3 \left(\frac{C}{R} \right)^{-2} \Lambda. \tag{6.25}$$

$$\frac{k^v}{p_o C} \propto \left(\frac{L}{D} \right) \left(\frac{C}{R} \right)^{-1} \Lambda. \tag{6.26}$$

Also note that the hydrodynamic stiffness depends on the rotor speed (bearing number) whereas the hydrostatic stiffness does not.

Modeling the rotor again as a simple disk with mass m the damping ratio for the hydrostatic journal bearing-rotor system is defined as $\zeta = \tilde{C}^{sys}/(2m\Omega_{\mathrm{N}})$, where $\tilde{C}^{sys}$ is the system damping and $2m\Omega_N$ is the critical damping above which the response of the system to an impulsive load is no longer oscillatory. Using the analytical model discussed in Section 6.5.3 the system damping can be determined from the pressure field that is setup as the flow is squeezed in the bearing gap when the rotor moves radially toward the wall. It can be shown that in the limit $e \to 0$ the system damping coefficient becomes $\tilde{C}^{sys} = \pi \mu R L^3 / C^3$. Combining this with the expression for natural frequency Eq. (6.18) and the mass of the disk $m = \rho_d \pi R^2 L$, the damping ratio becomes

$$\zeta \propto \left(\frac{C}{R} \right)^{-3} \left(\frac{\Delta p}{p_o} \right)^{-1/2} \left(\frac{\rho_o}{\rho_d} \right)^{-1/2} \frac{\sqrt{\Lambda Re_C} \mu L^2}{\rho_d C^3 (\Omega R)} \tag{6.27}$$

After rearranging terms the scaling law for the damping ratio yields

$$\zeta \propto \left(\frac{L}{D}\right)^2 \left(\frac{C}{R}\right)^{-1} \left(\frac{\Delta p}{p_o}\right)^{-1/2} \left(\frac{\rho_o}{\rho_d}\right)^{1/2} \Lambda^{1/2} Re_C^{-1/2} \tag{6.28}$$

6.5.5.4 Summary of Scaling Laws and Experimental Validation

The scaling laws for ultra-short hydrostatic gas journal bearings are established and are summarized in Table 6.2. The table entries contain the exponents of the bearing parameters in the scaling laws derived in the previous section.

Table 6.2 Scaling laws for ultra-short hydrostatic gas journal bearings.

	$\left(\frac{L}{D}\right)$	$\left(\frac{C}{R}\right)$	$\left(\frac{\Delta p}{p_o}\right)$	$\left(\frac{\rho_o}{\rho_d}\right)$	Λ	Re_C
$\left(\frac{K^{hs}}{p_o C}\right)$	1	−2	1	0	0	0
$\left(\frac{k^p}{p_o C}\right)$	3	−2	0	0	1	0
$\left(\frac{k^v}{p_o C}\right)$	1	−1	0	0	1	0
Λ_N	0	−1	1/2	1/2	1/2	−1/2
ζ	2	−1	−1/2	1/2	1/2	−1/2

Experiments were conducted to validate the scaling laws. The details of the experimental setup and data reduction schemes can be found in Section 6.9. Two silicon turbine rotor geometries supported by hydrostatic gas journal bearings with different L/D and C/R were tested at various rotor speeds over a range of bearing supply pressures $\Delta p/p_0$. The bearing parameters and geometries of the experimentally tested micro-devices are summarized in Table 6.3.

The experimentally measured natural frequency of the hydrostatic gas journal bearings of the micro-devices is plotted in non-dimensional form in Fig. 6.10 as a function of bearing pressure difference $\Delta p/p_0$. All experiments were conducted with the same gas (nitrogen). The scaling law is shown as the dashed and solid lines for the device geometries described in Table 6.3. The results show that the experimentally obtained trend with bearing pressure setting matches well with the suggested scaling law. Note that the natural frequency scales with the square root of bearing pressure difference $\Delta p/p_0$.

Table 6.3 Bearing and rotor geometries of experimentally tested micro-devices

	Si Rotor 1	Si Rotor 2
C (μm)	15	15
L (μm)	300	300
R (mm)	2.1	3
L/D	0.07	0.05
C/R	0.007	0.005

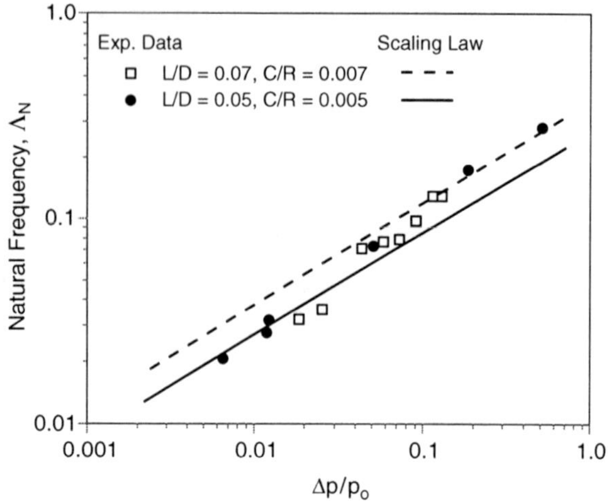

Fig. 6.10 Scaling of natural frequency with bearing pressure difference $\Delta p/p_0$ and bearing geometry [31]

Only little data exist on damping ratio because of the difficulty to measure the full unbalance-driven response curves required to deduce ζ. Figure 6.11 depicts the measured damping ratio of silicon rotor 1 as a function of bearing pressure difference $\Delta p/p_0$. The scaling law is plotted as the dashed line. Although only little data are available, the trend with bearing pressure difference seems to be captured. The result suggests that for a given geometry and lubricant, the damping ratio can be increased by reducing the bearing pressure difference: ζ is inversely proportional to the square root of $\Delta p/p_0$.

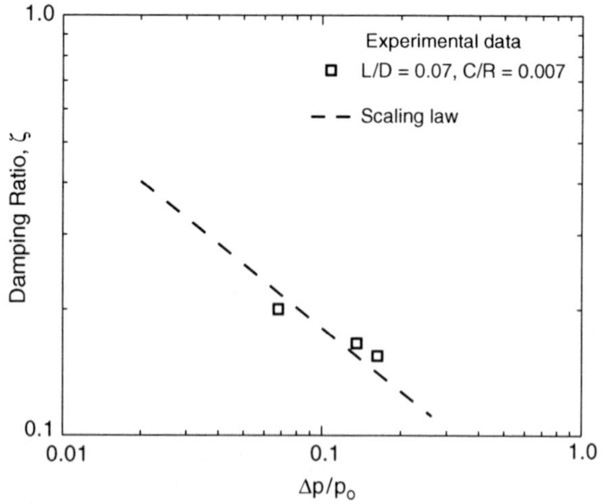

Fig. 6.11 Scaling of damping ratio with bearing pressure difference $\Delta p/p_0$ [31]

6.6 Whirl Instability in Ultra-Short Gas Journal Bearings

The pumping action of the rotor motion in the bearing journal generates a hydrodynamic force which is a cross-coupled force that can potentially contribute to whirl instability. The flow in the short bearing land is unsteady and inherently 3-D which complicates the description of the hydrodynamic force. However, at low rotor speeds appropriate simplifications can be made. This enables a first principles-based investigation of the mechanisms responsible for rotor whirl instability. The goal of this section is to first establish a simplified criterion for the onset of rotor whirl useful in the design and operation of ultra-short gas journal bearings and to detail the limitations of the theory. Then bearing anisotropy is introduced to extend the stable bearing operating range and to relieve fabrication tolerance requirements. Next, the theory is expanded to include unsteady flow effects at higher rotor speeds, and a generalized description in terms of a reduced frequency parameter is introduced.

6.6.1 A Simplified Criterion for Whirl Instability Onset

Using the analytical descriptions of the hydrodynamic and hydrostatic stiffness coefficients and the damping coefficients detailed in Section 6.5, a rotordynamic analysis can be carried out to determine the whirling motion of the rotor. Solving an eigenvalue problem, the whirl instability limit is determined by the rotor frequency $\Omega = \Omega_w$ at which the least stable system mode becomes unstable. A commonly used parameter that characterizes the limit of stable operation is the whirl ratio $\mathfrak{R} = \Omega_w/\Omega_N$, where Ω_N is the natural frequency.

The question arises how the whirl ratio $\mathfrak{R}$ scales with the bearing parameters and what sets the limit of whirl instability. The analytical model is used in a parametric study to analyze the effects of different bearing parameters on the onset of whirl instability. Figure 6.12 depicts the whirl ratio $\mathfrak{R}$ as a function of clearance to radius ratio C/R for a fixed L/D of 0.07, a fixed $\Delta p/p_0$ of 0.14 and for different levels of rotor unbalance a non-dimensionalized by the nominal bearing clearance $C_0 =$ 15 μm.

There is a range of C/R for which the whirl ratio is very large. For a small rotor unbalance of $a/C_0 = 0.07$ an infinite whirl ratio is obtained at a clearance to radius ratio C/R of about 0.01. Note that the trend of whirl ratio with C/R resembles singular behavior such that the clearance to radius ratio for infinite whirl ratio is denoted as "singular C/R." Further analysis of the system modes and mode shapes shows that for C/R less than the singular value (left branch of curves in Fig. 6.12) the whirl frequency is the natural frequency and the whirling motion is in the direction of rotor rotation (forward whirl). The opposite occurs for C/R greater than the singular value and the whirling motion is against the direction of rotor rotation (backward whirl).

In order to investigate the physical mechanism behind this behavior and to determine the bearing geometry that yields an infinite whirl ratio, the following first principles-based analysis is conducted. Assuming zero static eccentricity (no

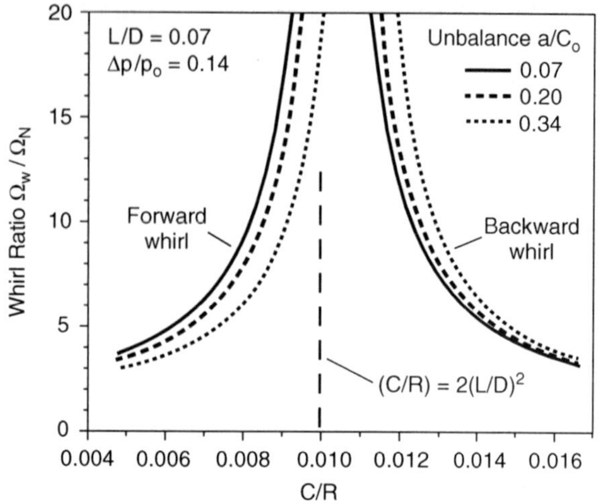

Fig. 6.12 Whirl ratio as a function of C/R for L/D = 0.07, $\Delta p/p_0 = 0.14$, and different levels of rotor unbalance a/C_0 [31]

unbalance or bearing side load) and introducing a small dynamic eccentricity $e = x + j\,y$, where x and y are small deflections of the rotor from the center of the journal in the X- and Y-directions, the whirling motion can be described using the journal bearing stiffness and damping coefficients:

$$m \cdot \frac{\mathrm{d}^2 e}{\mathrm{d}t^2} + \tilde{C}^{\mathrm{sys}} \cdot \frac{\mathrm{d}e}{\mathrm{d}t} + \left(K^{\mathrm{hs}} - jK^{\mathrm{hd}}\right) \cdot e = 0 \tag{6.29}$$

Note that the hydrodynamic stiffness is cross-coupled. Assuming a purely oscillatory solution of form $\mathrm{e}^{j\omega t}$ and setting the real and the imaginary parts of the resulting equation to zero yields

$$-\,\omega^2 + \frac{K^{\mathrm{hs}}}{m} = 0, \text{ and } \quad -\,K^{\mathrm{hd}} + \omega \cdot \tilde{C}^{\mathrm{sys}} = 0 \tag{6.30}$$

The first equation determines the frequency of oscillation which is the natural frequency $\omega = \Omega_N = \sqrt{(K^{hs}/m)}$. Using expressions for damping in terms of hydrodynamic stiffness due to pumping $\tilde{C}^{sys} = \pi\,\mu\,R\,L^3/C^3 = 2k^p/\Omega$ (first term in Eq. 6.22) and the hydrodynamic stiffness $K^{\mathrm{hd}} = k^p - k^v$, the whirl ratio can be written as

$$\Re = \frac{\Omega_\omega}{\Omega_N} = \frac{2}{1 - \frac{k^v}{k^p}}, \tag{6.31}$$

where $\Omega = \Omega_{\mathrm{w}}$ is the rotor speed at onset of whirl instability. We denote the second term in the denominator as the "whirl number" $\mathfrak{W}$ which is the ratio of hydrodynamic stiffness due to viscous drag (k^v) to hydrodynamic stiffness due to pumping (k^p). Using the expressions for k^v and k^p from Eq. (6.24), the whirl number yields in terms of bearing geometry

$$\mathfrak{W} = \frac{k^{\nu}}{k^{p}} = \frac{1}{2}\left(\frac{L}{D}\right)^{-2}\frac{C}{R} \tag{6.32}$$

Recalling that the cross-coupled hydrodynamic force due to pumping and the cross-coupled hydrodynamic force due to viscous drag act in *opposite* directions, the following rotor whirl criterion can be established. If the whirl number $\mathfrak{W}$ is less than one, or $C/R < 2(L/D)^2$, the hydrodynamic stiffness due to pumping k^p is larger than the hydrodynamic stiffness due to viscous drag k^{ν}. As a consequence the whirl ratio $\mathfrak{R} = 2 / (1 - \mathfrak{W})$ is positive and the net hydrodynamic force acts in the direction of rotor rotation such that the rotor tends to *forward whirl*. The opposite occurs for a whirl number greater than one, yielding a negative whirl ratio and inducing *backward whirl*. If the whirl number $\mathfrak{W}$ is equal to one, or $C/R = 2(L/D)^2$, the hydrodynamic stiffness due to pumping is in balance with the hydrodynamic stiffness due to viscous drag such that the system is *neutrally stable*. In this case, the whirl ratio $\mathfrak{R}$ is infinite.

Returning to Fig. 6.12 and assuming zero unbalance, the singular clearance to radius ratio is $C/R = 2(L/D)^2$ and yields 0.01 for an L/D of 0.07. This is marked by the vertical dashed line. Using the criterion of $\mathfrak{W} = 1$ for small unbalance, the conditions for an infinite whirl ratio is well predicted. The analysis also illustrates the key difference in the dynamic behavior between ultra-short gas journal bearings and large-scale, long ($L/D \sim 1$) gas journal bearings. Using Eq. (6.32), for long gas journal bearings with an L/D of typically 0.8 and comparable clearance to radius ratios (C/R ranging from 0.001 to 0.005) the whirl number $\mathfrak{W}$ is of order 0.001. This implies that the viscous drag effects are negligible compared to the pumping effects. With this the whirl ratio $\mathfrak{R}$ is estimated to be about 2. Larson and Richardson [15] present an experimental study of whirl instability of a large-scale pressurized gas bearing. An experimentally measured whirl ratio near 2 is reported for low-bearing pressure settings which agrees well with the estimate made here.

High-speed experiments were conducted on a large number of MEMS air turbines to investigate the whirl instability in ultra-short hydrostatic gas bearings [33]. More detail on the experimental setup can be found in Section 6.9. The damping ratio and natural frequency were measured and the whirl ratio at the highest speeds achieved was estimated. The test results in Fig. 6.13 demonstrate whirl ratios up to 40.

In summary, the first principles-based approach suggests that the viscous drag force is not negligible compared to the hydrodynamic pumping force in ultra-short gas bearings. Since the viscous drag force and the hydrodynamic pumping force act in opposite directions, the net destabilizing cross-coupled force is reduced and leads to relatively high whirl ratios. For long bearings with an L/D of order 1 the hydrodynamic pumping force is dominant compared to the viscous drag force such that the whirl ratio is relatively low. In other words, ultra-short hydrostatic gas journal bearings potentially yield very high whirl instability speeds compared to their long bearing counterparts. Furthermore, the simplified theory suggests that there exists an

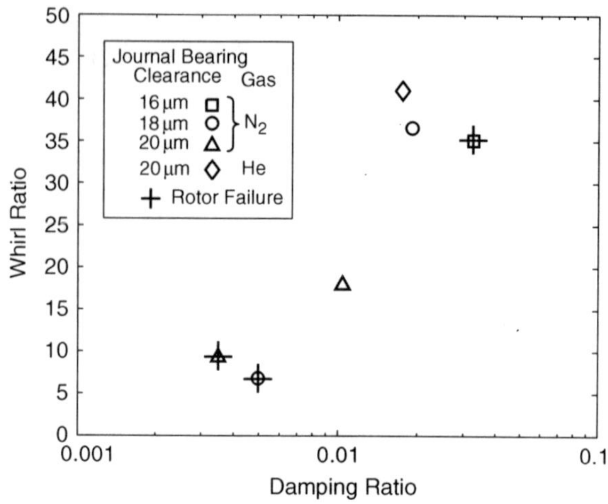

Fig. 6.13 Experimentally measured whirl ratio of ultra-short hydrostatic gas bearings as a function of damping ratio (data by [33])

optimum bearing geometry governed by $C/R = 2(L/D)^2$, which yields an infinitely high whirl instability speed.

6.6.2 *Unsteady Flow Effects on Gas-Bearing Dynamic Behavior*

In the previous analysis, it was assumed that the relatively low reduced frequency[7] during transcritical operation allows the decoupling of the axial flow due to hydrostatic effects from the circumferential flow due to hydrodynamic effects. It was argued that the whirling motion of the rotor drags the fluid around the circumference and sets up a fully developed Couette flow. Although this is valid at low rotational speeds and during journal bearing transcritical operation, near the design speed, the reduced frequency β is of order unity, and decoupling of the hydrostatic and hydrodynamic flow fields is no longer valid.

This section addresses the effects of unsteady flow in the bearing gap and it will be shown that, if the flow is not fully developed, the hydrostatic differential pressure and fluid properties strongly affect the dynamic behavior. This implies that, for design purposes, a higher fidelity computation of the bearing flow needs to be carried out. There is indication that in view of the short axial flow-through times governed by the hydrostatic differential pressure Δp, there is insufficient time for axial vorticity to diffuse radially outwards from the rotating surface. The circumferential flow thus does not fully develop which calls into question the validity of treating the circumferential and axial

[7] The characteristic axial flow-through time is much shorter than the characteristic flow-change time around the circumference such that the reduced frequency becomes $\beta = \Omega L/\overline{U} \sim 0.05$.

flow fields independently. The key implication is that the developing circumferential flow affects the magnitudes of the hydrodynamic forces acting on the rotor.

A two-pronged approach is adopted to investigate the effects of unsteady flow on the bearing dynamic behavior. The complicated flow field through the ultra-short journal bearing is first modeled using the impulsive starting of a Couette flow to gain insight on the essential physical mechanisms that govern the evolution of the circumferential flow. A reduced frequency parameter is introduced that determines the development of the circumferential flow field in the journal bearing. Then, detailed 3-D CFD calculations are performed to accurately quantify the destabilizing cross-coupled hydrodynamic stiffness components. A refined stability criterion is established and the criterion is used to delineate journal bearing operating conditions for stable high-speed operation.

6.6.2.1 Flow Field Development – Impulsively Started Couette Flow

Consider the impulsive startup of a Couette flow between two infinite parallel plates. The problem setup is illustrated in Fig. 6.14. The fluid is bounded by two rigid plates at $y = 0$ and $y = C$ and is initially at rest. The lower plate is suddenly brought to a steady velocity V in its own plane, with the upper plate held stationary.

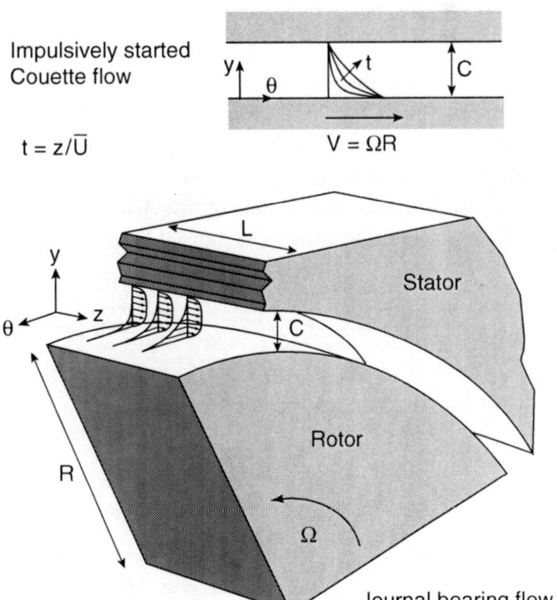

Fig. 6.14 Diffusion of vorticity in an impulsively started Couette flow (*top*) and across journal bearing gap in an ultra-short journal bearing (*bottom*) [38]

The unsteady velocity distribution is governed by the diffusion equation[8]

$$\frac{\partial u}{\partial t} = \nu \frac{\partial^2 u}{\partial y^2} \tag{6.33}$$

With boundary conditions

$$\begin{aligned} &u(0,t) = V \; u(C,t) = 0 \text{ for } t > 0, \\ &u(y,0) = 0 \text{ for } 0 < y \leq C, \end{aligned} \tag{6.34}$$

The above problem admits an analytical Fourier series solution given by

$$\frac{u\,(y,t)}{V} = \left(1 - \frac{y}{C}\right) - \frac{2}{\pi} \sum_{n=1}^{\infty} \frac{1}{n} \exp\left(-n^2\pi^2 \frac{\nu t}{C^2}\right) \sin \frac{n\pi y}{C}. \tag{6.35}$$

The flow is unsteady and the vorticity due to the moving wall takes a finite time to diffuse across the channel. The characteristic time is proportional to C^2/ν.

Both the impulsive startup of the Couette flow and the evolution of the circumferential flow through the journal bearing share a common concept: in order for the flows to develop, the vorticity put into the flow from the moving surface has to diffuse outwards through the flow field towards the stationary surface. The transient time of the impulsively started Couette flow is equivalent to the time it takes a fluid particle to reach an axial location z along the journal bearing such that

$$t = \frac{z}{\overline{U}} \tag{6.36}$$

where $\overline{U}$ is the mean axial velocity of the flow through the bearing. As shown schematically in Fig. 6.14, the circumferential velocity profile across the bearing gap at an axial location z then corresponds to the velocity profile of the impulsively started Couette flow at time instant t. Substituting the time-space transformation Eq. (6.36) into Eq. (6.35), and setting $V = \Omega R$, the circumferential velocity profile at an axial location z along the journal bearing becomes

$$\frac{u_\theta\,(y,z)}{\Omega R} = \left(1 - \frac{y}{C}\right) - \frac{2}{\pi} \sum_{n=1}^{\infty} \frac{1}{n} \exp\left(-n^2\pi^2 \frac{\nu z}{\overline{U}C^2}\right) \sin \frac{n\pi y}{C} \tag{6.37}$$

The above equation can be expressed non-dimensionally as

$$\frac{u_\theta}{\Omega R} = \psi\left(\frac{y}{C}, \frac{z}{L}, \frac{\nu L}{\overline{U}C^2}\right) \tag{6.38}$$

[8] Equation (6.33) has the form of the heat diffusion equation, the corresponding heat transfer problem being the sudden increase in temperature of a plate. The analogy is that vorticity spreads like heat.

By inspection, there are inherently two time scales that govern the evolution of the circumferential flow. The first time scale is the characteristic viscous diffusion time, τ_d, which scales with the kinematic viscosity ν and the journal bearing clearance C

$$\tau_d \sim \frac{C^2}{\nu} \tag{6.39}$$

The second time scale is the flow-through time of the hydrostatic axial flow, τ_f, scaling with the mean flow-through velocity $\overline{U}$ and bearing length L

$$\tau_f \sim \frac{L}{\overline{U}} \tag{6.40}$$

The dimensionless variable in Eq. (6.38),

$$\frac{\nu L}{\overline{U}C^2} = \frac{\tau_f}{\tau_d} \equiv \beta_{\mathrm{FD}} \tag{6.41}$$

can thus be interpreted as the ratio of the flow-through time to the characteristic viscous diffusion time. This ratio is denoted the reduced frequency β_{FD} [38]. For values of β_{FD} much greater than unity, the axial flow-through time is significantly longer than the viscous diffusion time. The axial vorticity from the moving wall has sufficient time to diffuse radially across the journal bearing clearance, and the circumferential flow can be assumed to be fully developed over a significant length of the journal bearing. On the contrary, if β_{FD} is much less than one, the viscous diffusion time is significantly longer than the axial flow-through time. The axial vorticity of the moving wall does not have sufficient time to diffuse radially outwards across the journal bearing clearance before the axial flow leaves the journal bearing. In this case, the circumferential flow is not fully developed and exhibits a significant amount of flow non-uniformity. Furthermore, two different journal bearings with different clearances C and lengths L are anticipated to have the same normalized circumferential flow profile if they have the same values of β_{FD}.

CFD calculations using the commercial solver FLUENT were conducted first for a radially centered rotor. The computations show that the axial velocity profile develops rapidly: for axial locations normalized by the journal bearing clearance $z/C > 2$, the axial velocity profile is independent of axial location. The circumferential velocity profile is shown in Fig. 6.15, for various axial locations z/L (now normalized by bearing length L). In contrast to the axial velocity profile, the flow in the circumferential direction is not fully developed at the exit of the journal bearing passage and exhibits a significant departure from a linear velocity profile over most of the bearing length.

Using Eq. (6.37), the analytical results for the circumferential velocity distribution are also plotted non-dimensionally at different journal bearing axial locations in Fig. 6.15. Close agreement exists between the analytical solution and the CFD calculations demonstrating similarity between the impulsive startup of a Couette flow and the 3-D flow development in the bearing gap.

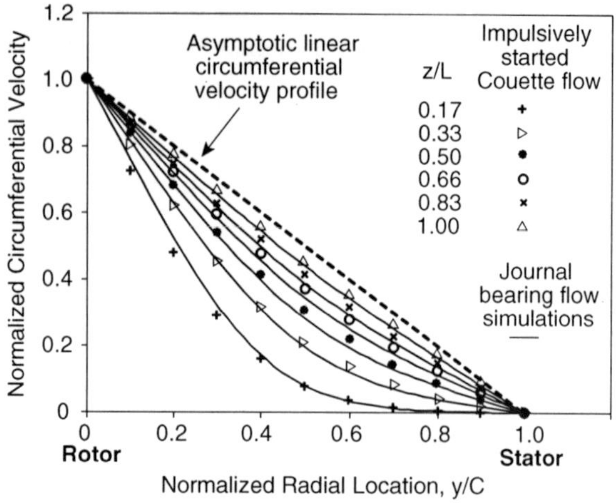

Fig. 6.15 Development of normalized circumferential velocity profile at various axial locations *z/L*: impulsively started Couette flow (*symbols*) and numerical simulation results (*lines*) [38]

6.6.2.2 Effects of Developing Circumferential Flow on Journal Bearing Whirl Stability

One of the shortcomings of the impulsively started Couette flow model is that inertia effects are not accounted for. If $\beta_{\mathrm{FD}} \leq O(1)$, inertial forces are no longer negligible such that classical hydrodynamic lubrication theory becomes invalid [2]. Since details of the kinematic flow field significantly influence the magnitudes of the hydrodynamic forces acting on the rotor, high-fidelity 3-D CFD calculations are necessary to determine the hydrodynamic stiffness for small radial eccentricities in the presence of a developing circumferential flow.

FLUENT calculations were performed on two rotor configurations with a bearing length over diameter ratio L/D of about 0.075, one with journal bearings of 2.1 mm radius and 330 μm length, and the other with rotor radius of 6 mm and journal bearing length of 900 μm. The first configuration considered a bearing clearance C ranging between 10 and 24 μm and a hydrostatic differential pressure Δp ranging between 1 and 8 psi. The second configuration was assessed for bearing clearances C ranging between 26 and 30 μm and pressure differentials of 1–5 psi. The flow domain within the journal bearing was represented using a structured grid with 331 grid points in the axial direction, 360 grid points in the circumferential direction and 21 grid points in the radial direction yielding a computational grid of over 2 million grid points. Dirichlet type boundary conditions were imposed for static pressure at the inlet and the exit of the journal bearing, and the flow was assumed to be incompressible[9] and laminar ($Re_c = \Omega RC/\nu < 500$). Note that the characteristic

[9] The axial Mach number M_z is of order 0.3. For the circumferential flow, numerical studies show that compressibility effects are insignificant for small radial eccentricities, as previously reported in [19].

viscous diffusion time $\tau_d = C^2/\nu$ does not depend on the axial vorticity magnitude diffusing across the journal bearing gap and is thus independent of rotor speed.

The rotor was subjected to a normalized radial perturbation (or radial eccentricity) of 0.1 and the viscous pumping and viscous drag components of the hydrodynamic forces acting on the spinning rotor were extracted from the converged CFD solution. The pressure field acting on the rotor was integrated to determine the viscous pumping component F^p, whereas the viscous shear stress field was integrated to yield the viscous drag component F^v. The cross-coupled hydrodynamic force F^{hd} can only arise from pressure forces and shear stresses due to the whirling motion of the rotor. The splitting of F^{hd} into F^p and F^v (Eqs. 6.24 and 6.26) is thus valid, no matter whether the circumferential flow is fully developed or not. The hydrodynamic stiffness components due to viscous pumping k^p and viscous drag k^v can thus be evaluated and the whirl-ratio $\Re$ can be obtained from Eq. (6.31).

Figure 6.16 (a) plots the computed whirl-ratio as a function of journal bearing clearance. The different points associated with each clearance correspond to different values of journal bearing Δp. The question arises whether it is possible to collapse the data by invoking first principles-based scaling arguments. In this regard, the impulsive startup of a Couette flow offers insight pertaining to the appropriate methodology for scaling the available CFD data. The key non-dimensional parameter that governs the evolution of the circumferential flow is the reduced frequency $\beta_{FD} = \nu L/(\overline{U}C^2)$, the ratio of the flow-through time of the axial hydrostatic flow to the characteristic viscous diffusion time. Note that the mean axial velocity through the journal bearing gap $\overline{U}$ depends on the hydrostatic differential pressure Δp and the bearing geometry, and can be readily determined from an analytical hydrostatic flow model.

For the same set of data, whirl-ratio $\Re$ is plotted as a function of β_{FD} in Fig. 6.16(b). The reduced frequency β_{FD} collapses the data for various combinations of journal bearing clearance and hydrostatic differential pressure Δp. Note also that singular behavior is observed at a critical value of $\beta^*_{FD} \approx 0.43$. This suggests the following criterion for whirl stability: if the axial hydrostatic flow-through time is approximately half of the characteristic viscous diffusion time or, if

$$\left(\frac{\nu L}{\overline{U}C^2}\right)^* \approx 0.43 \tag{6.42}$$

the whirl-ratio exhibits singular behavior and the value becomes an order of magnitude larger than in long bearings, where $\Re \approx 2$. Using the impulsively started Couette flow solution in Eq. (6.35), the criterion can be stated in an alternate form: since the characteristic time required for fully developed flow (within 1% of a linear profile) is $t/\tau_d = t/(C^2/\nu) \approx 0.5$, singular whirl-ratio behavior occurs when the flow-through time is approximately equal to the characteristic time required for a fully developed circumferential flow to be established.

To assess the effect of flow development on whirl instability, the computed rotor speed corrected cross-coupled hydrodynamic stiffness components k^p and $k^v$are compared to the analytical descriptions for fully developed flow (from Section

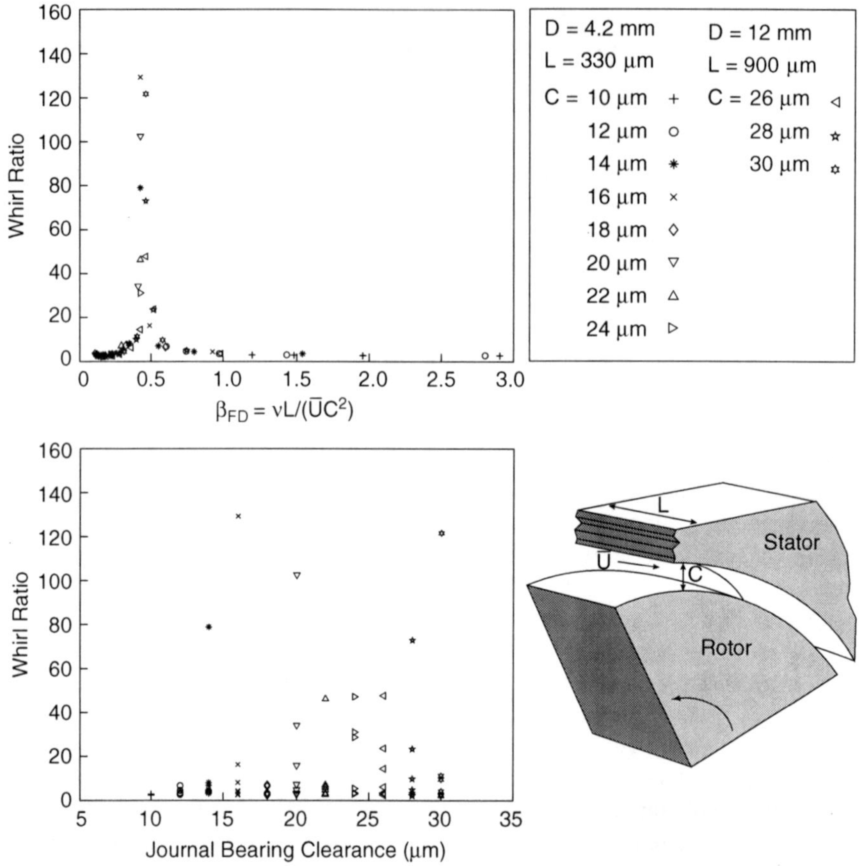

Fig. 6.16 Whirl-ratio as a function of: journal bearing clearance (*top*) and reduced frequency (*bottom*). Note singular behavior in whirl-ratio at a reduced frequency $\beta^*_{FD} \approx 0.43$ [38]

6.5.2). Figure 6.17 depicts k^p and $-k^v$ plotted against the journal bearing clearance for a 2.1 mm radius rotor. Also plotted is the net rotor speed corrected cross-coupled stiffness $k^p - k^v$ (thick lines), which indicates the direction of whirl tendency. A vanishing value of cross-coupled stiffness determines the conditions for singular behavior with an infinite whirl ratio. The journal bearing length L was kept constant at 330 μm and a hydrostatic differential pressure, Δp, of 1 psi was applied across the journal bearing. Two sets of data are plotted: one set corresponds to k^p and $-k^v$ obtained from the analytical journal bearing model with fully developed flow and the second data set is obtained from the CFD calculations for developing flow discussed above. Both cases show similar trends over most of the clearance range; however, the values for the cross-coupled stiffness are quantitatively different between the CFD results and the fully developed flow model, especially at larger bearing clearances.

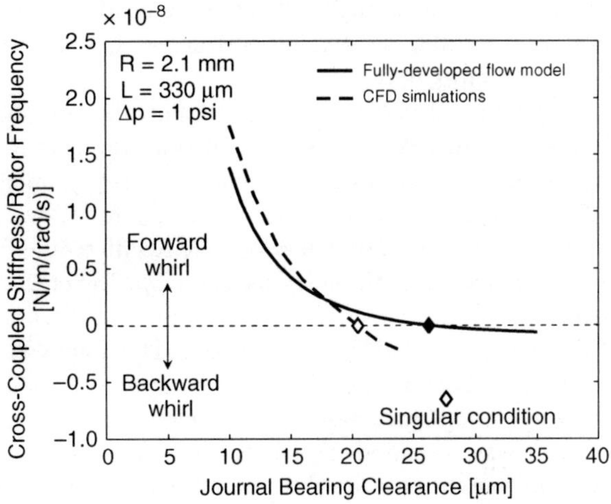

Fig. 6.17 Comparison of fully developed flow model and numerical simulation of developing flow – rotor speed corrected cross-coupled hydrodynamic stiffness as a function of journal bearing clearance [38]

The key difference between the two cases is that the net cross-coupled stiffness, $k^p - k^v$, vanishes at a larger bearing clearance for the fully developed flow case: at a journal bearing clearance of 26 μm, the magnitude of k^p matches the magnitude of k^v (marked by the solid diamond). This corresponds to a whirl number $\mathfrak{W}$ of unity, an infinite whirl-ratio $\mathfrak{R}$ and an infinite whirl instability speed. Note that the journal bearing clearance of 26 μm at this singularity point can also be obtained from the simple criterion $C/R = 2(L/D)^2$ in Eq. (6.32). In contrast to the fully developed flow model, the CFD results suggest equal magnitudes in k^p and k^v at a significantly smaller journal bearing clearance of 20 μm (marked by the open diamond). The implication is that the developing nature and the details of the circumferential flow field have a significant impact on the hydrodynamic forces. One direct consequence is the shift in location of the critical journal bearing clearance which yields an infinite instability speed.

In conclusion, the CFD results for developing journal bearing flow are in qualitative agreement with the trends obtained by the analytical model for fully developed flow but show differences in the quantitative magnitudes of the rotor speed corrected cross-coupled hydrodynamic stiffness k^p and k^v. The fully developed analytical flow model overestimates the critical journal bearing clearance for infinite whirl ratio. For design purposes, high-fidelity fluid dynamic simulations are necessary to accurately account for the effects of a developing circumferential flow field on k^p and k^v.

6.6.2.3 Effect of Hydrostatic Differential Pressure on Whirl Instability

Recall that the simplified criterion for singular behavior in whirl instability limit $C/R = 2(L/D)^2$ only depends on bearing geometry. This suggested that a certain combination of L/D and C/R can yield an infinite whirl ratio independent of the fluid dynamic characteristics of the bearing flow. Given that a developing journal bearing flow can strongly influence the bearing dynamic behavior as seen above, it is expected that the bearing pressure setting and bearing fluid can influence the whirl instability limit.

In the fully developed flow model, the diffusion of axial vorticity from the rotating wall of the rotor across the journal bearing gap is neglected. The rate at which axial vorticity diffuses towards the stationary journal bearing sidewall is in turn related to the flow-through time of the axial flow governed by the hydrostatic differential pressure Δp. For a developing circumferential flow, the critical journal bearing clearance therefore must depend on Δp. The preceding section showed that for a developing journal bearing flow an infinite whirl ratio occurs at the critical reduced frequency $\beta_{FD}^* = \nu L/(\overline{U}C^2)^* \approx 0.43$. This implies that, for a given bearing fluid and journal bearing length L, there is a direct relationship between the mean axial velocity through the journal bearing gap $\overline{U}$ and the journal bearing clearance C. Put another way, variations in the critical journal bearing clearance C^* are related to variations in the critical value of the axial velocity $\overline{U}^*$ according to

$$\frac{d\overline{U}^*}{\overline{U}^*} = -2\frac{dC^*}{C^*}. \tag{6.43}$$

Approximating the total hydrostatic pressure drop across the journal bearing to be the inlet loss arising from the Lomakin seal effect, one can write

$$\frac{d\,(\Delta p)^*}{p_0} = \gamma\omega_{\text{inlet}} M_z dM_z, \tag{6.44}$$

where γ is the ratio of the specific heats, ω_{inlet} is the journal bearing inlet stagnation pressure loss coefficient, M_z is the axial Mach number and p_0 is the ambient pressure. Combining Eq. (6.43) and Eq. (6.44) one finds

$$\frac{d\,(\Delta p)^*}{p_0} = -2\gamma\omega_{\text{inlet}} M_z^2 \frac{dC^*}{C^*}. \tag{6.45}$$

Thus an increase in hydrostatic bearing pressure setting Δp^* yields a decrease in critical journal bearing clearance C^* at which infinite whirl ratio occurs.

To assess the above hypothesis, a number of FLUENT simulations were conducted at various journal bearing pressure settings and bearing clearances for a given rotor radius and bearing length. Figure 6.18 depicts the CFD simulation results for the net rotor speed corrected cross-coupled hydrodynamic stiffness $k^p - k^v$ as a function of journal bearing clearance for various hydrostatic differential pressure settings. The results suggest that, for each journal bearing differential pressure, Δp, there exists a critical journal bearing clearance for which singular whirl-ratio behavior occurs. An increase in Δp reduces the critical bearing clearance. This trend is in agreement with the relationship captured in Eq. (6.45). Note that during typical journal bearing operation in a micro-turbomachinery application the hydrostatic journal bearing differential Δp increases from 1 to 5 psi from low speed to high speed (see

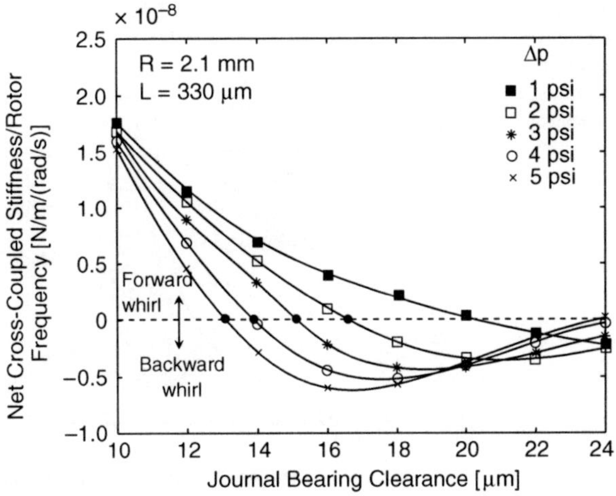

Fig. 6.18 Rotor speed corrected cross-coupled hydrodynamic stiffness as function of journal bearing clearance and hydrostatic differential pressure Δp [38]

Section 6.9.3 for details). For the example above, this yields a 35% change in critical bearing clearance from approximately 20 μm for a Δp of 1 psi, to approximately 13 μm at a Δp of 5 psi.

In summary, a key implication of the results presented in this section is that singular whirl-ratio behavior does not depend solely on the journal bearing geometry. For any given journal bearing geometry, the appropriate selection of hydrostatic bearing pressure difference Δp can yield singular whirl-ratio behavior and hence stable high-speed operation. High-fidelity numerical assessment of the developing journal bearing flow is necessary in the design of high-speed ultra-short gas bearings.

6.6.3 Journal Bearing Stiffness Anisotropy for Enhanced Whirl Stability

As seen in the preceding sections, the ultra-short high-speed bearings exhibit whirl instability limits and a dynamic behavior much different from conventional hydrostatic gas bearings. The design space for stable high-speed operation is confined to a narrow region and involves singular behavior. Apart from dynamic instabilities, the main challenges and limitations encountered with micro-scale gas bearings are the manufacturing constraints and fabrication tolerances and uncertainties using deep-reactive-ion-etching (DRIE). For example, the bearing clearance is typically a 350 μm deep and 20 μm wide circular trench which is cut at a radius of 2.1 mm. For such a bearing etch, the fabrication uncertainty can be as much as 1–2 μm in bearing clearance and 10–20 μm in bearing length. Given the high-speed bearing design requirements depicted in Fig. 6.19, the bearing clearance must be 21.5 μm in order to achieve a stable design speed of 2.4 million rpm. The misalignment of

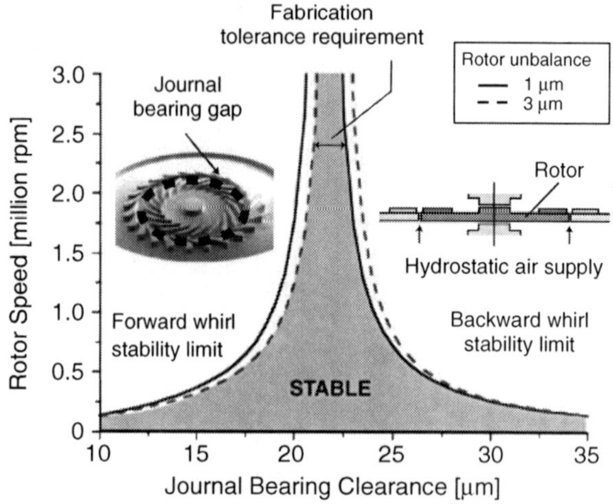

Fig. 6.19 Singular behavior of whirl instability limit imposes stringent fabrication tolerance requirements for high-speed stable bearing operation [18]

the geometric center of the ring of turbomachinery blades and the center of the journal bearing as well as the etch non-uniformity in forming the side of the disc at the base of the blades yield a rotor unbalance of typically 2 μm (expressed as displacement of the rotor's center of gravity from its rotational centerline). Returning to Fig. 6.19, the allowable tolerance on the bearing clearance then becomes $\pm$ 0.5 μm. This tolerance on the bearing clearance design specification is clearly below the fabrication capabilities and limits the maximum achievable rotor speed to roughly half the design speed.

6.6.3.1 Breaking Symmetry – Stiffness Anisotropy

A possible modification to plain cylindrical bearings to improve their stability range is to introduce multiple lobes or pads. In conventional, large-scale gas bearings this can improve the dynamic performance by creating a pre-load (the dimensional difference in clearance between the lobe and the bearing) which provides direct stiffness at the bearing centered operating condition. For higher speeds and light loads, multiple-lobe bearings are known to offer a slightly more stable behavior than plain cylindrical journal bearings as reported by Fuller [10] and Garner et al. [11]. Lobed journal bearings were also investigated as a possible means to improve micro-bearing dynamic behavior. Piekos [24] reports numerical simulations of multi-lobed hydrodynamic gas journal bearings and concludes that for the very short micro bearings a lobed configuration deteriorates the stability. More recently Kim et al. [14] discuss the modeling of micro-gas journal bearings with a stepped casing. The authors conclude that plain gas journal bearings are more stable than the stepped gas journal bearings and report that implementing axial grooves in addition to the steps, or reducing the number of steps, improves the bearing stability.

Smith [30] suggests that when there is considerable lack of symmetry in the bearing support stiffness, the transition speed from stable to unstable operation is very high compared with critical speeds even if there is no stationary damping. The asymmetry in support stiffness, commonly referred to as anisotropic support stiffness, was further investigated by Gunter and Trumper [12] to mitigate hysteretic whirl. More recently Ehrich [7] analyzed the stability limit of a Jeffcott rotor model for a representative range of values of support stiffness anisotropy and of system damping ratio. Ehrich showed that anisotropy is effective in suppression of instability for lightly damped systems but less so for systems with higher damping.

The modification to the micro-gas journal bearing configuration discussed here is based on the idea of bearing stiffness anisotropy. The concept is to break symmetry in hydrostatic direct stiffness without much affecting the hydrodynamic cross-coupled stiffness. This is in contrast to the lobed bearings reported by Piekos [24] where both direct and cross-coupled stiffness are affected by the non-uniform geometry such that the stability limit is deteriorated. In essence, the direct consequence is the elimination of the singular behavior of the whirl instability limit through asymmetry in hydrostatic stiffness. This extends the geometric design space for stable operation and yields a higher whirl instability limit and a major relief in the fabrication tolerance requirements, enabling an increased bearing top-speed. The impact of bearing stiffness anisotropy on the stable design space is shown in Fig. 6.20 in comparison to a micro-gas bearing with symmetric hydrostatic stiffness. The latter is from now on called isotropic bearing stiffness, implying that the bearing stiffness is independent of the direction of rotor displacement.

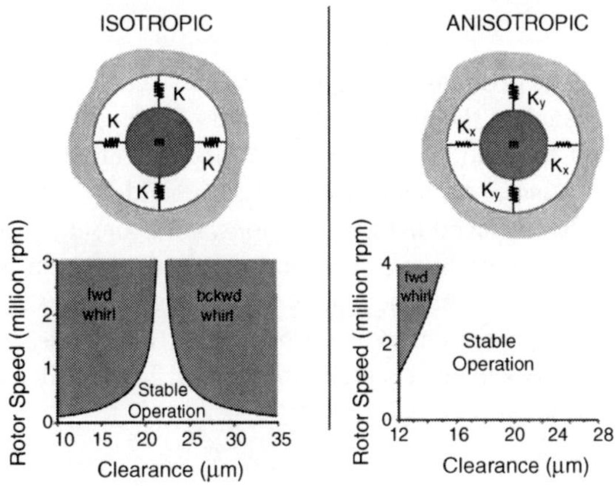

Fig. 6.20 Elimination of singular behavior of whirl instability limit and extension of stable design space through hydrostatic stiffness anisotropy [18]

6.6.3.2 Journal Bearing Architecture with Stiffness Anisotropy

The design intent is to break the symmetry in hydrostatic stiffness without a substantial increase in the hydrodynamic cross-coupled stiffness that gives rise to whirl instability. Since the bearing feed pressure and hence the flow rate govern the hydrostatic stiffness[10], anisotropy can be introduced by reducing or increasing the amount of flow through a circumferential sector of the bearing gap. The simplest configuration is to block the flow through two circumferential sectors as depicted in Fig. 6.21.

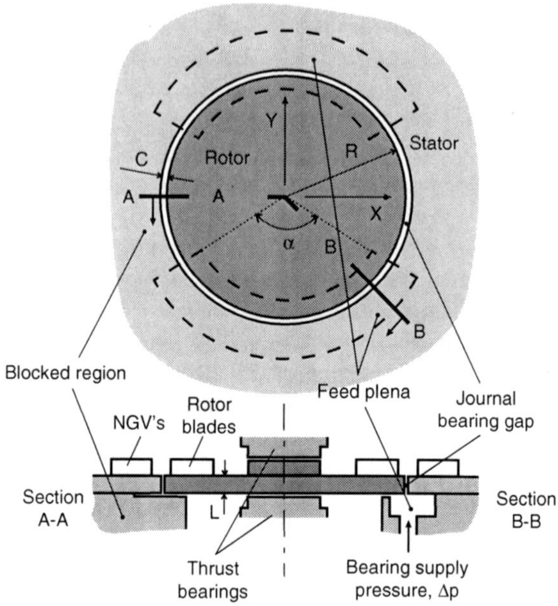

Fig. 6.21 Journal bearing architecture with hydrostatic stiffness anisotropy. Dimensions are not to scale [19]

The two remaining circumferential sectors are fed with pressurized air from plena located on the aft side of the rotor. Since the axial clearance gap between the rotor and the static structure in the blocked region is small and yields very high flow resistance, the hydrostatic pressure build-up is limited and the hydrostatic stiffness is thus reduced in these locations. As a consequence, the hydrostatic stiffness in the X and Y directions are different and anisotropy in journal bearing stiffness is introduced. Note that the journal and rotor geometries are still circular and axisymmetric, and that asymmetry is introduced only in the hydrostatic feed air system (without the blocked feed regions the journal bearing is axisymmetric and the hydrostatic stiffness is isotropic).

6.6.3.3 Anisotropic Gas-Bearing Theory

Using the modeling framework described in Section 6.5, the dynamic behavior and whirl instability limit of micro-gas journal bearings with stiffness anisotropy can be analyzed. The analytical model consists of two parts, a fluid dynamic model to describe the bearing stiffness and damping coefficients and a rotordynamic model

---

[10] The scaling law for hydrostatic stiffness is $K^{hs}/(p_0 C) \sim (L/D) \cdot (C/R)^{-2} \cdot \Delta p/p_0$.

of a simple micro-turbomachinery rotor supported by ultra-short gas bearings. The fluid dynamic model in Section 6.5 is modified to account for the flow blockage in the blocked sectors which affects the pressure field acting on the rotor. Integration of the pressure field and viscous stress field on the rotor yields the hydrostatic force F^{hs}, the hydrodynamic force F^{hd}, and the damping force F^{dp} which, in a stationary coordinate system, govern the motion of the rotor according to

$$\begin{aligned} m\ddot{x} &= F_x^{hs} + F_x^{hd} + F_x^{dp} + m\Omega^2\, a\cos(\Omega t) \\ m\ddot{y} &= F_y^{hs} + F_y^{hd} + F_y^{dp} + m\Omega^2\, a\sin(\Omega t). \end{aligned} \tag{6.46}$$

The excursions of the geometric center of the rotor in the X and the Y direction are denoted by x and y, respectively, and the last term denotes the harmonic forcing at rotor frequency Ω due to rotor unbalance a. For low unbalance, the motion can be viewed linear and the above equations become

$$\begin{aligned} m\ddot{x} + K_{xx}x + K_{xy}y + C_x\dot{x} &= O(x^2) + O(y^2) \\ m\ddot{y} + K_{yy}y + K_{yx}x + C_y\dot{y} &= O(x^2) + O(y^2), \end{aligned} \tag{6.47}$$

where K_{xx}, K_{yy}, C_x and C_y are the direct-coupled hydrostatic stiffness and damping coefficients in the X and Y direction, respectively. K_{xy} and K_{yx} denote the cross-coupled hydrodynamic stiffness coefficients. Note that from Section 6.5.5 for an *isotropic* bearing configuration the bearing coefficients become in the limit of small eccentricity:

$$K_{xx} = K_{yy} = K^{hs} \approx 8\frac{RL}{C}\cdot\Delta p \tag{6.48}$$

$$K_{xy} = -K_{yx} = K^{hd} \approx \pi\frac{\mu RL}{C}\left(\frac{L^2}{2C^2} - \frac{R}{C}\right)\cdot\Omega \tag{6.49}$$

$$C_x = C_y = C^{sys} \approx \pi\frac{\mu RL^3}{C^3}. \tag{6.50}$$

Returning to the anisotropic bearing configuration, the dynamics of the linearized equations of motion Eq. (6.47) can be solved in the frequency domain using the Laplace transform $\mathcal{L}$: $z(t) \rightarrow Z(s)$. The characteristic equation for the rotordynamic behavior of the anisotropic journal bearing then yields

$$\begin{aligned} &m^2s^4 + m(C_x + C_y)s^3 + m((K_{xx} + K_{yy}) + C_xC_y)s^2 \\ &+ (K_{xx}C_y + K_{yy}C_x)s + K_{xx}K_{yy} - K_{xy}K_{yx} = 0, \end{aligned} \tag{6.51}$$

where $s = j\Omega_N$ is the Laplace variable and Ω_N is the natural frequency of the oscillation. The hydrodynamic stiffness is proportional to the rotor frequency Ω such that one can write $K^{hd} = k^{hd}\cdot\Omega$ and similarly $K_{xy} = k_{xy}\cdot\Omega$ and so forth. Setting the real and the imaginary parts of the above equation to zero yields

$$\Omega_N^2 = \frac{K_{xx}C_y + K_{yy}C_x}{m(C_x + C_y)}, \tag{6.52}$$

and, using the rotor frequency at whirl instability onset $\Omega = \Omega_w$,

$$\Omega_w^2 = -\frac{C_x C_y}{k_{xy}k_{yx}}\left[\Omega_N^2 + \frac{(K_{xx} - K_{yy})^2}{(C_x + C_y)^2}\right]. \tag{6.53}$$

The first equation determines the natural frequency Ω_N of the rotor system and the second equation defines the whirl instability limit Ω_w. Note that generally k_{xy} and k_{yx} are of opposite signs[11] such the right-hand side of Eq. (6.53) is positive and Ω_w has a real root.

Similar to the case of isotropic bearing stiffness, as discussed in Section 6.6.1, the whirl instability limit depends on the ratio of the damping coefficients to the cross-coupled hydrodynamic stiffness. In fact, in the isotropic limit (Eqs. 6.48, 6.49, and 6.50), the above equations reduce to the classic whirl ratio result for isotropic bearing configurations

$$\mathfrak{R} = \frac{\Omega_w}{\Omega_N} = \frac{C^{sys}}{K^{hd}}. \tag{6.54}$$

The key difference between the isotropic and anisotropic bearing cases is the additional term on the right-hand side of Eq. (6.53), $(K_{xx} - K_{yy})^2/(C_x + C_y)^2$, which depends on the difference or asymmetry of the direct-coupled hydrostatic stiffness in the two orthogonal directions X and Y. This term carries the dynamic effect of anisotropy and constitutes the main mechanism that enhances the stability margin and increases bearing top speed. For the micro-gas bearing systems considered, the magnitude of this term is generally two orders of magnitude larger than the first term related to the isotropic bearing case.

To investigate the mechanisms underlying the improved stability margin due to hydrostatic stiffness anisotropy, the following simplifications can be made. Since the second term on the right-hand side of Eq. (6.53) is much larger than the first, the latter can be neglected. Furthermore, the asymmetric bearing system can be simplified by considering anisotropy only in the hydrostatic stiffness while keeping the hydrodynamic stiffness and damping coefficients isotropic. Equation (6.53) then reduces to

$$\Omega_w = \frac{|K_{xx} - K_{yy}|}{|2\,k^{hd}|}. \tag{6.55}$$

This illustrates that the bearing design should encompass as large a difference in hydrostatic stiffness between the two orthogonal directions $|K_{xx} - K_{yy}|$ as possible

[11] It can be shown that if k_{xy} and k_{yx} were of the same sign, the resulting root occurs at even higher frequency and shows a different dynamic behavior that does not constitute whirl instability.

while keeping the cross-coupled hydrodynamic stiffness coefficient k^{hd} as low as possible. This can be achieved by varying the hydrostatic feed air over parts of the circumference as outlined in Fig. 6.21.

The physical mechanism behind the enhanced stability margin and increased rotor speed at whirl instability onset is demonstrated in Fig. 6.22 which depicts the hydrostatic and hydrodynamic forces acting on the rotor for an anisotropic bearing configuration. Due to anisotropy in hydrostatic stiffness, $K_{xx} \neq K_{yy}$ as shown above, the hydrostatic force F^{hs} has a non-zero component in the tangential direction θ. The hydrodynamic force $F^{hd} = k^{hd}\Omega\varepsilon$ acts in the direction perpendicular to the rotor deflection and the total destabilizing force in the θ direction then becomes

$$F_\theta = \left(k^{hd} \cdot \Omega - \frac{K_{xx} - K_{yy}}{2} \sin(2\theta)\right) \varepsilon. \tag{6.56}$$

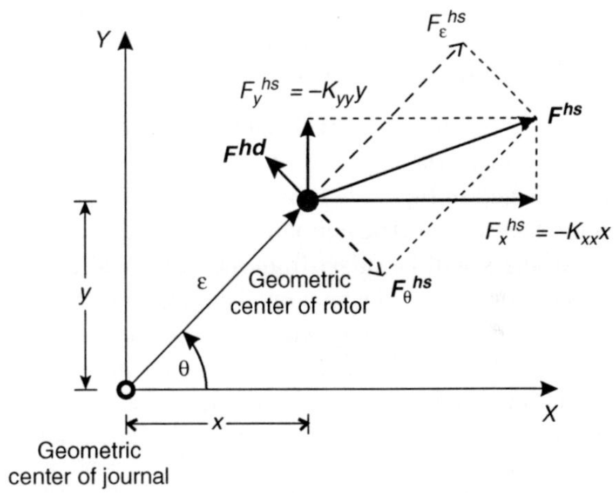

Fig. 6.22 Hydrostatic and hydrodynamic bearing forces for a bearing configuration with anisotropy in hydrostatic stiffness [18]

Non-dimensionalization of the above by the hydrodynamic force allows expressing the total destabilizing force as a function of whirl ratio and natural frequency

$$\begin{aligned}\frac{F_\theta}{F^{hd}} &= 1 - \frac{K_{xx} - K_{yy}}{2\,k^{hd}\Omega} \cdot \sin(2\theta) \\ &= -\frac{\Re}{\left(\frac{\Omega}{\Omega_N}\right)} \cdot \sin(2\theta).\end{aligned} \tag{6.57}$$

As seen in the preceding sections, the whirl ratio $\Re$ is much larger than one for ultra-short hydrostatic micro-gas journal bearings. For anisotropic bearing configurations, the whirl ratio becomes even larger (by one to two orders of magnitude) such that at the rotational speeds of interest (millions of rpm) the circumferential variation in the total destabilizing force becomes sinusoidal. The net destabilizing force over the full circumference then essentially vanishes, and as a consequence the whirl instability limit is shifted to very high rotational speeds. Put another way, introducing an asymmetry in hydrostatic stiffness yields a sinusoidally varying force component in the tangential direction, which, for ultra-short hydrostatic gas journal bearings, is much larger in magnitude than the cross-coupled hydrodynamic force. On a net basis, the whirl inducing tangential force component vanishes and the stable operating range is extended to very high rotational speeds.

For rotor speeds near the limit of whirl instability $\Omega \approx \Omega_w$, the magnitude of the sinusoidal variation in tangential force is of order one and the net whirl inducing force becomes positive. This suggests that, for anisotropic bearing configurations at whirl instability onset, the rotor tends to forward whirl only. This is in contrast to isotropic bearing configurations where both forward and backward whirl tendency is observed as depicted in Fig. 6.20. In summary, although cross-coupled hydrodynamic effects always induce destabilizing forces (a criterion to minimize this force was derived in Section 6.6.1), the whirl inducing forces can be balanced by asymmetry or anisotropy in direct-coupled hydrostatic stiffness.

6.6.3.4 Numerical and Experimental Assessment of Hydrostatic Stiffness Anisotropy on Bearing Dynamic Behavior

The analysis above assumed that the rotor unbalance is negligible and allowed a linearized treatment of the equations of motion around the equilibrium point x=y=0 as indicated in Eq. (6.47). Generally, there is a certain amount of unbalance of the rotor due to geometric non-uniformities and etch depth variations such that the center of mass will be offset from the geometric center of the rotor. At supercritical speeds, the rotor will spin around the center of mass, and as a consequence the fluid forces in the journal gap will vary with time as the rotor precesses in the journal. In addition, if the journal bearing supply pressures of the two feed plena are different, the rotor will experience a side-load and the journal bearing clearance will decrease on one side introducing non-linear behavior of the bearing fluid forces at larger eccentricities.

Since both rotor unbalance and rotor side-load introduce nonlinear effects and thus affect the whirl instability limit, the full non-linear set Eq. (6.46) must be solved. A numerical solution scheme was set up to directly solve the non-linear unsteady equations of motion. As such, for a selected rotational speed and bearing pressure supply and a fixed level of rotor unbalance and side-load, the non-linear hydrostatic, hydrodynamic, and damping forces are calculated for every time step according to the rotor position and velocity. The time trend of the orbit of the center of the rotor is then analyzed and permits the determination of whether the system is dynamically stable or unstable at the given rotor speed and selected bearing supply pressure.

Figure 6.23 depicts the whirl ratio $\Re$ as a function of clearance to radius ratio C/R for a fixed L/D= 0.07, a fixed bearing supply pressure $\Delta p/p_0 = 0.34$, and for different levels of rotor unbalance a marked by the solid, dashed, and dotted lines. The rotor unbalance is non-dimensionalized by the nominal bearing clearance $C_0 =$ 13 μm and calculations are conducted for both an anisotropic ($K_{xx} \neq K_{yy}$) and an

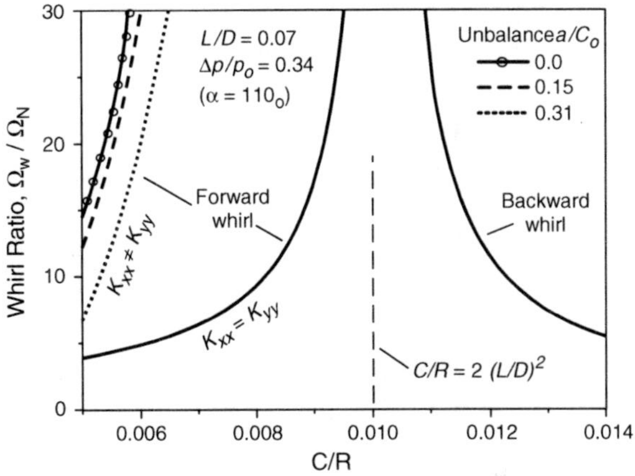

Fig. 6.23 Whirl ratio for isotropic and anisotropic bearing configurations as a function of C/R and different levels of unbalance a/C_0. Anisotropic case: numerical solution result (*solid line*) compared to simple analytical solution (*circles*) [18]

isotropic ($K_{xx} = K_{yy}$) bearing configuration. In the anisotropic case, the bearing pressure in the two supply plena, each covering 110° of the circumference, are at all times kept the same to avoid rotor side-load. The results indicate that, even with significant amounts of rotor unbalance, the whirl ratios are much higher compared to the isotropic case. Furthermore, the singular behavior in whirl instability limit[12] vanishes in the anisotropic case (forward whirl tendency branch only) such that the journal bearing top speed is increased and the stringent tolerance requirements are relieved.

In addition to the numerical simulations, the simplified, linearized analysis for bearing anisotropy is carried out for the same bearing conditions assuming zero rotor unbalance. The analytical results for the whirl instability limit are plotted as circles in Fig. 6.23. The results show that the analytical solution compares well with the numerical simulation for zero rotor unbalance (solid line) and thus validates the simplified approach for low rotor unbalance.

Furthermore, the simulation results suggest that the optimum circumferential extent of the bearing feed plena for a maximum extension of stable operating range is in the vicinity of $\alpha = 110°$. Figure 6.24 compares the isotropic bearing performance to the one of an optimized anisotropic bearing design. Both cases have the same geometric bearing characteristics $R = 2.1$ mm, $C = 13$ μm, and $L = 300$ μm. The effect of hydrostatic stiffness anisotropy greatly enhances the stable operating range: i.e., at a bearing supply pressure of $\Delta p = 5$ psig the bearing top speed of 500,000 rpm in the isotropic case is increased to a maximum stable rotor speed of 2.4 million rpm using the anisotropic bearing configuration sketched in Fig. 6.21.

Experiments in a micro-turbine test rotor supported by hydrostatic gas-journal bearings with stiffness anisotropy were carried out. Details of the test setup are

[12] The vertical dashed line indicates the simple whirl instability criterion for isotropic bearing configurations, $C/R = 2(L/D)^2$, established in Section 6.6.1.

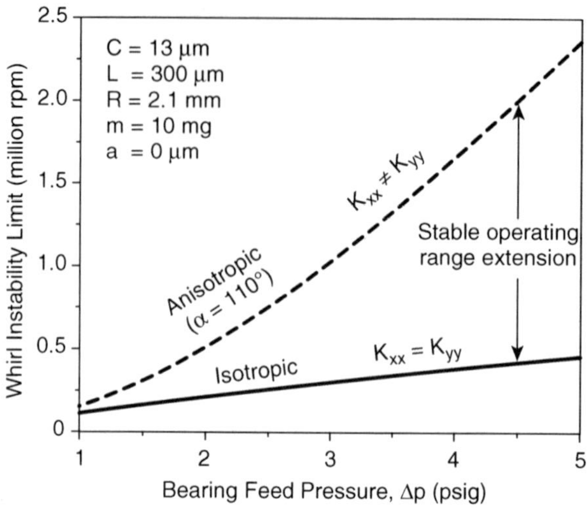

Fig. 6.24 Numerical simulation of stable operating range extension using bearing stiffness anisotropy [18]

discussed in Section 6.9.1. It is important to mention that it is difficult and not practical to experimentally determine the rotor speed at whirl instability onset for the following reasons: (1) a failure at high speed destroys the device and the experiment cannot be repeated on the same device, (2) the number of microfabricated devices is limited and can involve significant variations between different devices, and (3) a device can fail due to other phenomena which are not related to whirl instability (i.e., thrust-bearing resonance, thrust balance issues, and effects of conical modes – these effects are investigated in Section 6.7). Thus, an alternate approach is suggested to validate the theory of hydrostatic stiffness anisotropy and its impact on whirl instability limit. Since the micro-bearing design with stiffness anisotropy yields different stiffness in the two orthogonal directions, the rotordynamic system will exhibit two distinct natural frequencies associated with the two orthogonal directions. The natural frequencies of the micro-device can be measured using fiber optic sensors and pressure transducers and can be determined using system identification schemes as discussed in Section 6.9.1.

Figure 6.25 compares the experimentally determined natural frequencies of the micro-turbine test rotor with the model predictions using anisotropic gas-bearing theory. The experimentally measured first and second natural frequencies of the micro-device are plotted as open circles and solid squares. The model prediction of the natural frequencies for this device is marked as the dashed and the solid lines, respectively. The results show good agreement between the experimentally measured natural frequencies and the predictions validating the implementation of bearing anisotropy. Successful high-speed tests at rotor frequencies exceeding the estimated isotropic whirl instability limit and design implications are discussed in Sections 6.9.3 and 6.10, respectively.

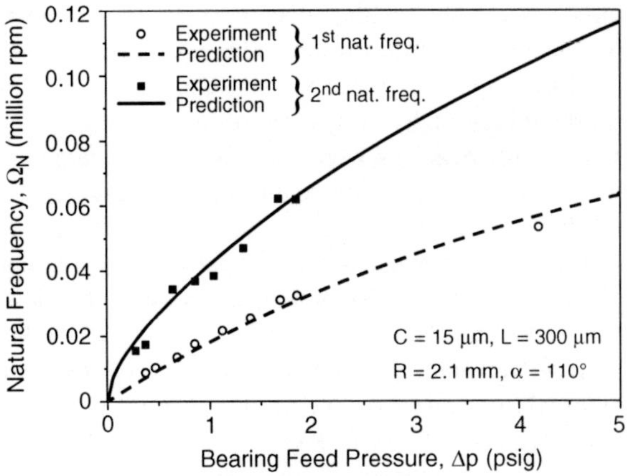

Fig. 6.25 Experimental validation of natural frequencies in an anisotropic bearing configuration [18]

In summary, a novel axial-flow hydrostatic micro-gas bearing concept was introduced which yields anisotropy in bearing stiffness. Anisotropy in hydrostatic stiffness is achieved by blocking the bearing supply air over two circumferential sectors located opposite to each other. The two remaining circumferential sectors are fed with pressurized air from plena located on the aft side of the rotor. As a consequence, a difference in hydrostatic stiffness in the two orthogonal directions is introduced without substantially altering the cross-coupled hydrodynamic stiffness. By breaking symmetry in hydrostatic stiffness, the bearing design increases bearing top speed and relieves fabrication tolerance requirements. This enables stable high-speed bearing operation and extends the bearing operating range.

6.7 Analysis of Hydrostatic Gas Thrust Bearings

In addition to radial support, all high-speed rotating machinery requires thrust bearings to provide axial support and stiffness. Similar to the journal bearings, there are two major categories of thrust bearings. Hydrostatic thrust bearings possess several advantages over their hydrodynamic counterparts. For example, hydrodynamic thrust bearings are incapable of developing a sufficiently high stiffness at low rotational speeds, since the magnitude of the stiffness arises from the relative motion between the stationary and rotating bearing surfaces. In contrast, hydrostatic thrust bearings depend on a controllable, external supply pressure to generate stiffness, which depends only weakly on the rotational speed of the rotor. Furthermore, hydrodynamic thrust bearings in general require smaller clearances posing two major challenges for micro-turbomachinery applications. The drag penalty associated with hydrodynamic thrust bearings outweighs that of hydrostatic thrust bearings, and the requirement for smaller running gaps in hydrodynamic thrust bearings renders them more susceptible to failure due to geometric non-uniformities. Hydrodynamic thrust bearings are also known to exhibit poor stability characteristics at high speeds, as discussed by Constantinescu and Galetuse [5]. In view of the above-mentioned shortcomings of hydrodynamic thrust bearings, inherent restrictor-orifice hydrostatic thrust bearings were selected to provide axial support and stiffness for the microturbomachinery applications discussed here.

Unlike conventional axial-flow turbomachines, where the engine architecture includes long slender shafts supporting the rotating structure, the rotors of the MEMS devices consist of thin disks with very low thickness-to-diameter ratios (L/D) and thrust bearings located near the center of the disk. Hence, the rotors can be prone to unwanted tilting and the journal bearing system may be unable to exert the required torque to resist the tilting. Therefore, the thrust bearings must be designed with adequate tilting stiffness.

There are a number of challenges associated with the development of micro-hydrostatic thrust bearings. The geometric dimensions of these bearings are typically one to two orders of magnitude smaller than conventional, large-scale hydrostatic thrust bearings (the thrust-bearing gap of the powerMEMS devices is of order 3 μm) and need to be controlled to higher levels of precision. The low Reynolds numbers and high pressure ratios encountered in the bearing land give rise to increased interactions between viscous and compressible flow effects, necessitating the development of new models to accurately capture the bearing performance and stability. Thrust-bearing performance may be compromised due to the presence of geometric non-uniformities such as orifice diameters and profiles. Occasionally, one has to contend with the presence of more severe defects, such as the failure of a number of orifices to etch through completely. Another anomaly which has been observed pertains to the profile of the orifices, where the diameters of the orifices are found to vary considerably along their lengths, thus resulting in tapered rather than uniform orifice profiles.

In summary, desirable attributes of hydrostatic gas thrust bearings include large axial and tilting stiffness, low mass flow and power consumption, as well as sufficient static and dynamic stability. Figure 6.26 illustrates three possible thrust-bearing configurations for powerMEMS applications. Configuration (a) refers to a commonly used baseline architecture with a circular thrust-bearing pad and typically 20 orifices uniformly distributed around the circumference. Configuration (b) is a novel circular pad architecture with 37 orifices distributed on different radial locations so as to form a hexagonal array. Configuration (c) refers to an annular thrust bearing, with a single row of 20 orifices arranged uniformly at the mean radius. This configuration is typically used in turbopumps where the annular pad acts as a thrust bearing and also as a seal to separate turbomachinery operating in different fluids. For example, a pump rotor could reside inside the annular ring pressurizing liquid whereas a turbine could be located on the outer periphery of the thrust-bearing operating on a gas.

In order to assess the performance of the various thrust-bearing configurations shown above, it is necessary to establish a modeling methodology capable of dealing with a number of effects. The thrust-bearing model should be able to estimate

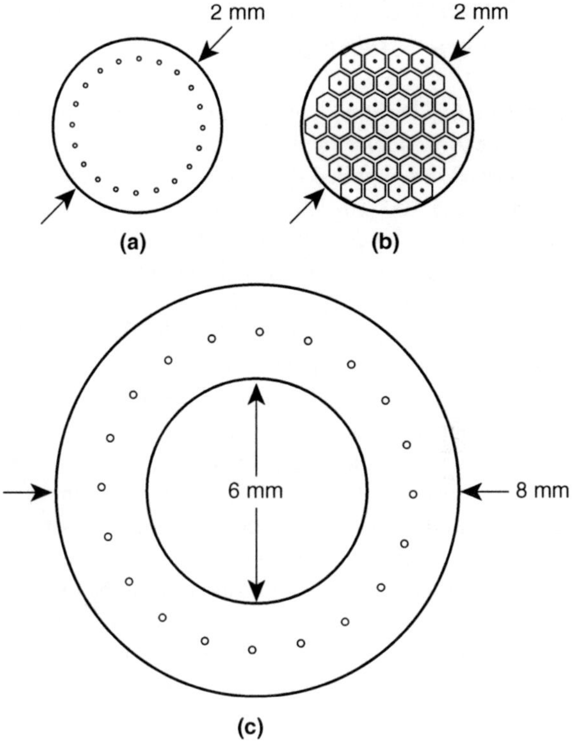

Fig. 6.26 Inherent-restrictor orifice type hydrostatic thrust-bearing architectures for powerMEMS: (**a**) baseline circular pad, (**b**) array of hexagonal pads, and (**c**) annular pad (dimensions are not to scale) [35]

the steady-state performance characteristics of micro-hydrostatic thrust bearings. The quantities of interest are axial stiffness, natural frequency, and mass flow rates as a function of thrust-bearing hydrostatic supply pressure. To evaluate the thrust-bearing tilting stiffness, the model needs to be able to handle operating conditions with non-uniform thrust-bearing clearances. Moreover, the model should have the attribute of simultaneously coupling the flow field through all orifices and the bearing land. Depending on the micro-turbomachinery application, the inner and outer radii of the thrust-bearing pad may be subjected to different static pressure boundary conditions, and the model should have the capability to deal with a variety of boundary conditions. In addition, the micro-thrust-bearing model should be able to analyze static and dynamic stability to characterize potential resonance and vibration problems.

This section is organized in three major parts. The first part introduces an analytical thrust-bearing model capable of characterizing the steady-state performance under uniform operating conditions. The second part extends this model to account for tilting effects and geometric non-uniformities using a Green's function approach. The last part deals with an unsteady flow model to capture the dynamic behavior and stability of the thrust-bearing-rotor systems used in powerMEMS applications.

6.7.1 Steady-State Performance Model for Hydrostatic Gas Thrust Bearings

Consider the baseline configuration of an inherent-restrictor orifice hydrostatic thrust bearing depicted in Fig. 6.27. The bearings rely on an external source of pressurized gas supplied through a series of orifice restrictors (typically of 10 μm diameter and 100 μm length), which together with the thrust-bearing gap (nominally 2–3 μm), generate the axial stiffness. The key advantage of hydrostatic thrust bearings is their ability to provide an almost constant stiffness independent of the rotational speed of the rotor. The steady state operating characteristics of the thrust bearings are modeled using a fluid resistance model. Its key components are outlined in the inset of Fig. 6.27. The station numbers are also labeled in the figure, and the mechanism for the generation of axial stiffness is illustrated schematically. The rotor is supported axially by means of a forward and an aft thrust bearing. When the rotor is given a slight perturbation in the axial direction and approaches the aft thrust-bearing pad, the hydraulic resistance in the outflow region of the aft thrust-bearing pad increases, necessitating a larger pressure difference between the outlet of the orifices and the thrust-bearing exhaust (vented to atmosphere). The increase in static pressure over the aft thrust-bearing pad yields a larger restoring axial force acting on the rotor. The opposite occurs on the forward thrust bearing, and the consequence is the generation of a net upward force on the turbomachinery disk restoring the axial equilibrium position of the rotor.

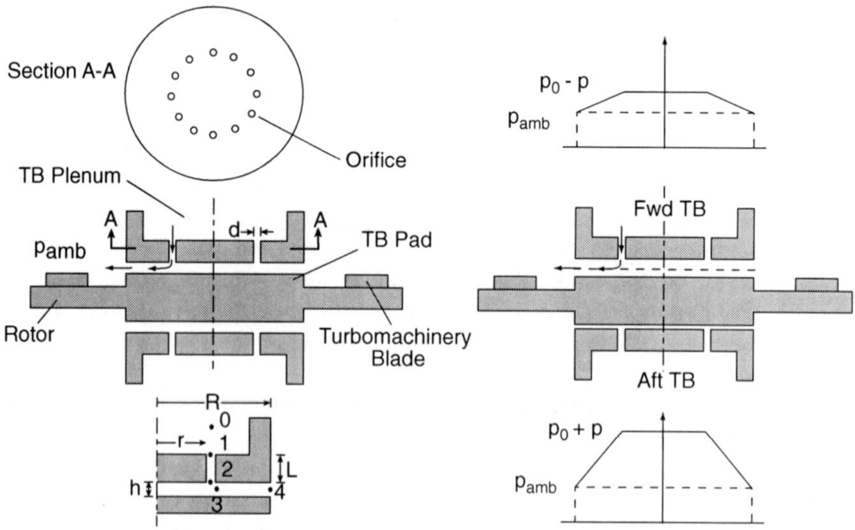

Fig. 6.27 Schematic of dual-thrust bearing arrangement for micro-turbomachinery using inherent-restrictor orifice type hydrostatic thrust bearings (dimensions are not to scale) [34]

A drop in stagnation pressure occurs at the inlet of the orifice, from station 0 to 1. The stagnation pressure loss is

$$\frac{\Delta p_t}{p_t - p} = \omega = 0.5, \tag{6.58}$$

where a typical loss coefficient of $\omega = 0.5$ can be found in handbooks of hydraulic resistance [13] and has been verified by CFD calculations. The flow through each of the circular orifices (from station 1 to 2) is modeled as a compressible, fully developed viscous flow. Due to the high thermal conductivity of silicon, the flow is isothermal. The relevant equations governing such a flow are given in Shapiro [29] and can be summarized by the following expression for the corrected mass flow

$$\frac{\dot{m}\sqrt{RT_{t2}}}{Ap_{t2}\sqrt{\gamma}} = \Phi\left(\mathrm{Re_d}, \frac{p_2}{p_{t1}}, \frac{L}{d}\right) = \Psi\,(M_2) \tag{6.59}$$

The corrected mass flow at the exit of each orifice depends on the Reynolds number Re_d based on orifice diameter d

$$\mathrm{Re}_d \equiv \frac{\rho V d}{\mu} = \frac{\dot{m} d}{A\mu} = \frac{4\dot{m}}{\pi d\,\mu}, \tag{6.60}$$

the ratio of the static pressure at the exit of the orifice p_2 to the total pressure at the inlet p_{t1}, and the aspect ratio of the orifice L/d, where L and d are the length and diameter of the orifice, respectively. It should be noted that although the Reynolds number of the flow is sufficiently low so as to render the flow laminar, the effects of friction and the associated decrease in stagnation pressure tend to increase the Mach number of the flow to appreciable values, necessitating the effects of compressibility to be taken into account. However, in contrast to the familiar Fanno flow for adiabatic ducts, where choking occurs at a local Mach number of one, the isothermal flow through ducts chokes at a local Mach number of $1/\sqrt{\bullet\gamma}$ [29]. Furthermore, since the flow is laminar (Re_d= O(10^2)), viscous effects, quantified by the friction coefficient C_f, can be related analytically to the Reynolds number according to $C_f = 16/\,Re_d$. Although the flow is dominated by viscous effects, compressibility has to be accounted for since $C_f M^2 (L/d)$ is of order one.

The turning losses at the exit of the orifices (from station 2 to 3) are modeled using the following empirical correlation by Vohr [39]

$$\frac{\Delta p_t}{p_t - p} = 0.31\left(\frac{\mathrm{Re}_h}{2000}\right) - 0.122\left(\frac{\mathrm{Re}_h}{2000}\right)^2 + 0.282\left(\frac{\mathrm{Re}_h}{2000}\right)^3 \tag{6.61}$$

where the Reynolds number here is defined based on the thrust-bearing clearance h

$$\mathrm{Re}_h = \frac{2\dot{m}}{\pi\, d\, \mu} \tag{6.62}$$

After leaving the orifices, the fluid flows radially outwards through the bearing land, before finally being discharged into the turbomachinery gas path. This occurs from station 3 to 4. In the radial outflow region, the static pressure distribution, which varies both radially and circumferentially, is computed numerically using a finite element method. The flow in this region is assumed to be isothermal and the mass flow $\dot{m}$ is driven by the difference in the squares of the static pressure Δp^2 between the exit of the orifices and the thrust-bearing exhaust. This can be expressed non-dimensionally as

$$\frac{\Delta p^2}{\left(\mu RT\dot{m}/h^3\right)} = \phi\left(N, \frac{r}{R}\right). \tag{6.63}$$

The ratio $\Delta p^2/\left(\mu RT\dot{m}/h^3\right)$ can be interpreted as a dimensionless hydraulic resistance between the exit of the orifices and the thrust-bearing exhaust, and the magnitude of this resistance depends on the layout and geometrical configuration of the orifices, which in turn depends on the number of orifices N, and the normalized radial location of the orifices r/R. Assuming the value of h to be constant, the static pressure p in the outflow region can be obtained by solving the Laplace equation for p^2

$$\nabla^2 p^2 = 0. \tag{6.64}$$

Having identified and modeled the individual resistances leading from the thrust-bearing supply plenum[13] to the thrust-bearing exhaust, the objective is to iteratively solve for the mass flow rate $\dot{m}$ given a thrust-bearing supply pressure and fixed thrust-bearing geometrical configuration.

The static pressure distribution in the outflow region between the stator and rotor can then be integrated to yield the axial force acting on the rotor. Figure 6.28 shows the finite-element solution of the static pressure distribution in the circular bearing land of the baseline hydrostatic thrust-bearing configuration depicted in Fig. 6.26(a). The model is not limited to circular pads and is capable of predicting key steady-state performance parameters of annular thrust bearings as well. The bearing mass flow, axial stiffness, and natural frequency can be readily obtained from the solution.

To validate the hydraulic resistance model, experimental results obtained from a micro-turbocharger device [28] are compared to the model predictions. Figure 6.29 displays static flow test results obtained experimentally by varying the supply pressure to the aft thrust bearing and monitoring the variations in mass flow through the forward thrust bearing. In the latter the supply pressure was kept constant. The modeling results are in agreement with the experimental data both in trend and in magnitude, validating the flow resistance approach.

---

[13] The pressure supply to the thrust bearing supply plenum is held at a constant value, whereas the thrust bearing exhaust is nominally at atmospheric pressure.

Fig. 6.28 Finite element solution of normalized bearing land pressure distribution p/p_{amb} [34]

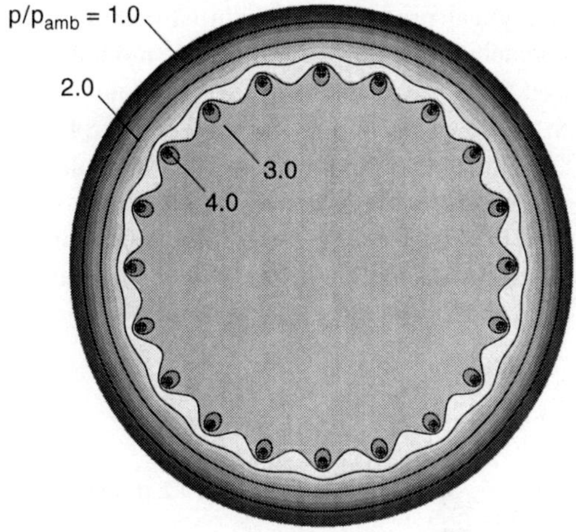

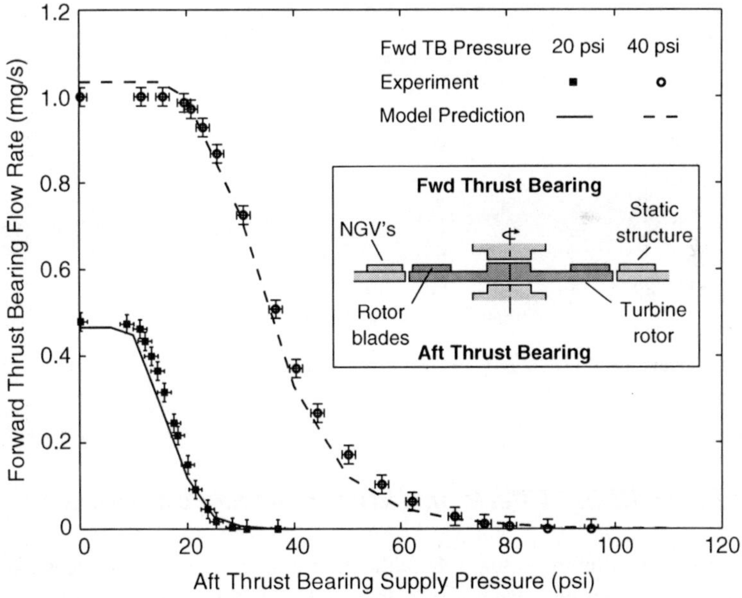

Fig. 6.29 Experimental measurement and model predictions of forward thrust-bearing mass flow rate as a function of aft thrust-bearing pressure setting [34]

It is important to quantify the natural frequency of the thrust-bearing system relative to the rotor design speed to avoid potential resonance phenomena. The goal is to either operate the micro-turbomachinery below the thrust-bearing natural frequency (subcritical operation) or to accelerate through the natural frequency to supercritical speeds with high enough damping to avoid resonance in the transcritical regime.

Analytical predictions for thrust-bearing natural frequency of a typical microturbomachinery application [28] are plotted against thrust-bearing supply pressure in Fig. 6.30. The results show a monotonic increase in axial stiffness and thus natural frequency with supply pressure. The rotor design speed is 1.2 million rpm indicated by the horizontal dashed line such that for supply pressures in the forward and aft bearings of less than 95 psi the thrust bearings will operate supercritically. The required thrust-bearing mass flow rates for various thrust-bearing supply pressures are also shown. For this particular application, the thrust-bearing mass flow rate is about 2% of the turbomachinery mass flow at design speed.

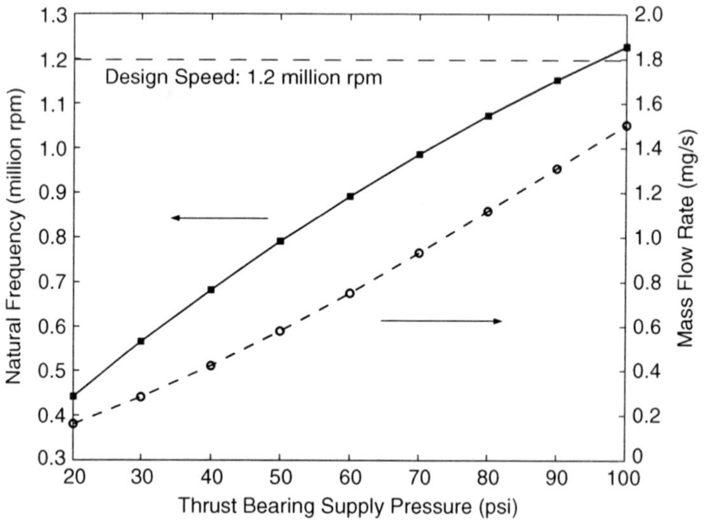

Fig. 6.30 Thrust-bearing natural frequency in a microturbocharger application (data by Savoulides [28])

6.7.2 Rotor Tilting Effects on Thrust-Bearing Performance

Tilting effects play an important role in the microturbomachinery applications as the rotor architecture is comprised of thin disks with low thickness to diameter ratio as compared to long, slender shafts in typical gas turbine engines. This section introduces a method to analyze the tilting response of the rotor-thrust-bearing system in terms of yaw and pitch dynamics. Referring to Fig. 6.31, in the absence of rotation, the vertical axis through the geometric center and fixed to the rotor disk is OB.

This is termed the static axis. The axis fixed to the static structure, which coincides with the static axis OB in this configuration is OZ. The dynamic imbalance χ, for example, due to a non-uniform disk thickness or a missing blade, is the angular misalignment between the static axis OB and the principal axis OA which corresponds to the polar moment of inertia of the rotor. When the rotor is set into motion,

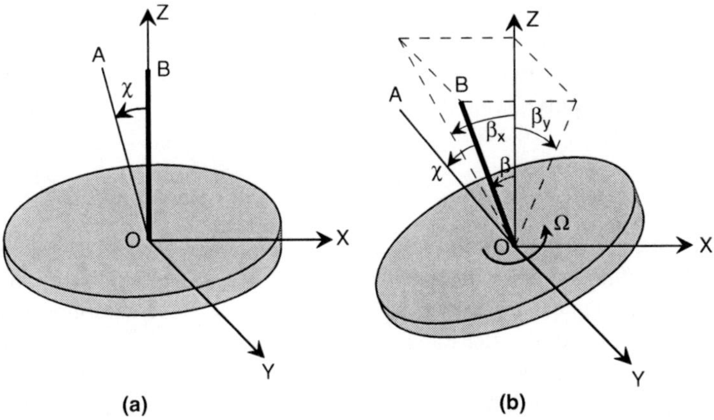

Fig. 6.31 Schematic of low *L/D* turbomachinery rotor in tilting motion: (**a**) rotor at rest, and (**b**) rotor spinning at rotation Ω [35]

the static axis OB, which is fixed to the rotor, exhibits a coning motion about OZ at a small inclination angle β. Denoting the projections of β onto the x-z and y-z planes by β_x and β_y, respectively, the linearized equations of motion governing β_x and β_y are given by

$$I\ddot{\beta}_x + I_p\Omega\dot{\beta}_y + k_{xx}\beta_x + k_{xy}\beta_y + c\dot{\beta}_x = \left(I - I_p\right)\chi\Omega^2\cos\Omega t \tag{6.65}$$

$$I\ddot{\beta}_y - I_p\Omega\dot{\beta}_x + k_{yy}\beta_y - k_{xy}\beta_x + c\dot{\beta}_y = \left(I - I_p\right)\chi\Omega^2\sin\Omega t \tag{6.66}$$

The notation for the Euler angles β_x, β_y, and ϕ (where $\Omega = \dot{\phi}$) follows the same convention as that used in Childs [4], where β_x is the yaw angle and β_y is the pitch angle (β_x,β_y << 1). Here, k_{xx} and k_{yy} are the thrust-bearing direct-coupled hydrostatic tilting stiffness, c is the thrust-bearing angular damping and k_{xy} is the cross-coupled stiffness due to hydrodynamic effects. The cross-coupled stiffness k_{xy} couples angular displacements β_x and β_y and vice versa. Note that the direct-coupled stiffness $k_{xx} = k_{yy} = K^{hs}$ and damping coefficient c can be calculated as a function of β. In addition to $k_{xy}= K^{hd}$, which couples β_x and β_y, an additional source of coupling arises due to the gyroscopic terms $I_p\Omega\dot{\beta}_y$ and $I_p\Omega\dot{\beta}_x$ in Eqs. (6.65) and (6.66). The non-homogeneous driving terms on the right-hand-side of each equation depend on the diametral and polar moments of inertia, denoted by I and I_p, respectively, as well as the dynamic imbalance χ. In the limit of infinite rotational speeds, the inclination angle β of the conical motion approaches the dynamic imbalance χ. Physically, the polar principal axis OA aligns itself with the inertial axis OZ as a consequence of gyroscopic stiffening which becomes dominant at higher speeds. If β becomes excessively large, physical contact between the rotor and static structure can occur.

To analyze the linear dynamic stability of the system, the rotor is subjected to small perturbations in inclination angles. This involves linearizing the above governing equations and solving them as an eigenvalue problem. The resultant fourth order characteristic polynomial then yields

$$\left(Is^2 + k + cs\right)^2 + \left(I_p\Omega s + k_{xy}\right)^2 = 0. \tag{6.67}$$

The real part of each of the eigenvalues carries information pertaining to the stability of the inclination motions to small perturbations about the equilibrium condition. The direct-coupled and cross-coupled tilting stiffness and damping coefficients need to be determined in order to calculate the dynamic behavior. This is achieved using a fluid dynamic model for the hydrostatic force, hydrodynamic force, and damping force. For the micro hydrostatic gas thrust-bearing architectures analyzed here, the time taken for the hydrostatic fluid to flow through the thrust-bearing gap is small compared to the period for one complete revolution of the rotor. Even at the maximum design speed, the reduced frequency $\Omega L/U$ is of order 10^{-1}, where Ω is the rotational frequency of the rotor and L/U is the mean flow through time for the hydrostatic fluid. In view of the very low reduced frequencies, it is possible to decouple the flow induced by the motion of the rotor from the flow arising due to the externally applied hydrostatic pressure drop across the thrust bearing. This is shown schematically in Fig. 6.32.

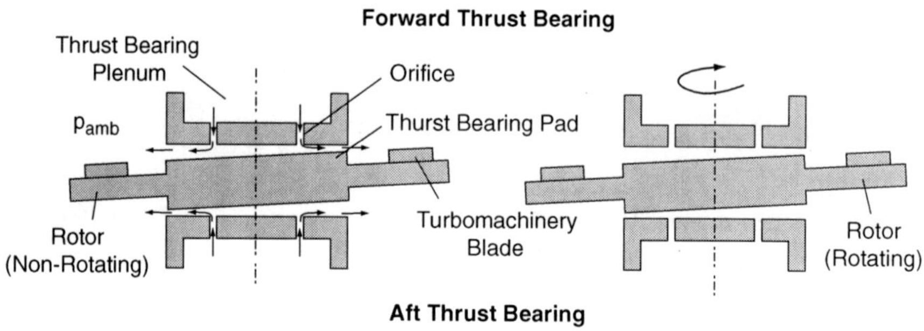

Fig. 6.32 Decoupled models for hydrostatic moments (*left*) and hydrodynamic moments (*right*) [35]

Based on these assumptions, a modeling approach is described for each of the three primary force components acting on the rotor, namely the hydrostatic force, the hydrodynamic force, and the damping force.

6.7.2.1 Hydrostatic Force

In evaluating the forces and moments acting on the rotor due to the externally supplied hydrostatic pressure, the rotor is assumed to be stationary, as depicted on the left in Fig. 6.32. In cylindrical coordinates, the continuity equation along the thrust-bearing pad can be expressed as

$$\nabla \cdot (\rho \vec{q}) = \frac{\partial}{\partial r}(\rho q_r) + \frac{1}{r}\frac{\partial}{\partial \theta}(\rho q_\theta) = \sum_{i=1}^{N} Q_i \delta(\vec{r}_i), \tag{6.68}$$

where $\vec{q} = q_r \hat{e}_r + q_\theta \hat{e}_\theta$ is the volumetric flow rate per unit length normal to the local direction of flow. The right hand side accounts for hydrostatic mass injection of Q_i at discrete locations $\vec{r}_i$ along the thrust-bearing gap. Assuming the flow to be locally of Poiseuille type, the pressure field p in the thrust-bearing gap is related to the volumetric flow rate $\vec{q}$ via

$$\vec{q} = -\left(\frac{h^3}{12\mu}\right)\nabla p, \tag{6.69}$$

where μ is the dynamic viscosity of the fluid. Substituting Eq. (6.69) in Eq. (6.68) thus yields

$$\nabla \cdot \left(\frac{\rho h^3}{12\mu}\nabla p\right) = -\sum_{i=1}^{N} Q_i \delta(\vec{r}_i). \tag{6.70}$$

Assuming an ideal gas and the flow to be isothermal, the above becomes

$$\nabla \cdot \left(h^3 \nabla p^2\right) = -24\mu RT \sum_{i=1}^{N} Q_i \delta(\vec{r}_i). \tag{6.71}$$

In the steady-state performance model discussed in Section 6.7.1, the thrust-bearing gap h was assumed to be uniform. Here, this assumption is relaxed and the general situation where $h = h(r, \theta)$ is analyzed. Away from the orifices

$$\nabla \cdot \left(h^3 \nabla \left(p^2 - p_{\text{amb}}^2\right)\right) = 0, \tag{6.72}$$

and letting

$$-\tilde{p} = \left(p^2 - p_{\text{amb}}^2\right) \Big/ (24\mu RT), \tag{6.73}$$

the governing equation for the pressure field becomes

$$\nabla \cdot \left(h^3 \nabla \tilde{p}\right) = \sum_{i=1}^{N} Q_i \delta(\vec{r}_i). \tag{6.74}$$

The above is a linear equation for $\tilde{p}$. To solve this equation, the method of Green's functions is employed, where the Green's function is obtained by solving for the impulse response along the thrust-bearing pad

$$\nabla \cdot \left(h^3 \nabla G \left(\vec{r}, \vec{r}' \right) \right) = \delta \left(\vec{r} \right). \tag{6.75}$$

Yabe and Watanabe [40] and Yabe and Yamamoto [41] employed the Green's function approach to analyze the tilting effects of hydrostatic thrust bearings using analytical expressions. This limited the analysis to simple geometrical configurations. Here, the Green's function is obtained numerically, permitting the assessment of a wide variety of thrust bearings with unconventionally shaped pads. A finite-element method with the appropriate boundary conditions is used to compute the Green's function $G\left(\vec{r},\vec{r}'\right)$ numerically.

The Green's function $G\left(\vec{r},\vec{r}'\right)$ determines the pressure at position $\vec{r}'$ within the thrust-bearing pad for a unit mass flow through an orifice of diameter d placed at $\vec{r}$. Hence, at the exit of the mth orifice (located at $\vec{r}_m$), the pressure $G\left(\vec{r}_n,\vec{r}_m\right)$ due to unit mass flow through orifice n (located at $\vec{r}_n$) is given by

$$\nabla \cdot \left(h^3 \nabla G \left(\vec{r}_n, \vec{r}_m \right) \right) = \delta \left(\vec{r}_n \right). \tag{6.76}$$

By superposition, the resultant pressure at the exit of the mth orifice due to the combined flows through all the N orifices is given by

$$\tilde{p}_m = \sum_{n=1}^{N} G\left(\vec{r}_n,\vec{r}_m\right) Q_n, \; m = 1,\ldots,N, \tag{6.77}$$

where Q_n is the mass flow rate through the nth orifice. Note that the above is a set of N equations for $2N$ unknowns, $\tilde{p}_m$ and Q_m, where $m = 1,\ldots,N$. In order to close the system, another set of N equations is required. These are obtained from the pressure-drop-flow-rate relationship across each of the N orifices:

$$Q_m = f\left(\tilde{p}_m\right), \; m = 1,\ldots,N \tag{6.78}$$

For fixed thrust-bearing supply pressure and temperature Eq. (6.78) represents the corrected flow function [29]. Since the flow through the orifice chokes for sufficiently low values of p_m and sufficiently high values of thrust-bearing supply pressure p_{TB},$\partial Q_m / \partial \left(p_m / p_{TB}\right)$ approaches zero at low values of p_m. Note that Eq. (6.78) is non-linear and one ends up with a set of coupled non-linear algebraic equations, for which the values of $\tilde{p}_m$ and Q_m are to be determined.

One common approach to solve such non-linear algebraic equations is to employ the multidimensional Newton's method with the vector of unknowns denoted by

$$\vec{w} = \left[\tilde{p}_1,\ldots,\tilde{p}_N,Q_1,\ldots,Q_N\right]^T. \tag{6.79}$$

The system of Eqs. (6.77) and (6.78) can then be expressed as

$$F(\vec{w}) = \begin{bmatrix} Q_1 - f(\tilde{p}_1) \\ \cdots \\ Q_N - f(\tilde{p}_N) \\ -p_1 + \sum_{n=1}^{N} G(\vec{r}_n,\vec{r}_1)\, Q_n \\ \cdots \\ -p_N + \sum_{n=1}^{N} G(\vec{r}_n,\vec{r}_N)\, Q_n \end{bmatrix} = \begin{bmatrix} 0 \\ 0 \\ 0 \\ \vdots \\ 0 \\ 0 \end{bmatrix} \tag{6.80}$$

The multidimensional Newton's method is given by

$$J_F(\vec{w})\left(\vec{w}^* - \vec{w}\right) = -F(\vec{w}), \tag{6.81}$$

where the Jacobian $J_F(\vec{w})$ is

$$J_F(\vec{w}) = \begin{bmatrix} -f'(\tilde{p}_1) & 0 & 0 & 1 & 0 & 0 \\ 0 & \cdots & 0 & 0 & \cdots & 0 \\ 0 & 0 & -f'(\tilde{p}_N) & 0 & 0 & 1 \\ -1 & 0 & 0 & G(\vec{r}_1,\vec{r}_1) & G(\vec{r}_2,\vec{r}_1) & \cdots \\ 0 & \cdots & 0 & \cdots & G(\vec{r}_n,\vec{r}_m) & \cdots \\ 0 & 0 & -1 & \cdots & \cdots & G(\vec{r}_N,\vec{r}_N) \end{bmatrix}. \tag{6.82}$$

Once the values of the $2N$ unknowns are found, the static pressure at any location $\vec{r}$ along the thrust-bearing pad can be determined using

$$\tilde{p}(\vec{r}) = \sum_{n=1}^{N} G(\vec{r}_n,\vec{r}) Q_n. \tag{6.83}$$

Note that the above equation is a discrete version of the Green's formula. Knowledge of the pressure field $\tilde{p}(\vec{r})$ facilitates the evaluation of the hydrostatic forces F^{hs} and moments M^{hs} acting on the rotor. The resulting hydrostatic tilting stiffness $K^{hs} = \partial M^{hs}/\partial\beta$ can then be evaluated for varying rotor tilting angles.

6.7.2.2 Hydrodynamic Force

As a consequence of rotor tilt, the spinning motion of the rotor drags the flow along a circumferentially varying clearance, thus giving rise to hydrodynamic pressures. This can be modeled using the Reynolds Equation. For incompressible flow in cylindrical coordinates, the Reynolds equation becomes:

$$\nabla \cdot \left(h^3 \nabla p\right) = 6\mu\Omega \frac{\partial h}{\partial \theta}. \tag{6.84}$$

The forces F^{hd} and moments M^{hd} due to the hydrodynamics can thus be evaluated for varying rotor tilting angles β and rotational speeds Ω. The hydrodynamic torque M^{hd} acts in a direction which is orthogonal to the rotor tilt, and thus contributes a cross-coupled torque and hence hydrodynamic tilting stiffness $K^{hd} = \partial M^{hd}/\partial \beta$.

In order to assess compressibility effects on the modeling results, fully compressible CFD calculations using the commercial flow solver FLUENT were performed. Figure 6.33 depicts the non-dimensional hydrodynamic moment $\zeta = M^{hd}/(p_a R^3)$ as a function of the bearing number $\Lambda = (6\mu\Omega R^2)/(p_a h^2)$ for varying normalized tip deflections (and hence rotor tilting angles). The incompressible model results compare well to the compressible CFD results. At high bearing numbers the assumption of incompressible flow over-predicts the cross-coupled hydrodynamic torque and stiffness. This in turn gives rise to more conservative modeling results, since larger magnitudes of cross-coupled hydrodynamic torques increase the system's tendency towards instability.

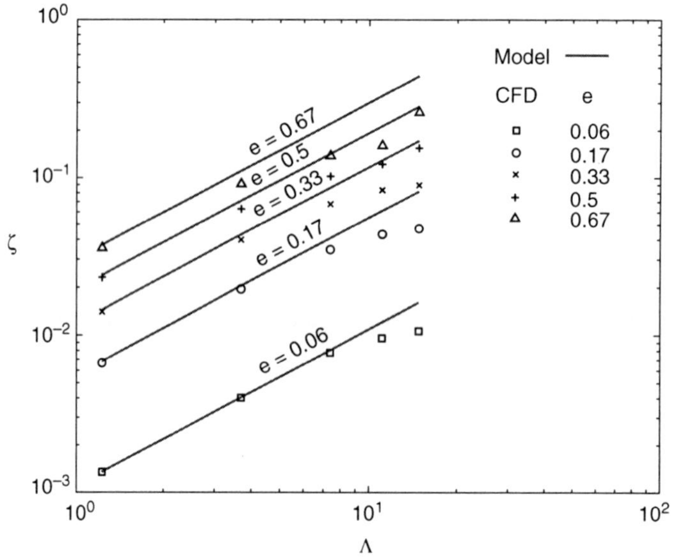

Fig. 6.33 Cross-coupled hydrodynamic torque: incompressible model vs. compressible CFD calculations [35]

In addition to the above-mentioned cross-coupled component, fluid compressibility results in hydrodynamic moments with a direct-coupled component. This direct-coupled hydrodynamic torque acts in a direction which opposes the rotor tilt, and hence contributes to a direct-coupled hydrodynamic tilting stiffness. However, at the rotational speeds and tilting angles of interest, the direct-coupled hydrodynamic torque is negligible compared to the direct-coupled hydrostatic torque. For example for a 1 mm radius thrust bearing with a nominal clearance of 3 μm rotating at a design speed of 1.2×10^6 rpm, a typical rotor tip deflection of 1.5 μm gives rise to a direct-coupled hydrodynamic torque which is approximately 20%

of the direct-coupled hydrostatic torque at a nominal supply pressure of 4 atm. In summary, the incompressible model enables the calculation of the resulting hydrodynamic torque which only contributes to a cross-coupled hydrodynamic tilting stiffness. As a consequence, the direct-coupled tilting stiffness arises solely due to hydrostatics.

6.7.2.3 Damping Force

The damping force and associated torque is determined from the pressure distribution established when the flow is squeezed through the gap and the thrust-bearing orifices due to the tilting motion of the rotor. This pressure field in the gap can be obtained by solving the Reynolds equation of the form

$$\nabla \cdot \left(h^3 \nabla p\right) = 12\mu \frac{\partial h}{\partial t} \tag{6.85}$$

together with the appropriate homogeneous boundary conditions $p = 0$ at the outlet of the thrust-bearing pad and at the entrance of the thrust-bearing orifices. However, the tilting motion of the rotor induces different mass flow rates through the individual orifices, thus necessitating the Green's function formulation to evaluate the varying mass flows and pressures at the exit of the individual orifices. The damping forces and torques are dependent on the rotor tilting angle, the tilting rate, and the bearing geometry. The damping coefficient for the tilting motion, c, can thus be determined by evaluating the pressure forces and concomitant tilting torques arising from a rotor tilt rate of $\dot{\beta}$.

6.7.2.4 Direct-Coupled Tilting Stiffness Assessment

The above model formulation is first used to compute the direct-coupled tilting stiffness of the baseline thrust-bearing configuration shown in Fig. 6.26(a). The radius of the thrust-bearing pad is 1 mm, the supply pressure to the thrust-bearing orifices is held constant at 60 psi (corresponding to a static pressure ratio $p_{supply}/p_{ambient} = 5$), and the nominal thrust-bearing gap is 3 μm. The computed bearing pad pressure distribution is depicted in Fig. 6.34 for an untilted rotor (a) and for a rotor tip deflection of 2 μm (b). The static pressure on the thrust-bearing pad becomes non-axisymmetric with a larger static pressure building up at smaller axial gaps.

This non-axisymmetric pressure distribution gives rise to a restoring torque which tends to oppose the tilting of the rotor. The monotonic increase in magnitude of the restoring torque with tilting angle thus results in a tilting stiffness. The non-dimensionalized tilting stiffness is marked by the solid line for a range of dimensionless rotor tip deflections. The dimensionless direct-coupled tilting stiffness is defined as $k^+ = k\Big/\left(\sqrt{A} p_{supply}\, r^2\right)$, where k is the direct-coupled tilting stiffness, A is the thrust-bearing surface area, and r is the radial location of the orifices. The rotor tip deflections are normalized by the maximum allowable tip deflection which is 3 μm for this configuration.

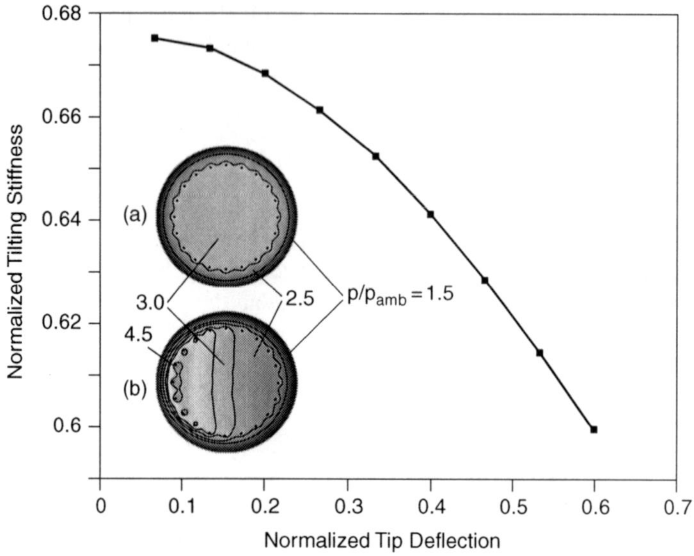

Fig. 6.34 Direct-coupled hydrostatic tilting stiffness as a function of rotor tip deflection – baseline thrust-bearing configuration [35]

The potential improvement in direct-coupled tilting stiffness of a hexagonal thrust-bearing configuration (see Fig. 6.26(b)) was also assessed using the Green's function formulation. The modeling results for the static pressure distribution on the thrust-bearing pad are shown in Fig. 6.35 for an untilted rotor (left) and a tilted rotor with a tip deflection of 2 μm (right). The supply pressure to the thrust bearings was kept constant at 60 psi for all calculations. In the untilted case the static pressure distribution is nearly axisymmetric. When the rotor tip deflection is 2 μm, the high-pressure "pocket" shifts outwards toward regions where the thrust-bearing gaps are smaller, and provides a restoring torque and direct-coupled tilting stiffness in response to angular perturbations on the rotor. Table 6.4 summarizes the

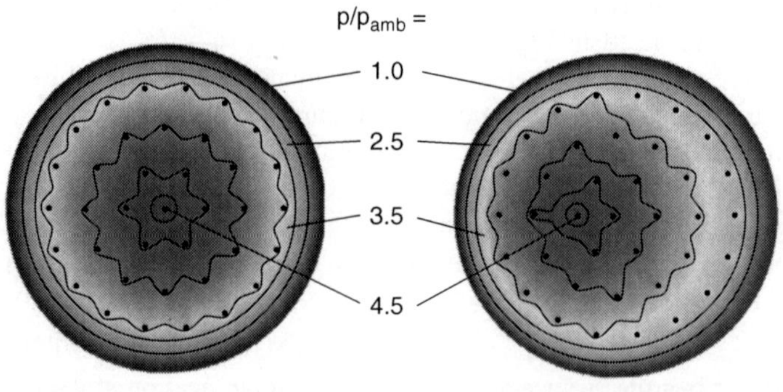

Fig. 6.35 Hydrostatic pressure distribution for untilted (*left*) and tilted rotor (*right*) – hexagonal thrust-bearing configuration [35]

Table 6.4 Comparison of key operating characteristics of baseline thrust-bearing configuration with hexagonal and annular configurations

Total clearance = 5.5 μm			
Type	***Axial stiffness (N/m)***	***Mass flow (g/s)***	***Tilting stiffness (N-m/rad)***
Baseline	2.2×10^5	1.00×10^{-3}	0.089
Hexagonal	1.7×10^5	1.19×10^{-3}	0.087
Annular	8.3×10^5	1.15×10^{-3}	5.0
Total clearance = 7 μm (Optimized for axial stiffness)			
Type	***Axial stiffness (N/m)***	***Mass flow (g/s)***	***Tilting stiffness (N-m/rad)***
Hexagonal	1.9×10^5	1.73×10^{-3}	0.075

results and compares the two configurations plus an annular thrust bearing to one another.

Two sets of results are presented for the hexagonal thrust-bearing configuration. The total thrust-bearing clearance of 5.5 μm, which optimizes the axial stiffness for the baseline case, is found to be sub-optimal for the hexagonal case. A total clearance of 7 μm maximizes the axial stiffness for the case with a hexagonal arrangement of orifices. It is apparent from the table that the hexagonal thrust bearing has a marginally lower axial and direct-coupled tilting stiffness, as compared to the baseline case. However, the mass flow requirement for the hexagonal thrust bearing is perceptibly higher. The hexagonal thrust bearing is thus seen to be inferior to the baseline configuration in terms of steady-state performance, even though the hexagonal thrust bearing has 37 orifices, whereas the original configuration has only 20 orifices.

A useful simplification in the modeling of the baseline configuration was to assume that the flow through each orifice contributes equally to the overall direct-coupled tilting stiffness of the system, thus permitting all the orifices to be modeled as springs with equal spring constants. When the number of orifices is increased and located at different radial locations, one might hypothesize that the overall stiffness of the system will increase, since there is an increase in the total number of springs. However, as seen from the above results, the orifices do not contribute equally to the overall axial stiffness of the system due to the interaction of the local pressure and flow fields at different radial locations. Even after optimizing the axial gap for maximum stiffness, the hexagonal thrust bearing is still inferior in performance, since the gap is optimized for the *overall* thrust-bearing system, and not with respect to individual orifices.

6.7.3 Dynamic Stability Considerations

Although steady-state operating characteristics are important, one issue in the design of fluid film bearings is stability. Before describing the details of the dynamic stability model, it is useful to highlight the essential difference between static and dynamic stability. Static instability refers to a pure divergence from the system's initial equilibrium state, and can be analyzed solely from the static- or steady-state characteristics of the system at the perturbed state. On the other hand, dynamic instability corresponds to oscillations of exponentially growing amplitude, which usually have to be treated using a dynamic analysis that accounts for the temporal evolution of perturbations in the system. As an example, for a classical mass-spring-damper system with mass m, spring constant k, and damping coefficient c, static instability occurs for negative values of k, whereas dynamic instability corresponds to negative values of c.

The dynamic stability of hydrostatic thrust bearings and the so-called "pneumatic hammering effect" have previously been studied by Licht et al. [16] and Roudebush [26] for circular thrust bearings with a circular central recess, by Allen et al. [1] and Stowell [32] for annular thrust bearings, and by Chen and Ho [3] for capillary compensated multi-pocket thrust bearings. Here, the dynamic stability of one thrust-bearing operating in isolation is first analyzed to elucidate the underlying mechanisms before extending the approach to study the dynamic stability of both the forward and aft thrust bearings operating in tandem, as is commonly the case in micro-turbomachinery applications.

The state variables chosen for this analysis are the static pressure at the exit of the orifices, p, and the thrust-bearing gap, h. Referring to Fig. 6.36, small perturbations in mass flow into and out of the thrust-bearing gap arise due to small perturbations p' and h' about the equilibrium states p_0 and h_0. The time rate of change of fluid mass in the bearing gap is given by the difference in the perturbed mass flow rates into and out of the thrust-bearing gap:

$$\frac{dm'}{dt} = \dot{m}'_{in} - \dot{m}'_{out} = -\left(C_1 + C_2\right) p' - C_3 h', \tag{6.86}$$

where the coefficients are

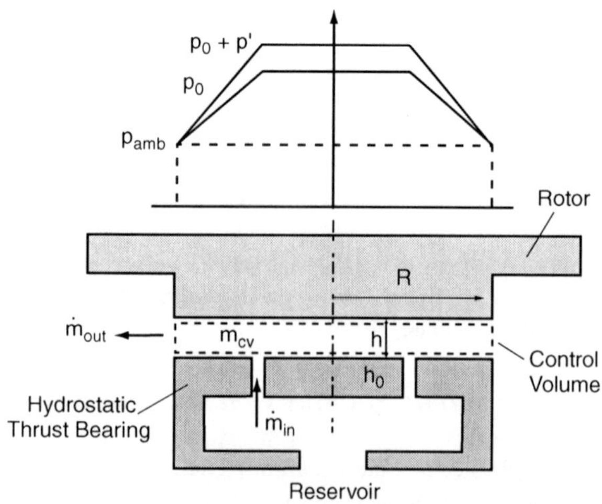

Fig. 6.36 Dynamic stability model for hydrostatic thrust bearings [34]

$$C_1 = \left(\frac{d\dot{m}_{in}}{dp}\right)_0, \quad C_2 = \left(\frac{\partial \dot{m}_{out}}{\partial p}\right)_0, \quad \text{and} \quad C_3 = \left(\frac{\partial \dot{m}_{out}}{\partial h}\right)_0 \tag{6.87}$$

respectively. It is useful to note that $\dot{m}'_{in}$ only depends on p' and does not depend on h'. This is due to the fact that the mass flow into the thrust-bearing gap has to match the mass flow out of the orifices. The perturbed mass flow out of the orifices, given by Eq. (6.59), is solely a function of the static pressure perturbations at the exit of the orifices since the thrust-bearing supply pressure, p_{supply}, the geometry of the orifices, and the fluid medium are assumed constant. Under steady-state conditions the mass of gas m_{CV} between the bearing surfaces is

$$m_{CV} = \int_0^{2\pi}\int_0^{R} \rho h_o r dr d\theta = \frac{h_0}{RT}\int_0^{2\pi}\int_0^{R} p r dr d\theta, \tag{6.88}$$

which can be evaluated once the pressure at every location along the thrust-bearing pad is known. Corresponding to perturbations p' and h', the perturbed time rate of change of the fluid mass between the bearing surfaces becomes

$$\frac{dm'_{CV}}{dt} = \left(\frac{\partial m_{CV}}{\partial p}\right)_0 \dot{p}' + \left(\frac{\partial m_{CV}}{\partial h}\right)_0 \dot{h}' = C_4\dot{p}' + C_5\dot{h}'. \tag{6.89}$$

From conservation of mass considerations, one can write using Eq. (6.86)

$$\frac{dm'_{CV}}{dt} = \dot{m}'_{in} - \dot{m}'_{out}. \tag{6.90}$$

Hence, one obtains

$$C_4\dot{p}' + C_5\dot{h}' + (C_1 + C_2)p' + C_3h' = 0. \tag{6.91}$$

The perturbations h' and p' are also coupled by Newton's second law governing the axial motion of the rotor. This yields

$$m_R\ddot{h}' = \left(\frac{\partial F_p}{\partial p}\right)_0 p' = \frac{RT}{h_0}C_4p' = C_6p', \tag{6.92}$$

where m_R is the mass of the rotor. Coupling Eqs. (6.91) and (6.92), the final ODE which governs the perturbed axial motion of the rotor takes the form

$$\dddot{h}' + \left(\frac{C_1 + C_2}{C_4}\right)\ddot{h}' + \frac{C_5C_6}{m_RC_4}\dot{h}' + \frac{C_3C_6}{m_RC_4}h' = 0, \tag{6.93}$$

which is a third-order differential equation in h'. The stability of the system is described by the above equation and is readily analyzed by applying the Routh–Hurwitz stability criterion. This yields the following criterion for dynamic stability:

$$\frac{C_1 + C_2}{C_3} > \frac{C_4}{C_5}. \tag{6.94}$$

The values of the coefficients C_1, C_2, C_3, C_4, and C_5 can be obtained using the steady state thrust-bearing model discussed in Section 6.7.1. Equation (6.94) reveals that one of the main factors which drives the system towards instability is a low value of coefficient C_1, which in fact becomes zero when the flow through the orifices of the thrust-bearing chokes. When this occurs, the mass flow out of the orifices becomes independent of the back pressure, thus

$$C_1 = \left(\frac{d\dot{m}_{in}}{dp}\right)_0 = 0. \tag{6.95}$$

It is also of interest to note that for an incompressible fluid with constant density, C_4 becomes zero and Eq. (6.93) reduces to a second-order ODE in h', which can then be expressed in a more familiar fashion as

$$m_R\ddot{h}' + c\dot{h}' + kh' = 0. \tag{6.96}$$

This is the equation of motion for a linear mass-spring-damper system, with spring constant k given by $C_3C_6/(C_1 + C_2)$ and damping constant c given by $C_5C_6/(C_1 + C_2)$. As discussed previously, static instability ensues when k becomes negative. However, static stability is necessary but not sufficient, and in practice dynamic instability is often encountered first. In the presence of dynamic effects, the phase difference between the unsteady perturbations in p' and h' may be such that in one cycle or period of the perturbations, the net disturbance energy input into the system is positive. This causes the amplitude of h' to experience exponential growth and results in dynamic instability.

Typically the micro-turbomachinery rotors are axially supported by a forward and an aft thrust bearing and it is necessary to extend the preceding analysis and to couple the disturbance of the rotor axial location to the dynamics of a dual-thrust bearing system. This eventually leads to a coupled system of equations:

$$\begin{aligned} &m_R\dddot{h}' + \left(\tfrac{C_{1a}+C_{2a}}{C_{4a}}\right) m_R\ddot{h}' + \left(\tfrac{C_{5a}C_{6a}}{C_{4a}} + \tfrac{C_{5f}C_{6f}}{C_{4f}}\right)\dot{h}' + \\ &\left(\tfrac{C_{2a}C_{6a}}{C_{4a}} + \tfrac{C_{2f}C_{6f}}{C_{4f}}\right) h' + C_{4f}\left(\tfrac{C_{1a}+C_{2a}}{C_{4a}} - \tfrac{C_{1f}+C_{2f}}{C_{4f}}\right) p_f' = 0 \end{aligned} \tag{6.97}$$

and

$$C_{5f}\dot{h}' + C_{2f}h' - C_{4f}p_f' - (C_{1f} + C_{2f})p_f' = 0. \tag{6.98}$$

The subscripts 'f' and 'a' in the above equations denote the forward and aft thrust bearings, respectively. In contrast to Eq. (6.93) the order of the characteristic polynomial or the number of poles for the system is now increased from three to four.

The system is dynamically stable if, and only if, all the roots of the characteristic equation have negative real parts, indicating exponential decay of the oscillations.

6.7.3.1 Dynamic Instability in Dual Hydrostatic Thrust Bearings

To illustrate the nature of thrust-bearing instability, a dynamic stability analysis is carried out on the thrust-bearing system of a micro-turbocharger using the analytical model described above. The details of the geometry can be found in Savoulides [28]. The analysis predicts that the thrust bearings are dynamically stable (the roots of the characteristic equations Eqs. (6.97) and (6.98) are either real or have negative real parts) when the supply pressure to both thrust bearings is kept below 60 psi and the rotor is axially centered. Under these conditions the orifices of both thrust bearings are unchoked. Increasing the supply pressure to both thrust bearings above 60 psi for an axially centered rotor, the analytical model predicts the onset of dynamic instability.

To further investigate the nature of dual thrust-bearing dynamic instability, the roots of the characteristic equation can be evaluated for different rotor axial positions and higher thrust-bearing supply pressures. At a bearing pressure setting of 120 psi, two of the roots are real and negative, while the remaining two roots are complex for all axial rotor positions. It is the pair of complex roots that govern dynamic instability. For nearly centered rotor axial positions (± 0.3 non-dimensional clearance), the flow through the orifices of *both* thrust bearings is choked. As a consequence, the two complex roots have positive real parts and the system is dynamically unstable. For larger positive and negative rotor axial positions, the flow through the orifices of one of the thrust bearings unchokes while the other side remains choked. The real part of the two complex roots becomes negative, indicative of a dynamically stable system. In conclusion, even though an isolated single thrust bearing by itself is dynamically unstable at choked flow conditions, the overall dual-thrust bearing system remains stable. The unchoked thrust bearing is capable of stabilizing the entire bearing system.

To test the above model prediction, experiments were performed in a micro-turbocharger device [28]. Both the rotational speed and axial position of the rotor were measured using an ultra-high frequency response fiber-optic sensor overlooking a series of speed bumps etched on the surface of the rotor. When the supply pressure to both thrust bearings was increased above 60 psi, the fiber-optic sensor detected a high-frequency axial oscillation of the rotor (this frequency was within the audible range of the human ear and could actually be heard). The frequency of the axial oscillations for various thrust-bearing supply pressures was quantified by passing the real-time voltage signal through a spectrum analyzer. The measured frequency of oscillations is shown in Fig. 6.37 together with analytical predictions for various thrust-bearing pressure setting. Unstable oscillations occur for thrust-bearing supply pressures in excess of 60 psi when the flow through the orifices of

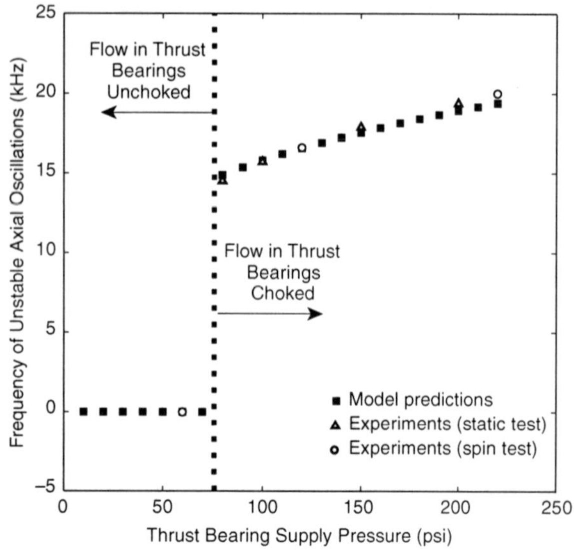

Fig. 6.37 Axial oscillations in hydrostatic thrust bearings – measured instability frequency compared with analytical model results [34]

both thrust bearings is choked. The analytical model predictions are in good agreement with the experimental measurements validating the ability of the simple model to accurately predict the frequency of the instability.

Based on this analysis, the following simple criterion for thrust-bearing dynamic instability can be established: if the flow through both forward and aft thrust bearings is choked, the system is dynamically unstable.

6.7.3.2 Pitch and Yaw Dynamics of a Rotor with Dynamic Imbalance

Dynamic imbalance induced pitching and yawing motion (see Fig. 6.31) can arise for example due to a non-uniform disk thickness or a missing blade. In this section, the pitch and yaw dynamics of a rotor-thrust-bearing system are analyzed solving Eqs. (6.65) and (6.66). For an isotropic system with $k_{xx} = k_{yy} = K^{hs}$, the yaw and pitch angles β_x and β_y are equal and the tilting angle of the rotor is simply quantified by a single parameter β. The Green's function formulation discussed in Section 6.7.2.1 can be applied to evaluate the variation of K^{hs} and c as a function of β. The values of K^{hd} are determined using the Reynolds equation. Two primary ways of reducing the values of β at high speeds are: (1) to increase the direct-coupled thrust-bearing tilting stiffness K^{hs} and (2) to maintain low levels of dynamic imbalance χ by reducing the rotor etch non-uniformity and potential bond misalignments between the turbomachinery disks. Figure 6.38 illustrates the variation of the inclination angle β with rotational speed for a microturbocharger rotor with three different values of dynamic imbalance χ subjected to the same thrust-bearing pressures and therefore direct-coupled tilting stiffness.

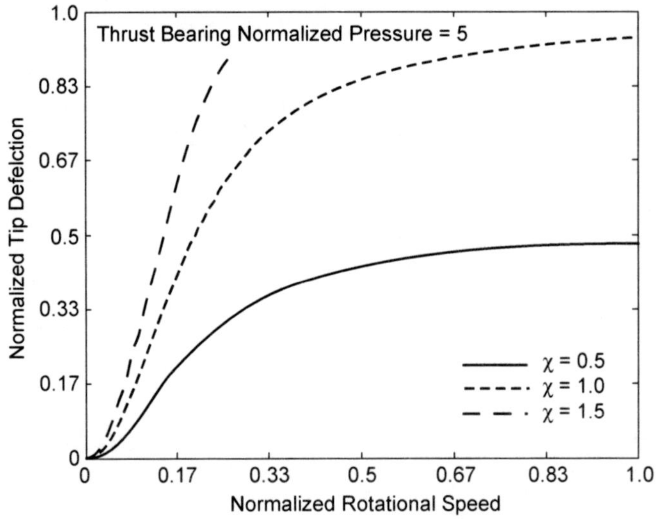

Fig. 6.38 Effect of dynamic imbalance on rotor disk tip deflection [35]

At a fixed rotational speed, the inclination angle β is greater for rotors with larger dynamic imbalance χ. Assuming the maximum allowable tip deflection is 3 μm,[14] the rotor with a dynamic imbalance of $\chi = 1.5$, which corresponds to an asymptotic tip deflection of 4.5 μm, is suggested to fail at a quarter of the design speed. Without considering the effects of dynamic stability, the maximum allowable tilting angle is limited by the total thrust-bearing clearance or the minimum clearances imposed by the presence of labyrinth seals. The above analysis thus implies that, in order to reduce the likelihood of physical contact between the rotor and static structure as the rotor accelerates, it is necessary to ensure that the rotor possesses a sufficiently low normalized dynamic imbalance χ of about 0.5.

The effect of thrust-bearing direct-coupled tilting stiffness is shown in Fig. 6.39 for rotors with a normalized dynamic imbalance of $\chi = 1$. An increase in thrust-bearing supply pressure and therefore direct-coupled tilting stiffness results in smaller rotor inclination angles β for the same rotational speed, reducing the likelihood of physical contact between the rotor and the static structure.

The analysis implies that if it is possible to quantify the dynamic imbalance of the rotor during the microfabrication process, the two degree-of-freedom model for the yaw and pitch dynamics can be used to predict the steady-state inclination behavior of the rotor at different operating speeds and thrust-bearing conditions.

The dynamic stability of the rotor-thrust-bearing system subjected to small perturbations in rotor inclination angles can by assessed by solving the eigenvalue

[14] This corresponds to a normalized dynamic imbalance $\chi = 1.0$.

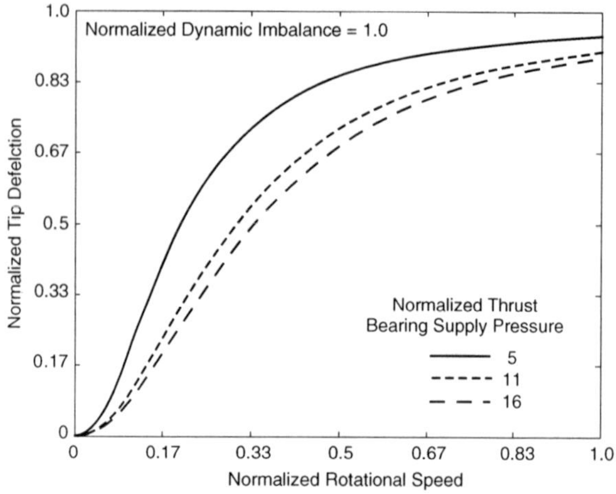

Fig. 6.39 Effect of thrust-bearing direct-coupled tilting stiffness on rotor disk tip deflection [35]

problem in Eq. (6.67). If any of the eigenvalues has a positive real part, the system is dynamically unstable in its pitching and yawing motion. For the specific case of the microturbocharger rotor with a large dynamic imbalance of $\chi = 1.0$ and supported by baseline-type thrust bearings with supply pressures of 60 psi, the eigenvalues are found to be complex conjugate pairs. The real parts of each complex conjugate pair are found to be negative, implying linear stability of the inclination or coning motion. Although the real parts of the eigenvalues are all negative, some of them have magnitudes very close to zero (corresponding to neutral stability). Coupling effects from other components of the system, such as the gas journal bearings located at the outer periphery of the rotor can potentially render them positive and hence unstable. Based on this analysis, a multi-degree of freedom rotordynamic analysis including coupling effects between radial, axial, pitching, and yawing motion can be found in Liu [17].

6.8 Annular Seals

Gas seals play an important role in the design of powerMEMS devices as the turbomachinery and gas-bearing components are closely coupled. Thrust and journal bearing feed systems are located next to each other and often exhaust to common plena or mix with the primary turbomachinery flows. Returning to the gas-bearing supported MEMS air-turbine in Fig. 6.2, an inner plain seal separates the aft exhaust from the thrust balance plenum, which is in turn isolated from the journal bearing feed chambers by an outer labyrinth seal consisting of row of annular teeth and grooves. Unlike in stationary devices, static or dynamic instability may arise due to rotordynamic coupling between the annular seal flows and other integral components of the powerMEMS device. A key challenge in the design of such seals is to ensure that the leakage flows do not deteriorate the performance and the dynamic stability of the high-speed gas bearings.

6.8.1 Negative Radial Stiffness in Annular Plain Seals

In annular plain seals, a negative radial stiffness can develop. Consider the support system of the MEMS air-turbine in Fig. 6.2. The hydrostatic flow to the journal bearing is supplied axially via feed chambers on the aft side of the rotor as shown in Fig. 6.40 (a). To enhance bearing dynamic stability and to relieve the stringent microfabrication tolerance requirements as discussed in Section 6.6.3.1, stiffness anisotropy in the journal bearing is introduced by blocking circumferential portions of the journal bearing feed via annular seals shown in Fig. 6.40(b). Under normal operating conditions, the static pressure difference between the exit of the journal bearing and the annular seal in the blocked region drives a leakage flow from the turbine side into the journal bearing through the annular seal towards the balance plenum.

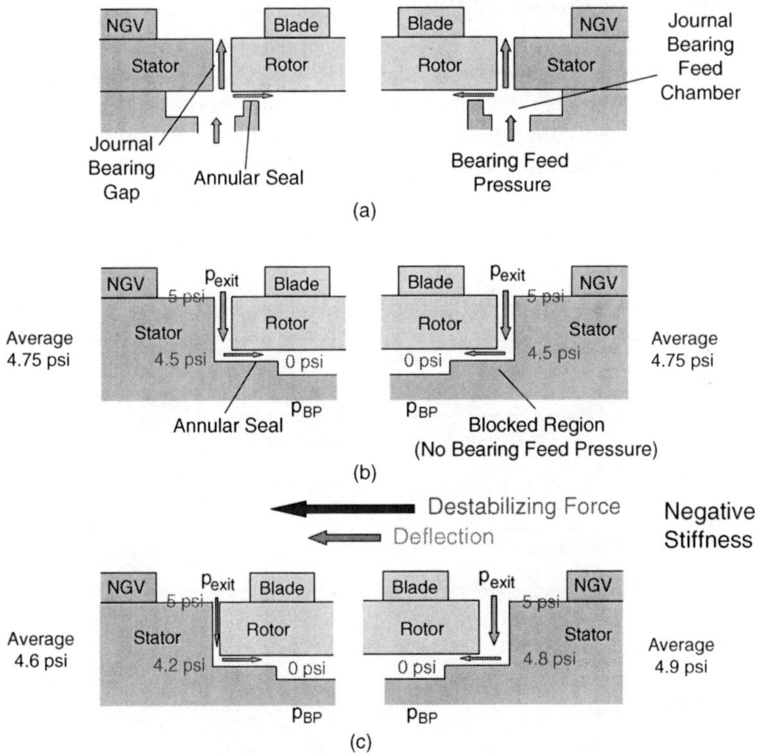

Fig. 6.40 Mechanism for negative (destabilizing) radial stiffness due to coupling of reverse flow through journal bearing land and annular seal in anisotropic bearing configuration [36]

A negative radial stiffness is generated because the reverse flow through the journal bearing occurs in series with the flow through the annular seal. This mechanism is illustrated above. Consider a nominal overall static pressure difference of 5 psi across the journal bearing and seal in Fig. 6.40(b): the rotor is radially centered and there is no net radial force acting on the rotor. However, when the rotor is deflected by a small radial perturbation, as shown in Fig. 6.40(c), the pressure drop or hydraulic loss across the journal increases over the region of reduced journal bearing clearance and decreases over the enlarged clearance on the opposite side. This yields a higher average static pressure in the bearing land on the enlarged clearance side and reduces the average pressure on the opposite side. Thus, a net radial force acts in the direction of rotor displacement inducing a destabilizing negative stiffness. The cause for the negative radial stiffness is the coupling of the flow through the journal bearing and the annular seal. To avoid this undesirable coupling, the flow resistance in the journal bearing land must not be directly connected in series with any other hydraulic resistance.

6.8.2 Plain Seal Radial Stiffness Model

The magnitude of the negative stiffness can be quantified using an analytical fluid resistance model. The key components and station numbers are outlined in the insert in Fig. 6.41. A small radial perturbation e of the rotor in the y-direction yields a circumferential variation in the journal bearing clearance, given by $C(\theta) = C_0 - e\cos\theta = C_0(1 - \varepsilon\cos\theta)$, where $\varepsilon = e/C_0$ is the radial eccentricity normalized by the nominal journal bearing clearance C_0.

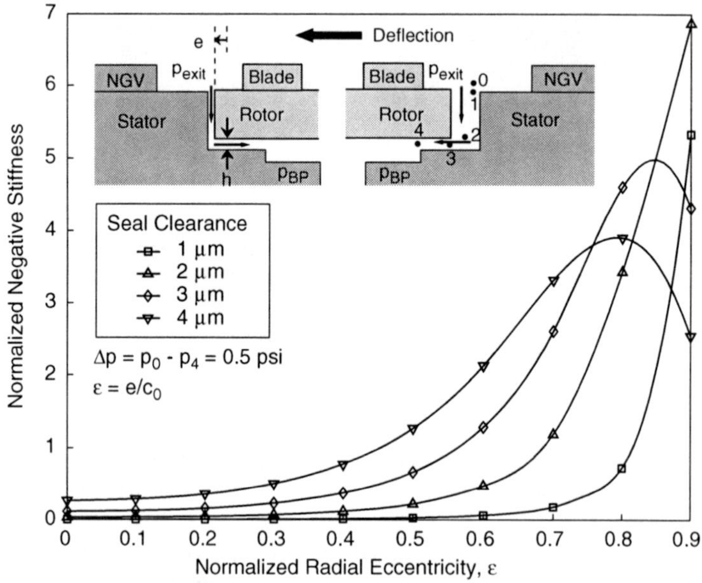

Fig. 6.41 Negative radial stiffness in plain annular seals normalized by journal bearing radial stiffness

Consider a small circumferential segment $C(\theta) \cdot Rd\theta$ over the blocked region. The inlet loss from stations 0 to 1 of the leakage flow is manifested as a drop in stagnation pressure as the flow enters the journal bearing. This loss can be described as

$$\frac{\Delta p_t}{p_t - p} = 0.5, \tag{6.99}$$

where the loss coefficient of 0.5 for this situations can be found in handbooks for hydraulic resistance [13]. The axial leakage flow through the journal bearing from stations 1 to 2 is modeled as a fully developed,[15] incompressible Poiseuille flow, and the local flow rate is given by

$$q(\theta) = \frac{C(\theta)^3}{12\mu} \frac{p_1 - p_2(\theta)}{L}, \tag{6.100}$$

where L is the journal bearing length and μ is the dynamic viscosity of the fluid. The turning losses experienced by the flow from stations 2 to 3 are modeled using the following empirical correlation by Vohr [39]

$$\frac{\Delta p_t}{p_t - p} = 0.31\left(\frac{Re}{2000}\right) - 0.122\left(\frac{Re}{2000}\right)^2 + 0.282\left(\frac{Re}{2000}\right)^3 + 0.179 \tag{6.101}$$

where $Re = 2\rho q(\theta)/\mu$, and ρ is the fluid density. The radially inward leakage flow through the annular seal from stations 3 to 4 is modeled as a fully developed, incompressible, viscous flow, and the local flow rate is given by

$$q(\theta) = \frac{\pi h^3}{6\mu (Rd\theta)} \frac{p_3(\theta) - p_4}{\ln(r_3/r_4)}, \tag{6.102}$$

where h is the seal clearance, r_3 and r_4 are the inner and outer radial locations of the annular seal, respectively. For a given static pressure difference $p_1 - p_4 = p_{exit} - p_{BP} = \Delta p$ driving the leakage flow, values of the local leakage flow rate $q(\theta)$ and static pressures $p_2(\theta)$ and $p_3(\theta)$ can be determined. The resultant radial pressure force acting on the rotor and hence the magnitude of the negative stiffness can be evaluated by circumferentially integrating the pressure distribution over the blocked regions along the journal bearing land.

Figure 6.41 shows the magnitude of the normalized negative radial stiffness generated for different values of rotor normalized radial eccentricity ε, corresponding to different values of seal clearance h, and for a fixed $\Delta p = p_{exit} - p_{BP}$ value of 0.5 psi driving the leakage flow. The negative stiffness is normalized by the journal bearing hydrostatic stiffness in the y-direction for a constant hydrostatic differential pressure of 1.0 psi. The nominal journal bearing clearance C_0 and length L

[15] Since the ratio of the axial flow-through time of the leakage flow to the viscous diffusion time is much larger than unity, the flow can be assumed fully developed.

are 18 and 300 μm, respectively. The inner and outer radial locations of the annular seal are $r_3 = 1.99$ mm and $r_4 = 2.1$ mm, respectively. The magnitude of the negative stiffness is sensitive to the seal clearance h, and is comparable with the journal bearing radial stiffness. For values of rotor eccentricity ε greater than 0.5, the negative stiffness overwhelms the journal bearing stiffness and can become three to four times larger. Peaks in the negative stiffness are observed for seal clearances h of 3 and 4 μm, which occur when the hydraulic resistance for the journal bearing land approximately matches the hydraulic resistance of the annular plain seal.

To avoid potential destabilizing effects, annular plain seals must be carefully designed. To significantly reduce the negative stiffness arising in situations where a variable resistance, such as the flow resistance through a bearing land, is coupled in series with the hydraulic resistance of another component, labyrinth seals can be employed. Such seals consist of a row of teeth and grooves. Compared to the teeth, the grooves yield negligible hydraulic resistance in the circumferential direction and mitigate possible circumferential static pressure non-uniformities. Thus, any possible net torque or destabilizing tilting stiffness on the rotor can be appreciably reduced.

6.8.3 Annular Labyrinth Seal Model

In view of the low Reynolds numbers and high pressure ratios across the seals, compressibility effects have to be taken into account to accurately estimate the flow characteristics of annular seals. A compressible flow model based on the method of influence coefficients [29] was developed to characterize the steady-state behavior of annular labyrinth seals in powerMEMS devices.

The main objective of the model is to quantify the dependence of the leakage mass flow on seal clearance distribution and static pressure ratio. The analytical fluid resistance model accounts for drops in stagnation pressure due to seal inlet losses and viscous losses over the seal face. The inlet stagnation pressure loss at the entrance to the teeth of the labyrinth seal is described by

$$\frac{\Delta p_t}{p_t - p} = 0.5. \qquad (6.103)$$

The quasi one-dimensional, viscous, compressible flow through each tooth is then computed using influence coefficients for effects of friction, heat addition, and flow area change. In view of the high thermal conductivity of silicon, the flow through the teeth is assumed to be isothermal. The change in static temperature dT/T due to changes in cross sectional area dA/A, stagnation temperature dT_t/T_t and a drop in stagnation pressure arising from viscous dissipation along a radial distance dr along the teeth yields

$$\begin{aligned}\frac{dT}{T} =& \frac{(\gamma-1)M^2}{1-M^2}\frac{dA}{A} + \frac{(1-\gamma M^2)\left(1+\frac{\gamma-1}{2}M^2\right)}{1-M^2}\frac{dT_t}{T_t}\\ &- \frac{\gamma(\gamma-1)M^4}{2(1-M^2)}4C_f\frac{dr}{D_h}\end{aligned} \tag{6.104}$$

where C_f is the local friction factor, D_h is the local hydraulic diameter, and γ is the ratio of specific heats of the fluid. The change in the square of the Mach number dM^2/M^2 can be written as

$$\begin{aligned}\frac{dM^2}{M^2} =& -2\frac{\left(1+\frac{\gamma-1}{2}M^2\right)}{1-M^2}\frac{dA}{A} + \frac{(1+\gamma M^2)\left(1+\frac{\gamma-1}{2}M^2\right)}{1-M^2}\frac{dT_t}{T_t}\\ &+ \frac{\gamma M^2\left(1+\frac{\gamma-1}{2}M^2\right)}{(1-M^2)}4C_f\frac{dr}{D_h}\end{aligned} \tag{6.105}$$

and the change in static pressure dp/p can be expressed as

$$\begin{aligned}\frac{dp}{p} =& \frac{\gamma M^2}{1-M^2}\frac{dA}{A} + \frac{\gamma M^2\left(1+\frac{\gamma-1}{2}M^2\right)}{1-M^2}\frac{dT_t}{T_t}\\ &- \frac{\gamma M^2(1+(\gamma-1)M^2)}{2(1-M^2)}4C_f\frac{dr}{D_h}\end{aligned} \tag{6.106}$$

In the above Eqs. (6.104), (6.105), and (6.106), the influence coefficients only depend on the local Mach number M and the specific heat ratio γ. Furthermore, the change in flow cross sectional area is given by

$$\frac{dA}{A} = \frac{dr}{r}, \tag{6.107}$$

and the local friction factor can be expressed as

$$C_f = \frac{\tau_w}{0.5\rho V^2} = \frac{12}{Re_h}, \tag{6.108}$$

where $Re_h = \rho V h/\mu$ is the Reynolds number based on the seal clearance h, and ρ and μ represent the density and dynamic viscosity of the fluid, respectively. Hence,

$$4C_f\frac{dr}{D_h} = \frac{48\mu\pi r}{\dot{m}h}dr. \tag{6.109}$$

Using Eq. (6.104) to eliminate dT_t/T_t in Eqs. (6.105) and (6.106) yields

$$\frac{dM^2}{M^2} = -\left[\frac{2\left(1+\frac{\gamma-1}{2}M^2\right)}{1-M^2} + \frac{(\gamma-1)M^2\left(1+\gamma M^2\right)}{\left(1-M^2\right)\left(1-\gamma M^2\right)}\right]\frac{dr}{r}$$
$$+\left[\frac{\gamma(\gamma-1)M^2\left(1+\gamma M^2\right)}{2\left(1-M^2\right)\left(1-\gamma M^2\right)} + \frac{\gamma M^2\left(1+\frac{\gamma-1}{2}M^2\right)}{1-M^2}\right]\frac{48\mu\pi r}{\dot{m}h}dr, \tag{6.110}$$

and

$$\frac{dp}{p} = \left[\frac{\gamma M^2}{1-M^2} + \frac{\gamma(\gamma-1)M^4}{\left(1-M^2\right)\left(1-\gamma M^2\right)}\right]\frac{dr}{r}$$
$$+\left[-\frac{\gamma^2(\gamma-1)M^6}{2\left(1-M^2\right)\left(1-\gamma M^2\right)} - \frac{\gamma M^2\left(1+(\gamma-1)M^2\right)}{2\left(1-M^2\right)}\right]\frac{48\mu\pi r}{\dot{m}h}dr. \tag{6.111}$$

Equations (6.110) and (6.111) are used to compute the changes in Mach number and static pressure across the entire tooth width. In view of the significantly larger depths of the grooves as compared to the teeth, the static pressure along each groove can be assumed constant, implying that the radial and circumferential hydraulic resistances in the grooves are negligible. In addition, it is assumed that all of the dynamic pressure is dissipated when the flow exits a tooth and exhausts into a groove. For a specified seal clearance, together with prescribed seal inlet and exit static pressure boundary conditions, the required mass flow rate through the seal can then be calculated.

Figure 6.42 compares experimental measurements and model predictions of the corrected seal mass flow as a function of static pressure ratio across the labyrinth seal of an MEMS-turbopump [6] for a given seal clearance. The labyrinth seal consists of two teeth with inner radial locations at 0.9 and 0.975 mm. The radial width of each tooth is 250 μm, and the two teeth are separated by a deep groove 500 μm wide. In contrast to an incompressible flow model, the results of the fully compressible flow analysis are in good agreement with the experimental measurements, confirming the importance of modeling compressibility effects.

The analytical model can also be used to determine the rotor axial position. This method was applied during high-speed testing of the same MEMS-turbopump [6]. Figure 6.43 depicts analytical results for the corrected mass flow plotted as functions of the static pressure ratio across the labyrinth seal for various seal clearances (and hence rotor axial positions). The maximum seal clearance is 5 μm at which the rotor contacts the forward thrust-bearing pad. Any given combination of corrected mass flow and static pressure ratio across the labyrinth seal corresponds to a unique seal clearance. The rotor axial location can thus be deduced by measuring the mass flow and pressure ratio across the labyrinth seal.

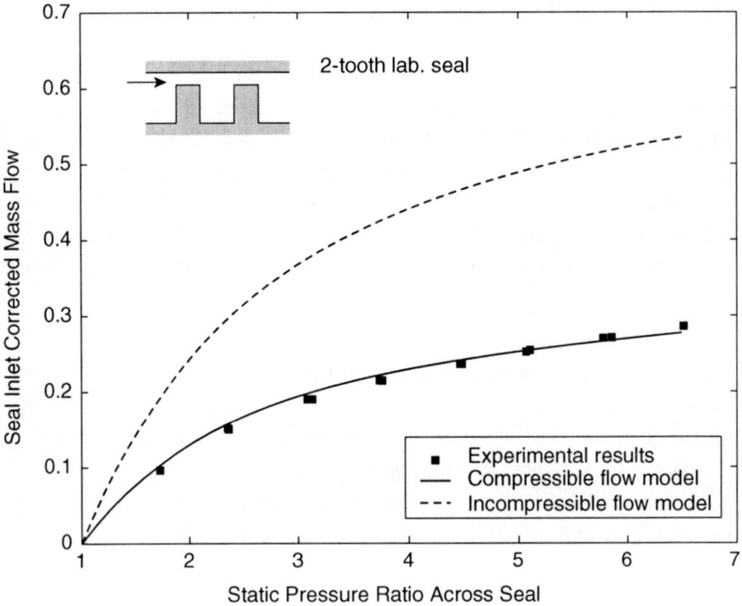

Fig. 6.42 Experimental measurements and model predictions of a two-tooth labyrinth seal in an MEMS-turbopump – seal inlet corrected mass flow as a function of static pressure ratio [36]

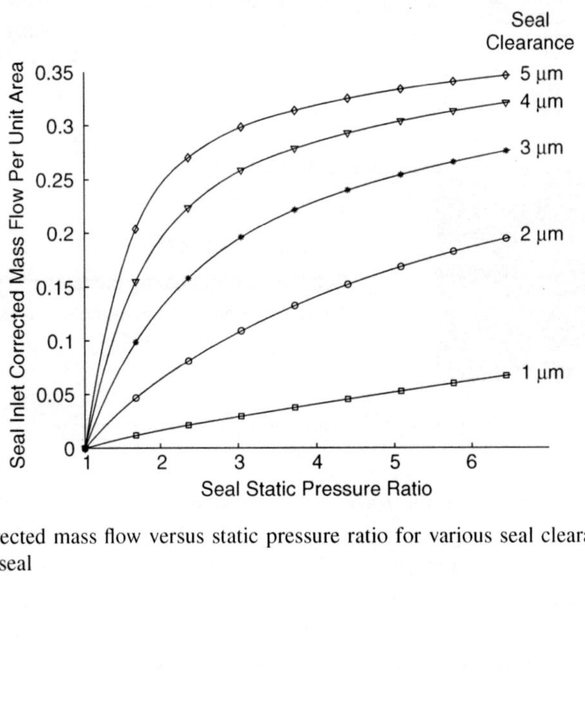

Fig. 6.43 Corrected mass flow versus static pressure ratio for various seal clearances in a two-tooth labyrinth seal

6.9 Experimental Considerations

Unlike for macroscale gas-bearing systems, where proximity and acceleration sensors can be readily implemented and placed along the rotor axis, experimental testing of the MEMS-based micro-gas bearings yields yet another set of challenges. Due to the small scale, new experimental methods and data reduction schemes have to be employed to deduce the rotordynamic whirl orbits, natural frequencies, and damping ratios. This section introduces instrumentation and diagnostic methods, and the experimental setup used in the high-speed testing of micro-gas bearings.

6.9.1 Instrumentation and Diagnostics

The experimental setup is schematically shown in Fig. 6.44. The turbine and bearing flow rates and pressures are regulated to achieve a particular rotational speed and a particular pressure difference setting across the journal bearing (the bearing pressure difference Δp sets the stiffness and thus natural frequency of the journal gas bearing).

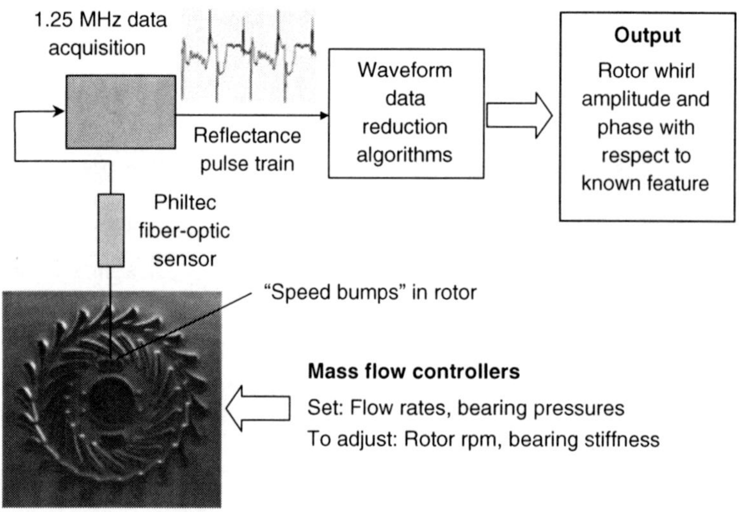

Fig. 6.44 Experimental setup for dynamic imbalance measurements [19]

The rotational speed of the rotor is determined using a Philtec fiber optic sensor that is inserted through an opening on top of the device, tracking the passage of speed bumps on the surface of the rotor. In order to obtain the whirl response curve for a particular Δp, fiber optic data are acquired over a range of rotational speeds with a sampling rate of 1.25 MHz. The bearing response is deduced using a waveform data reduction algorithm [34] which is briefly outlined below.

Experiments in the micro-bearing supported turbine test-rig were conducted using a high-resolution fiber optic sensor to measure rotor speed. A data reduction scheme was implemented to obtain the imbalance-driven whirl response curves of the rotor for different operating conditions. The stiffness, the natural frequency, and the damping ratio of the journal bearing and the imbalance of the rotor were deduced from the data. Additional experiments employed a flow control valve supplying the journal bearing with air. The natural frequency of the rotor bearing system was then deduced from bearing pressure measurements during rotor acceleration transients.

The imbalance of the rotor (i.e., the distance between the geometric center of the rotor and its mass center) induces a whirling motion superimposed on the rotational motion of the rotor as shown in Fig. 6.45. The precession frequency of precession-induced imbalance is known to be synchronous with the rotational frequency of the rotor, and the precession velocity yields an increase or decrease in the absolute

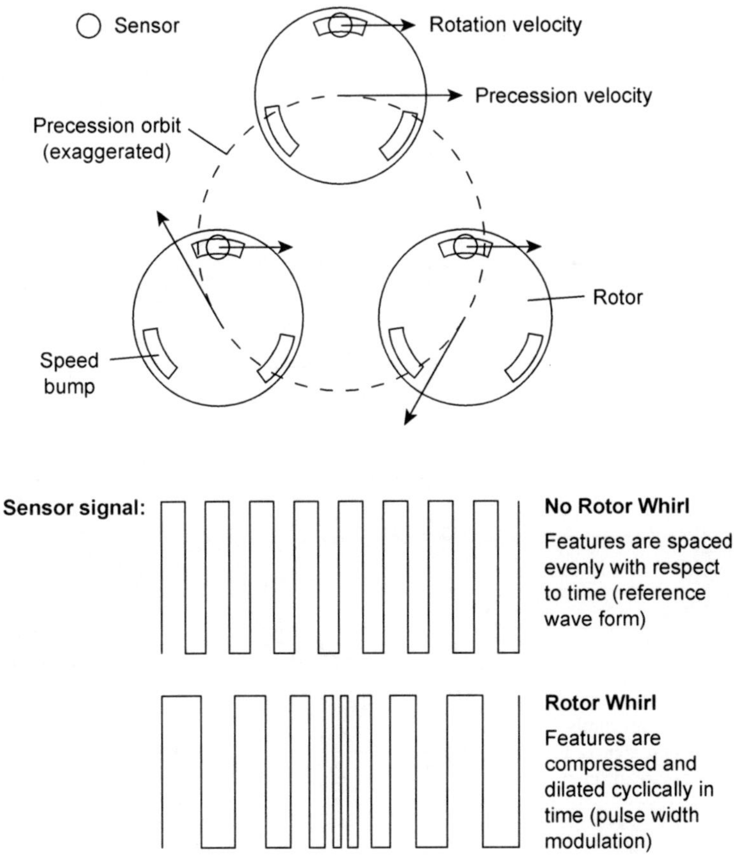

Fig. 6.45 Effect of whirling motion on speed bump absolute velocity and sensor signal wave form [33]

velocity of the speed bumps. This is reflected in the fiber optic signal where the features of the speed bumps are cyclically compressed or dilated in time. The resulting sinusoidal waveform is essentially a pulse-width modulation of the reference waveform obtained in the absence of rotor whirl.[16] A least-squares technique is used to deduce the optimum waveform dilation necessary to transform the whirl waveform to the reference waveform. The amplitude and the phase of this modulating function correspond to the size of the precession orbit and to the position or orientation of the orbit with respect to a particular feature on the rotor. The whirl-induced response curve is obtained by repeating this procedure for a range of rotational speeds at a fixed bearing pressure difference Δp.

Figure 6.46 depicts the response curve (amplitude and phase) of the micro-bearing rotor obtained from fiber optic measurements at a bearing pressure difference of $\Delta p = 2.0$ psi. The measurement shows a peak whirl amplitude of 7.7 μm at a natural frequency of 35,000 rpm. The transition from subcritical operation (rotation

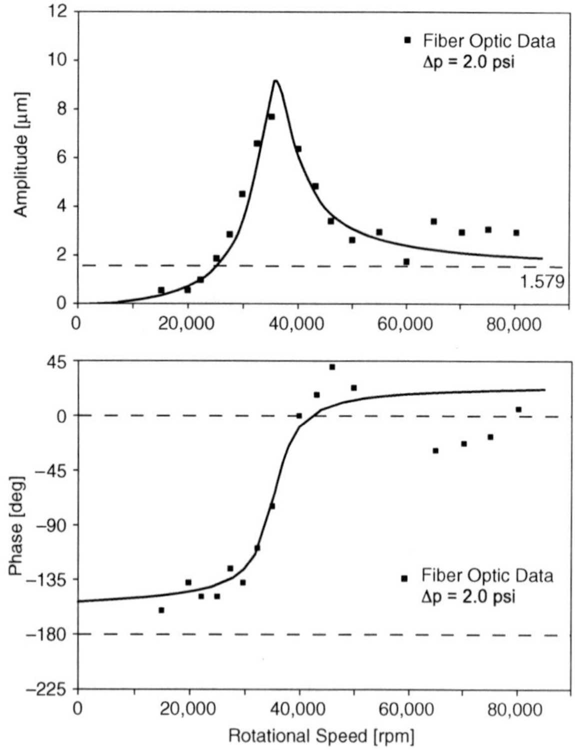

Fig. 6.46 Experimentally measured response curve for a bearing pressure difference $\Delta p = 2.0$ psi [19]

[16] The whirl amplitude of the rotor is negligible at very low speeds and the reference waveform is obtained from measurements at very low speed.

around the geometric center of rotor) to supercritical operation (inverted or self-centered rotation around mass center of rotor) is also visible in the phase shift of approximately 180° near the natural frequency. From this, the amplitude and location of the imbalance can be deduced. It can be seen from Fig. 6.46 that the mass center of the rotor is displaced by approximately 1.6 μm from the geometric center of the rotor, yielding an imbalance of about 1.6 μm. Note that there is an additional roll-off in phase and a change in amplitude at about 65,000 rpm (squares) since there is another resonance at play due to stiffness anisotropy as discussed in Section 6.6.3.1.

In general, the maximum peak response in the transcritical region increases with the level of imbalance, rendering the higher likelihood of the rotor colliding with the surrounding wall when accelerating or decelerating between subcritical to supercritical speeds. A high level of imbalance promotes instability, reduces the range of stable operation, and thus limits the maximum operational speed. Experience shows that devices with imbalance levels exceeding 5 μm may encounter difficulties crossing the natural frequency at bearing pressure settings of $\Delta p = 1$ psi. The above results provide the means to guide any dynamic balancing procedure that might be required using techniques such as laser ablation of the rotor blades to selectively remove mass from the rotor at a particular location.

Figure 6.47 shows the effects of bearing pressure difference Δp on the rotordynamic behavior of the micro-bearing rig. The natural frequency of the journal gas bearing increases monotonically with Δp, indicating that a higher hydrostatic pressure across the bearing results in a stiffer radial support. The trend of increasing stiffness as well as the magnitudes of the natural frequency compare well with the scaling laws discussed in Section 6.5.5.

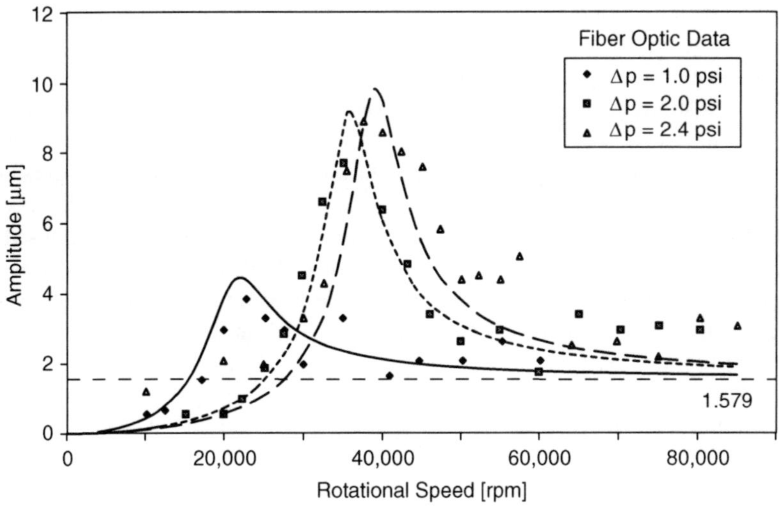

Fig. 6.47 Experimentally measured response curves for a bearing pressure differences $\Delta p = 1.0$, 2.0, and 2.4 psi [19]

An increase in Δp also affects the peak response amplitude and the damping of the rotor-bearing system. The distance to contact with the surrounding stationary wall decreases (the peak amplitude rises) as the bearing pressure difference is increased. Although a higher Δp results in stiffer bearings, the damping ratio $\zeta = C/(2m\,\Omega_N)$ decreases. The trend of decreasing damping ratio with increasing hydrostatic pressure across the journal bearing is consistent with the modeling results and scaling laws described in Section 6.5.5.

6.9.2 Operating Protocols – Strategies for Crossing Natural Frequencies and Acceleration to High Speed

The findings on the bearing dynamic behavior and the scaling laws yield important implications on the prospect of high-speed bearing operation. The following can be concluded for ultra-short gas bearings: (1) low values of bearing pressure difference Δp and thus higher damping ratios are necessary to cross the natural frequency without collision and contact with the journal wall. This was observed in tests performed on a number of different microturbomachinery devices where very low values of Δp were required to successfully cross the natural frequency and to invert the rotor [6, 28, 34]. (2) The level of imbalance obtained for different values of bearing pressure difference Δp is consistently about 1.6 μm. This value compares well to the static imbalance deduced from etch-variation or blade height measurements along the periphery of the rotor prior to wafer bonding[17] (for details see Miki et al. [22]). The maximum allowable imbalance for stable high-speed operation is discussed in more detail in Section 6.10.2.

One of the major challenges encountered in the experiments is the ability to smoothly transition from subcritical to supercritical operation when accelerating to high speed. Two different means to cross the natural frequency have been investigated and are briefly outlined. Figure 6.48 depicts the two strategies for acceleration from subcritical to supercritical speeds.

The first strategy involves an instantaneous acceleration of the rotor past its natural frequency at constant bearing pressure difference Δp (strategy A). This is carried out by a sudden increase in the main flow through the turbine and, as suggested by the model predictions and experimental data, the bearing pressure difference Δp needs to be set to low values to yield a higher damping ratio and an increased distance to contact. The second strategy is to accelerate from subcritical to supercritical speed by decreasing the pressure difference across the journal Δp at constant rotor speed as shown in Fig. 6.48 by strategy B. Both strategies have been successfully implemented and allowed the controlled operation of the micro-bearing rig from subcritical to supercritical speeds. The same strategies were applied when decelerating from supercritical speed to subcritical speed and showed a hysteresis

[17] For devices with single wafer rotors, wafer misalignment does not yield additional imbalance. Wafer misalignment effects become an issue for multi-wafer rotor designs such as the rotor design of a micro-turbocharger [28].

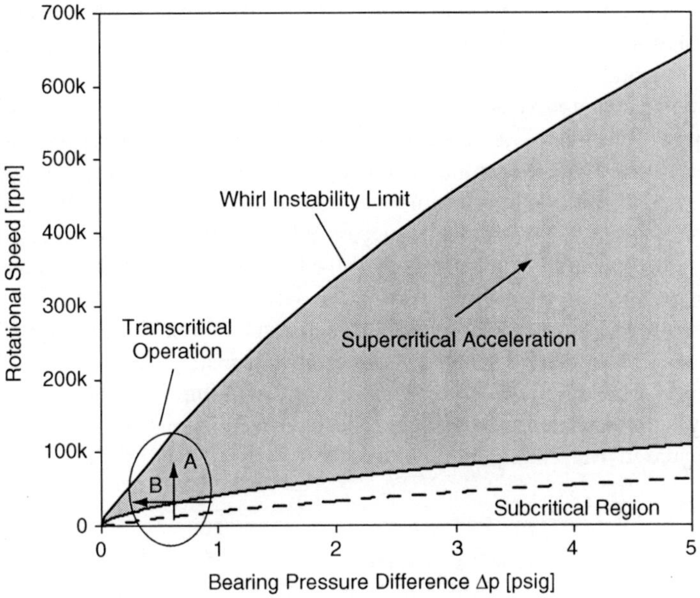

Fig. 6.48 Strategies to cross the natural frequency and bearing operating protocol for acceleration to high speed [19]

effect. The hysteresis effect is consistent with the modeling results and is conjectured to be due to the non-linear stiffening effect that results in a rightward leaning response curve as shown in Fig. 6.5.

From the above experimental data and the modeling results an operating protocol can be established to guide the stable operation of the micro-bearing rig. The approach to controlled acceleration to high speed is to cross the natural frequencies at low-bearing pressure differences Δp or low speeds and then to stay within the corridor bounded by the whirl instability limit and the highest natural frequency as illustrated in Fig. 6.48 by the shaded area. This type of operating protocol has been successfully implemented in a number of microturbomachinery devices. Attempts have also been made to automate the bearing operation by scheduling the journal bearing supply flow and the turbine main flow using a single pressure source and metering valves. The "single knob" automated operation of the micro-bearing rig has been experimentally demonstrated at low speeds and is under further investigation for high-speed operation.

6.9.3 Demonstration of High DN Bearing Operation

The gas-bearing supported MEMS micro-air turbine shown in Fig. 6.2 was tested to demonstrate repeatable, stable high-speed gas-bearing operation, and to validate the micro-gas bearing theories. The design features include a four-chamber journal bearing feed system to introduce stiffness anisotropy, labyrinth seals to reduce rotordynamic coupling effects of leakage flows, a reinforced thrust-bearing structural design, a redesigned turbine rotor to increase shaft power, a symmetric feed system to avoid flow and force non-uniformity, and a new rotor microfabrication methodology to improve uniformity and imbalance. Controlled high-speed operation up to 70% of design speed was demonstrated. This corresponds to a rotor rotation rate of 1.7 million rpm, a rotor tip velocity of 370 m/s and a bearing *DN* number of 7 million mm-rpm. Moreover, repeatable high-speed stable gas bearing operation above 1 million rpm was successfully demonstrated in a large number of devices. The operating schedule adopted for two of the MEMS air-turbine devices is presented in Fig. 6.49. Also shown in the same figure is the operating schedule for a previous-generation MEMS air-turbine (dashed line). The redesigned MEMS air-turbines are capable of achieving speeds approximately double of those reached by previous-generation MEMS air turbines and yield improved operability as indicated by only three to four necessary adjustments in journal bearing feed pressure as compared to over 20 required in the past.

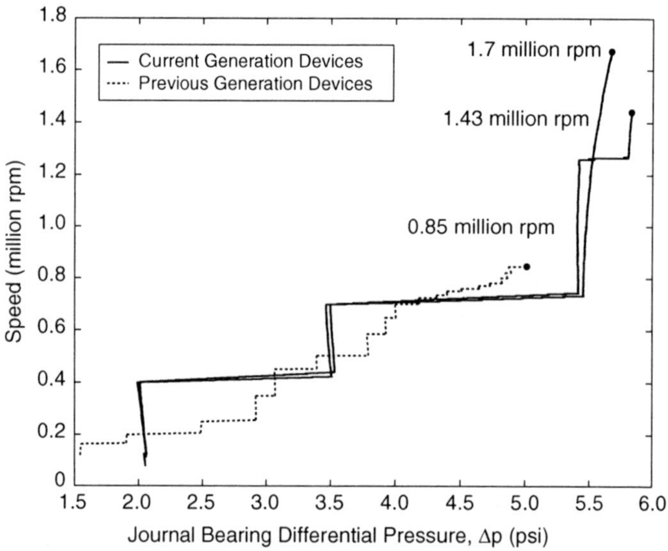

Fig. 6.49 Typical operating protocols for previous and current generation micro-gas bearing supported MEMS air-turbines. Maximum rotor speed for current generation devices is 1.7 million rpm equivalent to a DN of 7 million mm-rpm [37]

6.10 Design Implications

In summary, stable high-speed bearing operation demands the following design requirements from the hydrostatic gas-bearing system: (1) a sufficient amount of damping is required to safely cross the critical speeds of the rotor-bearing system; (2) the rotor speed at whirl instability must be well above design speed which requires a high whirl ratio together with high natural frequencies; (3) the bearing load capacity must meet the loads encountered during operation of the micro-turbomachinery device. In addition, the design of micro-gas bearing supported MEMS devices is strongly dictated by fabrication constraints and limitations of computer-chip manufacturing technology: (1) the bearing design should be limited to one or two wafers because of wafer alignment issues; (2) the bearing length L is limited by the maximum achievable etch depth using DRIE and the challenge of cutting straight, vertical side walls; (3) the lowest possible bearing diameter is set by the smallest bearing clearance that can be etched using DRIE. An in depth discussion of the fabrication aspects can be found in Chapter 3.

The following design implications are based on the assumptions that the bearing length and bearing clearance are constrained by a maximum length of $L = 500\ \mu\text{m}$ and a minimum width of $C = 10\ \mu\text{m}$, respectively. The bearing diameter is assumed unconstrained and can be chosen freely. Furthermore, the only bearing parameters that are independent of geometry are the bearing pressure difference $\Delta p/p_0$ and the properties of the bearing lubricant (gas). With the above design requirements and fabrication constraints, the scaling laws in Section 6.5.5 and the criterion for whirl instability limit in Section 6.6.1 are used to delineate design implications for ultra-short hydrostatic gas bearings.

One of the most critical issues apart from a high whirl instability limit is the requirement for a large enough damping ratio such that the rotor can be inverted safely from subcritical to supercritical speeds. Since the damping ratio ζ scales with $(L/D)^2$, $(C/R)^{-1}$, and $\Delta p/p_0)^{-1/2}$, a high ζ requires a low-bearing pressure setting $\Delta p/p_0$, a large L/D and a small C/R. Although the hydrostatic stiffness and the natural frequency rise when increasing L/D and decreasing C/R, both are reduced for lower $\Delta p/p_0$. The issue is that for a fixed bearing geometry with a small C/R and a large L/D, the damping ratio and the natural frequency change in opposite directions when raising the bearing pressure setting $\Delta p/p_0$: ζ decreases and Λ_N increases which makes crossing critical frequencies difficult.

It is desired to have the highest possible whirl ratio $\mathfrak{R}$ (typically 15–20), which means that the whirl number $\mathfrak{M}$ should be close to 1, or $C/R \approx 2(L/D)^2$. This means that for a small C/R, governed by the damping ratio and natural frequency requirements, the bearing length to diameter ratio L/D should be small and of order $\sqrt{C/R}$. Thus, the high whirl instability limit requirement demands a small L/D which compromises the damping ratio: for the same bearing pressure setting, clearance to radius ratio, bearing number and Reynolds number, an ultra-short hydrostatic gas bearing with an L/D of one order of magnitude smaller than a conventional hydrostatic gas bearing yields a damping ratio ζ of two orders of magnitude smaller, but a whirl ratio $\mathfrak{R}$ of one order of magnitude larger than a conventional gas bearing. Or put another way, ultra-high speed devices which require very high whirl ratios demand ultra-short (very small L/D) bearings with the drawback of significantly reduced damping ratios.

Possible ways to increase the damping ratio is to chose a lubricant with a higher viscosity (lower Reynolds number and higher bearing number for given rotor speed and geometry) or to add external damping independent of the internal damping of the gas bearing (e.g., foil-bearing type concepts). Another design strategy is to increase L/D in order to increase the damping ratio and hydrostatic stiffness and to remove the source of whirl instability by changing the bearing design. A potential design change is to introduce bearing anisotropy.

6.10.1 Effect of Bearing Anisotropy

Given the insight gained from the anisotropic gas-bearing theory described in Section 6.6.3 and the experimental results discussed above, the following bearing design implications are established.

In order to improve the dynamic behavior and to increase the whirl ratio, the difference in hydrostatic stiffness in the two orthogonal directions (anisotropy) should be maximized. This can be achieved (while maintaining adequate levels of average stiffness) by supplying hydrostatic feed air at different pressures over parts of the journal circumference. The optimum circumferential extent over which bearing air is supplied should be determined to maximize the effect of hydrostatic stiffness anisotropy. In general, each of the two feed plena at the higher supply pressure should cover about one third of the circumference.

In addition, the hydrodynamic stiffness of the journal bearing should be minimized. For a given bearing L/D, this can be done using the criteria established for isotropic bearing configurations by adjusting the clearance to radius ratio according to $C/R \approx 2\,(L/D)^2$.

The mechanical design of the bearing air supply system and piping should be kept bi-symmetric and the pressure losses in the opposite feed systems should be matched to avoid any rotor side-load.

In general, as for any rotordynamic design, rotor unbalance should be kept low. However, anisotropic bearing designs enable a relief in fabrication tolerance requirements and a relaxation in rotor balancing constraints compared to their isotropic counterparts.

6.10.2 Link Between Bearing Performance and Fabrication Tolerances

The main fabrication uncertainties for micro-bearings are journal geometry and unbalance. The lack of information on how well the manufactured micro-devices meet the design specifications demands a design strategy that accounts for fabrication tolerances and uncertainties in bearing geometry. The question arises how uncertainties in bearing geometry impact the prospects of stable high-speed operation. The analytical gas-bearing model together with the criterion for an infinite whirl ratio for isotropic journal bearing operation is used to establish the link between fabrication tolerances and optimal bearing geometry for stable high-speed operation.

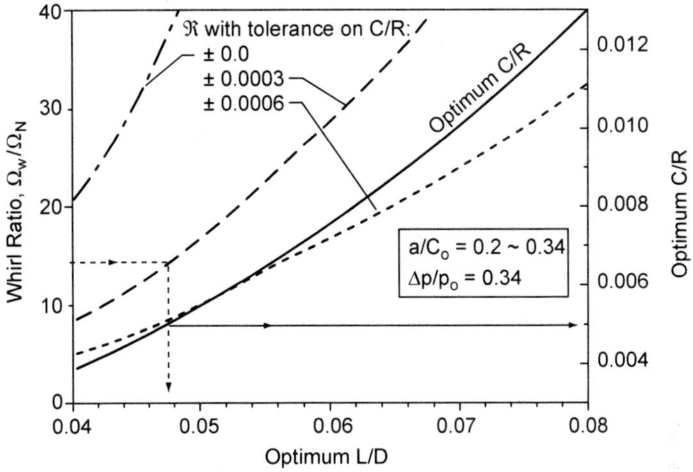

Fig. 6.50 Link between bearing performance and fabrication tolerances for isotropic bearing configurations [31]

The link between achievable whirl ratio, optimum *L/D*, and optimum *C/R* is shown in Fig. 6.50 for a bearing pressure setting of $\Delta p/p_0 = 0.34$ and a tolerable range of rotor-unbalance $a/C_0 = 0.02 \ldots 0.34$. Each of the dashed lines determines the optimum *L/D* for a given whirl ratio when the singular *C/R* is within a given tolerance. For example, the dashed line with a tolerance on *C/R* of ± 0.0003 allows an uncertainty of ± 1 µm in bearing clearance if the radius *R* is 3.1 mm. The solid line determines the corresponding optimum *C/R* for the choice of optimum *L/D*. Note that the solid line is parabolic and represents the quadratic behavior of the singular *C/R* with *L/D* according to the criterion $\mathfrak{M} = 1$ or $C/R \approx 2(L/D)^2$.

Figure 6.50 also shows that for a choice of optimum *C/R* and corresponding *L/D*, the achievable whirl ratio for isotropic bearing operation is very sensitive to the level of tolerance in *C/R*. For example, for an optimum *C/R* of 0.005, the whirl ratio is about 40 if the tolerance is 0. If the optimum *C/R* is missed by ± 0.0003 or ± 0.0006, the whirl ratio decreases to $\mathfrak{R} = 15$ or $\mathfrak{R} = 8$, respectively.

Another manufacturing uncertainty is rotor unbalance which mainly stems from turbomachinery blade etch depth non-uniformities when forming the disk surface and misalignment of rotor features with respect to the rotor centerline. In the light of stable high-speed bearing operation, it is desired to define the level of unbalance that is tolerable for a target whirl ratio. The level of unbalance can then be corrected by dynamically balancing the rotor. Figure 6.51 depicts the tolerable level of rotor unbalance for a range of clearance to radius ratios given a minimum required whirl ratio of $\mathfrak{R} = 15$, an *L/D* of 0.04 and a bearing pressure setting of $\Delta p/p_0 = 0.68$. For a chosen *C/R*, the vertical bars determine the range of rotor unbalance within which the given minimum whirl ratio is achieved. The optimum unbalance (solid line) corresponds to the singular clearance to radius ratio which yields an infinitely high whirl ratio.

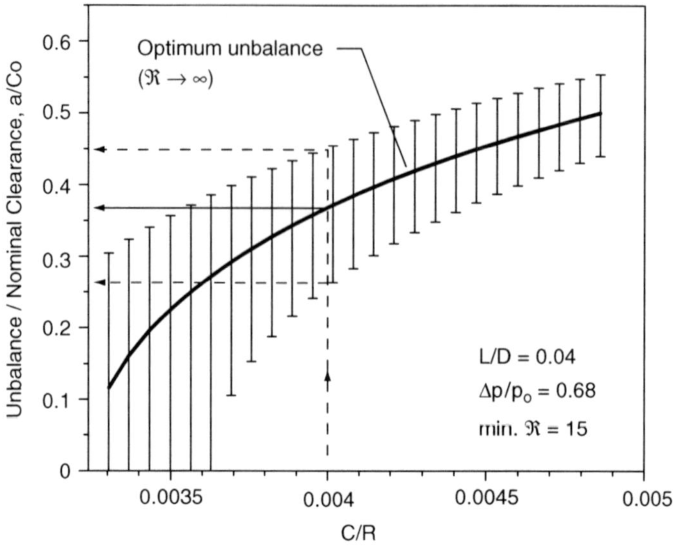

Fig. 6.51 Tolerable level of unbalance for a minimum whirl ratio $\Re = 15$, $L/D = 0.04$, and $\Delta p/p_0 = 0.68$ for isotropic bearing operation [31]

6.11 Summary

This chapter focused on high-speed gas bearings and rotordynamic considerations necessary to enable high-power density MEMS turbomachinery. The high rotating speeds (between one and two million rpm), the very low-bearing aspect ratios ($L/D < 0.1$), and the laminar flow regime in the bearing gap ($Re < 500$) place these micro-bearing designs into previously unexplored regimes in the parameter space. A theoretical basis for the dynamic behavior of ultra-short hydrostatic gas bearings and criteria for whirl instability were established from first principles. Theories for the dynamic behavior and stability of gas thrust bearings and annular seals were also introduced. Engineering solutions critical for stable high-speed bearing operation were implemented, and the first repeatable, stable high-speed silicon turbine rotor operation at 1.7 million rpm, equivalent to blade-tip speeds near 370 m/s and bearing *DN* numbers of 7 million mm-rpm, was demonstrated. A speed-to-whirl ratio of 30, roughly 15 times higher than conventional long bearing designs was achieved at this rotor speed. The experiments demonstrated the feasibility of microscale turbomachinery and validated key aspects of the micro-gas bearing theories.

Acknowledgments The work described in this chapter has resulted from contributions of a number of individuals, many of whom are co-authors of the referenced papers. It has been a pleasure working with the students and members of the bearing team of the microengine project. In particular the author would like to mention his gratitude to Dr. L. Liu, Dr. C.J. Teo, Dr. S. Jacobson, and Dr. F. Ehrich. The editorial help by Ms. D. Park is gratefully acknowledged.

Nomenclature

a	rotor imbalance
A, AR	area, area ratio
A_0	coefficient
c	angular damping coefficient
C	journal bearing clearance, coefficient
C, $\tilde{C}$	damping coefficient
C_f	friction coefficient
d	orifice diameter
D	rotor diameter, orifice diameter
D_h	hydraulic diameter
DN	bearing diameter times rotor speed in mm-rpm
e	dynamic eccentricity
$\hat{e}$	unit vector
F	force, function
G	Green's function
h	thrust-bearing gap or clearance
I	diametral moment of inertia
I_p	polar moment of inertia
k, K	stiffness
L	journal bearing length, orifice length
$\dot{m}$	mass flow
m	mass of rotor disk
M	Mach number, moment
n, N	harmonic number, number of
p	pressure,
Δp	hydrostatic differential pressure across journal bearing
q, Q	specific flow rate, flow rate
r	radial coordinate, radial location of orifices
R	rotor radius, radius of thrust-bearing pad, specific gas constant
Re	Reynolds number
$\mathfrak{R}$	whirl-ratio Ω_W/Ω_N
s	Laplace variable
t	time
T	temperature
u, $\overline{U}$	velocity, mean axial velocity due to hydrostatic flow
v, V	velocity
$\mathfrak{W}$	whirl number $= k^v/k^p$
z	axial location, parameter
α	plenum circumferential angle
β	reduced frequency $\omega L/$
β_{FD}	ratio of axial flow-through time and viscous diffusion time
β_x, β_y	Euler angles
δ	perturbation

$\varepsilon = e/C_0$	rotor normalized radial eccentricity
ϕ	angle
γ	ratio of specific heats
Λ	bearing number
χ	dynamic imbalance
μ	dynamic viscosity
ν	kinematic viscosity
ω	frequency, stagnation pressure loss coefficient
ω_{inlet}	journal bearing inlet stagnation pressure loss coefficient
Ω	rotor speed
Ω_N	natural frequency
Ω_W	rotor speed at onset of whirl instability
ψ	function
ρ, ρ_d	fluid density, rotor disk density
τ_w	wall shear stress
τ_d	characteristic viscous diffusion time
τ_f	flow-through time of axial hydrostatic flow
ζ	damping ratio, non-dimensional hydrodynamic moment
$\{\}^{dp}$	damping
$\{\}^{hd}$	cross-coupled hydrodynamic
$\{\}^{hs}$	direct-coupled hydrostatic
$\{\}^{sys}$	system
$\{\}^{p}$	hydrodynamic pumping action
$\{\}^{v}$	hydrodynamic viscous effect
$\{\}^{*}$	critical, singular condition
$\{\}_{\mathrm{a}}$	ambient
$\{\}_{\mathrm{o}}$	nominal, equilibrium condition
$\{\}_r$	radial
$\{\}_R$	rotor
$\{\}_t$	total or stagnation
$\{\}_x$	axial
$\{\}_\theta$	tangential

References

1. Allen DS, Stokes PJ, Whitley S (1961) The performance of externally pressurized bearings using simple orifice restrictors. ASLE Trans 4:181–196
2. Arghir M, Frene J (2001) A triangle based finite volume method for the integration of lubrication′s incompressible bulk flow equations. J Tribol-T ASME 123:118–124
3. Chen NNS, Ho KW (1981) Performance study of a hydrostatic air thrust bearing. Wear 70:207–217
4. Childs D (1993) Turbomachinery rotordynamics–phenomena, modeling and analysis. John Wiley and Sons, New York
5. Constantinescu VN, Galetuse S (1987) On the dynamic stability of the spiral-groove gas-lubricated thrust bearing. J Tribol-T ASME 109:183–188

6. Diez S (2003) Preliminary performance characteristics of a microfabricated turbopump. M.Sc. thesis, Massachusetts Institute of Technology
7. Ehrich FE (1989) The role of bearing support stiffness anisotropy in suppression of rotordynamic instability. ASME DTC-12th Biennial Conference on Mechanical Vibration and Noise, Montreal
8. Ehrich FE, Jacobson SA (2003) Development of high-speed gas bearings for high-power-density micro-devices. J Eng Gas Turb Power 125:141–148
9. Frechette LG, Jacobson SA, Breuer KS, Ehrich FE, Ghodssi R, Khanna R, Wong CW, Zhang X, Schmidt MA, Epstein AH (2005) High-speed microfabricated silicon turbomachinery and fluid film bearings. J Microelectromech S 14:141–152
10. Fuller D (1969) A review of the state-of-the-art for the design of self-acting gas-lubricated bearings. J Lubr Technol 91:1–16
11. Garner D, Lee C, Martin F (1980) Stability of profile bore bearings: infulence of bearing type selection. Tribol Int 13:204–201
12. Gunter E, Trumper P (1969) The influence of internal friction on the stability of high speed rotors with anisotropic supports. J Eng Ind:1105–1113
13. Idelchik IE (1994) Handbook of hydraulic resistance, 3rd edn. CRC Press, Boca Raton, FL
14. Kim D, Lee S, Bryant M, Ling F (2004) Hydrodynamic performance of gas microbearings. J Tribol-T ASME 126:711–718
15. Larson R, Richardson H (1962) A preliminary study of whirl instability for pressurized gas bearings. J Basic Eng-T ASME 84:511–520
16. Licht L, Fuller DD, Sternlicht B (1958) Self excited vibrations of an air-lubricated thrust bearing. Trans ASME 80:411–414
17. Liu L (2005) Theory for hydrostatic gas journal bearings for micro-electro-mechanical systems. Ph.D. thesis, Massachusetts Institute of Technology
18. Liu L, Spakovszky Z (2007) Effects of bearing stiffness anisotropy on hydrostatic micro gas journal bearing dynamic behavior. ASME J Eng Gas Turb Power 128(1): 177–184
19. Liu L, Teo CJ, Epstein AH, Spakovszky ZS (2005) Hydrostatic gas journal bearings for micro-turbomachinery. J Vib Acoust 127:157–164
20. Lomakin A (1958) Calculation of critical number of revolutions and conditions necessary for dynamic stability of rotors in high-pressure hydraulic machines when taking into account forces originating in sealings. Power Mech Eng (in Russian)
21. Lund J (1968) Calculation of stiffness and damping properties of gas bearings. J Lub Tech, Ser. F 90:793–803
22. Miki N, Teo CJ, Ho L, Zhang X (2000) Precision fabrication of high-speed micro-rotors using deep reactive ion etching (DRIE). In: Hilton Head Solid-State Sensors and Actuators Workshop, Hilton Head Island, SC
23. Orr DJ (2000) Macro-scale investigation of high speed gas bearings for MEMS devices. Ph.D. thesis, Massachusetts Institute of Technology
24. Piekos ES (2000) Numerical simulation of gas-lubricated journal bearings for microfabricated machines. Ph.D. thesis, Massachusetts Institute of Technology
25. Piekos ES, Orr DJ, Jacobson SA, Ehrich FE, Breuer KS (1997) Design and analysis of microfabricated high speed gas journal bearings. AIAA Paper 97–1966, 28th AIAA Fluid Dynamics Conference, Snowmass Village, CO
26. Roudebush WH (1957) An analysis of the effect of several parameters on the stability of an air-lubricated hydrostatic thrust bearing. NACA Technical Note 4095
27. San Andres L, Wilde D (2000) Finite element analysis of gas bearings for oil-free turbomachinery. Revue Européenne des Eléments Finis, 10 (6/7): 769–790
28. Savoulides N (2004) Development of a MEMS turbocharger and gas turbine engine. Ph.D. thesis, Massachusetts Institute of Technology
29. Shapiro AH (1953) The dynamics and thermodynamics of compressible fluid flow, vol I. Ronald Press, New York
30. Smith D (1933) Motion of a rotor carried by a flexible shaft in flexible bearings. Proc R Soc Lon Ser-A 142:92–118

31. Spakovszky Z, Liu L (2005) Scaling laws for ultra-short hydrostatic gas journal bearings. ASME J Vibr Acoust 127:254–261
32. Stowell TB (1971) Pneumatic hammer in a gas lubricated externally pressurized annular thrust bearing. ASME J Appl Mech 93:498–503
33. Teo CJ (2006) MEMS turbomachinery rotordynamics: Modeling, design and testing. Ph.D. thesis, Massachusetts Institute of Technology
34. Teo CJ, Spakovszky, Z (2006) Modeling and experimental investigation of micro-hydrostatic gas thrust bearings for micro-turbomachines. ASME J Turbomach 128:597–605
35. Teo CJ, Spakovszky Z (2006) Analysis of tilting effects and geometric non-uniformities in micro-hydrostatic gas thrust bearings. ASME J Turbomach 128: 606–615
36. Teo CJ, Spakovszky Z (2006) Analysis of annular seals for high-speed power-MEMS devices. Proc Power MEMS conference, Berkeley, CA
37. Teo CJ, Liu L, Li HQ, Ho L, Jacobson S, Ehrich F, Epstein A, Spakovszky Z (2006)"High-Speed Operation of a Gas-Bearing Supported MEMS Air Turbine. ASME paper IJTC2006-12173, STLE / ASME International Joint Tribology Conference, San Antonio, TX
38. Teo CJ, Spakovszky Z, Jacobson S (2008) Unsteady flow and dynamic behavior of ultra-short Lomakin gas bearings. ASME J Tribol 130(1): 1–9
39. Vohr JH (1966) An experimental study of flow phenomenon in the feeding region of an externally pressurized gas bearing. Proc ASME Lubrication Symposium, New Orleans, LA
40. Yabe H, Watanabe N (1988) A Study on the running accuracy of an externally pressurized gas thrust bearing (load capacity fluctuation due to machining errors of the bearing). JSME Int J Ser III 31:114–120
41. Yabe H, Yamamoto M (1989) A study on the running accuracy of an externally pressurized gas thrust bearing (bearing stiffness and damping coefficient). JSME Int J Ser III 32:618–624

Chapter 7
Thermofluidics and Turbomachinery

Stuart A. Jacobson

7.1 Introduction

The MIT MicroEngine rotor disk is less than 10 mm in diameter, which is more than two orders of magnitude smaller than that of a conventional aircraft gas turbine rotor disk. This significant length scale reduction impacts the underlying physics associated with the aerothermal design of the engine. Fluid dynamic and heat transfer parameters such as Reynolds number and Biot number in the MicroEngine are quite small, lying outside the commonly understood design space associated with gas turbine engines. The reduction in scale has a primarily negative impact on engine performance, although some benefits do arise as well.

While length-scale reduction plays a major role in the aerothermal performance of the MicroEngine, fabrication constraints play an equally important role. The MicroEngine is being constructed using MEMS microfabrication processes. Silicon is the primary material because high-quality single crystal silicon wafers are widely available.[1] There is a great deal of processing knowledge and tools available for silicon as well, including precision deep etch processes. However, current bulk MEMS fabrication processes provide limited control in producing 3D structures. Standard photolithography is a 2D process that images patterns on the surface of a wafer that are then etched essentially straight down into the wafer. Deep reactive ion etching (DRIE) is used to produce circularly symmetric rotors with blade heights spanning several hundred microns. The resulting 3D silicon structure is effectively a translation of the initial 2D shape into the silicon wafer. There are no standard processes currently available for achieving substantial etch depth variation when etching to depths of hundreds of microns, although the Ghodssi group at University of Maryland is developing gray scale masking processes that would provide the

S.A. Jacobson (✉)
Department of Aeronautics and Astronautics, Massachusetts Institute of Technology, Bldg 31-269, 77 Massachusetts Avenue, Cambridge, MA 02139, USA

[1] Although as this is being written in 2008, the solar cell industry is consuming a large fraction of the available supply of low TTV wafers, driving up material cost.

J.H. Lang (ed.), *Multi-Wafer Rotating MEMS Machines*, MEMS Reference Shelf,
DOI 10.1007/978-0-387-77747-4_7,

means for hundreds of steps of etch depth variation [4, 24]. The gray scale technology may ultimately play a major role in advancing the capabilities of Power MEMS, but at this point in time it still requires considerable development.

Without a means for precision 3D etching, high-performance axial-flow turbomachinery designs are not achievable, limiting the design space to centrifugal turbomachinery. Lack of etch depth variation control also means that turbomachinery blade heights cannot be varied with rotor radius, as they are for typical centrifugal turbomachinery. For instance, Fig. 7.1 shows a silicon turbine in which the blade heights are constant across the rotor. Without the ability to control flow passage size through blade height variation, separation control in the compressor becomes a substantial challenge, limiting the achievable compressor efficiency.

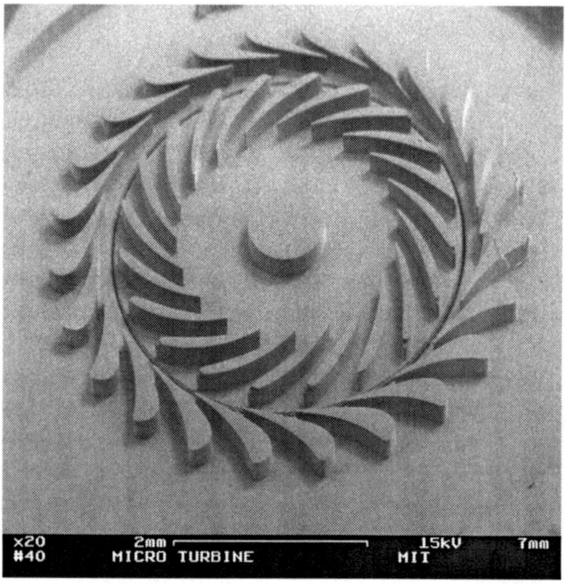

Fig. 7.1 Microscale silicon turbine. Blade height is 200 μm and rotor diameter is 4 mm

A conventional gas turbine is a highly complex machine consisting of thousands of parts necessary for achieving high levels of performance. The performance of a conventional gas turbine is so good (overall chemical to electric conversion efficiencies of order 50%), that one can sacrifice quite a bit of performance when scaling the device down to the microscale and still obtain an engine that is very competitive relative to a battery. At 10% conversion efficiency, a value we believe is achievable, the specific energy of the MicroEngine burning hydrocarbon fuels would still be over 1000 W-hr/kg, which is more than five times better than a standard lithium battery.

A relatively low-complexity (low performance) gas turbine engine design still requires a very high level of MEMS fabrication complexity, based on the current state of the art. So it is to be expected that the first MEMS gas turbines will have low efficiency relative to conventional engines. But as discussed in the previous paragraph, there is substantial room to give up performance and still have a competitive device. This argument is perhaps overly simplistic when applied to gas turbines, as small drops in component performances can lead to a precipitous drop in overall cycle performance. The net power obtained from a gas turbine is the difference between the power generated by the turbine and the power consumed by the compressor and other loss mechanisms in the cycle (bearings, bleeds, etc.). For a gas turbine engine operating at high-component efficiency (greater than 90% efficiency for both the turbine and compressor is typical for advanced aero engines), the turbine power is much larger than the compressor power, typically by a factor of about two. However, for gas turbines with lower efficiency components, the compressor power approaches the turbine power. The net power generated from such an engine will be the small difference between the much larger turbine and compressor values. Small changes to individual component performances can then have a profound effect on overall cycle performance. Cycle viability is extremely sensitive to component performance, so a deep understanding of the effects of scale and fabrication constraints on the MEMS engine system is critical for a successful design.

This chapter provides a detailed discussion of the impacts of length-scale reduction and fabrication constraints on the fluidic design and performance of Power MEMS devices, with a focus on the MIT MicroEngine. Section 7.2 offers a general discussion of the impacts of scale as they relate to physical effects and fabrication constraints. This is followed in Section 7.3 by a closer look at the performance of a microscale compressor, providing results from experimental measurements and computations. Section 7.4 presents a brief discussion on microturbines. Measurements from a microscale turbocharger designed, fabricated, and tested at MIT are discussed in Section 7.5. Finally, Section 7.6 summarizes the key issues raised in this chapter.

7.2 Implication of Scale

To maximize the exchange between kinetic and thermal energy in the cycle fluid (which then maximizes energy conversion efficiency and power density),[2] it is desirable for the engine flows to operate in a transonic regime [10]. Operating with rotational speeds approaching the limits set by the material strength generally achieves this desired flow velocity range. This is true for both conventional large-scale gas turbine engines (to be referred to as "conventional engines") as well as for small-scale engines, and the allowable maximum tip speed in both is comparable. As a result, one can expect that the internal flow velocities will also be comparable for both small-scale and conventional engines.

The MicroEngine is designed to operate with ambient air as the working fluid, comparable to a conventional engine operating at sea level (inlet and exhaust at one atmosphere of pressure, inlet at room temperature). Thus, the only scale that is

[2] The content of this section adapts and expands upon Jacobson [7].

necessarily different for the MicroEngine as compared to a conventional engine is the substantial decrease in length scale.

The initial engine designs pursued include a single stage centrifugal compressor and a single stage radial-inflow turbine. The complexity of adding wafers to the MicroEngine stack puts a severe constraint on the ability to add staging to the compressor and turbine. This limits the achievable pressure ratio to about 2–3 when heat transfer into the compressor flow is accounted for. As will be discussed later in this chapter, significant heat transfer into the compressor flow is one of the effects of the small length scale. This pressure ratio, about an order of magnitude smaller than what is achieved in multi-stage aircraft gas turbine engines today, also acts as a scaling issue that affects performance. But this scale effect is not inherent to the physics, but rather results from fabrication constraints. This section looks at the effects of scale from the perspective of both physical phenomena and fabrication constraints.

7.2.1 Physical Effects

Table 7.1 summarizes the effects of a reduction in length scale, x, on relevant parameters in a small-scale engine. The implications of the scale factors presented in Table 7.1 are discussed at length in this section. The third column of Table 7.1 includes data for the precise scaling of a baseline MicroEngine, which is about a factor of 100 smaller in length scale ($x \approx 0.01$) compared to a conventional engine.

When evaluating the impact of scale on engines, the critical issue is whether the change in the dimensionless parameters of Table 7.1 impacts the operating regime so as to alter one's fundamental understanding of that operating regime as gained from the extensive history of conventional engines. The impacts of the items in Table 7.1 are discussed in the following paragraphs related to major subsystems of the MicroEngine.

7.2.1.1 Power and Energy Densities

Several important parameters in Table 7.1 scale with the inverse of the scale factor, including area to volume. Since power scales with mass flow rate, and thus with flow area, the power per unit engine volume (power density) and the power per unit engine mass (specific power) also scale like the inverse of the scale factor. This power scaling is one of the few factors where the reduction in length scale benefits small-scale engine performance. Table 7.1 shows that the MicroEngine can have a power density that is 100 times larger than conventional engines, all else being the same. In reality, the power density improvement will not be quite this large as all else is not the same. Other scale effects increase losses in the engine, driving down the power density. However, conventional gas turbine engines already have exceedingly large power densities; this is why they are chosen as the prime movers for large aircraft. The increase in power density for an ideal microscale gas turbine engine provides significant margin that can be banked against other losses that decrease

Table 7.1 Parameter scaling. The scale factor is the ratio of the parameter in the MicroEngine to its value in a conventional engine. The factor x is defined as the ratio of small engine length scale(l) to conventional engine length scale (L)

Parameter	Scale factor	MicroEngine scale factor, $x \approx 0.01$	MicroEngine parameter magnitude
Scale:			
Length	x	0.01	$\sim$ 8 mm (rotor diameter)
Area	x^2	10^{-4}	
Volume	x^3	10^{-6}	$\sim 3\ cm^3$
Area/Volume	$1/x$	100	
Mass	x^3	10^{-6}	$\sim$ 4 g (silicon engine only)
Cycle power	x^2	10^{-4}	$\sim$ 10 W
Power ratios:			
Power/Mass	$1/x$	100	
Power/Volume	$1/x$	100	
Fluid mechanics scaling:			
Reynolds number (Re)	x	0.01	$\sim 10^4$
Mean friction coefficient (c_f)	$C/\sqrt{x}$ ($C \approx 0.3$)	3	$\sim 10^{-2}$
Boundary layer Displacement thickness/passage width	$C/\sqrt{x}$ ($C \approx 0.6$)		
Couette drag power/cycle power	$1/x$	100	$\sim 10^{-1}$
Heat transfer scaling:			
Mean Stanton number (St)	$C/\sqrt{x}$ ($C \approx 0.3$)	3	$\sim 10^{-2}$
Mean Biot number (Bi) (same material for macro and micro)	$C\sqrt{x}$ ($C \approx 0.3$)	0.03	
Mean Biot number (Bi) (stainless steel for macro, silicon for micro)	$\approx C\sqrt{x}/10(20°C)$ $\approx C\sqrt{x}/2(500°C)$ ($C \approx 0.3$)	0.003 0.02	$\sim 10^{-2}$ $\sim 10^{-2}$

Note: All correlations assume boundary layers with constant free stream velocity, along a constant temperature flat plate. Wherever relevant, laminar correlations are used for the small-scale device and turbulent correlations for the conventional scale device. The parameter C is calculated assuming the conventional scale device has $Pr \approx 0.7$ and $Re_L \approx 10^6$. C scales approximately like $Re_L^{-0.3}$.

cycle performance, resulting in a small-scale engine that can still have a large power density relative to competing small-scale power systems such as batteries, fuel cells, and other engine types.

The fuel mass will far exceed engine and fuel tank mass for typical applications. The specific energy (energy per unit mass) of the engine/fuel system is then primarily the specific energy of the fuel, reduced by the cycle energy conversion efficiency. Thus, the specific energy is dependent on scale only to the extent to which scale affects the achievable cycle efficiency. The specific energy of both hydrocarbon fuels and hydrogen is far superior to the attainable specific energy in batteries. A MicroEngine operating at about 2% chemical-to-electric conversion efficiency would have a specific energy comparable to current primary (non-rechargeable) lithium batteries, and would be about 50% better than current secondary (rechargeable) lithium batteries.

Our computational and experimental studies indicate that a chemical-to-electric conversion efficiency of 10% should be feasible. A MicroEngine operating at this efficiency level will have a fuel specific energy of about 1300 W-hr/kg burning a hydrocarbon fuel such as propane or JP-8. The mass of the balance of plant (generator electronics, fuel tank, accessories, etc.) must be accounted for as well, and it is estimated that these will drop the specific energy by about 20%. Even so, the resulting 10% efficient MicroEngine would still have a specific energy over five times larger than the best available primary batteries. With significant investment in PowerMEMS technology, MicroEngine efficiencies in excess of 10% should be achievable, although with an associated increase in device complexity.

7.2.1.2 Reynolds Number

The Reynolds number (Re) is a key parameter when considering the effects of scaling on the MicroEngine. Reynolds number, a measure of the relative importance of inertial forces compared to viscous forces, scales linearly with the length scale factor. Thus for the MicroEngine, Re is several orders of magnitude less than that of a conventional engine, indicating that viscous effects will be more pronounced. Put more simply, because the flow passage lengths within the MicroEngine are so small, boundary layers do not have the chance to develop much, resulting in thin boundary layers, which imply large skin friction and high-heat transfer rates. In particular, a factor of 100 decrease in Re places Microengine flows squarely in the laminar/transitional flow regime, whereas conventional engine flows fall well within the turbulent flow regime. This difference is significant. In performing the scaling comparisons in Table 7.1, laminar correlations are used for the MicroEngine scale flows and turbulent correlations are used for large-scale flows. The turbulence correlations are taken from Kays and Crawford [9]. Note that these correlations are approximate, and the scaling analyses performed with them are only meant to provide an order of magnitude feel for the effects of scaling. Since boundary layer parameters for laminar and turbulent flows scale differently with Re, the scaling effects are Re dependent.

Reynolds number impacts several physical processes within a gas turbine. These aspects are broken out below.

Turbomachinery Issues

Turbomachinery losses result from skin friction in the boundary layer along the blades and end walls. The friction coefficient for a laminar flat plate boundary layer with constant free stream velocity, averaged over a length l (chord length in the turbomachinery, for instance) is

$$c_{f,\mathrm{lam}} = \frac{1.328}{\sqrt{\rho u_\infty l/\mu}} = \frac{1.328}{\mathrm{Re}_l^{1/2}}, \tag{7.1}$$

while the correlation for a turbulent boundary layer, averaged over a length L is

$$c_{f,\mathrm{turb}} \approx \frac{0.0718}{\left(\rho u_\infty L/\mu\right)^{0.2}} = \frac{0.0718}{\mathrm{Re}_L^{0.2}}. \tag{7.2}$$

Combining the above two equations, and introducing the length scale factor $x = l/L$, one obtains:

$$\frac{c_{f,\mathrm{lam},l}}{c_{f,\mathrm{turb},L}} \approx \frac{18.5\mathrm{Re}_L^{-0.3}}{\sqrt{x}} = \frac{C}{\sqrt{x}} \approx \frac{0.3}{\sqrt{0.01}} = 3. \tag{7.3}$$

The values substituted into Eq. (7.3) assume $\mathrm{Re}_L \approx 10^6$(a typical value for conventional engines) and a length scale factor of 0.01. This scaling analysis indicates that skin friction drag losses will impact small-scale turbomachinery to a far greater degree than at conventional scale. Recall that flow density and velocity are expected to be equivalent for both the MicroEngine and conventional engines, so the scaling of the skin friction coefficient is equivalent to the scaling of wall shear. Although the relative increase in skin friction is substantial, one must still evaluate whether this increase in skin friction contributes to a significant drop in MicroEngine performance.

Casey [1] looked at the effects of Reynolds number on centrifugal compressor efficiency, extending the work of several earlier researchers. Casey showed that reducing the Reynolds number in a compressor can significantly reduce its polytropic efficiency. Casey also showed that the increased loss correlates with friction losses in the boundary layer. However, the lowest value of Reynolds number data measured by Casey (when converted to the form discussed in this section) is about a factor of 10 larger than the MicroEngine compressor Reynolds number. In addition, the Casey data remain in the turbulent regime, whereas the MicroEngine flow is in the laminar/transitional regime, so the Casey results cannot be simply extrapolated to the MicroEngine scale.

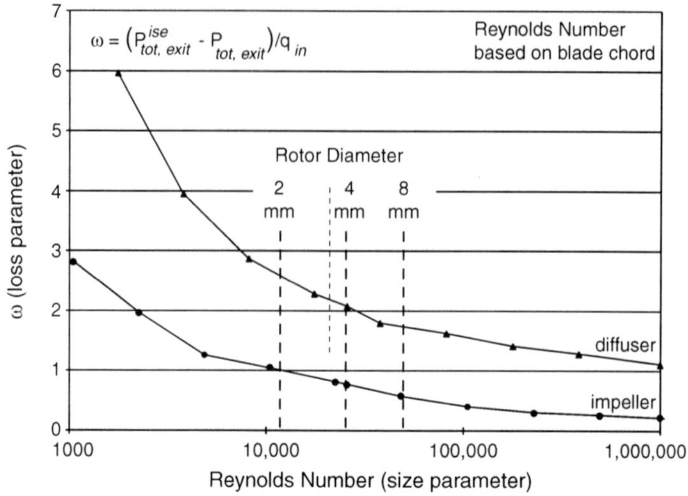

Fig. 7.2 Compressor loss as a function of size, MISES CFD results. Loss parameter is a measure of the difference between the isentropic total pressure and the actual total pressure at the impeller exit, normalized by the inlet dynamic head (all quantities calculated in the rotating frame)

Figures 7.2 and 7.3 show computational results of total pressure loss factor ω as a function of Reynolds number for a given 2D compressor and turbine geometry. The turbomachinery were designed for microscale operation, and then scaled in size to vary the Reynolds number for the analysis. These results were obtained using the MISES code developed by Drela and Youngren at MIT [3]. MISES is a 2D

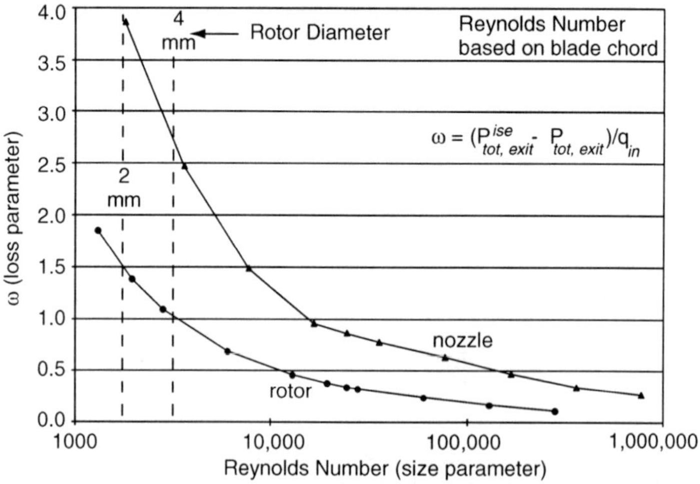

Fig. 7.3 Turbine loss as a function of size, MISES CFD results. Turbine Reynolds number is much lower than the compressor due to the high temperature of the gas at the turbine inlet

computational tool for multiple blade cascade analysis and design. The analyses were done with fixed turbomachine blade shape, and constant inlet flow properties, inlet Mach number and flow angle. Reynolds number is based on rotor blade chord and rotor inlet flow properties (in the relative frame). Varying the Reynolds number is then equivalent to changing the scale of the turbomachine. The loss factor ω is the difference between the isentropic and actual total exit pressure from the component, normalized by the inlet dynamic head of the rotating component.

The MicroEngine compressor impeller diameter of 8 mm is marked on Fig. 7.2. Losses are about a factor of three times higher at this Re than at the conventional scale Re value, which is of order 10^6. Although an increase in loss is observed, as predicted by the increase in skin friction, this loss increase is found to be modest, particularly for the impeller. Diffuser loss will be returned to later in a discussion of fabrication constraints. However, a decrease in scale by an additional factor of 10 over the MicroEngine scale increases the loss substantially over that at conventional scale, to a point where the loss is found to severely impact cycle performance.

The computed total-to-static stage isentropic efficiency using MISES at $Re = 10^6$ is 92%, whereas it is 85% at the MicroEngine design Re ($\mathrm{Re}_{\mathrm{des}}$). But at $\mathrm{Re}_{\mathrm{des}}/10$ the efficiency is 59%. Note that these efficiencies are based on a 2D flow analysis. A 3D computational analysis of the microturbomachinery flow was performed, resulting in losses that are about a factor of two larger than those predicted by the 2D analysis [12]. This increase in loss is due to the effects of endwalls and axial cross flow. As a general rule of thumb, we have found that 2D flow analyses account for only about half of the actual flow loss. Applying this rough rule-of-thumb to the 2D calculation of 85% efficiency (15% loss) results in an estimated efficiency for the true impeller at design of about 70%, a value that is relatively low for a gas turbine, but still adequate for the MicroEngine cycle. However, at $\mathrm{Re}_{\mathrm{des}}/10$ the 2D efficiency is 59%, indicating the 3D efficiency will likely be much less than 50%, a value that is too low for the MicroEngine to generate net power.

Results are presented later in this chapter from higher fidelity 3D computational fluid dynamics (CFD) that give more insight into the fluid mechanics at small scale. Results will also be presented from a scaled up compressor test facility built to verify the computational results, along with results from a microscale turbocharger, both of which show that the experimental results are consistent with the computational results.

Figure 7.3 shows the 2D loss factor for a microturbine computed with MISES. The turbine Reynolds number is considerably smaller than the compressor's because of the effects of elevated temperature on density and viscosity. The computed loss for the MicroEngine turbine (8 mm diameter) is more than ten times higher than at large Re. The predicted 2D total-to-static isentropic efficiency for the turbine is 80%. Doubling the loss gives an estimated 3D efficiency of 60% for the turbine, a value that is marginally sufficient for the MicroEngine thermodynamic cycle.

Thus, for the particular design regime of the MicroEngine, the small size is of concern for both the compressor and turbine. The 8 mm diameter compressor appears to be close to the borderline of suitability; some reduction in scale could be accepted, but a factor of 10 reduction would be untenable. The 8 mm diameter turbine has a Reynolds number an order of magnitude lower than the compressor, and appears to sit on the borderline of suitability. Note that Figs. 7.2 and 7.3 should be viewed as only suggestive of trends since they are based on specific turbomachine designs. In particular, the compressor and turbine designs adhere to the constant blade height constraint from microfabrication (to be discussed later), and removing this constraint would be expected to improve the results. Sirakov (2004, personal communication) showed a 5–10% improvement in isentropic efficiency for a microscale impeller when blade height was allowed to decrease with increasing radius. While early MicroEngine designs were based on a 4 mm diameter rotor, more recent designs are based on an 8 mm diameter rotor, which doubles the turbomachinery Reynolds number to improve engine performance.

Boundary Layer Issues

Table 7.1 includes the results of a scaling analysis of the boundary layer displacement thickness, normalized by the flow passage width. A flat plate boundary layer with constant free stream velocity is assumed. The relative displacement thickness scaling is about a factor of six times larger in the MicroEngine compared to conventional engines. Large boundary layers are a concern in turbomachinery flows as they limit the amount of flow through the blade passages (which are not generally fully developed) for a given rotor size, decreasing the obtainable power. This effect is known as blockage.

The boundary layer displacement thickness is about 7% of the flow passage width at the MicroEngine compressor impeller trailing edge. This fraction is not small, and some impact on the mass flow rate is expected. In decelerating flows, such as in a compressor, the relative displacement thickness increase with scale decrease will not be as large as that for a flow with constant free-stream velocity, although the thick boundary layers in such flows still result in blockage concerns. MISES computations were performed on the impeller at design and at large scale Re, with pressure ratio matched. The results showed a reduction in normalized mass flow of 30% for the MicroEngine impeller, compared to the same impeller geometry run at larger Re. In accelerating flows such as in a turbine, the relative displacement thickness increase with scale reduction becomes stronger. However, these types of flow typically have thin boundary layers, so the increased displacement thickness many not be significant. In fact, MISES computations showed the normalized mass flow rate between micro- and macroturbines to be virtually identical.

7.2.1.3 Gaps and Viscous Drag

There are many regions in the MicroEngine where a rotating wall is in close proximity to a stationary wall, including the bearings and the turbomachinery blade tips. The gap between rotating and stationary components tends to be on the order of several microns, and the flow developed in these gaps is predominantly Couette flow, with an associated viscous wall shear drag. While cycle power scales with the square of the length scale factor, the Couette drag power scales linearly with the length scale factor. Thus the ratio between drag and cycle power is inversely proportional to the length scale factor, making Couette drag an increasing concern as engine scale is reduced.

In fact, at the MicroEngine scale, the Couette drag in the bearings is expected to be of the same order of magnitude as the net cycle power output. Clearly one wishes to keep gaps as large as possible and areas between stationary and rotating parts as small as possible. But increased gap and decreased area translate into increased leakage for seals and decreased load capacity for bearings. The resulting system trade-offs that guide the design of a MicroEngine can thus be quite different from those of a conventional engine.

7.2.1.4 Heat Transfer

Heat Transfer Coefficient

The reduced scale of the MicroEngine has a strong effect on heat transfer characteristics as well. The Stanton number varies like the inverse of the square root of the length scale (Table 7.1), for the case of constant free stream velocity and constant wall temperature. As fluid properties and flow velocity are essentially unchanged by the scale reduction, the heat transfer coefficient scales like the Stanton number. The heat transfer coefficient is a measure of heat flow into the wall per unit area. Since area scales like cycle power, the heat transfer coefficient scaling can be viewed as equivalent to the scaling of a relative heat transfer parameter, defined as heat flux into the wall normalized by cycle power. Unsurprisingly, the effects of scaling on Stanton number are essentially identical to the scaling of friction coefficient (since air has $\mathrm{Pr} \approx 1$). Thus per Table 7.1, we expect, on a relative basis, about three times more convective heat transfer to the structure in the MicroEngine than in a typical large scale gas turbine engine. Physically, this scaling effect occurs because of the extremely thin thermal boundary layers that exist in the MicroEngine due to the limited lengths for growth. The increase in relative heat transfer is of particular concern in the MicroEngine's combustor where increased heat loss to the wall makes it that much tougher to sustain combustion.

Biot Number

Another key parameter impacted by scaling is the Biot number, a ratio of convective heat transfer to structural thermal conduction. Table 7.1 shows two sets of data for the Biot number. The first shows a case in which the same structure material thermal conductivity is used for both the macro- and microscale device, isolating the effects of the length scale alone. The second shows a case in which the thermal conductivity of stainless steel is used for the macro device while silicon is used for the microdevice, thus including the effects not only of scale but also the effects caused by the change in materials. At 20°C, the thermal conductivity of silicon is roughly 150 W/m-K (300 K), which places it closer to copper (~400 W/m-K) than to stainless steel (~15 W/m-K). As Table 7.1 indicates, the length scale reduction causes a significant drop in Biot number, by a factor of about 30. And the change from stainless steel to silicon causes the Biot number to drop even more. The Biot number for the MicroEngine is more than 100 times smaller than that for a conventional engine, and is much less than 1. The implication is that heat conduction in the structure will be much faster than heat convection into the structure, so the structure will not have substantial temperature gradients, i.e., the structure will be generally isothermal.

The low Biot number has serious implications for the thermal characteristics and design of the structure. For instance, in a conventional engine, the first stage turbine is a critically stressed component as it sees the highest fluid temperature while being under large centrifugal stress. First stage turbines are generally made from exotic materials and may include internal cooling. However, for the MicroEngine, the low Biot number suggests that the turbine leading edge temperature will not be very high, but will rather be at some average structural temperature as the convected heat into the structure is rapidly conducted and spread to the rest of the structure. Likewise, the low Biot number suggests that the MicroEngine can be cooled by applying a heat sink to any easily accessible part of the structure rather than targeting hot spots.

On the other hand, with the engine essentially isothermal, the walls of the compressor will be hot and there is the potential for substantial heat transfer into the relatively low-temperature compressor flow. Compressing a hot gas requires more work than compressing a cold gas. The heat transfer into the compressor is one of the critical challenges faced in designing a small-scale gas turbine, and must be accounted for in the compressor design. So the cooling benefit gained in the turbine from the low Biot number is countered by a substantial heating detriment in the compressor. A material that can be batch machined with precision to depths of hundreds of microns, with high strength and very low thermal conductivity would be ideal in place of silicon, but such a material/technology combination does not currently exist.

A recuperator, a heat exchanger between turbine exhaust air and combustor inlet air, can help improve gas turbine cycle efficiency by preheating the combustor inlet air, reducing the amount of fuel required. However, the low Biot number raises a concern about the suitability of a recuperator (see Sullivan [22, 23]). Care must be taken to thermally isolate the heat exchanger from the engine structure. Otherwise the heat exchanger structure will tend toward the temperature of the rest of the engine structure rather than a temperature intermediate between the two fluid flows through it. In addition, the compressor exhaust will be heated by the structure prior to entrance into the combustor. This heat transfer will tend to reduce the loss associated with heat transfer in the combustor to the structure. However, since the compressor exhaust prior to the recuperator is now at an elevated temperature, it reduces the amount of heat that can be transferred to it from the turbine exhaust stream, thus limiting the benefits to cycle performance from recuperation. The reduction in Biot number results in both gains and losses for the engine performance, generating new challenges in the design process for the MicroEngine.

Boundary Layer Effects

The previous paragraphs showed that the MicroEngine turbomachinery will see relatively more heat transfer than conventional scale components.[3] This increase in heat transfer has a substantial impact on the turbomachinery boundary layer physics. A heated wall will decrease the fluid density near the wall, increasing the boundary layer displacement thickness, which, per the momentum equation, increases the shape factor (ratio of the boundary layer displacement thickness to momentum thickness). The larger shape factor increases the growth of momentum thickness for retarded flows, increasing the drag or loss. More importantly, the larger shape factor due to heating acts to accelerate the growth of shape factor in such flows (per the shape factor equation). Note that separated flows are characterized by large shape factors, and attached flows are characterized by small shape factors. Thus, by increasing shape factor, the effect of heating is to move the boundary layer closer to separation. Conversely, cooling the wall will have the opposite effect, reducing losses in retarding flows and increasing the margin from separation. So these non-rotating effects are beneficial for the turbine, but detrimental for the compressor.

Effects resulting from rotation must also be accounted for. The heating results in a density gradient through the boundary layer. The density gradient combined with the acceleration due to rotation produces a buoyancy term. The buoyancy term is generally negligible in standard turbomachinery flows since the density gradient is typically small. However, the density gradients can become large at the microscale with uncooled turbomachinery. The buoyancy terms tend to increase the shape factor for both the compressor (whose walls are being cooled by the flow) as well as the turbine (whose walls are being heated by the flow), tending to move the boundary layer closer to separation. This increasing trend in shape factor can be countered by accelerating the flow, which tends to decrease the shape factor. While an accelerating flow is the natural state in a turbine, it is not what one desires in a compressor, and will limit the attainable pressure rise. So while the buoyancy effect is problematic for both the turbine and compressor, the turbine has more capability to counter the detrimental effects.

7.2.2 Fabrication Constraint Impacts

Bulk micromachining is basically a 2D process. A planar pattern is produced in a photoresist mask on top of the wafer and then etched into the underlying wafer, producing 3D structures that are essentially extrusions of the initial planar pattern. The processing tools do not allow for substantial etch depth variation. The MicroEngine obtains its 3D character by stacking multiple wafers, each of which has been processed on both sides, allowing for flow paths to run both along wafer planes as well as through wafers. However, there are serious implications for components like the

[3]This section on boundary layer effects was written with assistance from Harold Youngren.

compressor whose optimal shape is three dimensional. Other microfabrication fluidic constraints result from etch depth and thickness limits. These fabrication issues will be discussed in relation to some of the key components of the MicroEngine as they relate to thermofluidic issues. Fabrication constraints also significantly impact other aspects of the engine, such as the journal bearing, combustor, and electric machinery performance. These issues are discussed in their associated chapters in this book.

7.2.2.1 Turbomachinery

Flow through a compressor sees an adverse pressure gradient, making it highly susceptible to separation. The laminar/transitional boundary layers in the MicroEngine make them even more susceptible to separation than in a conventional-scale, turbulent flow compressor. In a conventional centrifugal compressor design, the blade height decreases as radius increases to counteract the increase in circumferential dimension, limiting the flow area increase. With this ability to control flow area by varying blade height, the adverse pressure gradient seen by the flow can be reduced, and the blade shapes can then be chosen to provide high aerodynamic performance.

However, using standard microfabrication techniques, there is no simple way to vary the blade height as a function of radius. The impeller blades are formed during a single etch and thus have uniform blade height. Multiple etches could be used to introduce a few steps in the blade height that could then be smoothed, but this is limited to no more than a few steps when etching to depths of hundreds of microns. A group at the University of Maryland [4, 24] is developing gray-scale photolithographic technology that allows for out of plane etch depth variation. This technique has the potential to significantly broaden the range of geometries available for MEMS and will likely play an important role in the future of PowerMEMS. However, the gray-scale process complexity must be reduced and its robustness must be improved before it can be considered for integration into the MicroEngine fabrication process.

When designing turbomachinery with constant blade height, the blade shape must be used to both control flow area change and provide good aerodynamic performance. Our early attempts to design compressor blades with controlled area increase, while trying to maximize turning in the flow to obtain high pressure ratios, resulted in very thick blades that were highly swept with flow in the relative frame exiting nearly tangentially. The compressor shown in Fig. 7.4 achieves a pressure ratio of 4.4 (total–total) across the rotor, and a pressure ratio of 3.8 (total–static) from rotor inlet to diffuser exit based on a Mises 2D computation, spinning at a 500 m/s tip speed. A significant drawback of such a blade configuration is that it severely limits available flow area, reducing the available mass flow rate per unit blade height, reducing power density and overall performance. On a positive note, the large thickness of the blades relative to rotor radius stiffens the blades, reducing stress at the blade roots. It is important to minimize stress at the blade roots because it is difficult to fabricate large fillet radii, which would require etch depth variation. We are generally only able to obtain blade root fillet radii of 5–10 μm on blades

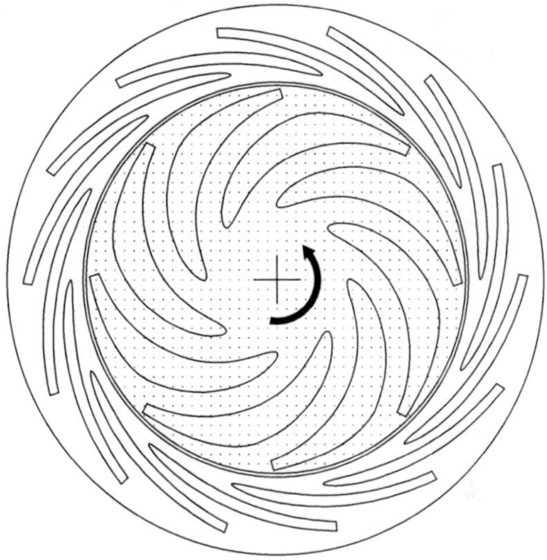

Fig. 7.4 Microfabrication compatible centrifugal compressor design, 3.8:1 pressure ratio (total-static) based on Mises 2D computation with 500 m/s tip speed. The centrally shaded region is the rotor. Flow inlet is at the *center*

with heights greater than 200 μm. The fillet radii are obtained by detuning the etch parameters as the etch is completed.

An additional drawback with highly swept designs is that the exit flow from the rotor contains considerable swirl, and thus a considerable portion of the pressure rise depends on good diffuser performance. However, diffuser performance is poor as it suffers substantially not only from the fabrication constraint of having constant blade height, but also from the additional fabrication result that the blade sides are nearly vertical. The flow angle of the impeller exhaust varies over a wide range in the axial direction due to end wall effects (see section 7.3.1), so a diffuser vane without twist will be off-design over most of its span.

Experience from running a series of 3D computations backed up by experimental results (discussed in section 7.3.1) has shown that for designs with large swirl (as in Fig. 7.4), the diffuser without twist is ineffective, providing a pressure recovery that is close to 0. The pressure ratio of 3.8 presented above is based on a (2D) MISES pressure recovery computation of about 60% across the diffuser. However, with a pressure recovery of 0, the pressure ratio across the full compressor becomes 2.9 (total:static).

This poor diffuser performance impacts efficiency as well. Using MISES, we computed an 85% isentropic efficiency for the compressor in Fig. 7.4, assuming a pressure ratio of 3.8. However, the efficiency drops to 64% if the pressure ratio is only 2.9. Note that this drop from 85 to 64% is consistent with the factor of two reduction rule-of-thumb discussed in the turbomachinery part of Section 7.2.1.2 for converting 2D efficiencies to 3D efficiencies.

Using 3D CFD tools, it is possible to design a constant height thin blade at this scale with efficiency close to the thick blade design described in the previous paragraph, but much larger flow rate. Figure 7.5 shows such a design developed by

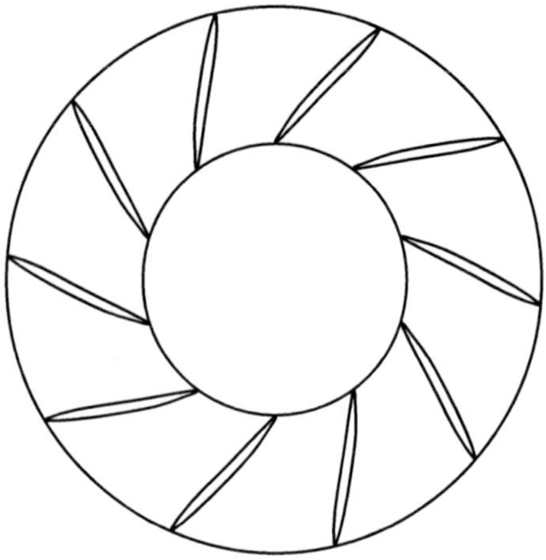

Fig. 7.5 Microscale compressor blading design (from Sirakov (2005) Fig. 3.6, portion reprinted with permission from B. Sirakov and Massachusetts Institute of Technology, Department of Aeronautics and Astronautics)

Sirakov [20]. The impeller of Fig. 7.5 suffers from separation and has lower pressure ratio than the impeller of Fig. 7.4. However, because of its larger flow rate, the exit flow angle (and Mach number) is less, placing less of a burden on the diffuser. Sirakov found that this type of impeller design provides a pressure recovery of about 60% based on 3D computations. While this is certainly not a good value when compared to conventional diffusers, it is a vast improvement over the ineffective diffuser of Fig. 7.4. The net result is a compressor design that has about the same efficiency and pressure ratio as that of Fig. 7.4, but much higher flow rate. As there are losses in the engine that are main flow independent (such as the bearings), a design with higher flow rate and thus higher power density is desirable.

An SEM of the microturbine analyzed in Fig. 7.3 is shown in Fig. 7.1. The 2D planar nature of the turbine blades is evident in this picture. As flow accelerates through the turbine, the microturbine design does not suffer from separation concerns, so flow area control is less of an issue than it is for the compressor. In addition, computational results show that the flow field remains essentially two dimensional in the microturbine. As a result, two-dimensional constant blade height designs are predicted to perform adequately, and fabrication constraints are not a large concern for a microturbine.

7.2.2.2 Flow Turning

The flow through the MicroEngine goes through sharp 90° turns at several locations as the flow direction changes between radial and axial. Rounding the walls of the flow passage at these points is desirable to reduce loss, but current planar microfabrication technology does not permit this. To keep loss from these turns small, one can try to keep the Mach number (and thus dynamic pressure) low through the turns.

This issue is particularly important at the MicroEngine inlet where a separation can result in significant blockage, sharply reducing the mass swallowing capacity of the engine. Mehra [12] studied the effects of inlet blockage, finding that mass flow could be cut in half compared to that obtainable with a smooth bell-mouth design. While such an inlet design is not consistent with standard MEMS processing, a close approximation to this may be possible using the gray-scale fabrication process discussed earlier. Alternately, a relatively simple conventionally machined small curved inlet could be fabricated and inserted into the MicroEngine. Mehra also found that a large improvement in the inlet flow could be obtained by increasing the inlet area as much as possible (decreasing Mach number) and by placing the outer diameter of the inlet at the same value as the compressor leading edge. This allowed the compressor to suck in the separation bubble, limiting its growth. With this scheme, the mass flow through the compressor approached that obtained with a smooth bell-mouth inlet, although a smooth inlet is still superior in terms of pressure loss.

Flow turning at the turbine exit is also a concern as it is a region of naturally high Mach number. Pressure loss at the exit reduces the available pressure ratio across the turbine. For a turbine generator design, all kinetic energy remaining in the turbine exhaust is a loss. One aims for a turbine design that leaves no swirl in the exit flow, and is thus radial in the absolute frame. Velocity is set by the mass flow rate (fixed), density (essentially fixed), and flow area. Again here it would be desirable to change flow area by varying blade height, but this is not an option. Flow area can be controlled somewhat through blade shaping, and one should design the exhaust geometry to have as large a radius to maximize exit area to reduce exit flow velocity.

7.2.2.3 Pressure Ratio Effects

The pressure ratio of the MicroEngine is limited by several factors. To reduce fabrication complexity, the engine turbomachinery design currently is single stage for both the compressor and turbine. Most gas turbine compressors, even those with a centrifugal turbomachinery stage, use more than one stage to obtain higher pressure ratios and increased power density for the engine. Lee et al. [11] explored MEMS turbomachinery designs with multiple stages, and we believe this may be feasible for the MicroEngine in later generation designs. Compounding the pressure ratio limitations from being single stage are the losses resulting from lack of three dimensionality in the blades and poor diffusion discussed above. The end result is an engine that is designed for a pressure ratio in the range of 2–3, an order of magnitude less than conventional scale gas turbines.

An important outcome of the relatively low pressure ratio of the MicroEngine is that it has a significantly lower fluid density in the "high" pressure sections of the engine, as compared to conventional engines. Thus, the local Reynolds number in the high-pressure sections of the MicroEngine is low not just from the lengths scale, but also from the low fluid density. The low Reynolds number results in relatively thick boundary layers, increased skin friction losses and increased heat transfer beyond the scale factors given in Table 7.1. A low pressure ratio is not inherent to the MicroEngine, and thus was left out of the earlier scaling discussions.

Rather it occurs from the choice to reduce complexity and stick with a single stage compressor in the baseline MicroEngine. This is then a fabrication issue that results in a scaling issue, and one that could be overcome in later generations of the device.

7.3 Microscale Compressor Design and Analysis

The previous section discussed the impacts of length scale and MEMS fabrication on the microengine design and performance. Several examples of component performance were presented based on computational models. In this section, we provide a more detailed discussion of the compressor design and analysis, along with experimental data to back up the computations. Only a few research groups besides ours have worked on the design of centrifugal compressors with diameters of order 1 cm (see for instance Isomura et al. [6], Kang et al. [8], Peirs et al. [14]), and these projects began several years after our effort. When we began, we planned to use CFD for designing the turbomachinery. However, since we were working in a relatively unexplored design space for high-speed turbomachinery, we felt it important to obtain experimental data for use in verifying the computational design codes.

This section begins with a presentation of experimental measurements obtained at MIT during this program, along with a comparison of these measurements to computational results. A scaled test facility was built for investigating the performance of microscale compressors under nominally adiabatic conditions. Many of these results have not been previously published. The second part of this section uses 3D CFD to explore the impact of heat addition on microcompressor performance. Results obtained at MIT by Sirakov et al. [21] and Gong et al. [5] are presented, focusing on those aspects of these studies that elucidate trends in phenomena of the microscale that are different from those typically observed at the conventional scale.

7.3.1 Macro-Compressor Test Rig – A Scaled Facility for Investigating the Performance of Adiabatic Microscale Compressors

Gregory B.Shirley, Khashif Khan, Christopher Cadou, Alan Epstein, and Stuart Jacobson[4]

Early in the planning efforts for the MicroEngine, it was deemed critical to generate experimental data for use in validating the computational codes in the appropriate flow regime as quickly as possible. There was then and there still remains relatively little related experimental work in the literature to which the microscale performance predictions can be compared. Developing an instrumented high-speed

[4]This section summarizes work completed by the listed authors that has not been previously published. Some of this work appeared in the Masters thesis of G. Shirley [19]. The authors acknowledge the assistance of Y. Gong.

prototype millimeter-scale compressor was not consistent with the time frame and resources available to this program. To allow for the use of conventional materials and instrumentation, a 75:1 scaled up *Macro-Compressor Test Rig* was built instead. This test rig was designed to be dynamically and kinematically similar to the microscale device by ensuring that all relevant non-dimensional parameters, such as Reynolds number and Mach number, are matched. A brief description of the facility follows. A more detailed discussion of the rig design and instrumentation can be found in Shirley [19].

7.3.1.1 Macro-Compressor Test Rig Design

The MAcro-Compressor Test Rig (MACTR) contains a 4 mm diameter microscale compressor design scaled up by a factor of 75, resulting in a 30 cm diameter rotor. For the compressor to operate in the desired physical regime, it is important that all relevant non-dimensional parameters be matched as closely as possible. The MACTR is operated so that the two most important physical parameters, the Reynolds number and the blade tip Mach number, are equivalent to their values in the actual microscale devices. Table 7.2 shows a dimensional comparison of the full-scale (fs) 4 mm diameter device to the MACTR. To achieve the proper Reynolds number scaling, the 75× increase in length scale is counteracted by a 75× decrease in operating pressure, which drops the air density (ρ) by 75×. The operating temperature is unchanged, as are fluid properties that depend mainly on temperature, such

Table 7.2 Scale comparison of the macro-compressor test rig to the full-scale microcompressor

Parameter	Units	Microscale device	Macro-compressor test rig	Scale factor
Gas properties (air)				
Gas constant	J/kg-K	287	287	1
Pressure, inlet	atm	1	0.0133	1/75
Temperature, inlet	K	300	300	1
Density, inlet	kg/m^3	1.176	0.0157	1/75
Geometry				
Impeller radius	mm	2.00	150.	75
Blade chord	mm	1.95	146	75
Blade height	mm	0.200	15.0	75
Dynamics				
Tip speed (100% speed)	m/s	400	400	1
Rotation rate	rpm	1,900,000	25,000	1/75
Mass flow rate, gas	g/s	0.11	8.25	75
Dimensionless parameters				
Reynolds number (blade chord)		50,000	50,000	1
Reynolds number (blade height)		5100	5100	1
Mach Number, tip		1.15	1.15	1

as dynamic viscosity (μ). Rotational and fluidic velocities (V) are also unchanged, keeping Mach number (M) fixed. Thus, keeping the product of length scale and density constant ensures that the Reynolds number (Re) will be unchanged from that in a comparable microscale device:

$$L_{\mathrm{MACTR}} = 75L_{\mathrm{fs}} \qquad \rho_{\mathrm{MACTR}} = \rho_{\mathrm{fs}}/75, \tag{7.4}$$

$$\mathrm{Re}_{\mathrm{fs}} = \frac{\rho_{\mathrm{fs}} V L_{\mathrm{fs}}}{\mu} = \frac{(75\rho_{\mathrm{MACTR}})V(L_{\mathrm{MACTR}}/75)}{\mu} = \frac{\rho_{\mathrm{MACTR}} V L_{\mathrm{MACTR}}}{\mu} = \mathrm{Re}_{\mathrm{MACTR}}, \tag{7.5}$$

$$M_{\mathrm{fs}} = \frac{V}{V_{\mathrm{s}}} = M_{\mathrm{MACTR}}, \tag{7.6}$$

where L represents a typical length scale and V_{s} represents the speed of sound,.

The MACTR impeller blade design, shown in Fig. 7.6 , is based on the impeller design for a motor-driven compressor that was being designed concurrently with the MACTR. This particular compressor was designed for 400 m/s tip speed and nominally adiabatic operation. MACTR impeller design, shown in Fig. 7.6, is similar

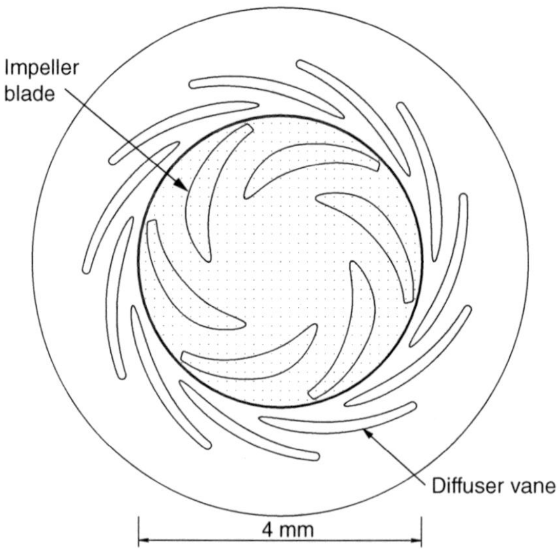

Fig. 7.6 Compressor design studied in the macro-compressor test rig. Note that the design is similar to that in Fig. 7.4. This compressor design was optimized in MISES for lower tip speed (400 m/s) operation. As modeled in MISES, the compressor has a diameter of 4.0 mm, tip speed of 400 m/s, resulting in a pressure ratio of 1.8 (total-static) across the impeller and 2.0 (total-static) across the combined impeller and diffuser. Mass flow rate at the design point is 0.11 g/s. The flow angle at the entrance to the diffuser is 83°. This geometry was scaled up by a factor of 75 for the MACTR

in design to the impeller shown in Fig. 7.4, which was discussed in the previous section.

Figure 7.7 shows a schematic of the full test rig and Fig. 7.8 shows a cross-section of the compressor section. The flow approaches the impeller along its axis of rotation. Just upstream of the compressor inlet, the flow is directed through a series of turbulence suppressing screens and a honeycomb straightener. Upon entering the compressor section, the flow turns 90° and passes radially outward through the blades of the impeller, through the diffuser, and continues radially outward before turning 90° again. The flow then passes through a ring of holes and enters a large annular chamber, designed to simulate the combustor in the MicroEngine. The flow exits the annular chamber through one of nine circumferentially located throttling valves into another annular chamber. The flow then exits the annular chamber

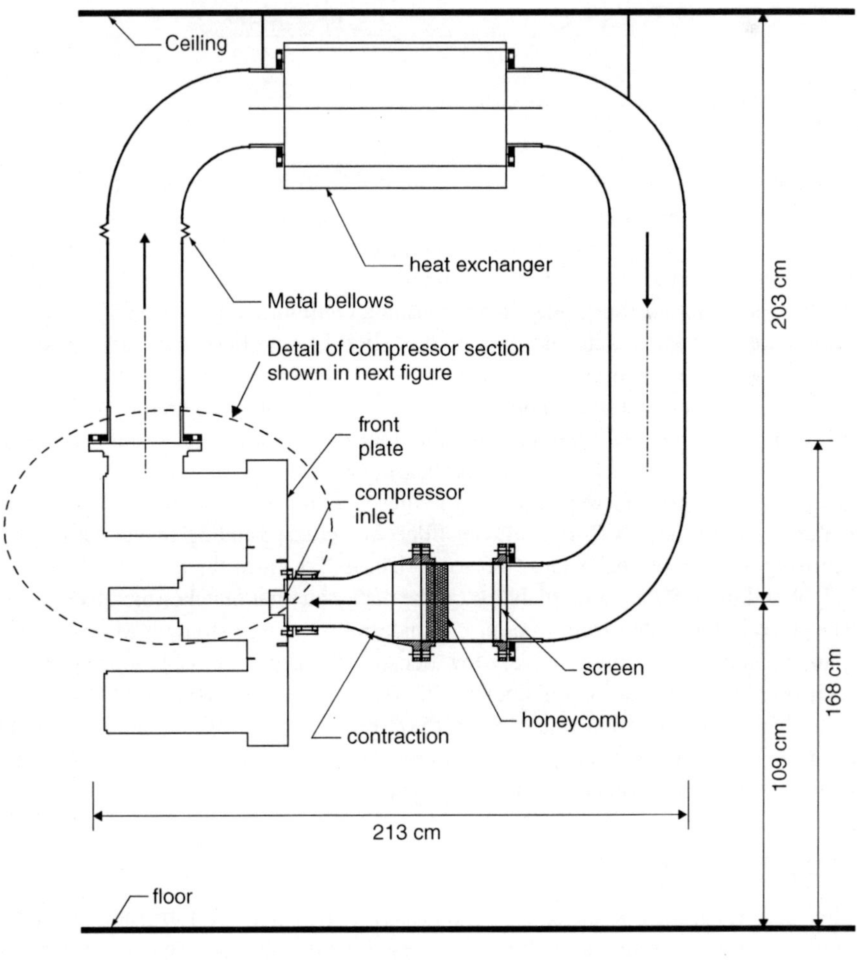

Fig. 7.7 Macro-compressor test rig schematic

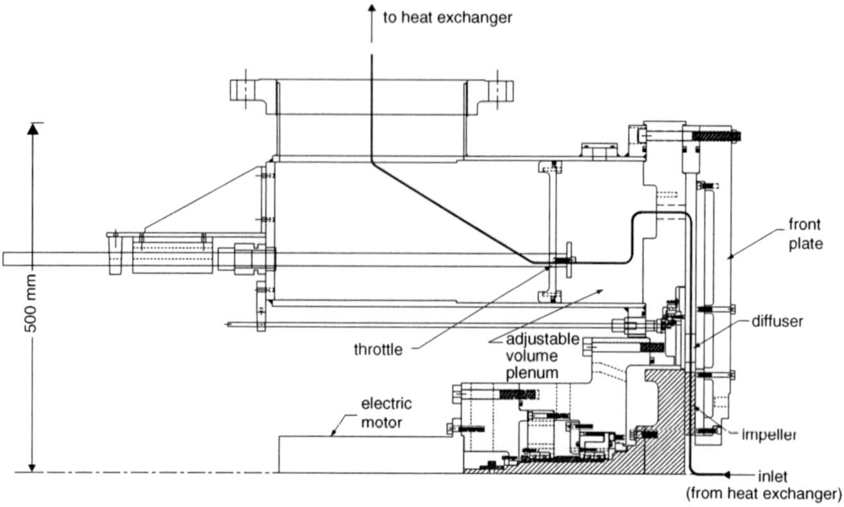

Fig. 7.8 Macro-compressor test rig compressor cross-section

vertically into a water-cooled heat exchanger before being returned to the impeller inlet. The throttling valves control the compressor pressure ratio and mass flow rate. The heat exchanger is used to maintain a constant fluid temperature at the compressor inlet through the course of an experiment.

The rotor support is designed in an overhung configuration, so no shaft obstructs the compressor inlet. Later designs of the MicroEngine have a centrally located thrust bearing that will cause some blockage to the inlet flow to the compressor. For the experiments presented here, the MACTR was not configured to include the thrust-bearing blockage; the computations that the experimental results are compared to also do not include the thrust-bearing blockage. The impeller blades are designed with minimal root fillets, as is the case in our microfabricated devices, which typically have a 5–10 μm root fillet. To reduce bending of the rotor from centrifugal loads on the blades, the rotor disk is 2.5 times thicker than the blade height and includes a counterbalance on the back side that matches the blade mass. The rotor is fabricated from titanium, a material with high strength to density (to manage blade stress) and relatively low density (to keep rotordynamic frequencies relatively high). With a rotor diameter of 0.30 m, operating at a tip speed of 400 m/s is equivalent to a rotation rate of over 25,000 rpm. Achieving this high rotational speed with a heavy overhung rotor requires supercritical rotordynamic operation, resulting in a considerable rotordynamic design challenge.

7.3.1.2 Instrumentation

The main quantities of interest for evaluating compressor performance are mass flow rate, pressure ratio, and efficiency. These quantities are determined from pressure, temperature, and rotation rate measurements. Three different types of pressure probes are used: wall orifices for measuring static pressure, a radially traversing total pressure probe in the inlet, and a spanwise traversing cobra probe located immediately downstream of the impeller discharge. When combined with an appropriate static pressure measurement, both the total pressure probe and the cobra probe permit measurement of the local flow velocity. The cobra probe also gives the flow direction in the plane of impeller rotation. Pressure signals from the various probes are routed via a manifold system to two absolute pressure transducers (MKS 220C 0–100 Torr). A single differential pressure transducer (MKS CD220CD 0–1 Torr) connected between the two legs of the manifold permits the measurement of very small pressure differences. The differential pressure transducer is used primarily with the inlet total pressure probe to measure the difference between the static and total pressure (and thus the velocity) at various radial positions in the inlet. The accuracy of the transducers is $\pm$ 0.15% of the reading while the resolutions are 0.01 and 0.0001 Torr for the absolute and differential transducers, respectively. The analog signals produced by the transducers are digitized using two Mark V Scanivalve Digital Interface Units (SDIUs) outfitted with 16 bit A/D converters.

All temperature measurements are made with K-type thermocouples connected to a Fluke Helios I thermocouple conditioner that amplifies and digitizes the signals. The wall temperature thermocouples are either attached directly to the surface of interest or to a washer that is bolted to the surface. Kiel-type total temperature probes are used for flow measurements. These probes contain three junctions: one to measure the gas temperature, one to measure the temperature of the radiation shield, and one to measure the temperature of the supporting stem. The latter two junctions permit the determination of appropriate radiation and conduction corrections to the gas temperature measurements, resulting in an uncertainty of $\pm$ 0.15 K.

Shaft rotation rate is monitored using a Hall effect sensor installed in the motor housing and a Hewlett Packard 3516A universal counter.

The voltage readings from the Helios, the counter timer, and both SDIUs are relayed via GPIB to a data acquisition computer running a custom program written in National Instruments' Lab Windows.

Efficiency is measured using a spindown method. Work done on the fluid by the compressor is determined by measuring how fast the impeller decelerates as a result of this work input and comparing this to how fast the impeller decelerates when pumped down to near vacuum to isolate non-aerodynamic loads (such as the bearings). This method is discussed in detail in Shirley [19].

7.3.1.3 Experimental Measurements

Diffuser Performance

The initial design of the microcompressor included a vaned diffuser downstream of the impeller. However, 3D CFD performed by Mehra [12] indicated that boundary layer growth along the diffuser end walls causes the flow area within the diffuser to remain nearly constant, which prevents any real diffusion. One of the first experiments designed for the MACTR was a study of the performance of vaned and vaneless diffusers.

Figure 7.9 shows measurements taken at both 100 and 42% speed for cases with vaned and vaneless diffusers. The pressure ratios shown are based on total inlet pressure to the impeller and static pressure measured downstream of the diffuser. Results from a MISES 2D computation vaned case are shown for both speed cases. The result of a design point 3D computation of a vaned case at 100% speed is also shown (Dawes code, see [12] and [2]). Focusing on the experimental results, it is clear that the vaned diffuser provides more diffusion than the vaneless diffuser at the low mass flow end of the speed lines in both Fig. 7.9a and b. However, at the high mass flow end, the vaneless diffuser is more effective at diffusion. Since the higher pressure ratio side of the speed line is more relevant for operation, these results indicate that a vaned diffuser is effective and should be included. Efficiency measurements show that the presence of the diffuser seems to improve the efficiency, further arguing for including a vaned diffuser.

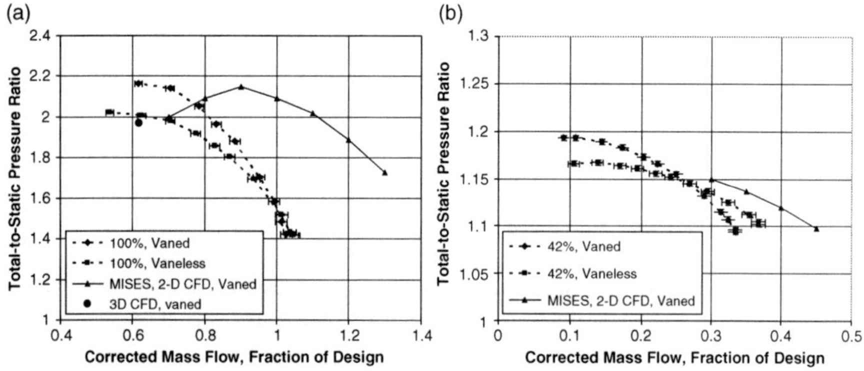

Fig. 7.9 Macro-compressor runs with and without diffuser vanes. Pressure ratio is across the combined impeller and diffuser. (**a**) Operating at 100% design speed (400 m/s tip speed); (**b**) Operating at 42% design speed (168 m/s tip speed). Mises 2D computational results are also shown. A 3D CFD (Dawes code) result is included in the 100% speed plot

The measured compressor performance is comparable in pressure ratio to the 2D MISES results. However, the MISES results predict about 25% more mass flow than is actually observed in the MACTR (vaned configuration) at the design pressure ratio of 2.1. This trend is expected as the MISES computations do not account for 3D effects such as the boundary layer growth on the end walls and the turning loss at the inlet (discussed further below).

On the other hand, the 3D DAWES code, computed at the design pressure ratio of about 2.0, underpredicts the measured mass flow rate. One reason for this discrepancy could be that the DAWES code assumes adiabatic flow, whereas the MACTR is not perfectly adiabatic. Heat loss from the flow to the MACTR structure, which is a massive thermal sink, could be acting to stabilize the boundary layer and delay separation leading to improved performance. Another hypothesis is that separation around the inlet where the flow makes a sharp 90-degree bend is not as severe as anticipated in the DAWES code. In addition, to make the analysis more tractable, the DAWES analysis was performed with a tip shroud, whereas the rig is unshrouded.

As discussed earlier, fabrication constrains the blades to have vertical sides, with no twist. Figure 7.10 shows flow angle measured just downstream of the diffuser for a run operating at 100% design speed with a vaneless diffuser. Note that the flow angle is relatively high and varies over 13°. At these high flow angles, a diffuser is typically sensitive to off-design conditions, only performing well within a few degrees of design. However, despite the large and varying flow angle, the diffuser appears to have performed with moderate success for this test.

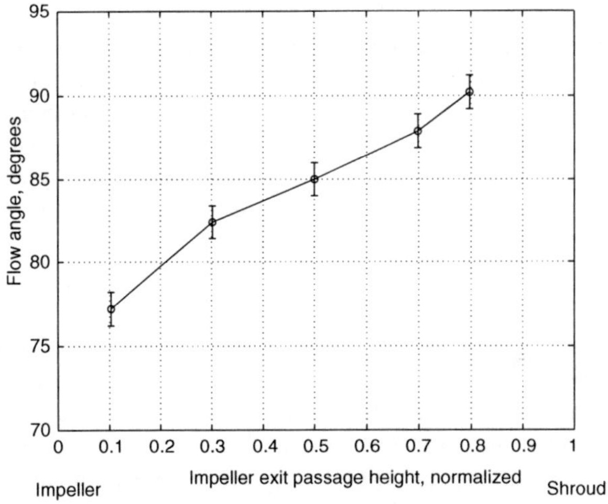

Fig. 7.10 Flow angles measured at the impeller exit in the compressor test facility for a run operating at 100% design speed with a vaneless diffuser (from G. Shirley [19] Fig. 5.4, portion reprinted with permission from Massachusetts Institute of Technology, Department of Aeronautics and Astronautics)

Inlet Effects

As flow enters the microcompressor, it must navigate a 90-degree turn from the axial inlet to the radial outflowing centrifugal compressor. Based on microfabrication constraints, the microcompressor has a sharp-turned square inlet, as sketched in Fig. 7.11. The potential for flow separation resulting from the sharp 90-degree turn was a major design concern for the microcompressor. Separation at this turn could lead to flow blockage and compressor performance degradation. So, in addition to the baseline sharp-turn inlet, a smoothed inlet was designed to assess the importance of separation at the inlet. Figure 7.11 also shows a sketch of the smooth inlet.

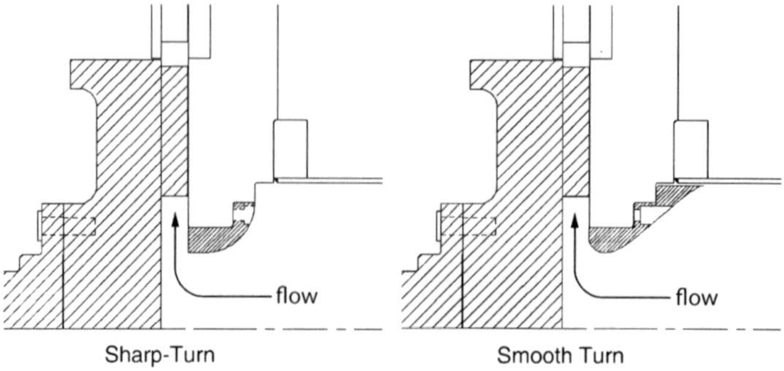

Fig. 7.11 Sharp-turn vs. smooth inlet geometries for the macro-compressor test rig

Figure 7.12 shows that the mass flow corresponding to a particular pressure rise increases for the smooth inlet relative to the square inlet, although the effect is more pronounced at higher mass flow rates, where one would expect increased flow separation. In the desired operating region of the compressor (lower mass flow rate and higher pressure ratio), there is almost no discernable effect. However, spindown efficiency measurements showed that the smoothed inlet improved the efficiency by 6 points, from 42 to 48% at 100% speed, which is a significant performance advantage. The MISES 2D result, which does not have an inlet turn section, is also shown in Fig. 7.12.

These results suggest that to improve efficiency, the inlet area of the microcompressor should be maximized to decrease inlet Mach number, which should reduce the size of the blockage resulting from the inlet turn. The measurements also suggest that a smoothed inlet will give better off-design performance capability. As a smoothed inlet is not possible with current microfabrication capabilities, we are exploring the possibility of attaching a conventionally machined smooth nozzle to the inlet section of the compressor.

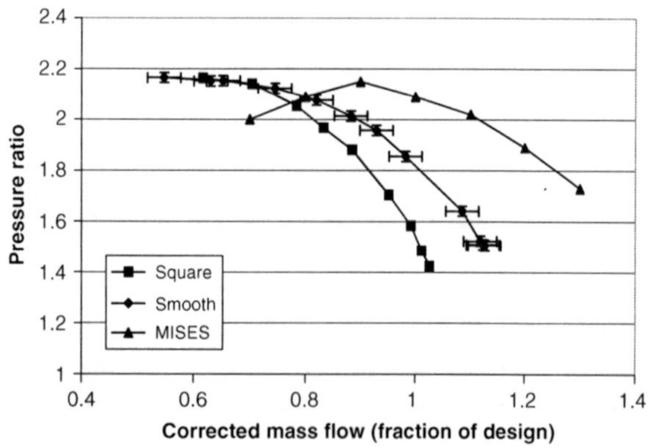

Fig. 7.12 Effect of inlet design on flow obtained from MISES and macro-compressor rig. All cases at 100% design speed with vaned diffuser

Compressor Map and Tip Clearance

Figure 7.13 shows a compressor map measured in the MACTR. Mass flow is corrected and normalized by the corrected mass flow, where corrected flow is defined as

$$\dot{m}_{\text{corrected}} = \frac{\dot{m}\sqrt{T_{t,in}/T_{\text{ref}}}}{P_{t,in}/P_{\text{ref}}}, \tag{7.7}$$

where $\dot{m}$ is the design microscale mass flow rate, $T_{t,in}$ is the measured inlet total temperature, $T_{\text{ref}} = 288$ K, $P_{t,in}$ is the measured inlet total pressure and $P_{\text{ref}} = 1$ atm. The pressure ratio shown is the ratio of the diffuser exit static pressure to the inlet (upstream of the turn) total pressure. Efficiency for each operating point was measured using the spindown method. The operating points are shown in Fig. 7.13. Result from a 2D MISES run at 100% speed is also shown, with the 2D results matching the pressure rise, but over-predicting mass flow (likely due to boundary layer growth on the end walls). Contours were computed based on the efficiency measurements at the operating points. Note that at design speed near the peak of the speed line, the compressor is operating with an efficiency approaching 50%. This relatively low efficiency results from the scaling and fabrication issues discussed above. A later 3D computational design effort by Sirakov [20] showed that, when

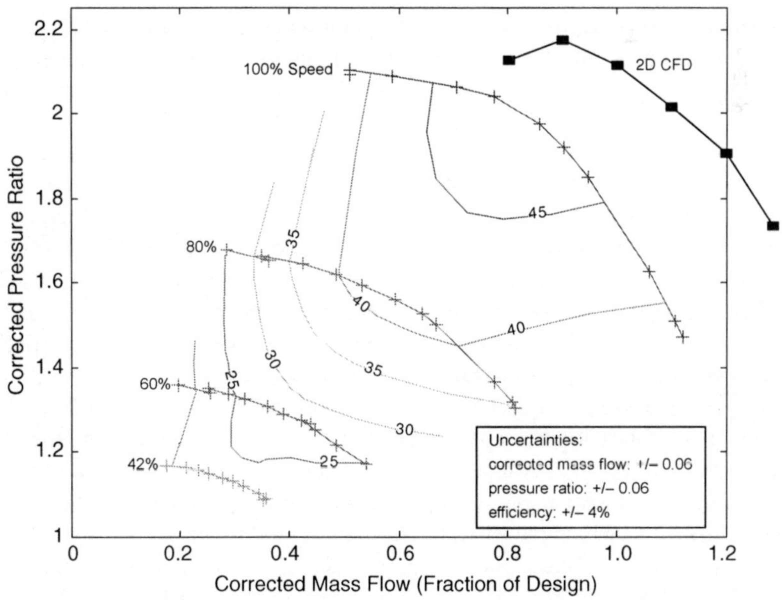

Fig. 7.13 Compressor map measurements from the macro-compressor test rig. Contours based on measured efficiency are shown. The pressure ratio is based on the diffuser exit static pressure and the inlet (upstream of the turn) total pressure. Tip clearance = 0.85% of blade height. Results from a 2D Mises run at 100% speed are also shown

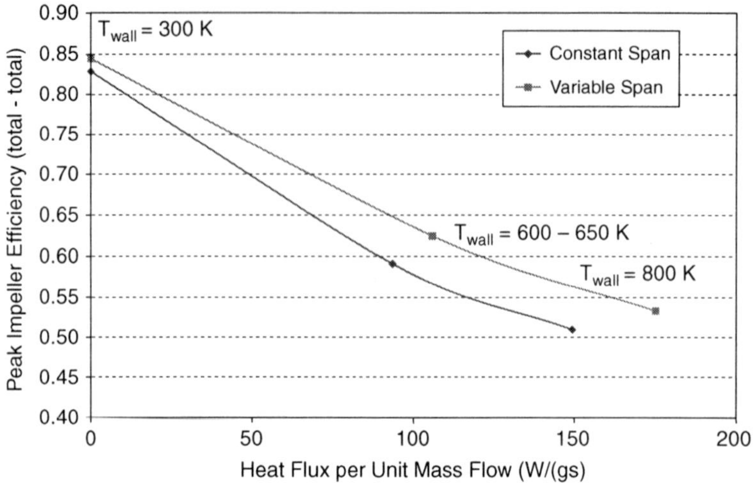

Fig. 7.14 Compressor performance improvement from a variable blade height impeller (20 blades, shrouded) under non-adiabatic conditions, computed using Fluent (3D CFD). Courtesy of Sirakov (2004, personal communication)

operating the compressor adiabatically (as we do in the macro-compressor rig), one could push the compressor efficiency to about 80% by doubling the microcompressor diameter to 8 mm (doubling the Reynolds number), along with some design optimization. However, for non-adiabatic operation as will exist in a microscale gas turbine, where heat transfer into the turbine is conducted to the compressor due to the high thermal conductivity of silicon, Sirakov found the efficiency dropped to about 50–60%. Clearly, maximizing thermal isolation between the turbine and compressor is critical for gas turbine operation, as compressor efficiencies below 60% make break-even operation a challenge. With the ability to vary blade height, Sirakov (2004, personal communication) showed that the compressor efficiency can increase by over 5% for cases with substantial heat transfer (see Fig. 7.14). As a result, gray-scale process development to enable variable blade height is considered an important technology for microscale engine viability [24, 4].

A tip shroud is expected to improve compressor efficiency. However, initial versions of the microscale gas turbine were designed without a tip shroud on the compressor as attaching a shroud adds an additional layer to the engine, which would increase fabrication complexity and impact yield and fabrication time. Without a tip shroud, the microcompressor performance becomes a function of the gap between the compressor blade tips and the stator tip gap. A series of experiments were performed in which the MACTR was configured to allow variation of the tip gap. Figures 7.15 and 7.16 show measurements taken with three different tip gaps: 6.75, 2.5, and 0.85% of blade height. At lower speed, there is little variation in the speed line pressure ratio performance as a function of tip gap; however, there is a drop off in pressure ratio with increasing tip gap at the higher speeds. At 100% design speed, however, there is also a substantial drop off in efficiency for the case

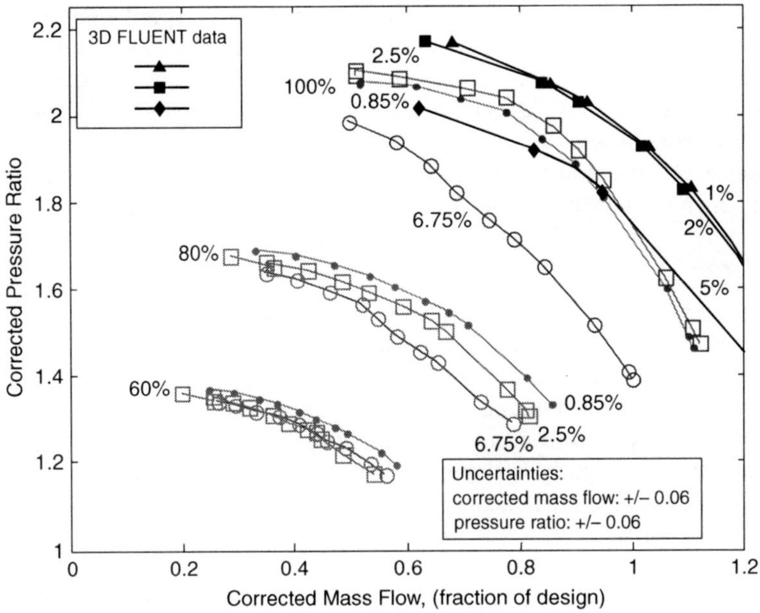

Fig. 7.15 Speed lines measured with three different tip gaps: 6.75, 2.5, and 0.85% of blade height. Results from 3D Fluent runs at comparable tip gaps (and 100% speed) are also shown. Fluent results courtesy of Gong (2000, personal communication)

with a 6.75% tip gap; the compressor performance remains comparable for the cases with tip gaps of 0.85 and 2.5%. At the microscale, designing for a tip gap of 2.5% (400 μm blade height with a 10 μm tip gap) is feasible as the thrust-bearing gap is of order 1–2 μm. To maintain the tip gap at high rotation rate, the compressor disk must have a counterbalance mass on its back side to reduce disk axial deflection that results from torque due to centrifugal forces developed by the blades.

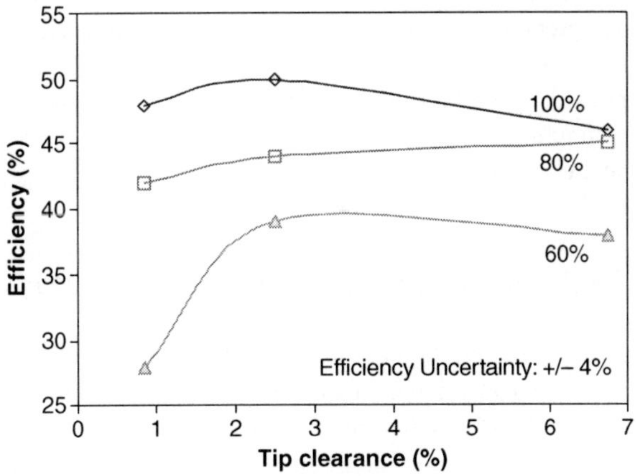

Fig. 7.16 Macro-compressor rig efficiency measurements as a function of tip clearance. The 3D Fluent computations at 100% speed estimated efficiencies of order 50%, comparable to the measurements obtained

Figure 7.15 also shows computational results at design speed from 3D Fluent runs. The Fluent mesh includes the inlet right angle turn. The Fluent results appear to do a good job of modeling the experimental results, lying just at the edge of the experimental uncertainty. All current design efforts, including those in Sirakov [20], were done using Fluent.

Gong et al. [5] also performed a comparison of Fluent results to compressor map measurements obtained from the MACTR, finding very close agreement.

Heat Transfer

Heat transfer is the most poorly scaled process in the MACTR. In a microscale gas turbine, the temperature of the structure is expected to be high resulting in a net heat flux to the flow. In the MACTR, however, the large thermal mass of the rig prevents it from heating up much during a run; the rig remains at a lower temperature than the flow and heat transfer occurs *from* the flow to the rig. In addition, the micro-compressor will be nearly isothermal because both blade and structure are silicon and Biot numbers are predicted to be small (< 0.1) (see Biot number discussion in Section 7.2.1.4). While the Biot number for the macro-compressor structure is similarly small ($\sim$0.03), the Biot number for the impeller is somewhat larger ($\sim$0.4) indicating that there may be larger temperature gradients within the impeller itself and between the impeller and the stationary structure than in the microscale compressor. Second, and possibly most importantly, heat addition to the boundary layer in a centrifugal compressor tends to be de-stabilizing (see Section 7.2.1.4). Indeed, it is possible that the effects of heat transfer on boundary layer instability and the resulting flow path blockage could overwhelm the purely thermodynamic effects considered for comparison. Therefore, the MACTR is best used for understanding and comparing to near-adiabatic cases only.

Macro-Compressor Rig Summary

Comparisons between MACTR measurements and computational results provide confidence in the ability of available CFD tools to correctly predict the adiabatic performance of microscale compressors. Differences between the measurements and predictions were generally small and within the uncertainty of the experimental measurements for cases that were directly comparable. As convective heat transfer physics is similar to boundary layer physics (Reynolds analogy), we believe that the computational tools should also perform well for cases in which there is substantial heat transfer.

7.3.2 Non-adiabatic Microcompressor Performance

Given some confidence in the performance of CFD in the microscale gas turbine design space, we next began a more complete exploration of the design space using 3D CFD tools, including the impacts of heat addition to the flow. The results of this study are presented in detail in Sirakov et al. [21], Gong et al. [5], and Sirakov

[20]. This section provides only an overview of the more important results obtained in these works. The first part of this section presents the results of Sirakov's analysis breaking down the efficiency loss in a microcompressor. The second part of this section summarizes a simple analytical model developed by Gong et al. [5] that can be used for representing and analyzing non-adiabatic compressors in a thermodynamic cycle.

7.3.2.1 Non-adiabatic Compressor Performance

The three main performance characteristics of a compressor are pressure ratio, mass flow rate, and efficiency. As presented earlier in this chapter, scale reduction alone does not strongly impact pressure ratio to first order, although microfabrication constraints such as constant blade height can limit pressure ratio. Scale reduction will impact mass flow rate as the relatively larger boundary layers create increased blockage relative to large-scale compressors. But the mass flow rate impact is not substantial, at least at the millimeter-scale being pursued for the devices of interest. The main impact of scale reduction as it relates to performance is a sharp reduction in efficiency.

As detailed in this chapter, small-scale physics and microfabrication constraints lead to relative increases in boundary layer losses and separation that will reduce turbomachinery efficiency. Scale reduction causes a drop in Biot number, which can lead to a substantial change in heat transfer characteristics. Significant heat transfer into turbomachinery flow is not part of the conventional design space. Accounting for the non-adiabatic conditions of the turbomachinery makes clear that this is the major impact of scale with respect to a gas turbine application, particularly on the compressor side. In a thermodynamic sense, compressing a hot gas requires more power than compressing a cold gas. So heat addition to the compressor reduces overall engine thermodynamic cycle efficiency. Heat addition to the compressor also decreases compressor efficiency and potential pressure ratio as it can destabilize the boundary layer.

Sirakov et al. [21] used Fluent, a 3D CFD analysis code for studying the efficiency impacts of different design choices. Table 7.3 shows results for five analysis cases, all of which used the same baseline impeller design. The centrifugal impeller

Table 7.3 Microcompressor performance under different operating conditions (taken from Sirakov et al. [21]). The efficiency is calculated across the impeller only; diffuser losses are not included

Case	Impeller temperature (K)	Tip clearance	Inlet	Isentropic total-total efficiency (%)
1	950	20 μm	90° turn	43
2	600	20 μm	90° turn	58
3	Adiabatic	20 μm	90° turn	68
4	Adiabatic	Shrouded	90° turn	85
5	Adiabatic	Shrouded	None	90

includes an 8 mm diameter rotor and 20 blades with a 20 μm tip gap (for non-shrouded cases). The blade height varies linearly from 400 μm at the inlet (2 mm radius) to 200 μm at the exit (4 mm radius). The blade height variation assumes a technology such as the gray-scale method discussed earlier. The impeller tip speed is 500 m/s.

Case 5 is a best-case operating scenario, consisting of the baseline compressor with no inlet loss, no casing (tip) loss, and no heat addition. For this case, an efficiency of 90% was obtained; this performance level is comparable to conventional scale impellers, indicating that at a scale of 8 mm, the Reynolds number is still large enough so that viscous losses are not yet substantial. For Case 4, an inlet is added to the computation that forces the flow to make a 90-degree turn (axial to radial outflow) at the impeller entrance, resulting in a 5% drop in efficiency. The previous two cases include a shroud on the blade tips, eliminating any loss between the blades and the stator. For Case 3, the shroud is replaced by a 20 μm tip gap. The tip gap causes a relatively costly 17% additional drop in efficiency. With variable blade height, the 20 μm tip gap is 5–10% of the blade height, depending on radial location. Recall that Fig. 7.16 showed that a tip gap greater than 2.5% would result in considerable efficiency loss at design speed. So the substantial efficiency loss obtained from a 5 to 10% tip gap is not surprising. For Cases 1 and 2, heat addition to the flow from the impeller structure is included. The impeller wall temperature is set to 950 and 600 K for Cases 1 and 2, respectively. Silicon has a considerable loss of strength and will begin to creep under modest stress at around 900 K. Thus in an MEMS engine design, the turbine is limited to a temperature of about 950 K. Case 1 assumes the compressor is directly coupled to the turbine with no thermal isolation. Case 2 assumes some level of thermal isolation between the turbine and compressor, such as a small diameter shaft. For Case 2, the efficiency drops an additional 10% from the adiabatic Case 3; for Case 1, the efficiency drops an additional 25% from the adiabatic Case 3.

This analysis clearly shows that the most important impact on compressor loss comes from heat addition. The boundary layer passage losses are not severe at this length scale. The tip gap and inlet turn resulted in a combined loss of 22%. This suggests that the tip gap should be held closer to 5% or less across the entire blade and the inlet should be sized as large as possible to minimize turning loss. A rounded inlet could also be considered as a post-fabrication addition. With these modifications, one could hope to gain back of order 10% in efficiency. The 25% efficiency loss associated with Case 1 indicates that some form of thermal isolation between the turbine and the compressor is absolutely necessary.

As was shown in Fig. 7.14, blade height variation accounts for just a few points of efficiency loss for an adiabatic compressor, but up to 5 points for cases with substantial heat addition. Keep in mind that the compressor analyzed in Fig. 7.14 was operated at a limited pressure ratio (~3), allowing for constant blade height operation. For a non-adiabatic case, as long as the constant blade height impeller is not pushed to high pressure ratio, its performance is not much different from the variable blade span case. However, the variable blade span impeller is more resistant to separation and thus can achieve higher pressure ratios.

The final device designed in this program at MIT was a self-sustaining microscale gas turbine engine. The compressor designed for this device was 8 mm in diameter, with constant blade height, 5% tip gap, smoothed inserted inlet and 900 μm diameter/660 μm long shaft between the compressor and turbine disk. Our computations and models predict a 57% isentropic efficiency for this design across the entire stage (impeller + diffuser), a value that should be sufficient for achieving self-sustaining operation.

7.3.2.2 Non-adiabatic Compressor Analytic Model

As discussed in Section 7.2.1.4, one of the key physical characteristics of an MEMS microscale engine, due to both small-scale and silicon's large thermal conductivity, is a relatively small Biot number. A small Biot number indicates that thermal conduction within the structure will occur more quickly than convection of heat from the flow to the structure, minimizing thermal gradients in the structure and making it difficult to thermally isolate any section of the engine. This is beneficial for the turbine, where design considerations require heat removal from the highly stressed turbine blades to keep temperature down and strength up. Great expense is incurred in the design of cooled blades for the hot turbine section in conventional gas turbines. In the MicroEngine, cooling comes cheaply by conduction through the structure as a result of the small Biot number.

The heat conducted away from the turbine moves of course to the cooler sections of the engine, namely the compressor. And dumping heat into the compressor flow can be catastrophic for gas turbine performance. Thermodynamically, the compressor is arguably the most critical component of a Brayton cycle based gas turbine. Compressing a gas is costly in terms of energy required per unit mass flow. Engineers strive to design high-efficiency compressors to minimize compressor power cost. But for a given pressure ratio, even for purely isentropic processes, compressing a hot gas requires more power than compressing a colder gas. Pressure ratio effectively sets a temperature ratio, but power scales with temperature difference, which increases for a given temperature ratio as the temperatures rise. It is perhaps more realistic to view the compression process as one where a given amount of work per unit mass flow is available. In this case, following the argument above, when the gas being compressed is at a higher temperature, the net pressure ratio obtained across the compressor will be reduced.

The optimal compressor pressure ratio (P_{opt}) for an idealized Brayton cycle is (per Reynolds and Perkins [16, equation 9–16])

$$P_{\text{opt}} = \left(\frac{T_C}{T_{\text{amb}}}\right)^{\gamma/[2(\gamma-1)]}, \tag{7.8}$$

where γ is the specific heat ratio, T_C is the combustor exit temperature and T_{amb} is the ambient temperature. Operating at this pressure ratio maximizes work only, not cycle efficiency; one must trade off work and efficiency in the design of a Brayton cycle gas turbine. Combustor exit temperatures in the range of 1200–1600 K are typically designed for to help maximize cycle efficiency and work output within allowable material limits. This range of combustor temperatures results in values of P_{opt} greater than 10. Conventional gas turbines typically use multiple turbomachinery stages to achieve this level of pressure ratio. In the current designs of the MicroEngine, however, we are planning for only a single stage for both the compressor and turbine to reduce fabrication complexity. With a single stage and constant blade height, the achievable pressure ratio is limited to about 2–3. Thus, the MicroEngine will be operating in the part of the design space that is well below the optimal pressure ratio. As discussed in the previous paragraph, heat input to the compressor will further reduce the achievable pressure ratio, which will then decrease the net work output per unit mass flow through the engine.

While heat addition to the compressor flow is then plainly bad for the cycle, analysis of its impact on design is not so straightforward. Heat is being transferred into the compressor flow from the inlet structure, the rotor and the stator. Gong et al. [5] suggest a simplified view of the physical process. Rather than computing the effects of heat transfer on the compression process through the compressor, they assume that all of the heat addition that takes place in the impeller section is added prior to the compressor. Gong et al. argue that in centrifugal compressors, such as those being pursued for the MicroEngine, most of the compression occurs in the highest circumferential speed section of the compressor, and thus in a small region near the impeller exit. Thus, in fact, most of the impeller heat transfer is happening prior to the bulk of the compression. Gong et al. make the additional assumption that the isentropic efficiency of the compressor is constant, regardless of the inlet stagnation temperature of the flow to the compressor. They first define a dimensionless heat addition parameter

$$\tilde{Q}_{\text{H}} = \frac{Q}{\dot{m} C_{\text{p}} T_{t1}}, \tag{7.9}$$

where Q is the total heat addition to the compressor flow, $\dot{m}$ is the mass flow rate, C_{p} is the fluid specific heat, and T_{t1} is the total temperature of the inlet flow to the compressor upstream of any heat addition. With this definition and the two assumptions presented above, they develop the following analytical equations for incorporating the impact of heat addition on compressor performance:

$$\pi = \left(\frac{1}{1+\tilde{Q}_{\text{H}}} \left[\pi_{\text{adb}}^{\gamma-1/\gamma} - 1 \right] + 1 \right)^{\gamma/\gamma-1}, \tag{7.10}$$

$$\eta = \eta_{\text{adb}} \frac{1}{1+\tilde{Q}_{\text{H}}}, \tag{7.11}$$

$$\dot{m} = \dot{m}_{\text{adb}} \frac{1}{1+\tilde{Q}_{\text{H}}}, \tag{7.12}$$

where π is pressure ratio, γ is the specific heat ratio, and η is efficiency. The subscript adb stands for the adiabatic value of that parameter.

From these equations, it is clear that increasing heat addition reduces pressure ratio, efficiency and mass flow rate. In the MicroEngine design space, mass flow rate is about 0.6 g/s and ambient inlet conditions are assumed. With a shaft between the turbine and compressor and the combustor exit temperature about 1200 K, the heat addition into the compressor flow is of order 25 W, resulting in a heat addition parameter ($\tilde{Q}_H$) of about 0.14. The analysis of Gong et al. then indicates that the heat addition will reduce the compressor efficiency and mass flow rate by about 12% and the compressor pressure ratio by about 10%. If there were less thermal isolation and the heat addition were to double, the impact on pressure ratio, mass flow rate, and efficiency would nearly double and quickly reach the point where the thermodynamic cycle was no longer sustainable.

Due to the low Biot number, the impacts of heat transfer within a microscale gas turbine are more significant than in a conventional gas turbine. The design process must consider and aim to minimize heat addition to the compressor flow. At the same time, it is still critical to remove heat from the turbine structure as it is a highly stressed region and silicon rapidly loses strength at temperatures above 900 K. The tradeoff between turbine temperature and compressor heat addition is one of the critical design considerations that must be accounted for. Ideally, advances in materials could help alleviate this tradeoff. At MIT we explored the deposition of silicon carbide into the turbine rotor to expand its temperature capabilities [13]. This process showed promising results, at the cost of added complexity to the fabrication process.

7.4 Microturbine Design

As the turbine generates the power to drive a gas turbine, good turbine performance is necessary. Fortunately, the favorable pressure gradient within the turbine helps to reduce the negative impacts of microscale operation and microfabrication. The favorable pressure gradient keeps the boundary layers thin and attached, even with uniform blade height. The heat flow experienced in the turbine is out of the fluid and into the structure. While this enthalpy loss decreases the power that can be generated by the turbine, some of this loss can be recouped if it is ultimately added back into the fluid upstream of the combustor in the cooler sections of the engine.

With thinner and attached boundary layers, a microturbine will generally perform better than a comparably scaled microcompressor. However, as was shown earlier in Figs. 7.2 and 7.3, the microturbine has a combination of small dimensional scale and low gas density (due to the high temperature in the turbine) resulting in a relatively small Reynolds number, even compared to the microcompressor. As a result, viscous loss will be a driving factor limiting the achievable isentropic efficiency.

There are several elements of the turbine design that can be optimized to improve efficiency (see also Gong et al. [5]). Due to fabrication constraints, the MicroEngine uses a radial inflow turbine. Thus, flow exiting the turbine is at relatively small radius, and must navigate a right angle turn to exhaust axially. Given the high temperature, low-density state of the turbine exhaust, and the small radius opening, the exit flow will be at high velocity, containing considerable lost kinetic energy (for turbine generator designs). Navigating the right angle exit with high velocity flow is also costly in terms of pressure loss. Thus, the turbine should be designed with as large an exhaust area as possible to reduce the exit velocity. The turbine should also be designed to minimize swirl in the flow at the exit of the turbine for the design condition.

While the previous paragraph argued for reducing axial exit velocity, the nozzle guide vane and turbine performance can be improved to some extent by designing for larger internal velocity. The higher the velocity in the nozzle guide vane and in the turbine rotors relative frame, the larger the Reynolds number, providing a reduction in viscous loss. Designing for large exit angles can increase the velocity in the nozzle guide vanes. Gong et al. [5] show that an exit angle of 70–80° from the nozzle guide vanes seems to be optimal for an 8 mm diameter turbine.

With proper design, 3D computations suggest that turbine isentropic efficiencies of order 70–80% are possible for 8 mm diameter rotor scale designs.

7.5 MEMS Turbocharger

In a gas turbine, compressor performance drives overall system performance. So it was considered important to demonstrate a microscale compressor and verify our design models at the earliest possible date. The initial device developed for this was the Macro-Compressor Test Rig (Section 7.3.1) as it did not rely on high-speed microscale bearings. Once microscale bearings capable of high-speed operation had been developed and proven (see Chapter 6), the turbomachinery development effort shifted to fabricating and testing high-speed MEMS turbomachinery.

The MEMS turbocharger was developed as a test vehicle for rotordynamics, turbomachinery, and microfabrication process development. This six-layer device is similar in geometry and process to our MEMS gas turbine engine. In fact, the turbocharger has all the components required of a gas turbine, including a chamber that can be used for combustion. The main exception is that the compressor flow path does not connect internally to the combustion chamber and turbine. The turbine and compressor flow paths are isolated from each other (making this a turbocharger), allowing us to evaluate their performance independently to some extent, although they are coupled mechanically by the silicon disk that supports them. Still, the turbocharger can be considered our first attempt at developing a demo-gas turbine engine since the compressor exhaust can be externally connected to the combustor, completing the engine flow path.

A complete discussion of the microturbocharger is included in Protz [15], Savoulides [17], and Savoulides et al. [18]. This section presents a summary of experimental results obtained from the turbocharger, focusing on turbomachinery results from both cold and hot operation. A MEMS turbocharger with an 8 mm diameter silicon rotor was spun to 480,000 rpm (200 m/s tip speed). Speed was limited by structural and rotordynamic issues that we now believe are understood and resolved. The measured turbocharger aerodynamic performance is consistent with our design models.

7.5.1 Design

The MEMS turbocharger is fabricated from six 100 mm single-crystal silicon wafers. Each wafer is etched on both sides to define flow passages and mechanical structures. During final assembly, the six-wafer stack is fusion bonded together. The multiwafer stack approach produces 3D structures and flow paths from what are essentially 2D lithographic and etch processes. The device fabrication is fairly complex, requiring over 20 different photomasks. For the initial builds of the device, four devices were fabricated on each wafer stack, allowing for large spacing between each device. While this spacing used less than 25% of the available area on the wafers, it improved the uniformity of the highly loaded blade etches. The development of the halo etch process (see Chapter 4) removed the need for highly loaded deep etches, allowing for 19 devices per wafer stack. However, the results presented in this chapter are for devices fabricated using the loaded, four-at-a-time etch process. All fabrication is done at the wafer level and diced into individual devices after the final bonding step.

The completed silicon turbocharger measures 23 × 23 × 2.9 mm. Figure 7.17 shows a sketch of a cross-section of the device. The rotor (compressor and turbine, shown with hash marks) is surrounded by a stator housing. A photograph of a diagonally sectioned device is included in Fig. 7.18. The rotor is formed from a 6.0 mm diameter turbine etched into the bottom of layer 4 and an 8.2 mm diameter compressor etched into the top of layer 3. The compressor and turbine rotors are bonded back-to-back as shown in Fig. 7.17. Note that the full disk area of the turbine is in contact with the compressor; there is essentially no thermal isolation between them in this design.

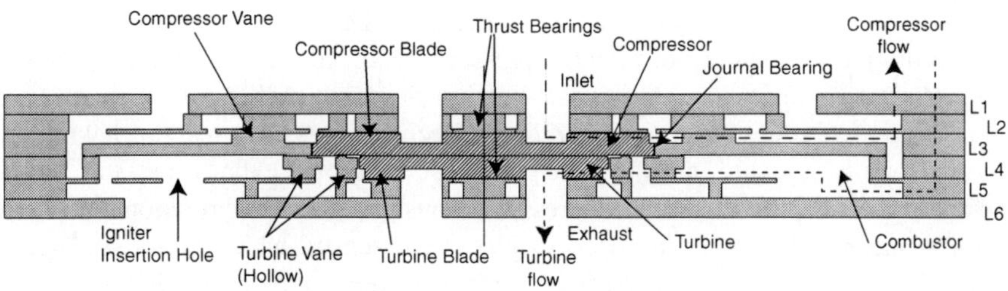

Fig. 7.17 Microscale turbocharger cross-section. Wafer layer numbers are indicated at right (courtesy of Savoulides et al. [18], ©IEEE)

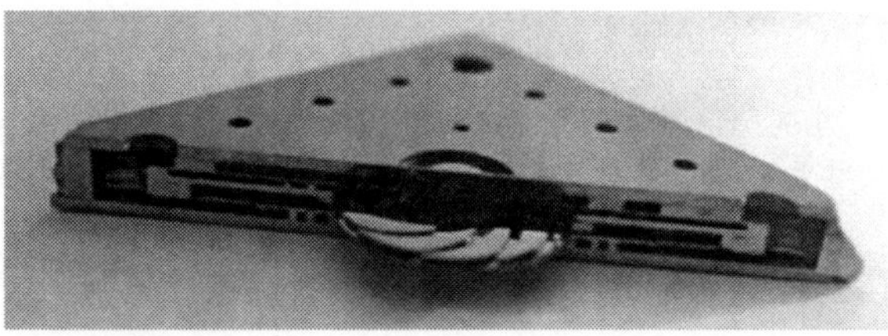

Fig. 7.18 Turbocharger device, diagonally sectioned, with rotor reinserted, compressor-side up. The device is 23 × 23 × 2.9 mm (courtesy of Savoulides et al. [18], ©IEEE)

The turbine inlet is supplied with pressurized nitrogen from a gas handling system. Flow through the turbine generates a torque that spins the rotor, inducing mass flow and pressure rise through the compressor. The turbine discharges through an external throttle valve to a vacuum pump to permit the turbine back pressure to be adjusted. The turbine back pressure is used to adjust the axial load balance on the rotor to avoid excessive loads on the thrust bearings. The compressor flow path is isolated from the turbine flow path with a rotating seal (narrow gap) that is in line with the journal bearing. The compressor ingests ambient air, and discharges through an external throttle valve, which is typically maintained at a fixed area during each test run. The compressor throttle valve is used to adjust the pressure ratio/mass flow rate characteristic of the compressor.

The gas handling system is also used to supply pressurized nitrogen to the hydrostatic thrust and journal bearings. Pressure regulators maintain a steady supply pressure to each thrust bearing. With supply pressure fixed, the flow rate through a thrust bearing becomes primarily a function of thrust bearing gap; rotation rate dependence is very weak. Based on this observation, the thrust bearings are used to track the rotor axial position and maintain centered axial operation. During operation, the thrust bearings are held at fixed inlet pressure and the mass flow rates supplied to them are monitored. Changes in the thrust bearing mass flow rates indicate axial motion, at which point the turbine back pressure is adjusted until the thrust bearing flow rates return to their initial values, indicating that the thrust bearing gaps are also back to their initial values.

The hydrostatic gas journal bearing consists of a nominally 300 μm long, 17 μm wide gap etched between the outer rim of the compressor disk and the stator. A pressure differential supplied across the journal-bearing gap induces radial stiffness. However, the bearing stiffness is not very large, requiring supercritical operation at design speed. For the device discussed below, operation at 480,000 rpm is about six times the natural frequency of the journal bearings. Device operation is very sensitive to the journal-bearing differential pressure, requiring fine differential pressure control. During operation, the journal-bearing flow rate and supply pressures are monitored at all times. The turbocharger also includes pressure taps for tracking the turbine supply pressure, turbine interrow pressure, turbine exhaust pressure, compressor interrow pressure, and compressor exhaust pressure. The rotation rate of the rotor is measured with a fiber-optic displacement sensor, which views six "speed bumps" etched into the compressor just outside of the compressor-side thrust bearing.

7.5.2 Results

Figure 7.19 shows measurements of pressure ratio across the compressor and turbine as a function of speed. At 480,000 rpm the mass flow rate through the turbine is 0.26 g/s, resulting from a pressure ratio of 1.7. At this speed, the compressor develops a pressure ratio of 1.2 with a flow of 0.13 g/s. Design speed for the turbocharger

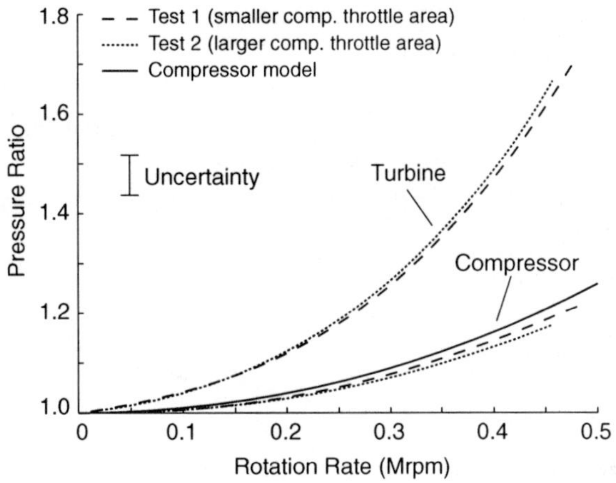

Fig. 7.19 Turbine and compressor measured pressure ratio as a function of rotation rate (courtesy of Savoulides et al. [18], ©IEEE)

is 1.2 million rpm; this device only achieved 40% of design speed and is operating in an off-design state for all data shown. Figure 7.19 actually shows results for two different runs that had slightly different compressor throttle valve settings. Also included in Fig. 7.19 is an operating line based on a compressor model that was calibrated with the design point compressor pressure ratio (~3) obtained from 3D CFD. The closeness of the model to the measurements indicates that the throttle setting was close to the design value (within measurement uncertainty).

Compressor and turbine maps are shown in Fig. 7.20. The maps are based on measurements taken from the two different runs shown in Fig. 7.19, with each run done at a different compressor throttle setting. As is expected for a compressor, when the compressor throttle area is decreased (at constant speed), the compressor pressure ratio increases and the compressor mass flow rate decreases. Within uncertainty, the turbine map shows no dependence on the compressor-side throttle area, as expected. The results of computations of two (off-design) impeller operating points are also shown on Fig. 7.20. The computations do not include a diffuser after the compressor impeller; however, the diffuser is not expected to provide significant pressure rise due to the off-design condition. The CFD data differs a small amount in terms of mass flow rate; this is likely due to not matching the compressor throttle area. If the CFD data is extrapolated to smaller throttle area, they appear to be consistent with the measured test results within experimental uncertainty.

Measurement of the individual compressor and turbine efficiencies would require precise internal enthalpy measurements, which would be very difficult in non-adiabatic devices this small and was not pursued. However, based on the measurements taken, it is possible to back out a spool efficiency that includes the coupled compressor and turbine performance. Based on measured pressure ratios and mass flow rates, the isentropic compressor and turbine power can be determined. We can also estimate the parasitic (bearing and seal) losses in the turbocharger. A power balance for the device can then be written with the introduction of isentropic

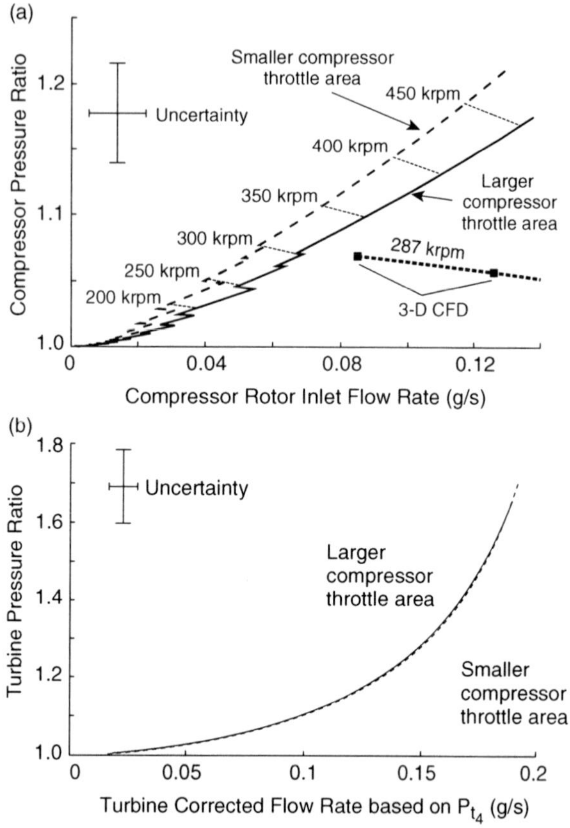

Fig. 7.20 (**a**) Compressor map, (**b**) Turbine Map (courtesy of Savoulides et al. [18], ©IEEE)

efficiencies:

$$\eta_t P_{ts} = \frac{P_{cs}}{\eta_c} + P_{loss}, \tag{7.13}$$

where η_t and η_c are the turbine and compressor isentropic efficiencies, P_{ts} and P_{cs} are the turbine and compressor isentropic powers, and P_{loss} represents the combined parasitic losses. Without internal enthalpy measurements, we cannot determine the isentropic efficiencies; however, Eq. (7.13) provides a functional relationship between them.

Four operating points at different speeds were chosen, and measurements at these points were used to determine the power values in Eq. (7.13). The result is four different lines that are functions of compressor and turbine isentropic efficiency, shown in Fig. 7.21. Each line is essentially a locus of possible compressor and turbine efficiency pairs associated with that operating point, although only one location on each line can contain the actual efficiency values.

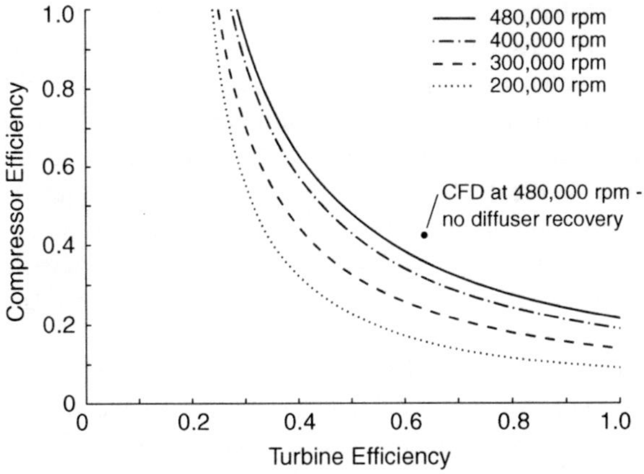

Fig. 7.21 Spool efficiency. Locus of possible compressor and turbine efficiencies is shown. Computational spool efficiency point is also plotted. As speed increases, spool efficiency improves (courtesy of Savoulides et al. [18], ©IEEE)

Several insights about the turbocharger performance can be learned from this figure. First, we can use experience to narrow down the possible values. It is highly unlikely that the compressor efficiency is greater than 0.6 or that the turbine efficiency is greater than 0.8. It is fairly likely that the turbine efficiency and compressor efficiency are close to each other, although we expect the turbine efficiency to be higher. So we can focus our view in the region straddling the symmetry axis in Fig. 7.21. Second, the turbomachinery efficiencies appear to be increasing with speed, as one would expect for off-design operation. Third, we can take points along each line and compute a product between the compressor and turbine efficiency. Along the 480,000 rpm line, this product is approximately 0.24 with little variance, providing a quantitative value that can be compared to computations. A 3D CFD (Fluent) computation at 480,000 rpm is shown on Fig. 7.21, with a compressor/turbine efficiency product of 0.27 (0.43 compressor efficiency and 0.64 turbine efficiency). The computational value is close to the measured value. Finally, while the spool efficiency at 480,000 rpm is still relatively low, if it continues to increase with speed in the same manner that the computed efficiency increases with speed, our engine model predicts that overall performance will be sufficient for gas turbine cycle operation.

7.5.3 Gas Turbine Tests

The turbocharger die can function as a gas turbine engine since it includes a combustion chamber upstream of the turbine inlet. The experiments described above did not include combustion, and the combustion chamber simply acted as a large volume prior to the turbine. Since the turbocharger was cold, fluidic connections to the silicon turbocharger were made with o-ring seals directly to the silicon surface. Combustion experiments require a different packaging scheme, as the silicon surface becomes too hot for o-rings. We thermally isolated the fluidic connections with a glass wafer anodically bonded to the turbocharger. The glass wafer has ultrasonically machined holes above the turbocharger fluidic holes. A silicon wafer with etched channels is anodically bonded to the top of the glass wafer. A new set of fluidic connection holes are obtained at the ends of the several centimeter long etched channels. The channel length and glass thickness provide sufficient thermal isolation from the hot turbocharger so that o-ring sealed fluid connections can then be made.

Several experiments were performed with hydrogen premixed into the turbine inlet flow to show that we could light and extinguish the combustor while spinning the rotor at high speed. Figure 7.22 shows results from one such run. Combustor temperature and rotor speed are shown as a function of time. The combustor temperature measurement is normalized by ambient temperature. There is combustion initially, but as equivalence ratio (a measure of the fuel-to-air mixture) drops, the combustor ultimately extinguishes, with a resultant temperature drop of about 110 K and rotor deceleration from about 63,500 to 52,900 rpm. The combustor temperature is measured with a K-type thermocouple that is simply inserted into the combustion chamber; because of radiation and conduction concerns, this thermocouple does not

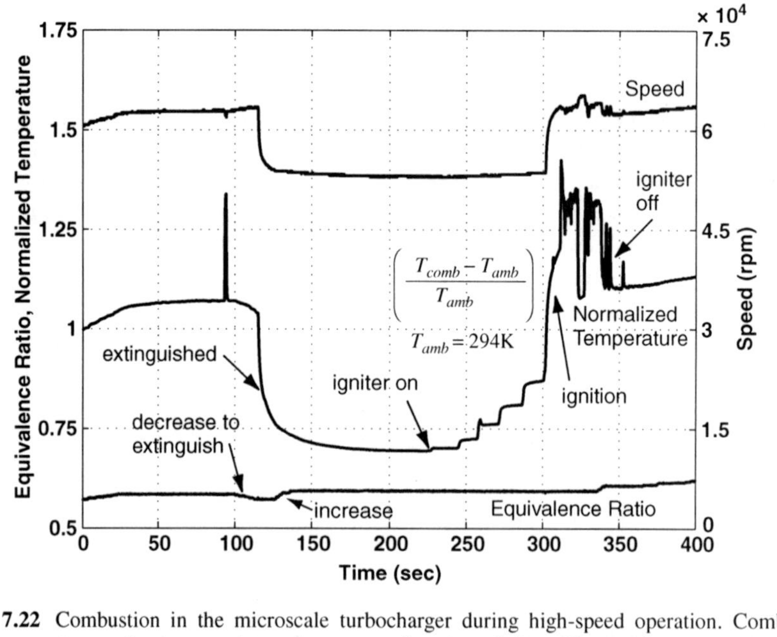

Fig. 7.22 Combustion in the microscale turbocharger during high-speed operation. Combustor temperature and rotor speed are shown as a function of time. The combustor temperature (T_{comb}) measurement is normalized by ambient temperature (T_{amb}). Note that the thermocouple is simply inserted into the combustion chamber and is not meant to give an accurate measure of temperature

provide an accurate measure of temperature. Rather, the thermocouple should only be interpreted as providing a sense of temperature change. After the combustor is extinguished, the equivalence ratio is increased back to its previous value and the igniter is turned on. The igniter is a small diameter Ni–Cr wire inserted into the combustor. The igniter voltage is stepped up (seen as a rise in the thermocouple temperature measurement in Fig. 7.22) until the combustion process restarts. At this point, a jump in combustion temperature and rotational speed is observed. The igniter is then turned off, and the rotor decelerates a bit. At around $t = 350$ s, the equivalence ratio is increased again and the rotor starts to accelerate as the combustor temperature rises. This experiment successfully demonstrates integration of microscale turbomachinery, bearings and combustion, which is an important step along the path to demonstrating our ultimate goal, a self-sustaining microscale gas turbine. Additional combustion test results are included in Savoulides et al. [18].

7.6 Summary

This chapter explored the impacts of reduced scale on the fluid mechanics and heat transfer within PowerMEMS devices. Both physical- and fabrication-related impacts were explored. There is no question that PowerMEMS devices will have increased viscous losses relative to conventional-scale devices. Thermal isolation in MEMS devices will also be far more difficult due to the low Biot number and large thermal conductivity of silicon. However, while the physical impacts of scale will limit the ultimate performance level of MEMS engine components, the experimental results presented here from a MEMS turbocharger indicate that the MEMS components can achieve sufficient efficiency to make a MEMS gas turbine viable. In fact, our work has not conclusively shown what the ultimate performance level could be for MEMS turbomachinery. The turbomachinery in conventional scale gas turbines have efficiencies in excess of 90%; however, these devices have also undergone many thousands of man-years of development and have had billions of research dollars invested in them. Such cannot be said for microturbomachinery. A more substantial development effort combined with material and microfabrication advances should allow for considerable performance improvements from the values presented here. At the same time, the scaling analyses presented here indicate that gas turbine designs significantly smaller than 1 cm are probably not feasible.

References

1. Casey, M.V., "Effects of Reynolds Number on the Efficiency of Centrifugal Compressor Stages," *Journal of Engineering for Gas Turbines and Power,* v 107, n 2, 541–548, 1985.
2. Dawes, W.N., "The practical application of solution-adaptation to the numerical simulation of complex turbomachinery problems," *Progress in Aerospace Science*, 29, 221–269, 1992.
3. Drela, M. and Youngren, H, "MISES 2.1," *MIT Computational Aerospace Sciences Laboratory,* June 1995.
4. Ghodssi, R., Morgan, B., and Waits, C.M., "Compensated aspect ratio dependent etching (CARDE) using gray-scale technology," *Microelectronic Engineering*, 77, 85–94, 2005.

5. Gong, Y., Sirakov, B., Epstein, A., and Tan, C.S., "Aerothermodynamics of micro-turbomachinery," *ASME Paper GT2004-53877, ASME Turbo Expo 2004*, Vienna, Austria, 2004.
6. Isomura, K., Murayama, M., Teramoto, S., Hikichi, K., Endo, Y., Togo, S., and Tanaka, S., "Experimental verification of the feasibility of a 100 W class micro-scale gas turbine at an impeller diameter of 10 mm," *Journal of Micromechanics and Microengineering*, 16, S254–S261, 2006.
7. Jacobson, S.A., "Aerothermal challenges in the design of a microfabricated gas turbine engine," *AIAA Paper 98-2545, 29th AIAA Fluid Dynamics Conference*, Albuquerque, NM, 1998.
8. Kang, S., Matsunaga, M., Johnston, J.P., Tsuru, H., Arima, T., and Prinz, F.B., "Micro-scale radial-flow compressor impeller made of silicon nitride – Manufacturing and performance," *Proceedings of the ASME Turbo Expo*, 3, 779–788, 2003
9. Kays, W.M. and Crawford, M.E., Convective Heat and Mass Transfer, McGraw-Hill Book Company, NY, 2nd edition, 1980.
10. Kerrebrock, J.L., *Aircraft Engines and Gas Turbines*, 2nd edition, The MIT Press, Cambridge, MA, 1992.
11. Lee, C., Liamini, M., and Fréchette, L., "Design, fabrication , and characterization of a micro-turbopump for a Rankine cycle micro power generator," *Proceedings of the Solid State Sensors, Actuators and Microsystems Workshop*, Hilton Head Island, SC, 2006.
12. Mehra, A., "Computational investigation and design of low Reynolds number micro-turbomachinery," Masters Thesis, Aeronautics and Astronautics, Massachusetts Institute of Technology, June 1997.
13. Moon, H.-S., Choi, D., and Spearing, S.M., "Development of Si-SiC hybrid structures for elevated temperature micro-turbomachinery," *Journal of Microelectromechanical Systems*, 13, 676–687, 2004.
14. Peirs, J., Reynaerts, D., and Verplaetsen, F., "A microturbine for electric power generation," *Sensors and Actuators, A: Physical*, 113, 86–93, 2004.
15. Protz, J., "An assessment of the aerodynamic, thermodynamic, and manufacturing issues for the design, development, and microfabrication of a demonstration micro engine," Ph.D. Thesis, Aeronautics and Astronautics, Massachusetts Institute of Technology, September 2000.
16. Reynolds, W.C. and Perkins, H.C., *Engineering Thermodynamics*, 2nd edition, McGraw-Hill Book Company, New York, NY, 1977.
17. Savoulides, N., "Development of a MEMS turbocharger and gas turbine engine," Ph.D. Thesis, Aeronautics and Astronautics, Massachusetts Institute of Technology, February 2004.
18. Savoulides, N., Jacobson, S.A., Li, H., Ho, L., Khanna, R., Teo, C-J., Protz, J.M., Wang, L., Ward, D., Schmidt, M.A., and Epstein, A.H., "Fabrication and testing of a high-speed microscale turbocharger," *Journal of Microelectromechanical Systems,* 17, no. 5, 1270–1282, 2008.
19. Shirley, G. B., "An experimental investigation of a low Reynolds number, high Mach number centrifugal compressor," Masters Thesis, Aeronautics and Astronautics, Massachusetts Institute of Technology, September 1998.
20. Sirakov, B.T., "Characterization and design of non-adiabatic micro-compressor impeller and preliminary design of self-sustained micro engine system," Ph.D. Thesis, Aeronautics and Astronautics, Massachusetts Institute of Technology, February 2005.
21. Sirakov, B., Gong, Y., Epstein, A., and Tan, C., "Design and characterization of micro-impellers," *ASME Paper GT2004-53332, ASME Turbo Expo 2004*, Vienna, Austria, 2004.
22. Sullivan, Shaun, "Development and testing of microscale silicon heat exchangers for the MIT micro gas-turbine engine," Masters Thesis, Mechanical Engineering, Massachusetts Institute of Technology, September 2000.

23. Sullivan, S., Zhang, X., Ayon, A.A., and Brisson, J.G., "Demonstration of a microscale heat exchanger for a silicon micro gas turbine engine," TRANSDUCERS '01. EUROSENSORS XV. 11th International Conference on Solid-State Sensors and Actuators. Digest of Technical Papers, p. 1606–1609 vol.2, 2001.
24. Waits, C.M., Morgan, B., Kastantin, M., and Ghodssi, R., "Microfabrication of 3D silicon MEMS structures using gray-scale lithography and deep reactive ion etching," *Sensors and Actuators A (Physical)*, 119, 245–253, 2005.

Chapter 8
Motors and Generators

Jeffrey H. Lang and Sauparna Das

This chapter describes the development of motors and generators as PowerMEMS machines. It also discusses aspects of the larger electromechanical energy conversion systems into which they fit. When coupled to microturbomachinery, typical applications of these machines might include motor/compressors, motor/pumps, and turbine/generators, all perhaps part of much larger systems such as a microengine or a microrocket.

Our development of PowerMEMS motors and generators takes place in the context of the system shown in Fig. 8.1. This system connects an electrical source or load to a mechanical load or source through an electric or magnetic machine and its attendant power electronics. If the machine acts as a motor, then power flows from left to right. If the machine acts as a generator, then power flows from right to left. The role of the machine in this context is the conversion of electrical to mechanical energy, or vice versa. The role of the power electronics is to commutate the machine; it is assumed here that the machine is an AC machine, while the electrical load or source is a DC device. The operation of the system takes place under the supervision of a local controller, which executes the desired regulation functions.

In every-day language, the terms "electric motor" and "electric generator" almost always refer to magnetic-field machines, rather than electric-field machines. This is understandable because electric-field machines are very rare in every-day life, while

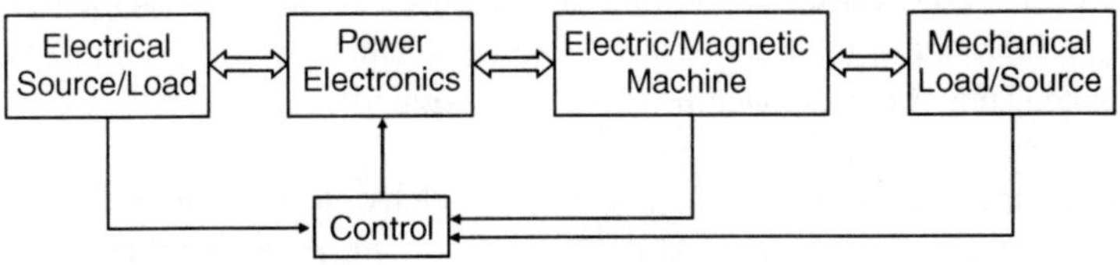

Fig. 8.1 An electric/magnetic machine system. Power flow is shown with *block arrows*. Signal flow is shown with *thin arrows*

J.H. Lang (✉)
Department of Electrical Engineering and Computer Science, Massachusetts Institute of Technology, Bldg 10-176, 77 Massachusetts Avenue, Cambridge, MA 02139, USA

J.H. Lang (ed.), *Multi-Wafer Rotating MEMS Machines*, MEMS Reference Shelf,
DOI 10.1007/978-0-387-77747-4_8,

magnetic-field machines are ubiquitous. The language of this chapter, however, must be more precise because both types of machines will be discussed. Thus, from this point forward the adjectives "electric" and "magnetic" will be used literally so as to avoid confusion. A non-specific machine will be refered to as "electromagnetic."

8.1 Background

Electromagnetic machine systems have been studied in great depth over the past two centuries, and their analysis, design and operation is thus very well understood [1–4]. Given this background, it is important to understand what is new and different in the context of PowerMEMS, beyond the obvious small size. It is also important to understand what is the same, and where electromagnetic PowerMEMS machines fit in the context of the prior art. That is the purpose of this section. Not surprisingly, there are important issues to reconsider, and these issues have a significant impact on the design and performance of electromagnetic PowerMEMS machine systems. At the same time, however, there is a very solid foundation on which to build their development. This too is discussed below.

8.1.1 What Is New?

Generally speaking, MEMS machines are fabricated through the selective addition (growth, deposition, electroplating, bonding) and subtraction (etching, lift-off, polishing) of materials to and from a wafer or wafer stack; see Chapter 2. This naturally leads to machines that are planar and disk shaped, which are uncommon, although not unseen, at the macroscale. Additionally, microfabrication is more limited in its capabilities than is macrofabrication, making it difficult to create microscale features that are common at the macroscale. For example, the aspect ratios of trenches and slots are more limited at the microscale. Also, useful features such as three-dimensionally smoothly rounded corners that would reduce breakdown in electric machines, or closed slots that would reduce losses in magnetic machines, are very difficult to fabricate at the microscale. Features on sidewalls are even more difficult to fabricate. Thus, MEMS machines must be designed within relatively constrained geometries.

Another important issue to reconsider is the relative torque and power density of electric and magnetic machines. At the macroscale, magnetic machines are considerably more power dense than electric machines, and this explains their ubiquitous presence. The stored energy density in the air gap of a magnetic machine is $B^2/2\mu_0$, where B is the magnitude of the magnetic field and μ_0 is the permeability of free space [5, 2]. This density is directly related to the magnetic shear stress acting on the rotor [2]. Generally, B is limited by magnetic saturation in the stator and/or rotor core. A typical air-gap B is 1.5 T, and so a typical magnetic air-gap energy density is 9×10^5 J/m^3. In contrast, the stored energy density in the air gap of an electric machine is $\epsilon_0 E^2/2$, where E is the magnitude of the electric field and ϵ_0 is the permittivity of free space [5, 2]. Generally, E is limited by electric breakdown, which in a large smooth air gap will occur as E approaches 10^6 V/m [6–8]. This limits the electric air-gap energy density to 4 J/m^3, which is considerably lower than that found in a magnetic machine. However, when the air gap becomes very small, it no longer contains enough gas molecules to support avalanching electric breakdown, and the practical limit to E can rise to near 3×10^8 V/m [6–9]. In this case, the electric air-gap energy density becomes 4×10^5 J/m^3, which is comparable to that found in a magnetic machine. At this point, other factors might become important. For example, an electric machine could be CMOS compatible, whereas a magnetic machine would most certainly not. On the other hand, the narrow air gap of an electric machine gives rise to increased viscous losses. In any case, electric machines should be considered along with magnetic machines at the MEMS scale.

Following [8], Fig. 8.2 details the dependence of the breakdown voltage of a smooth gap containing various gases on the product of the gap separation and the gas pressure. Note the 1 Pa·m is approximately 10 μm at 1 atm. From the figure, it is apparent that very small air gaps, for example 3 μm at 1 atm, can sustain voltages in excess of 300 V. This corresponds to an electric field strength in excess of 10^8 V/m. Even higher field strengths, reaching the field emission limit for electrons near 10^9 V/m, are possible for smaller air gaps. In any case, the field strengths are considerably greater than those achieved at the macro scale.

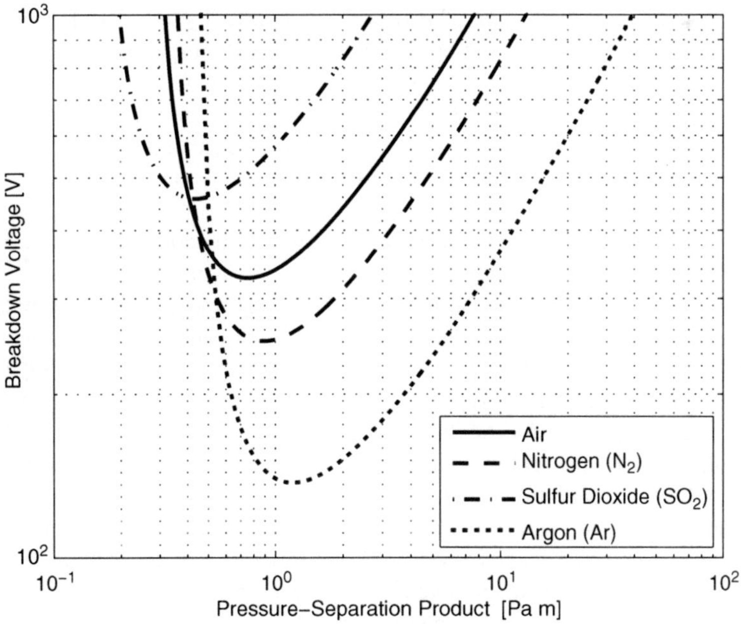

Fig. 8.2 Electric breakdown voltage for various gases in a smooth gap as a function of the gas-pressure-gap-separation product, following [8]. Note that 1 Pa·m is approximately 10 μm at 1 atm

Early micromotors, both electric and magnetic such as those described in [10–14] for example, applied electromagnetic stress to the circumferential edges of their rotors, as do most macroscale electromagnetic machines. Those micromotors converted energy through phases that exhibited a variable capacitance or a variable inductance. Such electromagnetic structures were very simple to implement by vertically etching salient stator and rotor poles simultaneously from a single planar material layer. However, when the rotor radius is larger than three times the rotor thickness, a greater torque can be achieved by applying the same stress to one planar surface of the rotor; here, the other planar surface is reserved for microturbomachinery. This, of course, requires more complex fabrication, but given the naturally planar geometry of MEMS machines it is the basic design pursued here in the interest of increased power throughput. Such designs have the added advantage that they free the gap at the edge of the rotor to become a journal bearing, as described in Chapter 6. However, they introduce the disadvantage of an unbalanced electromagnetic axial force on the rotor that must be balanced by an axial thrust bearing, again as described in Chapter 6. This added complexity is not common in macroscale machines, and should be managed through integrated electromechanical and bearing design.

As a further consequence of the way in which they are fabricated, it is rare to employ the assembly of many loose parts in an MEMS machine. Thus, the parts of MEMS machines are generally in intimate structural and thermal contact with one another over the broadest possible area, and this has two significant consequences. First, the rotors of MEMS machines have excellent structural integrity, enabling them to run at very high speeds. This offers the opportunity to design PowerMEMS machines with high power densities. Second, heat transfer through MEMS machines is relatively good, and this is enhanced by their planar geometries and the increasing ratio of surface area to volume of these machines as their size decreases. Thus, the traditional thermal limits observed in macroscale electric and magnetic machines are not observed in their PowerMEMS counterparts. This too opens up new design possibilities.

A simple rotating disk made from a homogenous material will burst when the speed of its peripheral tip reaches the burst speed u_B given by $\sqrt{8\sigma_F/(3+\nu)\rho}$, where σ_F is the fracture stress, ρ is the mass density, and ν is the Poisson's ratio of the material from which the disk is made [15]; see Chapter 2. This burst speed and the relevant material properties are given in Table 8.1 for silicon, 80/20 electroplated NiFe, a generic MnZn ferrite and a generic silicon steel; Young's modulus E is also given in the table for completeness. Two immediate conclusions can be drawn from the table. First, an electric machine having a silicon rotor can spin about two to three times faster than a magnetic machine having a solid steel rotor; note that solid steel rotors on the macroscale are rare. Thus, it is possible for the two machines to have comparable power densities even if the motive magnetic shear stress exceeds the motive electric shear stress. Again, electric machines should be considered along with magnetic machines in PowerMEMS applications. This becomes a stronger statement given the reduced burst speed of NiFe and MnZn ferrite, which are more probable materials for a PowerMEMS machine. Second,

Table 8.1 material properties and disk burst speed for silicon, 80/20 electroplated NiFe, a generic MnZn ferrite used for power applications, and a generic silicon steel used for power applications. Note that for silicon, Young's modulus varies from 130 to 196 GPa, and Poisson's ratio varies from 0.06 to 0.28, depending on crystallographic orientation. The properties of the other materials vary as well

Material	ρ [kg/m^3]	E [GPa]	ν	σ_F [GPa]	u_B [m/s]
Silicon	2330	160	0.18	4.1	2100
NiFe (80/20)	8500	140	0.3	1.6	675
MnZn Ferrite	4800	120	0.3	0.04	140
Silicon Steel	7700	230	0.25	2.0	800

macroscale magnetic machines exhibit maximum peripheral tip speeds more on the order of 100–150 m/s, and so PowerMEMS magnetic machines fabricated with more integral rotors can perhaps have greater power densities than their macroscale counterparts.

Because microscale PowerMEMS machines can operate at very high speeds they often do in order to obtain the highest possible power throughput. This in turn gives rise to very high electrical frequencies. As illustrated in subsequent sections, electrical frequencies ranging from many tens of kilohertz to several megahertz are not uncommon. Together, a high speed and high electrical frequency can lead to high material, bearing, aerodynamic and power electronics losses, and thus more complex design tradeoffs.

The importance of aerodynamic losses can be examined through Fig. 8.3, which shows a disk rotor of radius R spinning with tip speed u above a stator with an air-gap separation of G. Assuming laminar azimuthal air flow between the rotor and the stator, the viscous loss in the air gap is given by $\pi \mu_V R^2 u^2/2G$, where μ_V is the absolute viscosity of air, taken here to be 1.8×10^{-5} Ns/m^2. If at the same time, the rotor experiences a uniform electromagnetic shear stress of σ_{EM}, then the electromagnetic motive power is given by $2\pi \sigma_{EM} R^2 u/3$. The ratio of the viscous loss to the motive power is therefore $3\mu_V u/4G\sigma_{EM}$. As an example, consider an electric machine with a 3-μm air gap across which the rotor–stator potential difference varies sinusoidally in the azimuthal direction with a 300-V amplitude. In this machine the peak electric field is 10^8 V/m, and so the peak electric air-gap energy density is 4.4×10^4 J/m^3. Its spatial average is then 2.2×10^4 J/m^3. Typically, an electric machine might achieve an average electric shear stress equal to 10% of its average stored air-gap energy density, which averages to 2.2×10^3 N/m^2 in this example. Assuming that the rotor

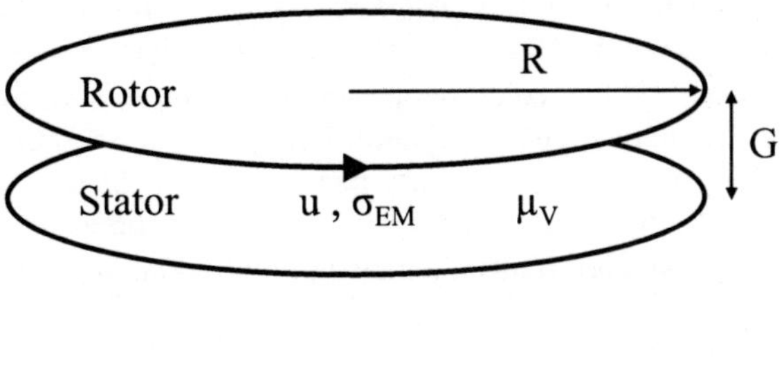

Fig. 8.3 Geometry for comparing viscous losses and electromagnetic motive power

tip speed u is 500 m/s, as allowed by Table 8.1 for silicon, the ratio of viscous loss to motive power is 1. Thus, it is important to account for viscous loss during the design of an electric PowerMEMS machine, particularly for motors as explained later in this subsection. Magnetic machines, which can successfully operate with much larger air gaps, need not exhibit such high viscous losses, but those losses can still be significant at the microscale.

In electric PowerMEMS machines, high-frequency operation has its most serious impact through losses in the power electronics; the typical power electronics for an electromagnetic machine is shown in Fig. 8.4 [3, 4]. The figure shows the switching branch for one phase, and includes two capacitors to model parasitic wiring and transistor capacitance. The power electronics operate by alternately turning the top and bottom transistors on and off, thereby applying a voltage square-wave to the corresponding phase of the machine; a varying duty cycle can be used to implement more complex time-average waveforms. If the switching frequency is f, then the power dissipated in this circuit due to the charging and discharging of the two parasitic capacitors is $2C_P V^2 f$ per phase. To illustrate the importance of this loss, let C_P be 10 pF, V be 300 V, and f be 1 MHz, as could be the case for an electric PowerMEMS machine. In this case, the power loss is 1.8 W per phase. Thus, the design of the power electronics and proper management of parasitics is critical to the performance of a PowerMEMS machine, particularly an electric PowerMEMS machine.

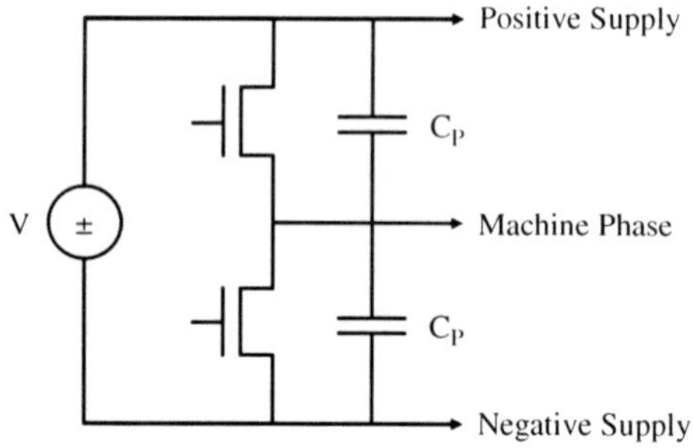

Fig. 8.4 Power electronics for an electric machine; only one phase branch is shown

In magnetic PowerMEMS machines, high-frequency operation introduces significant challenges through eddy currents and losses in the core. The role of the core in a magnetic machine is to guide magnetic flux through the machine without incurring a significant loss of magnetomotive force [1]. At high frequencies, eddy currents in the core act to exclude the flux, rendering the core ineffective. These eddy currents also generate loss. A core carries substantial magnetic flux only within a magnetic skin depth δ of its surface [5, 2]; the magnetic skin depth is given by $\sqrt{2/\omega\mu\sigma}$, where ω is the radian frequency of the flux, μ is the permeability of the core, and σ is the conductivity of the core. Therefore, planar cores, into which magnetic flux diffuses from both sides, carry flux only over a thickness of a few skin depths. Thicker cores exhibit wasted material from a magnetic viewpoint. To circumvent this problem, cores are thinly laminated, with the lamination thickness set to a skin depth or two, or less. This also helps reduce eddy-current losses by crowding the eddy currents into narrow regions. The skin depth at 100 kHz and the relevant material properties are given in Table 8.2 for electroplated NiFe, a generic MnZn ferrite and a generic silicon steel. Two immediate conclusions can be drawn from the table. First, if materials such as electroplated NiFe are to be used for high-performance magnetic PowerMEMS machines, their fabrication will be complicated by the need for very thinly laminated cores, on the order of 50 μm or less. Second, where possible, ferrite substrates might profitably be used in cores due to their greatly reduced conductivity. Unfortunately, their saturation flux density, B_S, is generally low, which will possibly require a thick substrate.

Table 8.2 magnetic properties and skin depth at 100 kHz for 80/20 electroplated NiFe, a generic MnZn ferrite used for power applications, and a generic silicon steel used for power applications; note that the permeability is relative to that of free space, $\mu_0 = 4\pi \times 10^{-7}$ H/m

Material	σ [S/m]	μ/μ_0	B_S [T]	δ [m]
NiFe (80/20)	6.7×10^6	500	0.8	2.5×10^{-5}
MnZn ferrite	0.2	3000	0.5	4.2×10^{-3}
Silicon steel	2×10^6	5000	2.2	1.6×10^{-5}

Thermal considerations are also very different at the microscale. In well-designed and well-constructed macroscale magnetic machines, it is typical to observe current densities ranging from about 10^6 A/m^2 for natural-air-cooled machines to about 10^7 A/m^2 for liquid-cooled machines. The current density is limited by the tolerable temperature rise, and so such current density limits are often used as substitutes for a thermal analysis. In the MEMS magnetic machines described here, current densities of 10^9 A/m^2 are common, as permitted by their excellent heat transfer. The immediate consequences of this can be seen through Fig. 8.5. The figure shows the rotor and stator of a simple magnetic machine having an air-gap separation of G. The stator supports windings carrying a current density J that are laid with alternating direction in slots each having an area A. The magnetic flux density created by the stator currents is B. Following Ampere's Law [1, 5], the peak magnetic flux density B in the air gap is approximately $\mu_0 JA/2G$, assuming that B is one-dimensional and the stator and rotor cores are highly permeable. From this, it is apparent that the achievable B will decrease linearly with dimensional scale as the size of the machine

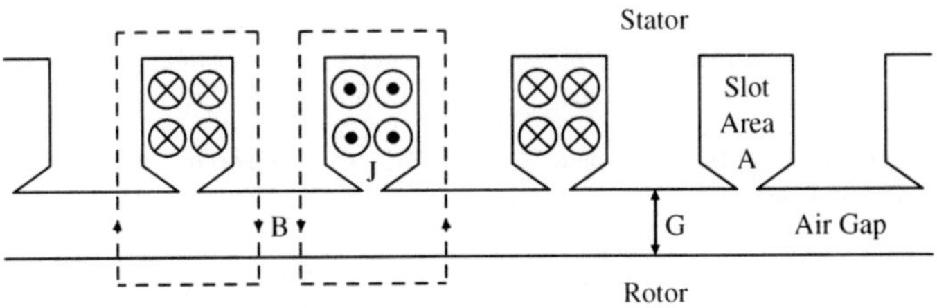

Fig. 8.5 The magnetic flux in a slotted magnetic machine

decreases for a given J. Thus, without some form of thermal relief, the performance of a MEMS magnetic machine will be relatively poor. Fortunately, the excellent heat transfer observed in MEMS machines provides the necessary thermal relief, and this must be taken into account during design.

Finally, the role of parasitics in general in PowerMEMS machine systems is more prominent than in macroscale machine systems. For example, parasitic capacitance in the stator wiring and power electronics of electric machines systems can be particularly significant, as can the parasitic winding inductance and core loss in magnetic machine systems. Aerodynamic and bearing losses can be significant in both machine systems. Moreover, the significance of these phenomena depends on the application as can be seen in Fig. 8.6. The figure shows the flow of power, including losses, through an electromagnetic machine system during both motoring (top) and generating (bottom) operation. In both sub-figures it is important to realize that the air-gap power is a bottleneck. This power is determined by the electromagnetic shear stress acting on the rotor, which is in turn limited by the available electromagnetic energy stored in the air gap. Therefore, losses downstream of the air gap not only lower efficiency, but more importantly limit the available output power. For a given machine, they cannot be compensated by an increased input power once the air-gap power limit is reached.

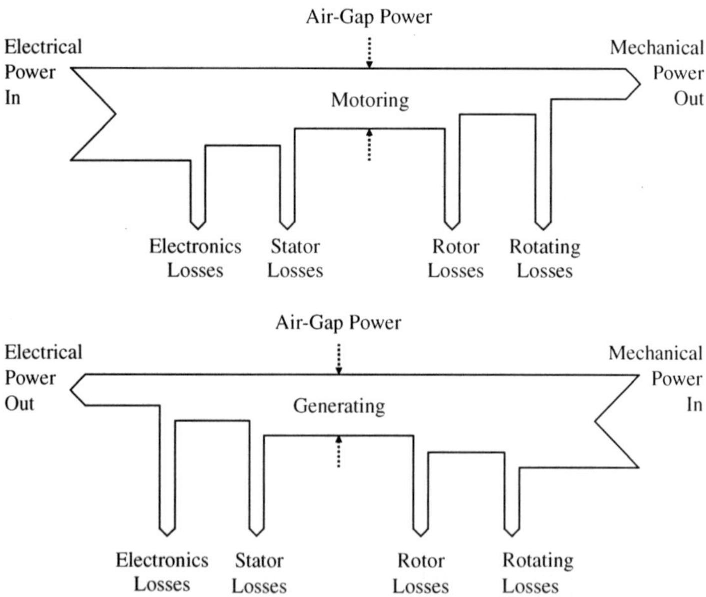

Fig. 8.6 Power flow in an electromagnetic machine system during both motoring (*top*) and generating (*bottom*) applications

By way of example, consider an electric PowerMEMS machine. These machines typically have relatively large parasitic capacitances between their stator electrodes and the surrounding silicon support structure or substrate. In an AC machine, these capacitances will cause reactive currents to flow through the stator wiring and power electronics, and the currents will in turn create significant losses. It is for this reason that thick dielectrics are used to separate the stator wiring and electrodes from the substrate, and thereby reduce the parasitic capacitances. In motoring operation, the losses are upstream of the air gap, as seen in Fig. 8.6, and so they act only to reduce efficiency. However, during generating operation, these losses are downstream of the air gap, and so serve not only to reduce efficiency, but also to limit output power. In fact, without the use of thick dielectrics to reduce parasitic capacitances, the associated losses can consume the majority, if not all, of the power that passes through the air gap. A similar example could be made of core loss in a magnetic machine given the typical high electrical frequencies involved. Alternatively, aerodynamic and bearing losses in both machines can have a limiting effect during motoring operation. Thus, parasitics in PowerMEMS machines must be managed early during design.

8.1.2 What Is Old?

While there are many issues that warrant reconsideration in the context of electromagnetic PowerMEMS machines, there are also important aspects of electromagnetic machine analysis, design and operation that are common across the macro- and microscales. For example, Maxwell's Equations are applicable at both scales. Further, since both macroscale and microscale machines make use of the bulk properties of their constituent materials, the relevant material laws are the same at both scales, even though the specific numerical values of their properties may differ due to material selection. Thus, the types and operating principles of electromagnetic machines are common to both scales, as are the approaches to analysis (analytical and numerical), design, and control.

8.1.3 Electromagnetic Machine Types Attributes, and Selection

Maxwell's Equations exhibit symmetry in the equations which govern the behavior of electric and magnetic fields. There is also a strong formulaic similarity between the equations which govern the behavior of electric and magnetic materials. Because of this, there is a duality in the basic types of electric and magnetic machines [16]. For example, Table 8.3 catalogs the dual pairs of some basic AC electric and magnetic machines [1]. Other dual machine pairs can be created by combining the features of those listed in the table. In principle, any such machine could be implemented as a PowerMEMS machine. In that case, which one is the best choice for a given PowerMEMS motoring or generating application? While there is no universal answer to this question, there are good guidelines for a selection. These guidelines are described next in the context of the Microengine Project.

Table 8.3 Some basic AC electric and magnetic machine duals. The first three machines in each column are synchronous machines, while the induction machines are asynchronous

Electric machine	Magnetic machine
Variable capacitance	Variable reluctance
Permanent-electret	Permanent-magnet
Excited-rotor	Wound-field
Induction	Induction

To begin, those machines with externally excited rotors, specifically the electric excited-rotor and the magnetic wound-field synchronous machines [1], require brushes and slip rings to bring the excitation to their rotating rotor phases. At the macroscale, this poses few challenges, and is common practice. However, at the microscale there is no proven brush and slip ring technology. Further, even if such components were developed, they would likely cause a significant mechanical loss. Thus, these machines are eliminated here as impractical at the microscale. Similar concerns eliminate the use of brushes and commutator bars as a means of commutating AC machines, which would result in a traditional DC machine as viewed from the outermost electrical terminals.

Consider next the electric variable-capacitance and magnetic variable-reluctance synchronous machines [1, 17], which have both been implemented as MEMS machines [10–14]. These machines usually use slots on both the stator and rotor to define salient poles that are excited by electrodes or windings on the stator. The poles then attract to produce a torque. In general, to be good torque producers, these machines require deep slots and very narrow air gaps in order to create a large difference between the maximum and minimum phase capacitance or inductance. Further, they require relatively accurate rotor position information to support the commutation required for good performance, both as a motor and as a generator [18, 19], which is challenging to measure at the microscale. While these machines exhibit excellent rotor integrity, they are nonetheless eliminated here due to their fabrication challenges, their control challenges, and the viscous loss they incur with a small air gap.

After elimination of the others, only the permanent-electret and permanent-magnet synchronous machines, and the two asynchronous induction machines, remain. Each is a viable candidate for the Microengine Project, and the selection among them is more complex. Consider first a comparison of the two electric machines, and a separate comparison of the two magnetic machines.

In an MEMS electric induction machine exhibiting a typical air-gap electric field magnitude E of 10^8 V/m, the induced surface charge on the rotor is approximately $\epsilon_0 E$, or 0.9×10^{-3} C/m^2. This is very close to the $0.5 - 1.2 \times 10^{-3}$ C/m^2 surface charge reported for MEMS permanent electrets fabricated from thin films of Teflon, similar materials and silicon dioxide [20–23]. Thus, the two machines appear electromechanically comparable at the microscale because it is the rotor surface charge which interacts with the stator excitation to produce a torque. However, the permanent electrets are not yet stable against leakage and decay over time, particularly at temperatures approaching 80°C and above. Further, being a synchronous machine, the permanent-electret machine requires position information to operate at its best, which is again difficult to provide at the microscale. Therefore, the electric induction machine is chosen here over the permanent-electret synchronous machine. The development of the electric induction machine is presented in Section 8.3.

The choice between the permanent-magnet synchronous machine and the magnetic induction machine is much less clear than in the case of the dual electric machines. Samarium-cobalt magnets are capable of high-performance operation even beyond 300°C. For this reason, both machines have been studied in detail as PowerMEMS machines. The primary conclusion from this work is that, when permitted by thermal considerations, the permanent-magnet synchronous machine appears to be the better choice, and so it is chosen here. The permanent-magnet synchronous machine is generally more power dense and efficient than the induction machine at the microscale due largely to the challenge of inducing substantial rotor currents at that scale. Permanent magnets, on the other hand, maintain their bulk performance at the micro scale, easily offering a good rotor field excitation. The development of the permanent-magnet synchronous machine is presented in Section 8.4. The development of the magnetic induction machine as a PowerMEMS machine is chronicalled in [24–28].

At this point, a comparison between the electric and magnetic machines chosen above is deferred until Section 8.5. As mentioned in Section 8.1.1, electric and magnetic PowerMEMS machines have very different properties, leading to very different advantages and disadvantage.

8.1.4 Summary

There are several important messages to take away from the discussion presented in this section. Perhaps the most important message is that while the practice of analysis and design of electromagnetic PowerMEMS machines may be no different than for any other electromagnetic machine, the end product likely will be quite different. This results from a combination of: (1) different geometries and fabrication constraints; (2) new materials; (3) revised thermal, speed and frequency considerations; (4) the need to consider both electric and magnetic machines; and (5) the increased importance of parasitics. More details may be found in Sections 8.1.1 and 8.1.2. Another important message is the interplay between microfabrication limitations and the performance of a PowerMEMS machine. The resulting design tradeoffs are significant, particularly given the need to manage such phenomena as parasitic capacitance and viscous loss in electric machines, and eddy currents in magnetic machines. Finally, from the viewpoint of PowerMEMS applications such as the Microengine Project, two electromagnetic machine systems stand out as being best suited to such applications. Those systems are based on the electric induction machine and the permanent-magnet synchronous machine. Both machine systems are discussed in detail below. The rationale for this conclusion is given in Section 8.1.3.

8.2 Modeling Simplifications

As indicated in Fig. 8.3, the electromagnetic machines studied here are disk-shaped. Thus, the fields between, within, and outside their stators and rotors are naturally analyzed in cylindrical coordinates. Such an analysis, however, leads to three-dimensional fields involving a mix of Fourier and Bessel functions [29]. While these fields are simple enough to derive, they are too complex to permit the most transparent discussion of machine performance. Therefore, approximate two-dimensional fields are derived here based on an approximate rectilinear machine geometry. The two-dimensional fields are simple enough to permit a transparent discussion of machine performance without a significant loss of accuracy, at least for the geometries typically seen in Power-MEMS electromagnetic machines.

The approximate two-dimensional fields are developed with reference to Fig. 8.7. Here, the annular rotor and stator of an electromagnetic machine are sliced along a radius, and unwrapped and stretched onto Cartesian coordinates. The resulting rectilinear machine is then analyzed using Cartesian coordinates, and its fields are wrapped back onto the original cylindrical geometry in the end. To further simplify matters, the fields are assumed to be uniform in x when analyzed using the Cartesian coordinates. That is, the fields are assumed to be two dimensional. This is justified for a typical Power-MEMS machine in part because the axial extent of the fields is much narrower than the distance from the inner radius of the machine to the outer radius, and so field fringing at the two extreme radii should be minor.

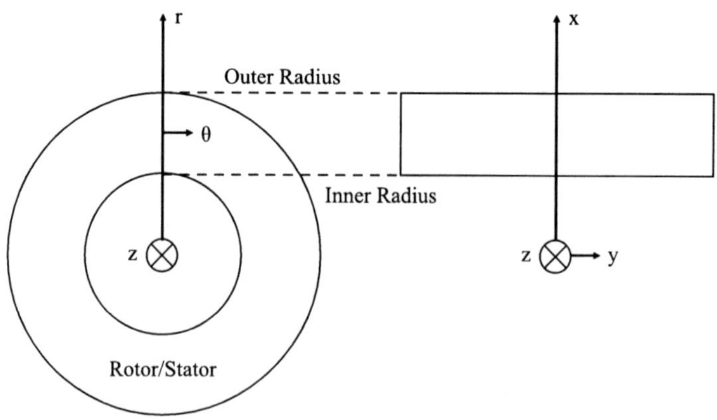

Fig. 8.7 The relation between the natural cylindrical coordinates (*left*) and approximate Cartesian coordinates (*right*) used to analyze the electromagnetic machines

The primary equation of interest here will be Laplace's Equation which is to be solved in the two-dimensional y-z plane. Since the electromagnetic machine is structurally periodic in θ, its fields are also periodic in θ, and hence in y. Thus, the corresponding solutions to Laplace's Equation will take the form $e^{\pm jky}e^{\pm kz}$ in Cartesian coordinates, where k is the wavenumber of the periodicity [5, 29]. This solution is mapped back to the cylindrical coordinates using the mapping

$$x \rightarrow r \quad , \quad y \rightarrow r\theta \quad , \quad z \rightarrow z \quad , \quad k \rightarrow \frac{mP}{r} \quad , \tag{8.1}$$

in which the wavenumber mapping guarantees that the solutions are periodic in θ for all r with fundamental periodicity P. The integer P is determined by the machine geometry, and will ultimately become the number of pole pairs in the machine. The integer m will be used to index higher space harmonics. Thus, the approximate solutions to Laplace's Equation in cylindrical coordinates take the form $e^{\pm jmP\theta}e^{\pm mPz/r}$. Such solutions essentially treat each cylindrical sliver of the machine as independent and two-dimensional.

To investigate the conditions under which the approximate solutions to Laplace's Equation are valid, those solutions can be substituted back into Laplace's Equation to yield

$$0 \approx \left(\frac{1}{r}\frac{\partial}{\partial r}\left(r\frac{\partial}{\partial r}\right) + \frac{1}{r^2}\frac{\partial^2}{\partial \theta^2} + \frac{\partial^2}{\partial z^2}\right) e^{\pm jmP\theta \pm mPz/r} \tag{8.2}$$

$$= \left(\pm\frac{mPz}{r^3} + \frac{m^2P^2z^2}{r^4} - \frac{m^2P^2}{r^2} + \frac{m^2P^2}{r^2}\right) e^{\pm jmP\theta \pm mPz/r} \tag{8.3}$$

in cylindrical coordinates. Note that the first two terms in (8.3) come from the partial derivatives with respect to r in (8.2), the third term comes from the partial derivatives with respect to θ, and the fourth term comes from the partial derivatives with respect to z. In order for (8.3) to evaluate approximately to zero, the dominant balance in the equation must be between the last two terms, with the first two terms being small. Since the product mP is a positive integer, the first two terms will be small in regions in which $|z| \ll r$. Most electromagnetic PowerMEMS machines exhibit fields that are essentially confined to narrow planar regions that are much shorter in the axial direction than the inner annular radius of their rotor and stator, and so this condition is easily met. When it is not, one must solve Laplace's Equation directly in cylindrical coordinates.

8.3 Electric Induction Machine

A cross-section side view of the electric induction machine studied here is shown in Fig. 8.8. Photographs of the machine fabricated as a PowerMEMS machine are shown in Fig. 8.9. The stator comprises radially directed electrodes, like spokes, supported by an insulator on a conducting substrate. The stator electrodes are wired

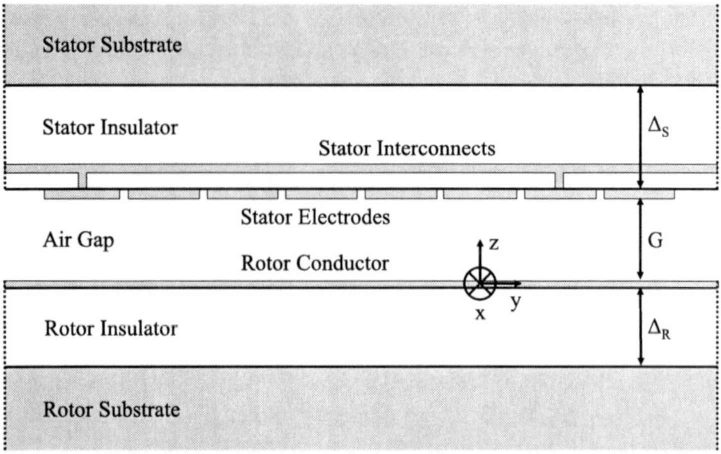

Fig. 8.8 Side view of the electric induction machine with the Cartesian coordinates used for analysis

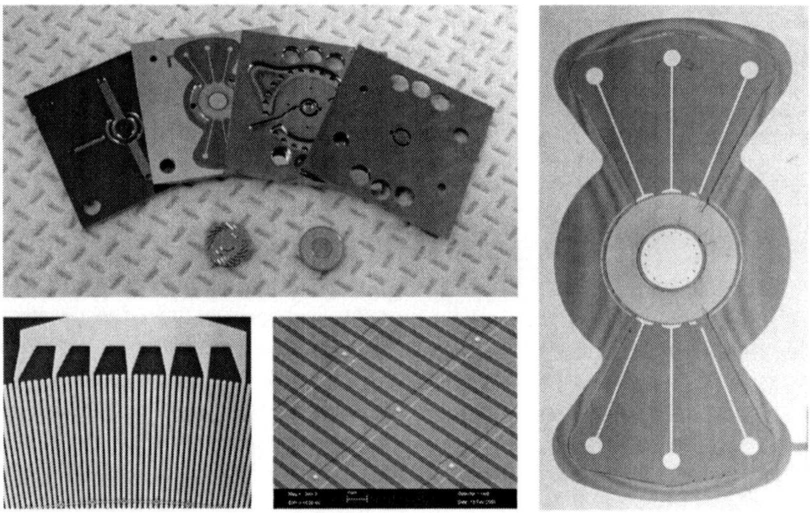

Fig. 8.9 Photographs of an MEMS electric induction machine from [35]. The *top-left* photograph shows a five-wafer set which, when fully bonded, forms a complete motor/compressor or turbine/generator depending on the design of the blades. The middle-right wafer is already a bonded wafer pair that contains the rotor. The middle-left wafer is the stator wafer. The two outer wafers are foundation wafers that contain many of the air passages connecting to the compressor or turbine, and the air bearings. Below the wafers are two separated rotors. The rotor on the left shows its turbine blades, while the rotor on the right shows its annular rotor conductor surrounding a thrust bearing pad. The *right*-most photograph shows a close-up view of the stator fabricated on top of buried silicon dioxide in the shape of a bow tie. The six phase leads that terminate on contact pads are also visible. The stator electrodes in the central annular region are not distinguishable, but the air nozzles around the periphery of the center thrust bearing are visible. The *lower-left* photograph shows a close-up view of the stator electrodes near their connection to one of the phase leads. One interconnect ring is visible at the very bottom of the photograph. The *lower-middle* photograph shows another close-up view of the stator electrodes which shows three interconnect rings (reprinted from Livermore et al. [35] © 2004 IEEE)

together through back-side interconnects to form N phases; every Nth electrode is connected to form a single phase. In both figures, N is 6. The interconnected electrode pattern is repeated P times around the stator. Thus, the total number of stator electrodes is NP. In Fig. 8.9, P is 131 for a total of 768 electrodes. The rotor consists of a thin uniform conductor also supported by an insulator on a conducting substrate. In the applications considered here, the bottom side of the rotor could be etched to form turbine or compressor blades. The blades are not shown in Fig. 8.8, but they are visible in Fig. 8.9. In the machines studied here, the substrates are silicon, the insulators are silicon dioxide, the stator electrodes and interconnects are either metal or heavily doped polysilicon, and the rotor conductor is moderately doped polysilicon.

The stator phases of an electric induction machine are generally excited with balanced sinusoidal voltages, having a temporal phase shift of $2\pi/N$ from one phase to the next, and hence one electrode to the next. In this way, the phases support a traveling wave of potential that circulates around the stator. The stator potential and the corresponding charges on the stator electrodes induce image charges on the rotor conductor, as shown in Fig. 8.10. In steady state, the image charges on the rotor must travel in synchronism with the stator excitation. Part of this travel is provided by convection arising from the motion of the rotor, and part is provided by conduction through the rotor. During motoring operation, as shown in Fig. 8.10, the stator potential travels at a speed that exceeds that of the rotor motion, and so the rotor charges lag behind. To stay in synchronism with the stator potential, the rotor charges conduct forward through the rotor conductor, and this conduction is driven by the horizontal component of the air-gap electric field. This same electric field component produces the motoring torque acting on the rotor. During generating operation, the speed of the rotor motion exceeds that of the stator potential wave, and the rotor charges lead their images on the stator electrodes. In this case, the charges conduct backward through the rotor to maintain synchronism, and the horizontal component of the electric field is reversed.

The development of the electric induction machine as a PowerMEMS motor is chronicalled in [30–36], with [30, 31] providing the greatest detail, and [35]

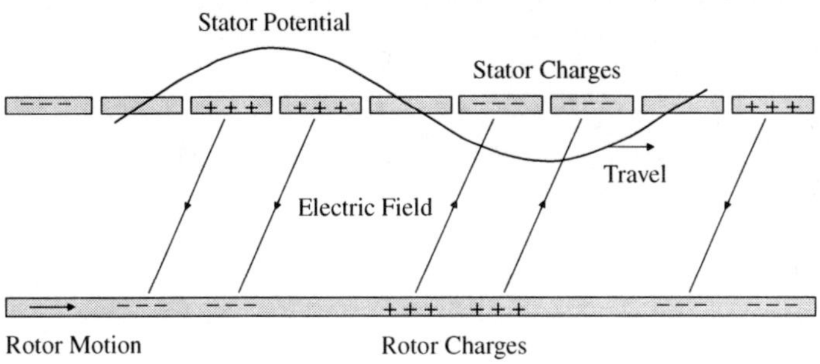

Fig. 8.10 A diagram of the traveling stator potential, the charges on the stator electrodes and the rotor conductor, and the electric field in the air gap, within an electric induction machine during motoring operation

providing additional experimental results. The development of the electric induction machine as a PowerMEMS generator is chronicalled in [37–39], with [37] providing the greatest detail. Additionally, the electric induction machine has a substantial development history as a macro-scale motor. Much of this early work is described in [29, 40–45]. Finally, the electric induction machine has been studied as a direct fluid pump both at the microscale [46, 47], and much earlier at the macroscale [29, 48]. The subsections that follow are based largely on this collective work.

8.3.1 Analysis

The analysis presented here is a steady-state analysis which closely follows that presented in [31–33, 36]. It is one in which the rotor moves with a constant angular velocity, and the stator electrodes are excited with balanced sinusoidal voltages. Its objective is to determine the steady-state forces and torques at work within the electic induction machine, and the steady-state power flow through it. Such an analysis is adequate to understand the fundamental behavior of the electric induction machine, and its power electronics, and to investigate design and performance tradeoffs. However, there are situations in which an unbalanced and/or transient model is required. One important situation is during the analysis of self-exciting resonant operation in which imbalances play a critical role at the microscale. In such cases, the reader is refered to [37, 38, 39] for an extended model.

The analysis of the electric induction machine begins with its stator excitation. This excitation is spatially piecewise constant, except for the small gaps between the stator electrodes which are ignored here. It is also periodic, and thus expressible in a Fourier Series over the angle θ defined in Fig. 8.7. Thus, if the N stator voltages v_n, $0 \leq n < N$, are given by

$$v_n(t) = V \cos(\omega t - 2\pi n/N) \quad , \tag{8.4}$$

where V and ω are the electrical stator excitation amplitude and frequency, respectively, then the stator potential ϕ_S at $z = G$ may be written as

$$\phi_S(r,\theta,z = G,t) = \sum_{m=-\infty}^{\infty} V_m \cos(\omega t - mP\theta) \tag{8.5}$$

$$V_m \equiv \begin{cases} V\frac{N}{m\pi} \sin\left(\frac{m\pi}{N}\right) & m = \cdots 1-N, 1, 1+N, 1+2N \cdots \quad ; \\ 0 & \text{otherwise} \quad . \end{cases} \tag{8.6}$$

For the case of $N = 6$, as shown in Figs. 8.8, 8.9, and 8.10, for example, $V_1 = 0.95\,V$, while $V_{-5} = 0.19\,V$ and $V_7 = -0.14\,V$. Thus, while the higher space harmonic amplitudes do fall off essentially as $1/m$, this decay may not be rapid enough to justify ignoring the higher harmonics altogether. The stator potential expressed in (8.5) becomes

$$\phi_S(y,z=G,t) = \sum_{m=-\infty}^{\infty} V_m \cos(\omega t - k_m y) \tag{8.7}$$

$$k_m \equiv \frac{mP}{r} \tag{8.8}$$

in the approximate Cartesian coordinates defined in Figs. 8.7 and 8.8, after the mapping in (8.1) is applied. It is a superposition of potential waves traveling in the y direction.

The next step in the analysis is to determine the electric potential within the stator, the air gap and the rotor, all driven by the stator potential at $z = G$. This requires the solution to Laplace's Equation within these three regions, which is developed here in the Cartesian coordinates shown in Fig. 8.8. To solve Laplace's Equation, boundary conditions of zero potential are applied at the axial extremes, namely at $z = G + \Delta_S$ and $z = -\Delta_R$. Further, the thin rotor conductor at $z = 0$, which separates the air gap from the rotor insulator, is assumed to behave as a surface conductor with surface conductivity σ_R, and the stator and rotor insulators are assumed to have permittivity ϵ. Finally, the rotor is assumed to rotate with angular velocity Ω. Under these conditions, the three potentials are given by

$$\phi_S(y,z,t) = \sum_{m=-\infty}^{\infty} \Re\left\{V_m \frac{\sinh(k_m(\Delta_S + G - z))}{\sinh(k_m \Delta_S)} e^{j(\omega t - k_m y)}\right\} \tag{8.9}$$

$$\phi_G(y,z,t) = \sum_{m=-\infty}^{\infty} \Re\left\{\left(V_m \frac{\sinh(k_m z)}{\sinh(k_m G)} + \hat{\Phi}_{R_m} \frac{\sinh(k_m(G - z))}{\sinh(k_m G)}\right) e^{j(\omega t - k_m y)}\right\} \tag{8.10}$$

$$\phi_R(y,z,t) = \sum_{m=-\infty}^{\infty} \Re\left\{\hat{\Phi}_{R_m} \frac{\sinh(k_m(\Delta_R + z))}{\sinh(k_m \Delta_R)} e^{j(\omega t - k_m y)}\right\} \tag{8.11}$$

where ϕ_S, ϕ_G, and ϕ_R are the potentials in the stator ($G \le z \le G + \Delta_S$), the air gap ($0 \le z \le G$), and the rotor ($-\Delta_R \le z \le 0$), respectively, and

$$\hat{\Phi}_{R_m} \equiv V_m \beta_m \frac{jS_m \omega \tau_m}{1 + jS_m \omega \tau_m} \tag{8.12}$$

$$\beta_m \equiv \frac{\epsilon_\circ / \sinh(k_m G)}{\epsilon_\circ \coth(k_m G) + \epsilon \coth(k_m \Delta_R)} \tag{8.13}$$

$$\tau_m \equiv \frac{\epsilon_\circ \coth(k_m G) + \epsilon \coth(k_m \Delta_R)}{\sigma_R k_m} = \frac{\epsilon_\circ}{\sigma_R \beta_m k_m \sinh(k_m G)} \tag{8.14}$$

$$S_m \equiv \frac{\omega - mP\Omega}{\omega} \quad . \tag{8.15}$$

As expressed in (8.12), $\hat{\Phi}_{R_m}$ is the complex amplitude of the electric potential in the rotor conductor for space harmonic m. Further, as defined in (8.13), (8.14) and (8.15), β_m is the capacitive potential division from the stator to the rotor, τ_m is the rotor charge-relaxation time, and S_m is the slip such that the electrical frequency of the traveling stator potential wave as viewed from the moving rotor is $S_m\omega$, again all for space harmonic m.

The electric field in the three regions, the free surface charge densities on the stator electrodes and the rotor conductor, the shear and normal stresses of electric origin acting on the rotor, and the power dissipation density in the rotor conductor, can all be determined from the potentials given in (8.9), (8.10) and (8.11). The electric field $\bar{E}$ is given in each region by

$$\bar{E} = -\nabla\phi \quad . \tag{8.16}$$

The free surface charge densities on the stator and rotor, ρ_S and ρ_R, respectively, are given by

$$\begin{aligned}\rho_S(y,t) &= \epsilon E_{S_z}|_{z=G} - \epsilon_0 E_{G_z}|_{z=G} \\ &= \sum_{m=-\infty}^{\infty} \Re\left\{V_m k_m\left(\epsilon \coth(k_m\Delta_S) + \epsilon_0 \coth\right.\right. \\ &\qquad \left.\left. - \frac{\epsilon_0\beta_m}{\sinh(k_m G)}\left(\frac{jS_m\omega\tau_m}{1+jS_m\omega\tau_m}\right)\right) e^{j(\omega t - k_m y)}\right\}\end{aligned} \tag{8.17}$$

$$\begin{aligned}\rho_R(y,t) &= \epsilon_0 E_{G_z}|_{z=0} - \epsilon E_{R_z}|_{z=0} \\ &= \sum_{m=-\infty}^{\infty} \Re\left\{\frac{-\epsilon_0 k_m V_m}{\sinh(k_m G)}(1+jS_m\omega\tau_m)e^{j(\omega t-k_m y)}\right\} \quad .\end{aligned} \tag{8.18}$$

The y-directed spatial average of the shear and normal stresses of electric origin acting on the rotor, $<f_y>$ and $<f_z>$, respectively, are given by

$$\begin{aligned}\langle f_y\rangle &= \langle \epsilon_0\, E_{G_y}\big|_{z=0}\, E_{G_z}\big|_{z=0}\rangle = \langle \epsilon_0 \left.\frac{\partial\phi_G}{\partial y}\right|_{z=0} \left.\frac{\partial\phi_G}{\partial z}\right|_{z=0}\rangle \\ &= \sum_{m=-\infty}^{\infty} \frac{\epsilon_0\beta_m k_m^2 V_m^2}{2\sinh(k_m G)}\left(\frac{S_m\omega\tau_m}{1+S_m^2\omega^2\tau_m^2}\right)\end{aligned} \tag{8.19}$$

$$\begin{aligned}\langle f_z\rangle &= \langle\frac{\epsilon_0}{2}\left(E_{G_z}^2\big|_{z=0} - E_{G_y}^2\big|_{z=0}\right)\rangle = \langle\frac{\epsilon_0}{2}\left(\left.\left(\frac{\partial\phi_G}{\partial z}\right)^2\right|_{z=0} - \left.\left(\frac{\partial\phi_G}{\partial y}\right)^2\right|_{z=0}\right)\rangle \\ &= \sum_{m=-\infty}^{\infty}\frac{\epsilon_0 k_m^2 V_m^2}{4\sinh^2(k_m G)}\left(\frac{1+S_m^2\omega^2\tau_m^2\left(1-2\cosh(k_m G)\beta_m+\beta_m^2\right)}{1+S_m^2\omega^2\tau_m^2}\right)\end{aligned} \tag{8.20}$$

where $\langle\cdot\rangle$ denotes the y-directed spatial average of its argument; the electric stress tensor has been used to determine $\langle f_y\rangle$ and $\langle f_z\rangle$ [2, 29]. The shear stress $\langle f_y\rangle$ is responsible for the motoring and generating operation of the electric induction machine. The normal stress, or pressure, $\langle f_z\rangle$ acts to collapse the air gap, and hence must be restrained by thrust bearings such as those developed in Chapter 6. Finally, the spatial average of the power dissipation density in the rotor conductor, $\langle p_\mathrm{R}\rangle$, is given by

$$\begin{aligned}\langle p_\mathrm{R}\rangle = \langle \sigma_\mathrm{R}\, E^2_{\mathrm{G}_y}\Big|_{z=0}\rangle &= \langle \sigma_\mathrm{R} \left(\frac{\partial\phi_\mathrm{G}}{\partial y}\right)^2\Bigg|_{z=0}\rangle \\ &= \sum_{m=-\infty}^{\infty} \frac{\sigma_\mathrm{R}\beta_m^2 k_m^2 V_m^2}{2}\left(\frac{S_m^2\omega^2\tau_m^2}{1+S_m^2\omega^2\tau_m^2}\right) \quad .\end{aligned} \tag{8.21}$$

The final step in the analysis is to use the free stator surface charge density, the electric stresses acting on the rotor, and the power dissipation density in the rotor conductor, to determine the electrical and mechanical terminal variables of the electric induction machine, and the power dissipated in it. To do so, the surface charge density, the stresses, and the power dissipation density are first mapped into the cylindrical coordinates of Fig. 8.7 using (8.1). Although these quantities are not functions of x, they do become functions of r through the mapping. This dependence occurs because k_m is a function of r, and so β_m and τ_m become functions of r as well. For typical MEMS geometries, the dependence of β_m on r will likely be weak, but the dependence of τ_m on r may be strong.

Because β_m and τ_m become functions of r, $\hat{\Phi}_\mathrm{R}$, and the electric potentials in general, become functions of r as well. The gradient of the rotor potential will therefore yield a radial electric field that drives radial currents in the rotor conductor. These currents in turn redistribute the rotor surface charges, and hence the operation of the electric induction machine, in ways that are not modeled in (8.9–8.21). However, if the conditions discussed in Section 8.2 are met, as is likely the case given typical MEMS geometries, then the unmodeled effects are inconsequential. Otherwise, Laplace's Equation must be solved in the cylindrical coordinates of Fig. 8.7.

After mapping back to cylindrical coordinates, the free stator surface charge density may be used to determine the total charge, q_n, on the electrodes of the nth phase. This charge is given by

$$q_n(t) = P\int_{r_\mathrm{IN}}^{r_\mathrm{OUT}}\int_{(2n-1)\pi/NP}^{(2n+1)\pi/NP} \rho_\mathrm{S}(r,\theta,t) r\,\mathrm{d}\theta\,\mathrm{d}r \quad , \tag{8.22}$$

where r_IN and r_OUT are the inner and outer radii, respectively, of the rotor as depicted in Fig. 8.7. The time derivative of q_n is the current into the nth phase of the electric induction machine. Similarly, the two stresses can be used to determine the torque of electric origin, T, and the normal force of electric origin, F_z, acting on the rotor. The torque and force are given by

$$T = 2\pi \int_{r_{\mathrm{IN}}}^{r_{\mathrm{OUT}}} \langle f_y \rangle(r) r^2 \mathrm{d}r \tag{8.23}$$

$$F_z = 2\pi \int_{r_{\mathrm{IN}}}^{r_{\mathrm{OUT}}} \langle f_z \rangle(r) r \mathrm{d}r \quad . \tag{8.24}$$

Finally, the power dissipation density can be used to determine to total power dissipated in the rotor conductor, P_{R}. This loss is given by

$$P_{\mathrm{R}} = 2\pi \int_{r_{\mathrm{IN}}}^{r_{\mathrm{OUT}}} \langle p_{\mathrm{R}} \rangle(r) r \mathrm{d}r \quad . \tag{8.25}$$

These quantities, in combination with the voltage applied to each phase, constitute the electrical and mechanical terminal variables of the electric induction machine, and the power dissipated in it. They directly control the power flow through the machine, as investigated in the next subsection.

8.3.2 Design and Performance

From (8.19) through (8.21), it is apparent that the electric stresses acting on the rotor, and the power dissipated in the rotor, all involve the superposition of terms proportional to V_m^2. Therefore, the behavior of the spatial fundamental, $m = 1$ in (8.6), is dominant even for machines constructed with only three phases. Correspondingly, this section investigates the design and performance of the electric induction machine as viewed primarily through its fundamental. In addition, a long-wave approximation is adopted to greatly simplify the investigation and improve its clarity. The long-wave approximation assumes that $k_1 G$, $k_1 \Delta_{\mathrm{R}}$, and $k_1 \Delta_{\mathrm{S}}$ are all much less than unity so that their hyperbolic sines and cosines are approximately themselves and unity, respectively. (As a practical matter, if each product is less than 0.4, then the error is less than 10%.) Physically, this approximation assumes that one wavelength of the stator excitation is much greater than the air gap separation and the insulator thicknesses. Also for simplicity, r_{IN} is taken to be zero, and $r_{\mathrm{OUT}} \equiv R$. As a result, the long-wave approximation will not be valid at the innermost radii, but this is not so important because phenomena at these radii make the least contribution to T, F_z, and P_{R}. When greater accuracy is required, (8.19), (8.20) and (8.21) should be used.

The objective now is to derive simple expressions for the phase charge, the torque and axial force, the input and output powers, and the efficiency of the electric induction machine. These expressions will then be used to assess performance and gain design insight. Again, these expressions will be approximations valid within the limitations of the long-wave approximation. Beginning with (8.13) and (8.14), β_1 and τ_1 simplify to

$$\beta_1 \approx \left(1 + \frac{\epsilon G}{\epsilon_0 \Delta_R}\right)^{-1} \tag{8.26}$$

$$\tau_1(r) \approx \frac{\epsilon_0 r^2}{\beta_1 \sigma_R P^2 G} \tag{8.27}$$

after use of the long-wave approximation, and substitution for k_1 using (8.1). Next, (8.26) and (8.27) are substituted into (8.17), and (8.19) (8.20) and (8.21), which are in turn substituted into (8.22) (8.23), (8.24), and (8.25) to yield

$$q_n(t) \approx V_1 \sin\left(\frac{\pi}{N}\right) e^{j(\omega t - 2n\pi/N)}$$
$$\left(\frac{\epsilon R^2}{\Delta_S} + \frac{\epsilon_0 R^2}{G}\left(1 - \beta_1 + \beta_1 \frac{\tan^{-1}(\gamma_1)}{\gamma_1} - \frac{j\beta_1}{2}\left(\frac{\ln(1+\gamma_1^2)}{\gamma_1}\right)\right)\right) \tag{8.28}$$

$$T \approx \frac{\pi P \beta_1 \epsilon_0 V_1^2 R^2}{4G}\left(\frac{\ln(1+\gamma_1^2)}{\gamma_1}\right) \tag{8.29}$$

$$F_z \approx \frac{\pi \epsilon_0 V_1^2 R^2}{4G^2}\left(\frac{\tan^{-1}(\gamma_1)}{\gamma_1} + (1-\beta_1)^2\left(1 - \frac{\tan^{-1}(\gamma_1)}{\gamma_1}\right)\right) \tag{8.30}$$

$$P_R \approx T\,\frac{S_1 \omega}{P} \tag{8.31}$$

$$\gamma_1 \equiv \frac{S_1 \omega \epsilon_0 R^2}{\beta_1 \sigma_R P^2 G} = S_1 \omega \tau_1 \,|_{r=R} \tag{8.32}$$

upon completing the indicated integrations. In deriving (8.28), (8.29), (8.30), and (8.31), all space harmonics except $m = 1$ have been dropped, the long-wave approximation has again been used, and the simplifications $r_{IN} \equiv 0$ and $r_{OUT} \equiv R$ have also been used. The mechanical power output from the machine, P_M, is the product of torque and rotor velocity, hence

$$P_M = T\Omega \approx \frac{\pi P \beta_1 \epsilon_0 V_1^2 R^2 \Omega}{4G}\left(\frac{\ln(1+\gamma_1^2)}{\gamma_1}\right) \quad . \tag{8.33}$$

Combining (8.31), the equality in (8.33), and (8.15) yields an alternative expression for the power dissipated in the rotor, namely

$$P_R \approx P_M \frac{S_1 \omega}{P\Omega} = P_M \frac{S_1}{1 - S_1} \quad . \tag{8.34}$$

Then by energy conservation, the electrical power into the machine, P_E, is given by

$$P_E = P_M + P_R \approx P_M \frac{\omega}{P\Omega} = P_M \frac{1}{1 - S_1} \quad . \tag{8.35}$$

Finally, the electromechanical motoring efficiency η_M and the electromechanical generating efficiency η_G are given by

$$\eta_M = \frac{P_M}{P_E} \approx 1 - S_1 \tag{8.36}$$

$$\eta_G = \frac{-P_E}{-P_M} \approx \frac{1}{1 - S_1} \quad , \tag{8.37}$$

in which $0 < S_1 < 1$ for motoring operation, and $S_1 < 0$ for generating operation. Note that the efficiencies are based on the power balance that occurs between the electrodes of the machine and its shaft. There will always be additional electrical and mechanical losses that occur outside this domain, and hence are not represented in (8.36) and (8.37). In summary, the approximate expressions derived above can be used to assess the performance of the electric induction machine, and gain insight into its proper design. To aid in this process, a graph of $\ln(1 + \gamma_1^2)/\gamma_1$ is provided in Fig. 8.11.

In general, a harmonic can operate to perform one of three functions: motoring, generating, or braking. Which function it performs depends on its slip S_m, as defined by (8.15); ω is taken here to be positive without loss of generality. When space harmonic m travels at the same velocity as the rotor, then its excitation through (8.5) is stationary as viewed by the rotor, and $S_m = 0$. Correspondingly, its steady-state contribution to $\langle f_y \rangle$ and p_R, and hence T and P_R, is zero, as can be seen from (8.19) and (8.21). This can also be seen through (8.29) and (8.31) for $m = 1$, since γ_1 is

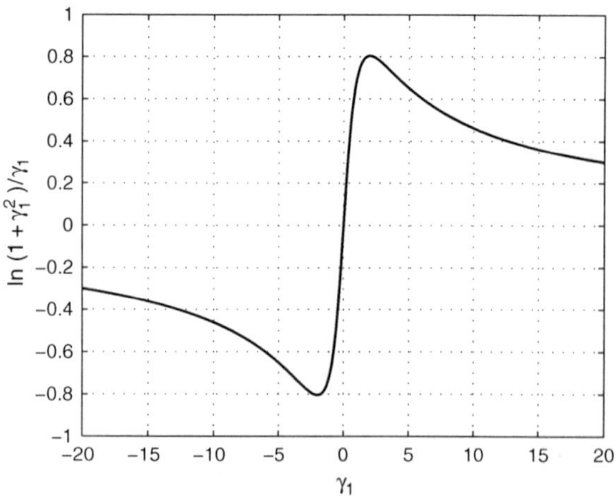

Fig. 8.11 A graph of $\ln(1 + \gamma_1^2)/\gamma_1$ as a function of γ_1

proportional to S_1. As the velocity of the rotor increases beyond synchronism, S_m becomes negative, as does its contribution to $\langle f_y \rangle$; its contribution to p_R, however, becomes positive. In this case, the harmonic generates, as power enters the machine through mechanical work, and exits through the electrical terminals, with some loss along the way. Conversely, as the velocity of the rotor decreases below synchronism, S_m becomes positive, as does its contribution to $\langle f_y \rangle$ and p_R. In this case, the harmonic motors, as power enters the machine through the electrical terminals, and exits through mechanical work, again with loss along the way. Finally, as the velocity of the rotor reduces to zero and below, the mechanical power reverses sign, and again enters the machine. In this case, the harmonic acts as a brake in which power enters both through the electrical terminals and through mechanical work. This power is dissipated entirely in the machine. This section will focus primarily on motoring and generating operation.

From (8.6) and (8.29), it is apparent that the torque produced by, and hence the power through, the electric induction machine increases as V^2. The principle limit to V is imposed by electrical breakdown inside the machine. As discussed in Section 8.1.1, and shown in Fig. 8.2, a realistic limit to V is approximately 300 V for an air gap. Such voltages have been successfully applied to a variety of electric MEMS devices from motors and generators, to valves to relays. In the electric induction machine, there are actually two gaps of importance to electrical breakdown, namely that between the rotor and the stator, and those between neighboring electrodes. Either gap can break down and hence limit machine performance. It is assumed here that the rotor and stator insulators have a greater breakdown strength.

From (8.6) and (8.29), it is apparent that $\beta_1 < 1$ and so $\left|V_1 - \hat{\Phi}_R\right|$ is maximized at V_1 for $S_1 = 0$. Thus, the worst case potential difference across the air gap of the electric induction machine is approximately V. At the same time, the maximum potential difference between two neighboring electrodes is given by $V\sqrt{2 - 2\cos(2\pi/N)}$, which equals V for $N = 6$, and exceeds V for smaller N. Thus, in order to maximize the permissible V, without requiring too many phases, and without losing too much to the space between them, N is chosen to be six for the electric induction machines shown in Fig. 8.9.

From (8.32), (8.33), (8.34), and (8.35) it is apparent that the power through the electric induction machine depends on the rotor surface conductivity σ_R only through the parameter γ_1. Thus, γ_1 can be considered an independent design parameter which in turn determines the required rotor surface conductivity. This is the viewpoint taken here. In the end, however, the rotor surface conductivity is restricted by fabrication and physical constraints, which may establish design trade offs. One means of constructing the rotor conductor is with a very thin layer of doped polysilicon which can exhibit a bulk resistivity ranging from 10^{-3} to 10^6 Ω·cm, depending upon the dopant and its concentration [49]. This variation, combined with permissible variations in the conductor thickness, allows for a wide range of effective rotor surface conductivities.

The dependence of torque, and hence power throughput, on γ_1 is shown in Fig. 8.11. For $\gamma_1 = 0$, the rotor and the stator excitation are in synchronism. That is, they travel at the same speed, and so the stator excitation appears to be stationary as viewed from the rotor. In this case, charge relaxation causes the rotor charges to settle directly beneath their images on the stator, and the electric field lines between the two charges, shown in Fig. 8.10, are axially directed. No torque is produced in this case. As the stator excitation begins to slip past the rotor for $\gamma_1 > 0$, the rotor charges begin to lag the stator excitation, and the electric field lines tip as shown in Fig. 8.10. This produces a positive torque on the rotor. Eventually, the slip becomes so great that there is insufficient time to induce charge on the rotor, and so the torque drops off. The competing phenomena of charge induction and alignment for $\gamma_1 > 0$ give rise to the positive peak in Fig. 8.11. Alternatively, as the rotor begins to slip past the stator excitation for $\gamma_1 < 0$, the rotor charges begin to lead the stator excitation, and the electric field lines tip in the opposite direction shown in Fig. 8.10. This produces a negative torque on the rotor. Again, eventually, the slip becomes so great that there is insufficient time to induce charge on the rotor, and so the torque again drops off.

Motoring and generating torque can be maximized with $\gamma_1 \approx \pm 2$, respectively. In this case, $\ln(1+\gamma_1^2)/\gamma_1 \approx \pm 0.8$, respectively. In practice, it is probably unwise to design an electric induction motor to operate in an open loop manner at the peak torque if the mechanical load can vary. For example, if the mechanical load increases beyond the peak torque, then the induction machine will slow down, which will increase S_1, thereby increasing γ_1 and further reducing torque. In this case, depending on the dependence of the load torque on speed, the motor could slow down significantly. Thus, some room for load variation is desirable from a practical viewpoint in an open-loop application.

From (8.33), it is also apparent that the power through the electric induction machine increases with decreasing G, and increasing P, R, and Ω. Beyond electrical breakdown concerns, the principle limits to decreasing G are viscous losses in the air gap, the planarity of the rotor, and the ability of the thrust bearings to hold the air gap open in the presence of F_z. Air gaps as small as 3 μm, for example, have been successfully implemented in the machines described in [30, 31, 37, 34–36, 38, 39], even with rotors having diameters approaching 1 cm. The principle limits to increasing P are the minimum feature size of the separation between electrodes, the electric field fringing that occurs as the long-wave approximation is violated, and practical limits to the resulting electrical frequency ω. Given a fixed minimum feature size, the fraction of the stator area that is allocated to the actual electrodes diminishes with increasing P. This will eventually result in a significant loss of torque as the width of the electrodes approaches the electrode separation. Generally, the electrode separation should be limited to at most 25% of the electrode width at $r = R$. The performance of machines having wider electrode separations between the electrodes must be examined with more detailed electric potential solutions than are given in (8.9), (8.10), and (8.11). Separately, as $k_1 G$ approaches 0.4, the electric fields lines emanating from the stator electrodes begin to fringe more towards the neighboring electrodes rather than being directed to the rotor. As this occurs, the torque of electric origin begins to decrease rapidly, and a further increase in P causes a reduction in torque and power throughput. Finally, the principle limits to raising R and Ω are the planarity and mechanical integrity of the rotor, and the resulting electrical frequency. The mechanical integrity of a silicon rotor becomes limiting as the tip speed of the rotor, $R\Omega$, approaches 2000 m/s, as observed in Table 8.1. However, if the rotor carries turbomachinery blades etched into the silicon, then a limit near 500 m/s is more realistic.

Following (8.15), the electrical frequency ω is given by

$$\omega = \frac{P\Omega}{1 - S_1} \quad . \tag{8.38}$$

Given the dependence of η_M and η_G on S_1 observed in (8.36) and (8.37), $(1 - S_1)$ will likely be near unity so that $\omega \approx P\Omega$. Thus, raising both P and Ω results in an increase in ω for a given efficiency. In this case, practical electronics limits must be observed so that ω is limited to perhaps the MHz range and below as discussed in Section 8.1.1.

Finally, consider Δ_R and Δ_S. A small Δ_R allows the rotor substrate to short circuit the rotor surface conductor. This results in a reduction of torque through β_1. A small Δ_S increases the stator charge, as observed in (8.28). While this does not directly reduce power throughput, the increase in stator charge results in an increase in reactive current to the electric induction machine. This current will pass through electrode and wiring resistances, and the power electronics, and thereby create additional losses. Thus, it is desirable to make both Δ_R and Δ_S as large as possible, though there are eventually diminishing returns as $k_1\Delta_R$ and $k_1\Delta_S$ become large enough so that the majority of the electric fields lines do not reach the stator and rotor substrates. This may occur, however, for Δ_R and Δ_S as large as 10 times G given typical PowerMEMS geometries.

To summarize, assume that: (1) Δ_R can be made large enough so that $\beta_1 \approx 0.5$; (2) that P can be made large enough so that k_1 at $r = R$ satisfies $k_1 = PG/R \approx 0.4$; and (3) that the operation of the electric induction machine can be controlled so that the machine operates at the peak of the curve in Fig. 8.11. In this case,

$$P_M \approx 0.5 \left(\frac{\epsilon_0 V_1^2}{4G^2} \right) R^2 U \tag{8.39}$$

where $U = \Omega R$ is the peripheral tip speed of the rotor. Note that the term inside the parentheses is the spatial average of the electric energy stored in the air gap. If the slip can be set near zero, then the electromechanical efficiency of the machine will be high, and P_M in (8.39) is a good measure of the power through the machine during both motoring and generating operation. By way of example, let: $V_1 = 300$ V; $G = 3$ µm; $R = 2$ mm; and $U = 500$ m/s. In this case, $P_M = 22$ W from (8.39), which is quite high for a small device. However, it is also the case that $P = 267$, and so $\omega \approx 11$ MHz. Thus, fabrication and power electronics issues will play a significantly limiting role. Note that by way of comparison, the evaluation of (8.19) and (8.23) for the same machine and operating conditions yields $P_M = 12$ W. As expected, the simple models presented in this section are optimistic, and are best used to understand performance trends and develop design guidelines. The results of Section 8.3.1 should be used for detailed analysis, or a Fourier–Bessel analysis of the electric potentials in cylindrical coordinates could be used for yet greater accuracy.

8.3.3 Power Electronics and Control

To achieve its best performance as a motor or a generator, the electric induction machine must be operated at both high voltage and high frequency. Voltage amplitudes near 300 V and frequencies near 1 MHz are typical for highest power throughput. In addition, the applied stator voltage waveforms should be sinusoidal. Non-sinusoidal waveforms contain time harmonics that can degrade both motoring and generating performance, and they will unnecessarily increase the peak voltage between neighboring electrodes making electrical breakdown a more limiting phenomenon. Thus, hard-switching power electronics such as those shown in Fig. 8.4 are inappropriate for the electric induction machine, and will most likely be too lossy as discussed in Section 8.1.1, particularly for generating applications. An alternative is required.

One alternative is to use resonant power electronics such as those shown in Fig. 8.12. The electric induction machine is essentially a capacitive machine. The effective capacitance at its electrical terminals, which includes any parasitic capacitance, can be used to form a resonator with an external inductor. The function of the resonator is to step the voltage from the transistors in the power electronics up to the phases of the machine. This is the approach taken in [30, 37, 34, 38, 39]. In [37–39], the low-voltage sources are themselves sinusoidal to minimize the presence of time harmonics and the associated loss. This is important during generation. However, it is possible, particularly in motoring applications, that the sources can be simple switching square-wave sources as shown in Fig. 8.4 because each resonator acts as a band-pass filter. This is the approach taken in [30, 34]. Whether or not this approach is acceptable will depend on the application. The advantage of resonant power electronics is that the transistors in the power electronics operate at a lower voltage, and so their switching losses can be greatly reduced. Voltage step-up ratios near 20 have been achieved, which lowers the transistor switching losses by a factor of 400 [30, 34].

Resonant power electronics also have important disadvantages. One disadvantage is the need to tune each phase so as to maintain a balanced excitation at the

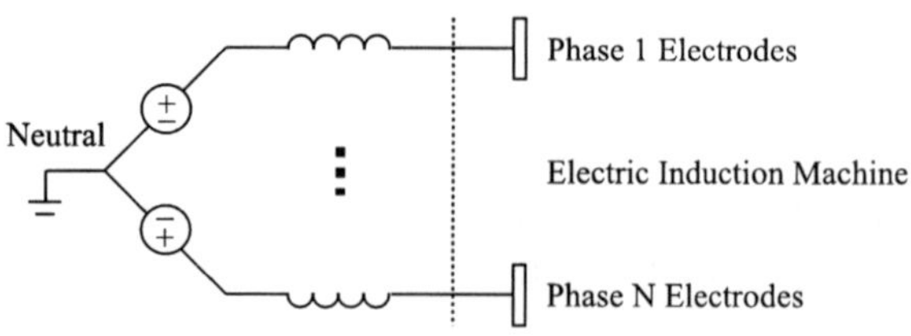

Fig. 8.12 Resonant power electronics for an electric induction machine; only two phase legs are shown

electrodes of the machine. Without a balanced excitation, the amplitudes of the space harmonics may be large enough to significantly reduce the performance of the machine, particularly when used as a generator. One source of imbalance is unbalanced parasitic capacitance in the machine itself. For example, the interconnect rings seen in Fig. 8.9 are concentric, with some rings inside others. The parasitic capacitance between these rings and the other phases will be different depending on the radius of the ring at least because the electrodes become wider as radius increases. Another source of imbalance is the inductors, no two of which will be the same. For this reason, it may be necessary to place tuning capacitors from each phase to neutral, and adjust them, and the voltage and temporal phase of the low-voltage sources, to achieve balanced excitation at the electrodes. This is done in [30, 37, 34, 38, 39]. Note that the presence of the tuning capacitors increases reactive currents and the associated losses, but the net result can be an improvement in performance. A second disadvantage is the presence of the resonating inductors, and the losses that they bring at high voltage and high frequency. Low-loss inductors having the appropriate inductance that operate at high voltage and high frequency can be physically large, and almost certainly larger than the electric machine itself. A final disadvantage is that the effective capacitance of the electric induction machine will change with operating conditions. For this reason, it may be necessary to change the excitation frequency and temporal phase shifts of the low-voltage sources so as to keep the resonator tuned at resonance.

A second alternative is to again use resonant electronics, but with a crystal as opposed to an inductor [37]. Crystals exhibit inductive behavior, although they do so only over a limited frequency range. The advantage to using a crystal over an inductor is the potential for much smaller size, and much lower loss. Preliminary experiments with crystals indicate that they can tolerate acceptably high-voltage operation.

A third alternative for providing a high-voltage and high-frequency excitation is to make use of multi-level power electronics as described in [50]. In [50], a multiphase inverter is implemented with a collection of identical single-phase inverters, one inverter per machine phase. Each single-phase inverter is in turn constructed from several identical full-bridge inverters whose outputs are transformer coupled and connected in series to produce a high net output voltage. By timing the switching of the individual inverters that constitute a single phase, the net output can be made nearly sinusoidal over a wide frequency range so that electrical breakdown between neighboring electrodes is not a limiting phenomenon. The advantages of such an inverter are again the low-voltage operation of the individual inverters that constitute a complete phase inverter. Additionally, the multi-level inverter is non-resonant, and so need not be tuned. Its primary disadvantages are the presence and size of the output transformers, and increased hardware complexity.

Finally, a fourth alternative for providing a high-voltage and high-frequency excitation is the use of piezoelectric transformers [51]. These transformers can also be used in resonant circuits to provided very-high-voltage sinusoidal outputs at very-high frequency. To date, however, they have not been pursued extensively in the context of inverters for MEMS electromagnetic machines.

The control of an electric induction machine is identical in spirit to that of a magnetic induction machine [3, 4] since the two are electromagnetic duals. As a motor, the electric induction machine may be successfully run open loop with a fixed electrical excitation, although care must be taken to anticipate the peak load so that the motor is capable of producing the necessary torque over the range of $0 \leq \gamma_1 \leq 2$. This is the manner in which most magnetic induction machines are operated today. Since it is an asynchronous machine, the electric induction machine will produce starting torque at zero speed, for which $S_m = 1$, even if it is excited at a high electrical frequency. Alternatively, a more complex control can be implemented with a controller that measures speed and adjusts the electrical excitation frequency ω so that the output torque follows the positive peak of Fig. 8.11. Yet more complex torque and speed controllers are discussed in [3, 4] in the context of magnetic machines. As a generator, the electric induction machine may also be run successfully open loop with an electrical frequency that yields negative slip, namely such that $P\Omega > \omega$. Again, more complex control is possible, such as following the negative peak in Fig. 8.11. Interestingly, the electric induction generator with resonant power electronics is capable of self excitation as discussed in Section 8.3.5, and in [37–39].

8.3.4 Special Fabrication Issues

To obtain high power throughput from an electric induction machine, the machine must run at very high speeds, perhaps hundreds of thousands to millions of revolutions per minute. In order to support such high-speed operation, air thrust and journal bearings of the type described in Chapter 6 are required. For this reason, a MEMS electric induction machine will look very much like one of the bearing test devices described in Chapter 6. The only difference will be the inclusion of stator electrodes and interconnects, and their supporting insulator, on one side of the stator, and the inclusion of a rotor conductor and its supporting insulator on the facing side of the rotor. Further, if the electric induction machine is to be integrated into a motor/compressor or a turbine/generator, then it will also look much like the turbomachinery described in Chapter 7. Thus, in order to understand the fabrication of the electric induction machine, it is first necessary to understand the fabrication of the bearing test devices and turbomachinery [52, 53]. The focus of this subsection is fabrication issues that are important to the electric induction machine, yet outside bearing test devices, turbomachinery and the main stream of Chapter 2. Additional fabrication details for specific electric induction machines may be found in [30, 31, 37, 34, 35, 38, 36, 39], which together describe the fabrication and operation of five separate machines.

8.3.4.1 Thick Rotor and Stator Insulators

As discussed in Section 8.3.2 it is desireable to fabricate the rotor conductor and stator electrodes on top of very thick insulators. That is, it is desirable to fabricate electric induction machines with large Δ_R and Δ_S. While there are diminishing returns beyond some thickness, it is likely that insulators as thick as 20 μm or more are desirable. Silicon dioxide is the insulator considered here, and such thicknesses are well beyond those that can be obtained with thermally grown silicon dioxide. Therefore, deposited silicon dioxide, primarily densified PECVD and PECVD TEOS silicon dioxide, is used in all electric induction machines discussed in this subsection.

Because the air gaps of the electric induction machine and its thrust bearings are likely to be on the order of several microns, rotor and stator planarity is critical. Consequently, the stress internal to any thick layer of deposited silicon dioxide must be properly managed, since this stress can lead to significant wafer bow. In the electric induction machines discussed here, the stress is managed primarily by limiting the deposited insulators to those regions of the rotor and stator that are directly beneath the rotor and stator conductors. Further, to maintain planarity, the insulators are embedded in their silicon substrates so that their top surfaces are nearly flush with that of the silicon. These embedded silicon-dioxide islands are the annular region beneath the rotor conductor, and the bow-tie-shaped region beneath the stator electrodes and wiring, shown in Fig. 8.9. Alternatively [31, 36], investigate the use of a quartz stator substrate to eliminate the conducting silicon substrate, and hence the associated parasitic capacitance, altogether. This also eliminates the need for an additional thick insulator.

While their fabrication flows are similar, no two electric induction machines discussed in this subsection employ the same fabrication flows to create embedded silicon-dioxide islands on the rotor or the stator. Nonetheless, there are significant similarities worthy of note. First, all flows use the same two essential steps to create an embedded silicon-dioxide island. These steps are the dry etch of deep recesses in the silicon substrates, typically 10 μm for the rotor and 20 μm for the stator, and the subsequent deposition of silicon dioxide so that the top surface of the silicon dioxide in the recess is flush with the silicon substrate, or slightly recessed. Second, all flows either remove or substantially thin the silicon dioxide in the field regions outside the embedded islands so as to limit wafer bow. There are several ways to accomplish this. The fabrication of the early machines described in [30, 31, 34, 36] uses dry or wet etching, and/or chemical–mechanical polishing (CMP) following a fabrication flow first described in [54]. Subsequent machine fabrication described in [37, 35, 38, 39] uses lift-off to simplify the fabrication as first described in [35]. The lift-off of silicon dioxide is carried out using a 1-μm layer of sacrificial aluminum beneath the thick silicon dioxide. Following a lengthy wet sacrificial etch in HCl, the silicon dioxide breaks off cleanly at the edges of the islands leaving behind micron-scale stubs that can be removed with light CMP. The resulting wafer bow is insignificant with respect to an air gap of several microns. Third, all fabrication flows integrate the creation of the embedded silicon-dioxide islands with the preparation of the rotor and stator wafers for fusion bonding. This often involves repeated CMP and silicon dioxide deposition until a suitable bonding surface is obtained.

Upon completed fabrication, the axial gaps between the rotor and the stator will likely be very small across the radial span of the thrust bearings, stator electrodes, and rotor conductor. The bearings gaps and the electrode-to-conductor gap are typically on the order of several microns on both sides of the rotor; as discussed in

Chapter 7, the gaps across any turbomachinery can be larger. The fabrication of the narrowest gaps is best integrated with the fabrication of the thick stator and rotor insulators, since the insulators can either be recessed or used as shims to create the narrow gaps with good control. While the various electric induction machines described in this chapter manage this integration differently, they all nonetheless integrate the fabrication of the narrow gaps and the thick insulators.

8.3.4.2 Stator Electrodes and Interconnects

The fabrication of the stator electrodes is complicated by at least four factors: (1) the need for two levels of conductors to accommodate the electrodes and their interconnect rings, (2) the need for good electrical conductivity, (3) concerns for electrical breakdown, and (4) thermal exposure during subsequent fabrication, particularly during fusion wafer bonding. Good electrical conductivity is required to minimize losses. This is particularly important for generators since their stator losses come after the limiting air gap energy conversion; see Section 8.1.1. It is also important to limit electric breakdown, which ultimately limits the air-gap energy conversion in any electric induction machine. Electrical breakdown tends to initiate at sharp corners. The sharpest corners in the machine are probably at the edges of the stator electrodes. Electric breakdown will also initiate at particles, and so cleanliness is extremely important during the fabrication of an electric induction machine.

In the electric induction machines described in this chapter, the stator electrodes are connected through back-side interconnect rings, as can be seen in Fig. 8.9. This requires two levels of conductors that are separately patterned and connected through vertical vias. The overlap between electrodes and the interconnect rings from other phases leads to a parasitic capacitance between phases that is unbalanced. The imbalance arises from the electrode geometry that widens with increasing radius. The parasitic capacitance leads to reactive currents that cause loss. The imbalance also leads to undesired potential space harmonics that further decrease performance. Both effects are particularly important during generation. To minimize the parasitic capacitance, it is desireable to use a thick insulator between the two conducting layers. However, this makes via fabrication difficult given the typically narrow electrodes and interconnect rings. Separating the conductors with approximately 1 μm of TEOS silicon dioxide has been found to a good compromise.

The early stator electrodes and interconnect rings for motoring applications described in [30, 31, 34, 36] are fabricated from highly doped polysilicon because that material easily withstands the high temperatures associated with fusion wafer bonding. Even highly doped polysilicon, however, is too lossy for generating applications. It can also be difficult to pattern with smooth edges using dry etching, and so the electrodes may break down at low voltages. Subsequent machines described in [31, 36], and other machines described in [37, 35, 38, 39] use stator electrodes fabricated from 0.3-μm-thick platinum using a 10-nm titanium adhesion layer and lift-off patterning as detailed in [35]. Platinum is selected because it exhibits good electrical conductivity, it has good chemical and thermal stability, and it can lift off smoothly leaving electrode edges that do not exhibit premature electrical breakdown. In particular, the platinum electrodes can withstand all necessary chemical cleans and wafer bonding anneals up to 800°C that came after their deposition and patterning [35]. Fusion wafer bonding at temperatures as low as 700°C is described in Chapter 2, and in [37–39].

8.3.4.3 Rotor Surface Conductor

The primary fabrication challenge posed by the rotor surface conductor is its typically high surface resistivity. In all electric induction machines described here, the target surface resistivity is 200 $\sqrt{\text{M}\Omega}$. Correspondingly, the target surface conductivity σ_R is 5 nS. To achieve this electrical behavior, the rotor conductor is fabricated as a thin layer of course-grain polysilicon, moderately doped with boron [31]. The rotor conductor thickness is designed to be 0.5 μm in order to be much thinner than the air gap, and boron is chosen as the dopant because it does not segregate at grain boundaries. The dopant dose is 5×10^{12} cm^{-2}, leading to a dopant density of 10^{17} cm^{-3}. With appropriate anneals, this leads to the desired bulk resistivity of 10^4 Ω·cm. However, at this doping level, the resistivity of polysilicon is highly dependent on the anneal history, dopant concentration and temperature [49], and so it is difficult to control during fabrication and operation. Therefore, considerable variation in the rotor surface conductivity is observed in the experiments reported here.

In addition to presenting fabrication and operating challenges, lightly to moderately doped polysilicon presents the possibility that its conductors can be significantly depleted at high air-gap field strengths. In this case, charge depletion will effectively increase the air gap, and so this phenomenon must be considered during design. To begin, the rotor surface charge density can be computed using (8.18). This charge density and the dopant concentration can then be used to compute the depth of the depletion layer. Rotor conductor depletion effects have been observed during generating experiments in [37–39].

There is perhaps a better approach to manufacturing a highly resistive rotor conductor that has not yet been explored. It may be possible to use highly conducting rotor electrodes that are connected to their neighbors by very thin bridges. In this case, the high apparent resistivity is achieved through the narrow geometry of the bridges, allowing the resistivity of the material to be much lower. If this option is pursued, then the number of rotor electrodes and the number of stator electrodes should probably be mutually prime to prevent cogging.

8.3.5 Experiments

Between them, [30, 31, 37, 34, 35, 38, 36, 39] describe the experimental operation of five separate electric induction machines. Motoring experiments are described in [30, 31, 34–36] and generating experiments are described in [37–39]. All machines have the same number of stator phases and electrodes, and approximately the same geometry. They differ primarily in their fabrication details, and the manner in which the experiments were conducted. For example, the motors described in [31, 36] were tethered. That is, their rotors were suspended by springs so as to eliminate the need for, and the effects of, air bearings. This made torque easier to measure. The focus of the motor described in [30, 34] was integration into a motor/compressor application, while the focus of the motor described in [35] was higher power operation. The focus of [37–39] was generation. This subsection focuses on the experiments described in [37, 35, 38, 39].

8.3.5.1 Motoring Experiments

The parameters for the electric induction motor described in [35] are given in Table 8.4. The air gap G was smaller than the 3 μm for which it was designed due largely to the operation of the air bearings. The rotor surface conductivity σ_R was smaller than the 3 nS for which it was designed due largely to the omission of several high-temperature bonding anneals that would have promoted grain growth in the polysilicon rotor conductor.

Table 8.4 Parameters for the electric induction motor described in [35]

Parameter	Value	Parameter	Value	Parameter	Value
N	6	r_{IN}	1.0 mm	Δ_R	10 μm
P	131	r_{OUT}	1.9 mm	Δ_S	20 μm
G	1.8 μm	σ_R	1.25 nS	ϵ	3.9 ϵ_o

Data from two motoring experiments are presented in [35]. In both experiments, the motoring torque was balanced only by the viscous losses in the air gaps. This permitted the torque to be measured via motor speed following a spin-down calibration experiment, also described in [35], and in [34]. In the first experiment, a balanced six-phase 45-V square-wave excitation was applied to the stator at electrical frequencies ranging from 50 to 600 kHz. The motor speed was recorded and converted to torque. The data from this experiment are reported in Fig. 7 of [35]. In the second experiment, a balanced six-phase 200-kHz square-wave excitation was applied to the stator with voltage amplitudes ranging from 25 to 95 V. Again the motor speed was recorded and converted to torque. The data from this experiment are reported in Fig. 8 of [35]. The data from both experiments are shown here in Fig. 8.13, although in a slightly different format. In Fig. 8.13, measured torque is normalized by the division of V_1^2, and it is plotted against the electrical slip frequency $S_1\omega$ of the fundamental. In this way, the data from both experiments can be shown in the same graph. Note that in the process, V_1 is multiplied by 0.92 to account for the spacing between the electrodes, which reduces the amplitude of the potential space harmonics.

Figure 8.13 also includes a graph of normalized theoretical torque, based on the fundamental alone as computed numerically from (8.23) using the parameters from Table 8.4. The fit between the data and the theory is good, but a better fit is

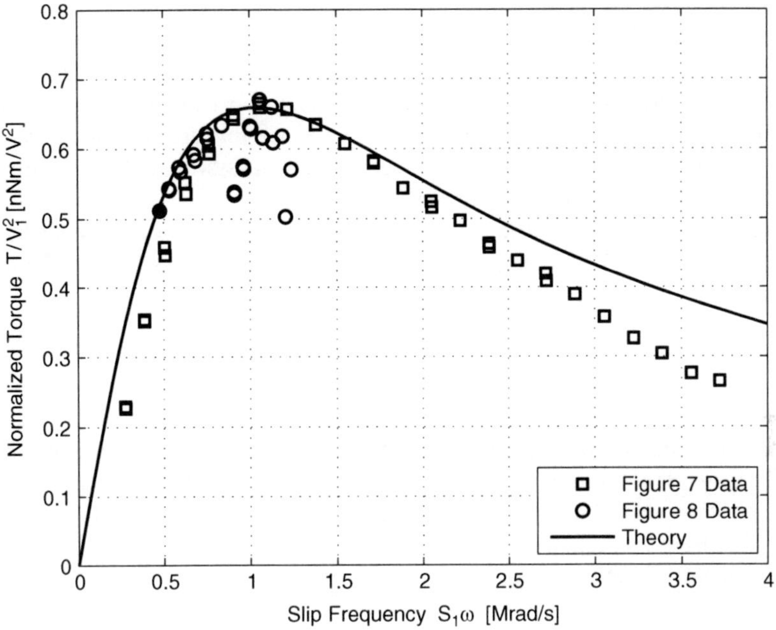

Fig. 8.13 A graph of the normalized torque T/V_1^2 versus the electrical slip frequency $S_1\omega$; the data is taken from [35]

obtained in [35] using additional space and time harmonics. The time harmonics, which are not represented in (8.4), arises from the square-wave excitation of the stator electrodes.

The greatest torque and power output from the electric induction motor was achieved during the second experiment at 95 V. The data corresponding to this experiment is identified by an solid circle in Fig. 8.13 near $S_1\omega = 0.5$ MRad/s. During this experiment, the measured torque was 3.51 μNm, the measured speed was 56.7 krpm, and the measured mechanical power was 20.8 mW. The efficiency, as determined by (8.36) was 62%. As measured, it was 61% with the additional loss coming as conduction loss in the stator wiring. In separate experiments, the stator electrodes were tested for electrical breakdown. As described in [35], the electrodes were capable of withstanding a 300-V excitation. Scaling the peak torque to full voltage would then increase it by a factor of $(300/95)^2$ to 35 μNm. At fixed speed, the power would scale up by the same factor to 0.207 W, with only 10% of that power being consumed by viscous losses in the air gap and bearings.

One interesting metric for comparison purposes is power density, computed on the basis of electromagnetically active materials. The electromagnetically active materials in the electric induction machine include the rotor and stator conductors and insulators, and the air gap. The axial length of these materials is approximately 34 μm after including the thickness of the conductors, and the insulator between the stator electrodes and the interconnect rings. The active volume of the electric induction motor is therefore 3.9×10^{-10} m^3, which yields a power density of 54 MW/m^3 at the peak power point. If this power density is scaled up to full voltage, then it becomes 540 MW/m^3. Both are greater than the power density of a 1-GW utility generator, which approaches 20 MW/m^3. Such generators are designed to run with a peripheral tip speed near 200 m/s, a mechanically stress-limited rotor, and maximum magnetic shear stress through forced cooling.

Another interesting metric for comparison purposes is electromagnetic shear stress. If f_z in (8.23) is replaced with the average shear stress $\bar{f}_z$, then the torque is given by

$$T = \frac{2\pi}{3}\bar{f}_z(r_{\mathrm{OUT}}^3 - r_{\mathrm{IN}}^3) \quad , \tag{8.40}$$

from which it may be concluded that $\bar{f}_z = 286$ N/m^2 at the maximum power point. If the demonstrated average stress is scaled up to full voltage by the factor of $(300/95)^2$, then the average shear stress would be 2850 N/m^2. This is not far from the shear stress in an ordinary integral-horsepower magnetic induction motor, approximately 7 kN/m^2 [17], but it is quite far from the shear stress in a 1-GW utility generator, which approaches 180 kN/m^2.

8.3.5.2 Generating Experiments

Producing net electrical power with a microscale electric induction generator is much more difficult than producing net mechanical power with a microscale electric induction motor. The primary reason for this is the electrical losses in the machine system, which are downstream of the air gap during generation; see Fig. 8.6. These losses occur in the stator wiring and in the power electronics; the inductors in Fig. 8.12 are particularly lossy. Nonetheless, successful generation with net power output has been achieved as described in [37–39].

Eliminating electrical losses from the machine system is critical to the production of net electrical power with a microscale electric induction generator. This requires attention to many details, including: (1) reducing the resistance of the electrodes and the stator wiring; (2) eliminating parasitic capacitance inside and outside the generator; (3) maintaining a balanced sinusoidal excitation; and (4) using low-loss inductors and transistors. Each of these issues is discussed further below. In this discussion, it is assumed that resonant power electronics like those shown in Fig. 8.12 are employed. Hard-switching power electronics like those shown in Fig. 8.4 will be too lossy.

To reduce losses, all stator electrodes and wiring must be fabricated with metal. Even heavily doped polysilicon is too lossy to permit net generation. Metal electrodes, however, lead to subsequent fabrication challenges, the most noteworthy of which is wafer bonding at temperatures low enough to preserve the mechanical and electrical integrity of the metal conductors. See [37, 35] and Chapter 2 for fabrication details.

To the extent possible, all parasitic capacitance must be eliminated. This includes parasitic capacitance inside and outside the generator. Such capacitance leads to reactive currents that produce loss in the stator electrodes and wiring, and more importantly in the power electronics. Thick insulators help eliminate parasitic capacitance inside the generator, and attention to the layout of the power electronics helps eliminate parasitic capacitance outside the generator. Inside the electric induction generator, it is the capacitance between the stator electrodes and the stator substrate, and the capacitance between the stator electrodes and the interconnect rings of other phases, that make up the dominant parasitic capacitance. Outside the generator, even oscilloscope probes at the generator terminals can present excessive capacitance.

Also to the extent possible, the stator excitation should be sinusoidal and balanced as it appears at the electrodes. A non-sinusoidal excitation excites potential time harmonics inside the electric induction generator, while an unbalanced excitation excites potential space harmonics. The reactive currents associated with these harmonics will almost surely create more loss than the harmonics can generate. Further, some of the harmonics will likely motor, in which case they could consume all generated power. The dominant source of electrical imbalance is physical imbalance in the parasitic capacitance inside and outside the machine, and differences between inductors. Since these physical imbalances cannot be avoided, it is necessary to tune the power electronics so that the excitation presented to the stator electrodes is balanced, despite the imbalance outside. Time harmonics can be eliminated by using sinusoidal sources in Fig. 8.12.

At least two convenient adjustments can be made to tune the power electronics so as to produce a balanced stator electrode excitation. Both are employed in [37–39]. First, the voltage amplitudes and phase shifts of the sources driving the generator phases can be adjusted relative to that of one source chosen as the reference. Second, tuning capacitors can be placed from the external terminal of each generator phase to neutral, and these capacitors can be adjusted to help balance the parasitic capacitance. The presence of the tuning capacitors will increase the reactive currents, but their ability to eliminate imbalance is more important [37]. Executing this tuning procedure for a six-phase electric induction machine results in the tuning of 18 variables: five voltage amplitudes, five voltage phase shifts, six tuning capacitors, the fundamental electrical frequency, and the speed of the generator. Such tuning is too complex to be performed in an ad hoc manner. Consequently, a model-based approach to tuning is employed in [37–39]. This approach follows three basic steps. The first step is to develop a model of the parasitic capacitances and the inductors. The second step is to perform a set of experiments to determine the parameters in the model. The third step is to use the model to determine how to tune the power electronics, and the operation of the generator, so as to maximize the electrical output power. This approach proved to be essential to successful generation.

Finally, it is important to make use of very-low-loss inductors and transistors in the power electronics since it is in these components that most of the loss actually takes place. If voltage step-up ratios of 20 or more can be obtained, then the losses in the transistors can be made small. However, low-loss inductors that can operate at high voltage and high frequency are considerably more challenging, and will almost certainly be bigger than the generator.

Following the approach outlined above, the electric induction generator has been successfully run to generate net electrical power. This is reported in [37–39], with [37] providing the most detail. The generator was identical to that described by Table 8.4, except for its air gap of 4.2 μm and rotor surface conductivity of 0.75 nS. With a speed of 250 krpm, an electrical frequency of 403 kHz, and sinusoidal source voltage amplitudes of 0.8 V, the electric induction generator produced 108 μW while very good correlation with the tuning model was observed. The tuning model further estimated that the efficiency of the electric machine system, from the input mechanical power to the electrical power delivered to the phase sources, was 14%; the associated power flow is tabulated in Table 8.5. However, the overall efficiency from input air power to output electrical power was estimated to be 0.036%, with turbine losses of 219 mW, air-gap viscous losses of 54 mW, and viscous bearing losses of 26.7 mW being dominantly responsible for the low overall efficiency.

Table 8.5 Electromechanical power flow in the electric induction generator described in [37–39]. The electrical output power is the total power delivered to the sources in Fig. 8.12

Phenomenon	Power [μW]
Rotor conductor loss	255
Stator wiring loss	75
Inductor loss	280
Dielectric loss	50
Proximity loss	22
Electrical output power	108
Mechanical input power	790

Experiments were also run at higher voltages in an attempt to generate greater electrical output power. However, these experiments actually yielded a lower output power due a nonlinearity in the rotor surface conductor. As explained in [37], the nonlinearity was caused by charge depletion in the rotor surface conductor.

Finally, it is also interesting to observe that the electric induction generator is capable of self excitation. To arrange for this, the sources in Fig. 8.12 are simply removed and replaced with a load. If the generator speed and the inductor inductance are properly chosen, the generator will self excite and deliver power to the load. Self excitation can be understood qualitatively from a circuit's viewpoint. From this viewpoint, each phase of the electric induction machine behaves as a (frequency dependent) capacitor in parallel with a (frequency dependent) resistor between the phase terminal and neutral, assuming that only the potential fundamental is active. When the electric induction machine operates as a generator, each phase resistor exhibits a negative resistance. If this circuit model is placed in parallel with an inductor, as is done for each phase in Fig. 8.12, and the inductor resonates with the phase capacitance at a frequency that excites the machine into generation, then the system becomes unstable due to the negative effective resistance, and the electric generator will self excite. This mode of operation is also demonstrated in [37–39]. In the experiments reported therein, the load connected to each phase was a pair of anti-parallel light-emitting diodes (LEDs).

The same model used to guide driven generation was used to guide self-excited generation in [37–39]. Using the model, the external capacitors were tuned, the inductor inductances were chosen, and the generator speed was set to maximize the self-excited electrical output power. This yielded a maximum output power of approximately 140 μW delivered to the LED load when the turbine/generator was run at 255 krpm. In that experiment, the phase terminal voltages grew to an amplitude of 14.5 V before being limited by the nonlinearity of the rotor surface conductor. Figure 8.14 shows a photograph of the self-excited electric induction generator driving the LED load. The turbine/generator is at the center of the top of the figure. Were it not for the nonlinear rotor surface conductor, this experiment should have produced a net electrical output power of almost 3 mW if the phase voltage amplitudes were to grow to 300 V.

Fig. 8.14 A photograph of a PowerMEMS electric induction turbine/generator operating under self-excitation. The photograph is taken from [37]

8.3.6 Summary

The focus of this section has been the electric induction machine, specifically its analysis, design, microfabrication, operation, and experimental performance. To begin, an analytical model of the machine is presented in Section 8.3.1. While this model is quite detailed, it is based on several limiting assumptions. First, the stator excitation is limited to a single time harmonic. Such excitation is desireable, but not required. In more complex cases, the model can be extended in a straight forward manner to involve a Fourier expansion having multiple time harmonics [31]. Second, for simplicity and transparency, the model is built on an analysis in Cartesian geometry. In cases where this is inappropriate, as described in Section 8.2, a Fourier–Bessel analysis can be employed. Third, the model ignores the spacing between the electrodes. If the spacing is wide, presumably to accomodate fabrication limitations, then a more complex electric potential solution is required. Finally, the model is concerned only with the core of the electric induction machine. Detailed models of the wiring between the core and the bonding pads, including parasitic resistance and capacitance, are developed in [31].

Section 8.3.2 develops a simplified model, and uses this model to design the electric induction machine within the limitation of MEMS fabrication. This design effort exposes several design trends. In an effort to maximize power throughput, a Power-MEMS electric induction machine will likely be designed with six phases, a narrow air gap, many hundreds of electrodes, and thick rotor and stator insulators. It will likely operate with phase voltages having amplitudes near 300 V and frequencies on the order of 10^5–10^6 Hz.

Section 8.3.3 discusses some of the power electronics that have been, or could be, developed for the electric induction machine. In order to avoid dominant switching losses and large power electronics, resonant power electronics are probably preferred. Resonant power electronics have so far been implemented with inductors. However, similar power electronics implemented with either crystals or piezoelectric transformers should also be explored.

Section 8.3.4 examines three aspects of the electric induction machine that require special attention during microfabrication. These are the stator electrodes, the rotor surface conductor, and the rotor and stator insulators. The stator electrodes require special attention because they should be metal for good conductivity, and smooth to avoid electrical breakdown. The presence of metal electrodes, however, complicates downstream fabrication since wafer–wafer bonding at elevated temperatures is desired. The rotor surface conductor requires special attention because it will likely be designed to be highly resistive. Further, attention should be paid to rotor conductor depletion to permit high-power operation. The stator insulators require special attention because they must be thick on an MEMS scale, and so their internal stress must be managed to preserve the planarity of the rotor and stator at the air gap.

Finally, Section 8.3.5 presents experiments involving both electric induction motors and generators. References to many more experiments are also given. The motoring experiments in particular show a very good match to the analysis provided in Section 8.3.1. Further, the experiments indicate that the electric induction motor, if run at full voltage, can produce a motive shear stress nearing that of ordinary macroscale magnetic machines. Further, Watt-level power output from a machine having a radius of 2–3 mm certainly appears possible. Even without a full-voltage stator excitation, the motor described in Section 8.3.5 achieves a power density exceeding that of most magnetic machines when computed from the volume of its electromagnetically active materials.

Generating is much more difficult than motoring due to the many electrical losses that occur downstream of the air gap. Indeed, the management of these losses is critical to successful generation, as explained in Section 8.3.5. With the losses under control, net generation is possible through both driven and self-excited operation. Such experiments are reported in Section 8.3.5.

8.4 Permanent-Magnet Synchronous Machine

Unlike the electric induction machines described in Section 8.3, the specific permanent-magnet synchronous machines described in this section have not been fully integrated into a silicon housing with self-contained air bearings. In order to achieve this integration, several challenges must be met. These challenges include: (1) the integrated fabrication of deep laminations in a silicon substrate, (2) the fabrication of multi-level three-phase windings on the stator surface, (3) the design and operation of bearings that can support a heavy magnetic rotor, and (4) the design and fabrication of a rotor that can hold together at speeds approaching 300 krpm to 400 krpm. While not described in this section, many to most of these challenges have now been met, as described in [55, 56].

The permanent-magnet synchronous machine (PMSM) consists of two-key components, a rotor with a multi-poled permanent magnet, and a stator with appropriately designed windings. During operation the rotating multi-poled permanent-magnet rotor creates a time-varying magnetic flux in the rotor–stator air gap. This flux induces AC voltages at the terminals of the stator windings, which forms the basis for power generation. The PMSMs considered in this section use an annular permanent magnet mounted on a ferromagnetic core. Interior PMSMs are not considered here due to the difficulty of embedding magnets into an electroplated ferromagnetic core. With the use of suitably thick magnets (500 μm) and high-performance magnetic materials in the rotor and stator core, the air gap can be made very large (300–500 μm) without appreciable loss of performance. This allows the stator windings to occupy space in the air gap, fabricated on the surface of a flat stator core as shown in Fig. 8.15. PMSM fabrication is thus greatly simplified. A fabricated stator is shown in Fig. 8.16.

Microscale magnetic machines can match the linear tip speeds of their macroscale counterparts by spinning much faster, reaching speeds of 100–1000 krpm. This, combined with the higher current densities allowed in the windings, enables microscale machines with very high-power densities. Microscale magnetic machines operate at electrical frequencies one to two orders of magnitude higher than typical macroscale machines, typically in the kHz to tens of kHz range, due to their higher rotational speeds. These high-operating electrical frequencies lead to smaller passive filtering components in their attendant switch-mode power electronics. The small size of microscale machines also leads to low stator inductances. This permits both generator and motor drives with fast transient responses, but also requires very high switching frequencies (hundreds of kHz to low MHz) in their power electronics.

Prior work has focused on planar MEMS PMSMs. Several axial-air-gap motors have been demonstrated [57–61], and speeds of up to 200 krpm with torques of up to 7.5 μNm have been achieved [59, 60]. Planar three-phase motors using magnetic thrust bearings have also been demonstrated [60]. Generators too have been demonstrated. One was a turbine/generator using NdFeB permanent magnets that produced 1.1 mW at 30 krpm with an air flow of 35 L/min [62]. The composite turbine/generator rotor was supported by conventional ball-race bearings, and

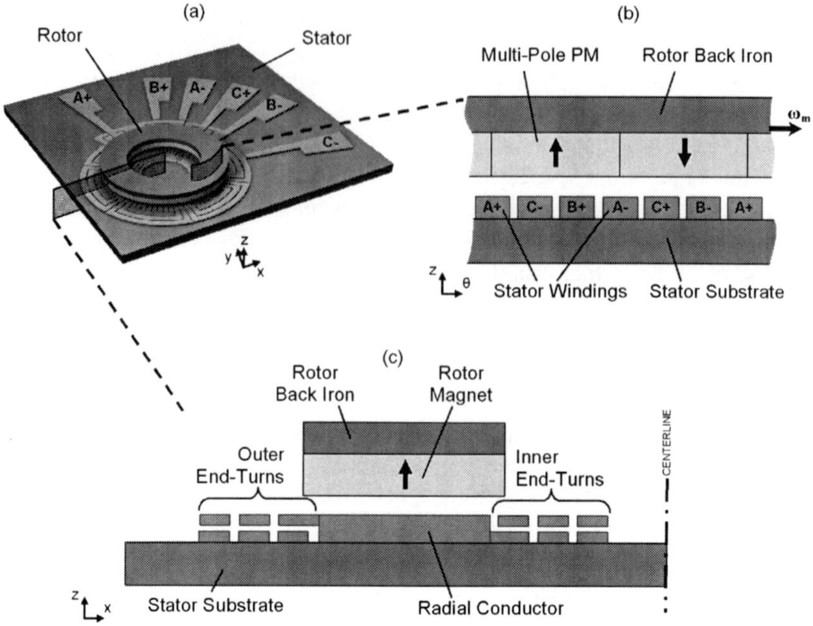

Fig. 8.15 Conceptual drawing of the **PMSM**: (**a**) perspective view, (**b**) angular cross section, and (**c**) radial cross section (Adapted from [64])

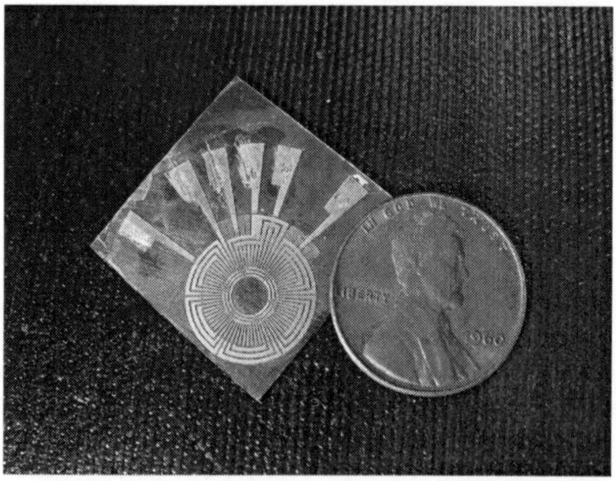

Fig. 8.16 Fabricated 3-turn/pole concentric wound stator is shown along side a penny for size comparison (Adapted from [64])

measured 1 mm in thickness and 6 mm in diameter. Another demonstrated high-speed rotation (260 krpm) by using a combination of air and magnetic bearings [63]. Using an 8-mm diameter $SmCo_5$ rotor, 14.6 mW of power was delivered to a wye-connected resistive load at 58 krpm. Using an air-powered drill and 8-mm-diameter

NdFeB permanent magnets, the same generator delivered 5 W into a wye-connected resistive load at 380 krpm.

The previous work on microscale PMSMs for power generation focused solely on the generator. This section focuses on the complete generating system, starting from the electric generator that performs the electromechanical energy conversion to the power electronics that provide usable power for portable applications. This section presents the modeling, design, and evaluation of high-speed (300+ krpm) microscale PM machines and power electronics capable of generating and delivering multi-Watt electric power to a load at a regulated DC output voltage. Unlike the electric induction machines described in Section 8.3, the PMSMs described in this section have only recently been fully integrated into a silicon housing with self-contained air bearings [55, 56]. In order to achieve this integration, several challenges were met beyond those described here. The additional challenges include: (1) the integrated fabrication of deep laminations in a silicon substrate, (2) the fabrication of multi-level three-phase windings on the stator surface, (3) the design and operation of bearings that can support a heavy magnetic rotor, and (4) the design and fabrication of a rotor that can hold together at speeds approaching 300–400 krpm.

8.4.1 Modeling

This section develops models for the planar surface-wound PMSM with laminated and unlaminated stators. The development closely follows that presented in [64, 65]. The analytic models developed here permit the quick determination of machine performance as a function of speed, physical dimensions, material properties, and input currents. The planar surface-wound PMSM is modeled by solving Maxwell's Equations in their magneto-quasistatic (MQS) form [29]. Correspondingly, the magnetic field intensity H, and magnetic flux density B, in the machine driven by the rotor magnet and stator currents are determined as functions of radius, and then integrated over the radial span of the machine to determine machine parameters such as open-circuit voltage, inductance, and torque. The field solutions are solved in the spatio-temporal frequency domain, which permits modeling of complex winding patterns and rotor magnetization profiles by decomposing them into their respective Fourier components. Performance parameters can then be determined on a harmonic-by-harmonic basis.

The continuum model presented here can be used to determine how thin the rotor and stator back iron, or cores, can be without becoming heavily saturated. It can also be used to determine hysteresis and eddy-current losses in both cores in order to predict machine efficiency as a function of speed, geometry, and material properties. The model supports these tasks by utilizing the nonlinear B–H relationship of the core materials. An iterative algorithm is first used to solve for the fields and effective permeabilities within the rotor and stator cores. Second, the field theory solutions are used to determine hysteresis losses from the nonlinear core B–H curve. Finally, the Maxwell Stress Tensor is used to find the eddy-current losses without the need to calculate current densities or evaluate volume integrals.

The field solutions are used to compute the inductance and open-circuit voltage for a single phase of the machine. Together with a simple model for the single-phase winding resistance, these parameters are summarized through an equivalent-circuit model of the PMSM. The proximity eddy-current losses in the stator windings can also be computed using the field solutions from the continuum model.

Because it contains no complicated axial structures such as poles or slots, the surface-wound PMSM can be modeled through a set of planar continuum layers as shown in Fig. 8.17. The use of two-dimensional Cartesian coordinates follows the discussion of Section 8.2. Note that Fig. 8.17 contains two parallel coordinate axes. The x-y-z axis is fixed to the stationary stator, while the x′-y′-z′ axis is fixed to the moving rotor which travels with velocity U_y with respect to the stator.

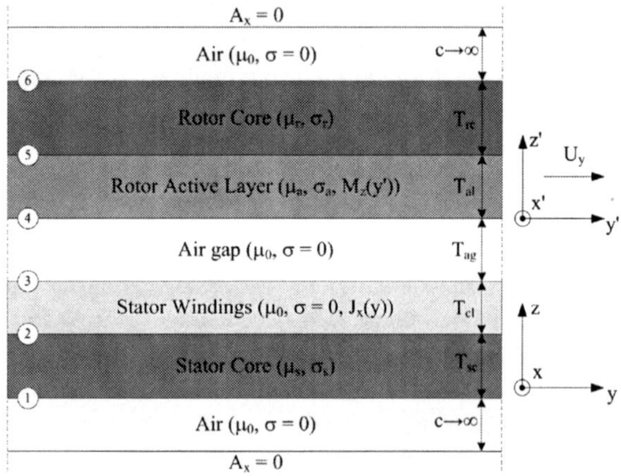

Fig. 8.17 The surface-wound PMSM is modeled as a series of planar continuum layers. The interfaces between the layers are designated 1–6. The rotor layers move in the y=y' direction with velocity U_y (Adapted from [64])

The rotor–stator air gap, as well as the space above the rotor and below the stator, are modeled as layers with permeability μ_0 and conductivity $\sigma = 0$. The rotor and stator cores are modeled as layers with permeabilities μ_r and μ_s, and conductivities σ_r and σ_s, respectively. For the purposes of modeling magnetic fields, the stator windings are modeled as a layer with zero conductivity and permeability μ_0, with an imposed current density distribution $J_x(y,t)$. The effects of proximity eddy currents within the coil layer are modeled separately. The active rotor layer has a spatially varying magnetization in the axial direction, $M_z(y')$, as well as a permeability μ_a and conductivity σ_a. It can be used to model a permanent magnet such as SmCo or NdFeB. The active layer can also model the rotor conductor in an induction machine by setting $M_z(y')$ to zero.

A two-dimensional Cartesian-coordinate convection-diffusion equation for the magnetic vector potential A, derived from the MQS Maxwell's Equations, can be solved to find the H and B fields at the radius r produced by the rotor active layer and the stator windings [29]. Each planar layer moving uniformly with velocity v_y, will have a vector potential with only one component, A_x. If no magnetization or independent current sources exist within the layer then the convection-diffusion equation for A_x is

$$\nabla^2 A_x = \mu\sigma\left(\frac{\partial A_x}{\partial t} + v_y\frac{\partial A_x}{\partial y}\right). \tag{8.41}$$

The normal B field and tangential H field can be computed from the vector potential according to

$$H_y = \frac{1}{\mu}\frac{\partial A_x}{\partial z}, \tag{8.42}$$

$$B_z = -\frac{\partial A_x}{\partial y}. \tag{8.43}$$

Equation (8.41) governs the behavior of the vector potential in the rotor core, the stator core, the air gap, and the regions above the rotor and below the stator. In each region, μ, σ and v_y may be different; in particular, v_y is U_y for the rotor core, and zero otherwise. Equation (8.41) also governs the vector potential in an open-circuited stator-winding layer in the PMSM. For the active rotor layer with magnetization $M_z(y')$, moving uniformly with velocity $v_y = U_y$ the convection-diffusion equation for A_x, derived in [64], is

$$\nabla^2 A_x + \mu_a\frac{\partial M_z}{\partial y} = \mu_a\sigma_a\left(\frac{\partial A_x}{\partial t} + U_y\frac{\partial A_x}{\partial y}\right). \tag{8.44}$$

The current density in the stator winding layer, J_x, will be a function of y and t. For such a planar layer with current density distribution J_x, the governing equation for the vector potential is [66]

$$\nabla^2 A_x = -\mu_0 J_x. \tag{8.45}$$

The vector potential in each layer of the machine will have the form

$$A_x(y,z,t) = \mathrm{Re}\left\{\sum_{n=-\infty}^{\infty} \hat{A}_{xn}(z)\, e^{j(m\omega t - nky)}\right\}, \tag{8.46}$$

where m indexes the time harmonics, n indexes the space harmonics, the lowest-order time frequency is ω, the lowest-order spatial wave number k is P/r, and P is the number of pole pairs. Note that the lowest-order wavelength λ of the magnetic periodicity is $2\pi/k$. For each spatial harmonic, the complex amplitudes of the normal B field and tangential H field will be

$$\hat{B}_{zn} = jnk\hat{A}_n, \tag{8.47}$$

$$\hat{H}_{yn} = \frac{1}{\mu}\frac{d\hat{A}_n}{dz}, \tag{8.48}$$

where

$$B_z\,(y,z,t) = \mathrm{Re}\left\{\sum_{n=-\infty}^{\infty} \hat{B}_{zn}\,(z)\,\mathrm{e}^{j(m\omega t - nky)}\right\}, \tag{8.49}$$

$$H_y\,(y,z,t) = \mathrm{Re}\left\{\sum_{n=-\infty}^{\infty} \hat{H}_{yn}\,(z)\,\mathrm{e}^{j(m\omega t - nky)}\right\}. \tag{8.50}$$

Each layer in Fig. 8.17 is the same except for the details of its material properties (μ and σ) and velocity. The field solution for the entire machine can be found by splicing together the solutions within each layer using the appropriate boundary conditions. This approach takes advantage of the fact that the vector potential A, the normal B field and tangential H field are all continuous across the interface of two layers. Using the linear superposition of the two separate solutions, the vector potential and magnetic fields due to the stator currents and rotor magnetization can be found everywhere in the machine. These field solutions are iterated using different values for the permeability of the rotor and stator cores until the correct effective permeabilities are found so that the fields in the machine lie on the B–H curves of the core materials.

8.4.1.1 Rotor Magnetization Profile

For rotors with an annular permanent magnet, the magnetization impressed using a magnetic poling process does not abruptly transition between north and south poles. Instead, the magnetic remanence transitions smoothly between $+B_r$ and $-B_r$. Normally, in a macroscale machine, discrete permanent magnets are used so this transition region is not present. However, for microscale machines this effect is important because of the small dimensions involved. Therefore, the annular-rotor permanent magnets are modeled here with a trapezoidal magnetization profile having a linear "transition region" of length c_m, as shown in Fig. 8.18. This linear transition is used as a first-order approximation for computational simplicity.

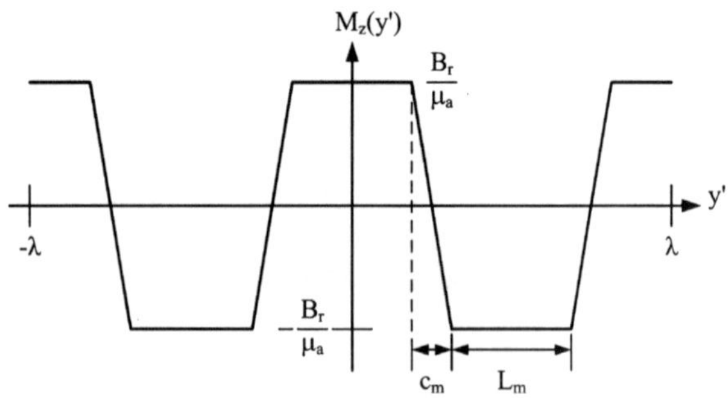

Fig. 8.18 Trapezoidal magnetization profile of the annular permanent magnets. B_r is the remanent flux density of the magnets, and c_m is the transition region length (Adapted from [64])

In the stationary coordinate frame, the time-varying magnetization profile can be decomposed into its Fourier harmonics according to

$$M_z\,(y,t) = \text{Re}\left\{\sum_{n=1,odd}^{\infty} \hat{M}_{zn} e^{jn(\omega_m t - ky)}\right\}, \tag{8.51}$$

where

$$\hat{M}_{zn} = \frac{1}{\lambda}\int_{\lambda} M_z\,(y')\, e^{jnky'}\, dy. \tag{8.52}$$

The linear velocity U_y is given by $\Omega_m r$, the rotor angular velocity multiplied by the radius. The electrical angular frequency ω_m is given by $P\Omega_m = kU_y$.

Defining the length L_m of the region in which the magnetization is uniform according to

$$L_m = \frac{\lambda}{2} - c_m = \frac{\pi r}{P} - c_m, \tag{8.53}$$

the magnitude of each magnetization harmonic becomes

$$\hat{M}_{zn} = \frac{4}{n^2\pi^2}\frac{\lambda}{c_m}\frac{B_r}{\mu_a} cos\left(nk\frac{L_m}{2}\right). \tag{8.54}$$

Given the magnetization in (8.54), the particular solution to (8.44) is

$$\hat{A}_{xn(m)} = \frac{8}{n\pi}\frac{B_r}{n^2k^2c_m} cos\left(nk\frac{L_m}{2}\right). \tag{8.55}$$

8.4.1.2 Stator Current Density Profile

The windings on the surface of the stator are arranged so that a traveling magnetic field is created by the winding currents. Like the rotor magnetization, the spatially varying current density in the stator windings can be represented in terms of its Fourier harmonic components according to

$$J_x\,(y,t) = \text{Re}\left\{\sum_{n=-\infty}^{\infty} \hat{J}_{xn} e^{j(\omega_e t - nky)}\right\}, \tag{8.56}$$

where the current in windings is assumed to operate at the single electrical frequency ω_e. For a three-phase surface-wound machine with balanced excitation, P pole pairs, N turns per pole, and κ winding layers, the harmonic amplitudes of the current density are

$$\hat{J}_{xn} = \frac{6}{n\pi}\sin\left(n\pi\frac{W_{cl}}{\lambda}\right) J_o \sum_{\substack{\nu=2N'+1 \\ \text{step2}}}^{4N'-1} e^{-j\frac{n\pi}{6N'}\nu}, \text{for } n = 1, -5, 7, -11, 13 \ldots \quad (8.57)$$

where

$$J_o = \frac{\kappa I_o}{W_{cl} T_{cl}}, \quad (8.58)$$

and $\lambda = 2\pi r/P$, $N' = N/\kappa$, W_{cl} is the width of the stator winding, and T_{cl} is the total height of all κ winding layers, and I_o is amplitude of the phase current. For the current density amplitudes given by (8.57) the resulting particular solution to (8.45) is

$$\hat{A}_{xn(J)} = \frac{6}{n\pi}\frac{\mu_0 J_o}{n^2 k^2}\sin\left(n\pi\frac{W_{cl}}{\lambda}\right) \sum_{\substack{\nu=2N'+1 \\ \text{step2}}}^{4N'-1} e^{-j\frac{n\pi}{6N'}\nu}. \quad (8.59)$$

8.4.1.3 Stator Laminations

The laminations are a series of thin concentric shells, as shown in Fig. 8.19. Adding laminations to the stator core creates a planar layer with non-uniform properties in the radial (x) direction which complicates the field analysis. Rather than solving Maxwell's Equations in three dimensions for a laminated stator core, a one-dimensional analysis is used to determine an effective conductivity $\sigma_{s,eff}$. The effective conductivity is then used in the two-dimensional continuum analysis derived above. The effective conductivity $\sigma_{s,eff}$ for the laminated stator core is given by [64]

$$\sigma_{s,eff} = \sigma_s\left(1 - \left|\frac{\tanh(Q)}{Q}\right|\right), \quad (8.60)$$

where

$$Q = \frac{1+j}{2}\frac{W}{\delta} \quad (8.61)$$

and

$$\delta = \sqrt{\frac{2}{\mu_s \sigma_s \omega_e}}. \quad (8.62)$$

The parameter Q describes how induced eddy currents reduce the net flux through a lamination. As the lamination width W becomes much larger than the skin depth, δ, the right-hand term inside the parenthesis in (8.60) approaches zero so that $\sigma_{s,eff} \approx \sigma_s$ and the solution in the stator core becomes that of an unlaminated stator. As W becomes much smaller than δ, this term approaches unity and the effective conductivity goes to zero.

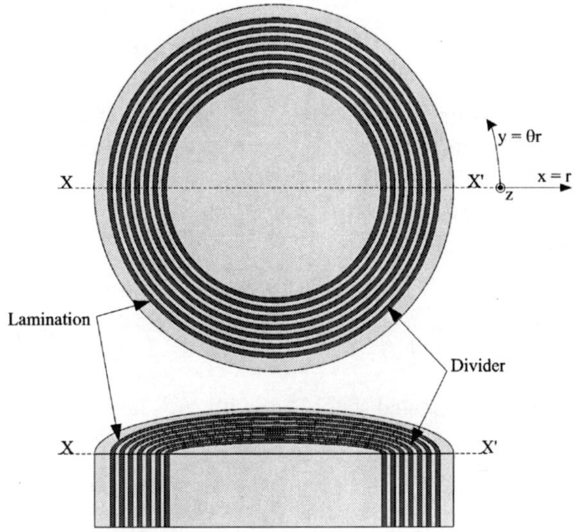

Fig. 8.19 Laminated stator for a planar surface-wound magnetic machine (Adapted from [64])

8.4.1.4 Saturation and Hysteresis Loss

Saturation effects in the machine are incorporated into the model by assigning effective permeabilities to the rotor and stator cores. The effective permeabilities are found by iteratively solving for the B and H fields until $B = \mu_{\text{eff}} H$ lies on the B–H curve of the magnetic material [64]. Since the B fields in the rotor and stator cores vary spatially over one wavelength, the root-mean-square (RMS) B field for each region is used to determine the effective permeability. For a stator consisting of laminations, the B field in the stator core will be higher than in the unlaminated case. To account for this effect, a packing factor, p_{f}, is introduced. The packing factor represents the fraction of the core that is taken up by ferromagnetic material. The RMS B field used to determine the effective permeability is increased by the factor $1/p_{\text{f}}$.

For an unlaminated stator, the fields are evaluated at the average radius $\frac{1}{2}(R_{\text{o}} - R_{\text{i}})$. The analysis of the laminated stator on the other hand is no longer continuous over the radial span of the machine. The iteration procedure is evaluated at the radius of each lamination to determine the permeability $\mu_{\text{s}}(r_{\text{lam,i}})$, where $r_{\text{lam,i}}$ is the radius of the ith lamination. Since the field solutions are computed over a continuous radial span, a permeability that varies smoothly with radius, $\mu_{\text{s}}(r)$, can be computed by interpolating between permeabilities at successive laminations, that is, between $\mu_{\text{s}}(r_{\text{lam,i}})$ and $\mu_{\text{s}}(r_{\text{lam,i+1}})$.

The hysteresis losses in the stator core are calculated using the B–H curve for the core material. The hysteresis-loss density is defined to be the energy density lost while traversing the static B–H loop. For a material with a peak flux density B_1, the hysteresis-loss density is [67]

$$\rho_{\text{core}}(B_1) = \int_{-B_1}^{B_1} H(B)\,\mathrm{d}B. \tag{8.63}$$

For an unlaminated stator, B_1 is evaluated at the average radius of the machine. For a laminated stator, B_1 is computed for each lamination at $r_{\text{lam,i}}$. The hysteresis loss is the sum of the total energy dissipated in the stator core over one hysteresis cycle times the frequency at which the B–H loop is traversed. For an unlaminated stator this yields

$$P_{\text{core}} = \frac{\omega_{\text{m}}}{2\pi}\int\!\!\int\!\!\int \rho_{\text{core}}(B_1)\mathrm{d}V = \frac{\omega_{\text{m}}}{2}\left(R_{\text{o}}^2 - R_{\text{i}}^2\right)\int_0^{T_{\text{sc}}} \rho_{\text{core}}(B_1(z))\,\mathrm{d}z. \tag{8.64}$$

For a laminated stator core with N_{lam} laminations the hysteresis loss is

$$P_{\text{core}} = \omega_{\text{m}} W \sum_{i=1}^{N_{\text{lam}}} r_{\text{lam},i} \int_0^{T_{\text{sc}}} \rho_{\text{core}}\left(\frac{1}{p_f} B_1(z)|_{r=r_{\text{lam},i}}\right)\mathrm{d}z. \tag{8.65}$$

8.4.1.5 Eddy-Current Losses in the Stator Core

The eddy-current losses in the stator core can be found by computing the drag torque on the rotor generated by these currents. The Maxwell Stress Tensor [29] is used to find the torque on the rotor using the field solutions found for the air gap in the continuum model. The time-average shear stress at the bottom surface of the rotor permanent-magnet surface (interface 4) is

$$\langle\tau_{\text{yz}}\rangle = \frac{\mu_0}{2}\sum_n \text{Re}\left\{\hat{H}_{zn}^{(4)}\hat{H}_{yn}^{(4)*}\right\} = \frac{1}{2}\sum_n \text{Re}\left\{\hat{B}_{zn}^{(4)}\hat{H}_{yn}^{(4)*}\right\}. \tag{8.66}$$

The total time-averaged torque is $2\pi\, r^2$ times this shear stress integrated over the radial span of the machine such that

$$T_{\text{drag}} = -2\pi \int_{r=R_{\text{i}}}^{r=R_{\text{o}}} \langle\tau_{\text{yz}}\rangle\, r^2 \mathrm{d}r. \tag{8.67}$$

Note that the minus sign in front of the integral in (8.67) is due to the fact that integration is performed over bottom surface of the rotor permanent magnets, which has a normal in the $-\hat{z}$ direction. The eddy-current losses in the machine are then

$$P_{\text{eddy}} = \Omega_{\text{m}} T_{\text{drag}}. \tag{8.68}$$

Physically, the drag torque T_{drag} manifests as eddy-current loss and stator-core heating. It can be reduced by using a ferrite or a laminated core.

8.4.1.6 Proximity Eddy Current Losses in the Coil Layer

The continuum model of the surface-wound PMSM does not model the stator windings as a layer with conductivity σ_c because the windings are made of discrete conductors. The width of each conductor is much smaller than a magnetic wavelength. The coil layer is instead treated as a layer with zero conductivity and an imposed current density $J_x(y,t)$ in the continuum model. However, eddy currents within each radial conductor will be present. These eddy currents are driven by the magnetic fields from the rotor permanent magnets, and the resulting losses are modeled separately using the field solutions from the continuum model. If the losses due to these eddy currents are too large, they can be reduced by laminating the radial conductors. There are 6PN radial conductors in a surface-wound stator. If each of these conductors has C_{lam} laminations, then the total loss due to proximity eddy currents is [64, 65]

$$P_{prox} = \frac{3\sigma_c\omega_m^2}{64\pi}(PNC_{lam})^2 T_{cl}\left[W_{cl}(R_o)^4 - W_{cl}(R_i)^4\right]\sum_{n=1,odd}^{\infty} n^2\left|\hat{B}_{zn,coil(m)}\right|^2,$$

where $\hat{B}_{zn,coil(m)}$ is flux density at half the thickness of the coil layer due the rotor PM.

8.4.1.7 Equivalent Circuit Model

The equivalent circuit for a single phase of a balanced PMSM is shown in Fig. 8.20. It consists of a back EMF source, or open-circuit voltage V_{oc}, due to the spinning permanent magnets, as well as an inductance L_s, and resistance R_s, due to the stator windings. The open-circuit voltage and stator inductance can be found using the vector potential solutions within the machine due the rotor magnetization and stator currents, respectively. The resistance is calculated from the dimensions of the radial conductors, and the inner and outer end turns of the stator windings.

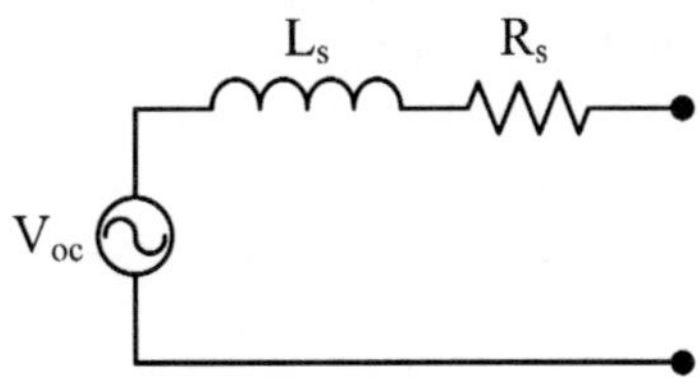

Fig. 8.20 Equivalent circuit for a single phase of a PMSM consisting of an open-circuit voltage V_{oc}, stator winding inductance L_s, and resistance R_s (Adapted from [64])

Open-Circuit Voltage

The spinning multi-pole permanent-magnet rotor creates a time-varying flux that induces the voltage V_{oc} on each phase of the stator windings governed by Faraday's Law according to

$$V_{\mathrm{oc}} = \frac{\mathrm{d}\lambda_{\mathrm{m}}}{\mathrm{d}t}, \tag{8.70}$$

where λ_{m} is the total flux from the rotor permanent magnets linked by a single phase. In order to determine the open-circuit voltage, the axial flux through a single phase of the stator windings must be found. This is achieved by integrating the incremental flux at a given radius over the radial span of the machine. The incremental flux from a single turn can be found by taking the difference between the vector potential of the two radial conductors in that turn [29].

For a three-phase surface-wound machine with P pole pairs, N turns per pole, and κ winding layers, the incremental flux per phase is [64, 65]

$$\hat{\Phi}_{zn,\mathrm{coil}(m)} = 2\kappa P \hat{A}_{xn,\mathrm{coil}(m)} \sum_{\substack{\nu=2N'+1 \\ \mathrm{step}2}}^{4N'-1} \mathrm{e}^{j\frac{n\pi}{6N'}\nu}, \tag{8.71}$$

where $\hat{A}_{xn,\mathrm{coil}(m)}$ is the vector potential evaluated at half the axial thickness of the stator winding layer, and $N' = N/\kappa$. Equation (8.71) works for both full-pitched and concentric-wound stators. Integrating over the radial span of the machine, the total flux linked by a single phase for each harmonic is

$$\hat{\lambda}_{n,\mathrm{coil}(m)} = \int_{r=R_{\mathrm{i}}}^{r=R_{\mathrm{o}}} \hat{\Phi}_{zn,\mathrm{coil}(m)} \mathrm{d}r. \tag{8.72}$$

The harmonic amplitude of the open-circuit voltage induced by this flux is then

$$\hat{V}_{\mathrm{ocn}} = \frac{\mathrm{d}}{\mathrm{d}t} \hat{\lambda}_{n,\mathrm{coil}(m)} = jn\omega_{\mathrm{m}} \hat{\lambda}_{n,\mathrm{coil}(m)}. \tag{8.73}$$

Stator Inductance

The stator inductance can be determined in the same manner as the open-circuit voltage. The total flux generated by the stator currents and linked by a single phase of the machine is found by integrating the incremental flux at a given radius over the radial span of the machine. The incremental flux linked by a single phase is P times the incremental flux per pole. For a three-phase surface-wound machine with P pole pairs, N turns per pole, and κ winding layers, the incremental flux per phase is [64, 65]

$$\hat{\Phi}_{zn,\mathrm{coil}(J)} = 2\kappa P \hat{A}_{xn,\mathrm{coil}(J)} \sum_{\substack{\nu=2N'+1 \\ \mathrm{step}2}}^{4N'-1} \mathrm{e}^{j\frac{n\pi}{6N'}\nu}, \tag{8.74}$$

where $\hat{A}_{xn,\mathrm{coil}(J)}$ is the vector potential due to the stator currents evaluated at half the thickness of the stator winding layer, and $N' = N/\kappa$.

The stator currents, which have the single temporal frequency ω_e, generate magnetic fields with multiple spatial harmonics. However, due to the large air gap, the higher-order harmonics are much smaller than the fundamental. Therefore, the flux linked by a winding due to the stator currents is dominated by the fundamental component of the magnetic field in the air gap. Integrating over the radial span of the machine, the flux linked by a single phase is then

$$\hat{\lambda}_{\mathrm{coil}(J)} = \int_{r=R_i}^{r=R_o} \hat{\Phi}_{z1,\mathrm{coil}(J)} \mathrm{d}r. \tag{8.75}$$

Note that this is the flux linkage generated by the sum of the three phase currents. The resulting inductance, L_d, is referred to as the synchronous inductance [1] and is given by

$$L_d = \frac{\mathrm{Re}\left\{\hat{\lambda}_{\mathrm{coil}(J)}\right\}}{I_o}. \tag{8.76}$$

In a balanced machine, the Phase-B and Phase-C currents produce a flux linked by Phase A that is scaled by $\cos(\pm 2/3\pi) = -\frac{1}{2}$. Since $i_b + i_c = -i_a$, the sum of the fluxes produced by Phases B and C equals half the flux produced by Phase A. Therefore, Phase A will link 1.5 times the flux produced by currents in only Phase A during balanced operation. Therefore, the stator phase inductance, L_s, in the single phase model will be

$$L_s = \frac{2}{3I_o}\mathrm{Re}\left\{\hat{\lambda}_{\mathrm{coil}(J)}\right\}. \tag{8.77}$$

Stator Resistance

For a surface-wound PMSM, the radial conductors of all three phases will occupy the same winding layer. Radial conductors in the same phase are connected by concentric end turns. Figure 8.21 shows a two-turn-per-pole stator with concentric windings. The width of the radial conductors, W_{cl}, is a function of radius r. Each outer end turn has a width W_{co} and each inner end turn has a width W_{ci}. The conductors are separated from each other by a width g_{cl}.

There are $2PN$ radial conductors per phase. Given C_{lam} laminations per radial conductor, the resistance of all the radial conductors in a single phase is [64]

$$R_{\mathrm{radial}} = \frac{6\,(PNC_{\mathrm{lam}})^2}{\pi\sigma_c T_{cl}} \ln\left(\frac{W_{cl}\,(R_o)}{W_{cl}\,(R_i)}\right), \tag{8.78}$$

where T_{cl} is the height of the radial conductor and σ_c is the conductivity of copper. If N is even, there will be $N/2$ short-pitched windings and $N/2$ long-pitched windings. If N is odd, their will be one full-pitched winding, $(N-1)/2$ short-pitched windings and $(N-1)/2$ long-pitched windings. The arcs spanned by the nth inner and outer end turns are, respectively,

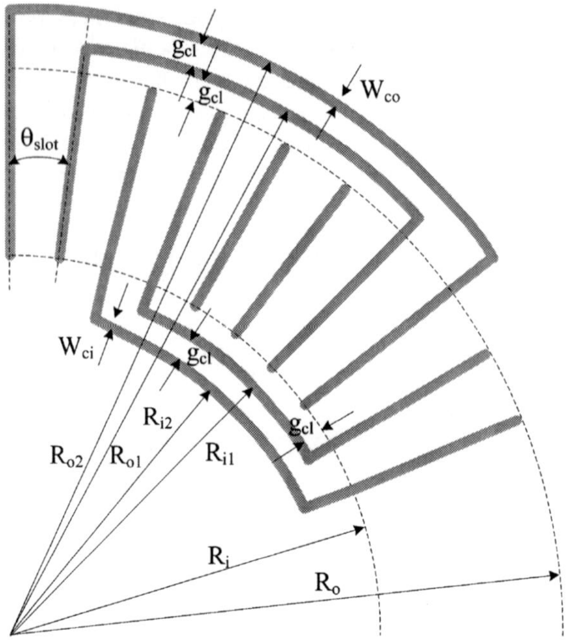

Fig. 8.21 End turn connections for a 2-turn-per-pole concentric winding (Adapted from [64])

$$L_{\text{inner}(n)} = \frac{[2(N+n)-1]\,\pi R_{\text{i}(n)}}{3PN}, \tag{8.79}$$

$$L_{\text{outer}(n)} = \frac{[2(N+n)-1]\,\pi R_{\text{o}(n)}}{3PN}, \tag{8.80}$$

where the radii of the arcs are

$$R_{(n)} = R_{\text{i}} - \frac{2n-1}{2}W_{\text{ci}} - ng_{\text{cl}}, \tag{8.81}$$

$$R_{(n)} = R_{\text{o}} + \frac{2n-1}{2}W_{\text{co}} + ng_{\text{cl}}. \tag{8.82}$$

The total resistance of the inner and outer end turns is then

$$R_{\text{inner}} = P\sum_{n=1}^{N}\frac{L_{\text{inner}(n)}}{\sigma_{\text{c}}T_{\text{ci}}W_{\text{ci}}} = \frac{\pi}{\sigma_{\text{c}}T_{\text{ci}}W_{\text{ci}}}\sum_{n=1}^{N}\frac{[2(N+n)-1]\,R_{(n)}}{3N}, \tag{8.83}$$

$$R_{\text{outer}} = P\sum_{n=1}^{N}\frac{L_{\text{outer}(n)}}{\sigma_{\text{c}}T_{\text{co}}W_{\text{co}}} = \frac{\pi}{\sigma_{\text{c}}T_{\text{co}}W_{\text{co}}}\sum_{n=1}^{N}\frac{[2(N+n)-1]\,R_{(n)}}{3N}. \tag{8.84}$$

The total stator resistance per phase is $R_s = R_{radial} + R_{inner} + R_{outer}$.

8.4.1.8 Simplified Open-Circuit Voltage Equation

The continuum model is fast and accurate, but also complex. To derive analytical insight into how various parameters affect the open-circuit voltage, the continuum model can be simplified. To do so, let the magnetization profile approach an ideal square wave ($c_m \to 0$). Also, assume that the rotor and stator cores and active rotor layer are ideal such that $\mu_r = \mu_s \to \infty$, and $\sigma_r = \sigma_s = \sigma_a = 0$. In addition, assume that the magnetic wavelength λ is much larger than the thicknesses of the air gap, stator windings, rotor permanent magnets, and rotor and stator cores. With the last approximation, the exponential decay of the fields in the machine can be ignored. In addition, the B and H fields are assumed to be independent of radius. The magnetic wavelength of the rotor is also independent of the radius, and the wave number k is evaluated at the mean radius. Given these approximations, the vector potential evaluated at half the thickness of the stator winding layer is

$$\hat{A}_{xn,\mathrm{coil}(m)} = -j\frac{(R_o + R_i)}{P}\left(\frac{T_{al}}{T_{al} + T_{cl} + T_{ag}}B_r\right)\frac{2}{n^2\pi}\sin\left(\frac{n\pi}{2}\right). \tag{8.85}$$

The corresponding incremental flux per phase reduces to

$$\hat{\Phi}_{zn,\mathrm{coil}(m)} = 2PN\hat{A}_{xn,\mathrm{coil}(m)}e^{j\frac{n\pi}{2}} = (R_o + R_i)\left(\frac{T_{al}}{T_{al} + T_{cl} + T_{ag}}B_r\right)N\frac{4}{n^2\pi}. \tag{8.86}$$

Since (8.86) is independent of the radius, the flux linked by a phase becomes

$$\hat{\lambda}_{n,\mathrm{coil}(m)} = (R_o - R_i)\hat{\Phi}_{zn,\mathrm{coil}(m)} = \left(R_o^2 - R_i^2\right)\left(\frac{T_{al}}{T_{al} + T_{cl} + T_{ag}}B_r\right)N\frac{4}{n^2\pi}. \tag{8.87}$$

while the harmonic amplitudes of the open-circuit voltage become

$$\hat{V}_{ocn} = jn\omega_m\hat{\lambda}_{n,\mathrm{coil}(m)} = j\left(R_o^2 - R_i^2\right)\left(\frac{T_{al}}{T_{al} + T_{cl} + T_{ag}}B_r\right)NP\Omega_m\frac{4}{n\pi}. \tag{8.88}$$

The voltage harmonics can be added to generate the total open-circuit voltage

$$V_{oc}(t) = \mathrm{Re}\left\{\sum_{n=1,\mathrm{odd}}^{\infty}\hat{V}_{n,\mathrm{coil}(m)}e^{jn\omega_m t}\right\} = V_o\sum_{n=1,\mathrm{odd}}^{\infty}-\frac{4}{n\pi}\sin(\omega_m t), \tag{8.89}$$

where

$$V_o = \left(R_o^2 - R_i^2\right)\left(\frac{T_{al}}{T_{al} + T_{cl} + T_{ag}}B_r\right)NP\Omega_m. \tag{8.90}$$

The term inside the summation of (8.89) is the nth Fourier component of a square wave. Therefore the open-circuit voltage due to a square wave magnetization is also a square wave with magnitude V_0. The first term in the parentheses in (8.90) shows how the open-circuit voltage scales with the active area of the machine. The second term in parentheses is the B field in the air gap, and indicates that the thickness of the rotor permanent magnets should be larger than the thickness of the coil layer and physical air gap combined in order to maximize the flux generated from the permanent magnets. Lastly, the open-circuit voltage scales linearly with B_r, N, P, and Ω_m.

8.4.2 Design

The surface-wound PMSM was chosen as the best candidate for efficient high power generation in the microengine. This section focuses on the design of a microscale PMSMs capable of generating and delivering Watt-level power at speeds of approximately 300 krpm. The emphasis will be on DC power generation via power electronics that rectify the AC voltages delivered by the generator phases. Since integration issues are not dealt with at this stage, silicon wafers and electroplated core materials will not be used. Instead, a ferromagnetic wafer will be used as the stator core. Since this wafer will not be laminated, eddy current losses will degrade efficiency but will not affect output power provided that the turbine that spins the rotor provides sufficient mechanical power to overcome the added losses. Integration issues are addressed in [55].

The PMSMs considered here are three-phase axial-air-gap machines. Each machine consists of a rotor with an annular multi-pole permanent magnet and ferromagnetic core, and a stator with multi-turn windings on the surface of a ferromagnetic substrate which serves as the stator back iron as shown in Fig. 8.22. During operation, the rotating multi-poled permanent-magnet rotor creates a time-varying magnetic flux in the rotor–stator air gap, which induces AC voltages at

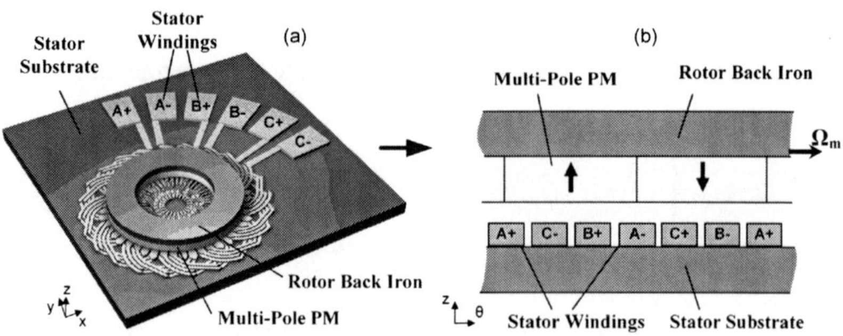

Fig. 8.22 Conceptual drawing of the PMSM: (**a**) perspective view and (**b**) cross-section (Adapted from [64])

the terminals of the stator windings. The magnetic flux generated by the permanent magnets can be concentrated in the rotor–stator air gap by using cores with very large permeabilities. By using rare earth permanent magnets, which have large remanences (~1 T), strong magnetic fluxes can be generated in the air gap. In fact, by using high-performance magnetic materials (permanent magnet, and rotor and stator cores) and suitably thick magnets (~500 μm), the actual rotor–stator air gap can be made relatively large (300–500 μm) to reduce windage loss while maintaining high magnetic flux in the air gap. Thus, the stator windings can be hundreds of microns thick and occupy space in the air gap on the surface of the stator substrate.

The simplified model of the PMSM will be used to derive the scaling laws that allow optimization of the physical and electrical parameters of the machine. The open circuit voltage scales as

$$V_{\mathrm{oc}} \propto \left(\frac{T_{\mathrm{al}}}{T_{\mathrm{al}} + T_{\mathrm{cl}} + T_{\mathrm{ag}}}\right) NP\Omega_{\mathrm{m}}. \tag{8.91}$$

The total resistance per phase [65, 68] scales as

$$R_{\mathrm{s}} \propto \frac{1}{T_{\mathrm{cl}}}\left[k_1(PN)^2 + k_2 PN\right]. \tag{8.92}$$

The k_1 term represents the resistance of the radial conductors which grows as PN squared, while the k_2 term represents the resistance of the inner/outer end turns which grow linearly with *PN*. Even though the end turn resistance grows more slowly, it can make up a large fraction of the over all resistance. In addition, the resistance is inversely proportional to the thickness of the radial conductors that make up the coil layer. Equation (8.92) assumes the thickness of the inner and outer end turns are a constant fraction of the coil layer.

Combining (8.91) and (8.92), the output power capability of a PMSM scales approximately as

$$P_{\mathrm{out}} \propto \frac{V_{\mathrm{oc}}^2}{R_{\mathrm{s}}} \propto \left(\frac{T_{\mathrm{al}}}{T_{\mathrm{al}} + T_{\mathrm{ag}} + T_{\mathrm{cl}}}\right)^2 T_{\mathrm{cl}}. \tag{8.93}$$

Equation (8.93) contains important information about the relative thicknesses of the active layer, air gap, and coil layer in order to maximize output power. First, the air gap should be made as small as possible. Second, for a given coil layer and air gap thickness, power is maximized as the term inside the parentheses approaches unity. Finally, taking the derivative of (8.93) with respect to T_{cl}, the maximum output power for a fixed air gap and active layer thickness is

$$T_{\mathrm{cl}} = T_{\mathrm{al}} + T_{\mathrm{ag}}. \tag{8.94}$$

Thus, for a fixed air gap and active layer thickness, the coil layer should equal the sum of the two. The maximum thickness for the rotor permanent magnets is 500 μm, set by microengine wafer-thickness constraints. The maximum height of the coil layer is 200 μm due to fabrication constraints. So, for a given coil thickness, the maximum power is achieved by setting T_{al} to its maximum value, while for a given rotor permanent magnet thickness, the maximum power is achieved by setting T_{cl} to its maximum. Thus, the maximum power is achieved when T_{al} is set to 500 μm and T_{cl} is set to 200 μm.

To optimize N and P, scaling laws for the power electronic losses are required. Some losses in the power electronics are independent of frequency, such as conduction loss in the on-state resistance of a power MOSFET. Others are proportional to the square root of the electrical frequency, such as skin depth losses in the windings of high-frequency transformers. Still others grow linearly with frequency, such as the switching loss of a power MOSFET. The frequency-dependent components are the most important because they scale with both the speed and number of pole pairs of the machine. To model these losses, the effective power-electronics resistance R_{pe}, will be assumed to be proportional to $P\Omega_{\mathrm{m}}$ because the fastest growing frequency-dependent component scales linearly. Thus

$$R_{\mathrm{pe}} \approx k_3 P \Omega_{\mathrm{m}}. \tag{8.95}$$

Since the load resistance can be expressed as some fraction of $R_{\mathrm{s}} + R_{\mathrm{pe}}$, the overall output power of the generator and power electronics scales as

$$P_{\mathrm{out}} \propto \frac{V_{\mathrm{oc}}^2}{R_{\mathrm{s}} + R_{\mathrm{pe}}} \propto \frac{(PN\Omega_{\mathrm{m}})^2}{k_1 (PN)^2 + k_2 PN + k_3 P\Omega_{\mathrm{m}}}. \tag{8.96}$$

By making N large, both the open-circuit voltage and stator resistance increase, while dissipation in the power electronics remains the same leading to higher output power. In addition, for large N, P_{out} will be independent of both P and N, and scale with the square of the rotational speed. In reality, the minimum feature size constraint will limit the gap between adjacent radial conductors, and so a large number of turns per pole will lead to thin inner and outer end turns. Both effects make R_{s} increase slightly faster than N^2. Therefore, to maximize output power, the number of turns per pole should be maximized subject to stator fabrication and surface area limitations.

The total generator system efficiency can be expressed as

$$\eta_{\mathrm{g}} = \frac{P_{\mathrm{out}}}{P_{\mathrm{out}} + P_{\mathrm{prox}} + P_{\mathrm{core}} + P_{\mathrm{cond}} + P_{\mathrm{pe}}}, \tag{8.97}$$

where P_{cond} is the loss in the stator windings and P_{pe} is the loss in the power electronics. P_{core} is the sum of the stator core eddy current and hysteresis losses. The eddy current loss is either proportional to frequency squared (skin depth > core/lamination thickness) or linear with frequency (skin depth < core/lamination thickness). The hysteresis loss is linear with frequency. In the worst case,

$$P_{core} \propto k_4(P\Omega_m)^2 + k_5 P\Omega_m. \qquad (8.98)$$

Thus, the generator system efficiency can be increased by making P small. However, doing so will make the rotor permanent-magnet poles wide. Consequently, the rotor and stator cores must be made thicker to prevent them from saturating. The proximity skin effect losses, P_{prox}, are approximately independent of P [64, 65] and scale as speed squared. However, proximity losses decrease as N squared. Therefore, to maximize efficiency, the number of poles should be minimized subject to rotor and stator core thickness limitations while the number of turns/pole should be maximized subject to stator fabrication and surface area limitations. Optimization of the output power and efficiency [64, 69] resulted in a stator with 8 poles and 3 turns per pole.

8.4.2.1 Stator Design

The stator uses interleaved, multi-layer electroplated copper windings that are dielectrically isolated from a 1-mm thick $Ni_{80}Fe_{15}Mo_5$ (Moly Permalloy) substrate by a 3-μm spin-on-glass layer and/or a 5-μm polyimide layer. The active area of the stator has an outer diameter of 9.5 mm, and an inner diameter of 5.5 mm. NiFeMo is selected as the stator core (substrate) for its combination of high permeability ($\mu_r > 10^4$), low coercivity ($H_c \sim 0.16$ A/m) to minimize hysteresis losses, and commercial availability in sheets of suitable thickness [70].

An optimized 2nd-generation 8-pole 3-turn-per-pole stator is shown in Fig. 8.23. The winding pattern is designed for minimal resistance to ensure optimal generator performance. These stators use a single layer for the radial conductors that are 200 μm thick. The end turns occupy two layers and are each 80 μm thick. There is a 40 μm thick SU8 insulation layer between the top and bottom end turns. The stators use concentric windings, shown in Fig. 8.24, in which a complete end turn occupies

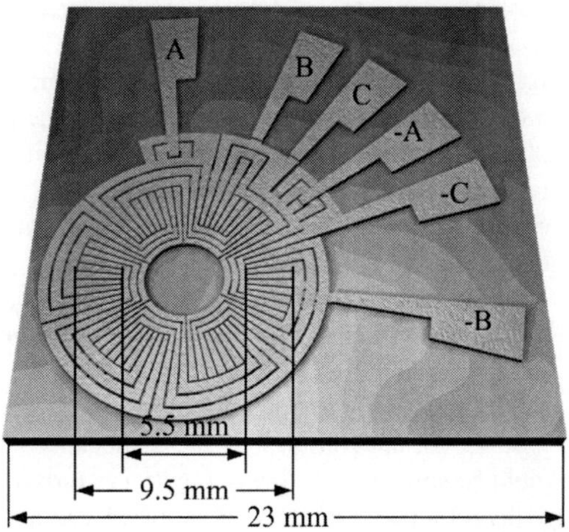

Fig. 8.23 8-pole, 3-turn/pole concentric wound stator (Adapted from [64])

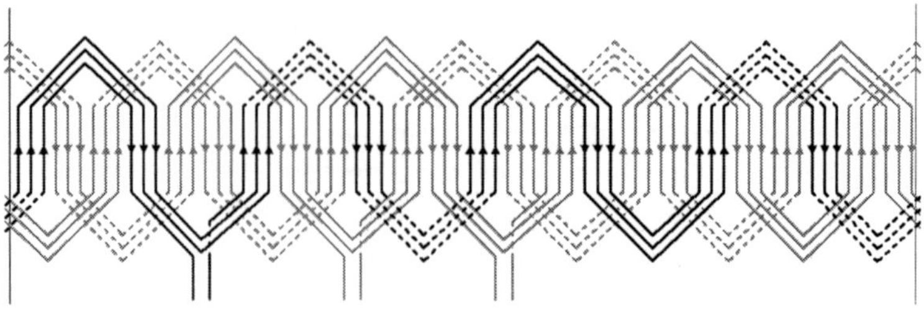

Fig. 8.24 Winding diagram for the 3-turn-per-pole concentric-wound stator. *Solid lines* represent end turns on the top layer while *dashed lines* represent end turns on the bottom layer. The radial conductors, marked with *arrows*, occupy both coil layers. Phase A is *darkened* for reference (Adapted from [64])

a single layer. This eliminates the need for cross-over conductors like the ones used in the 1st-generation stators [65, 68]. In fact, the only cross-over conductors needed are for connections to the pads. With fewer crossovers the number of vias is greatly reduced, from 96 down to 3. In addition, by using concentric windings, the end turns can be made to be circular arcs. This makes much more efficient use of the surface area of the inner and outer regions of the stator than the triangular end turns used in the 1st-generation devices [68]. Furthermore, variable-width coils are used with the gaps between adjacent radial conductors and end turns equal to 50 μm rather than 130 μm. The combination of the reduced number of vias, better end turn design, and more aggressive fabrication leads to a predicted stator resistance of 100 mΩ. This resistance is about the same as that of a 1st-generation 2-turn-per-pole machine. In fact, the predicted output power capability is increased by a factor of 2.25 over the 2-turn-per-pole machines presented in [68] due to the improved stator resistance.

8.4.2.2 Rotor Design

The rotor contains an 8-pole, annular, pressure-formed (sintered) Sm_2Co_{17} permanent magnet and a ferromagnetic $Fe_{49}Co_{49}V_2$ (Hiperco 50) core, as shown in Fig. 8.25. The rotor components are mounted in a titanium adaptor for assembly onto a 1.6-mm diameter shaft for testing. The permanent magnet and rotor core are 500 μm thick to be compatible with standard wafer thicknesses, and have an outer diameter of 10 mm and inner diameter of 5 mm. This outer diameter was chosen because it was the closest match to the 10-mm outer diameter of the microengine generator.

SmCo magnets were selected for their combination of high-energy product (BH_{max} ~240 kJ/m^3) for high-energy conversion, and high-operating temperatures (T_{max} ~300°C) [71]. Although NdFeB has a higher energy product (BH_{max} ~400 kJ/m^3), it does not provide the necessary operating temperatures (T_{max} ~150°C) for integration with the combustion-driven microengine [71]. However, it would be suitable for a low-temperature turbine generator. Hiperco 50 was selected for the rotor core for its combination of a reasonably high permeability ($\mu_r > 3000$),

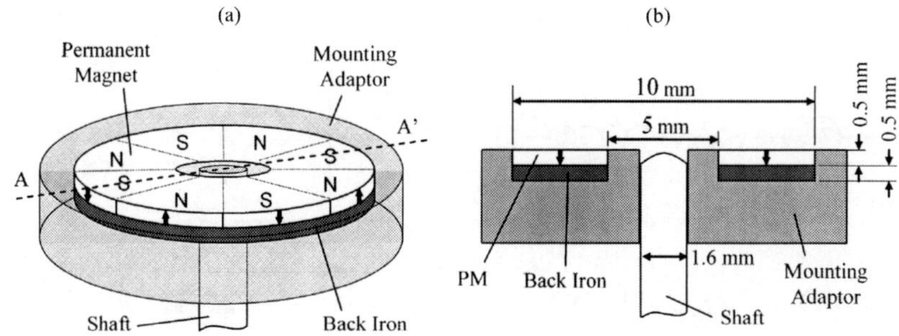

Fig. 8.25 Schematic (**a**) perspective view and (**b**) cross-section of the PM machine rotors. Note that the rotor is shown upside down (Adapted from [64])

and its high saturation flux density ($B_S \sim 2.4$ T) to minimize the thickness of the cores while avoiding magnetic saturation [70]. Note that the fields in the rotor core are nearly time-invariant, so the coercivity of the rotor back iron is not a major concern.

8.4.3 Power Electronics

For the microengine system to replace batteries, it must provide a constant DC output voltage. The PMSM presented in the previous section produces a three-phase AC voltage. A PMSM can work with both passive and active power electronics to create a DC output voltage. If the microengine is to provide a 12-V battery replacement, for example, then the power electronics must rectify and boost the AC voltages from the generator, while operating under closed-loop control to provide a regulated output voltage.

One way to passively step up and rectify the AC voltages is to use a Δ/wye-connected transformer and diode bridge rectifier as shown in Fig. 8.26. The generator windings are connected in a wye configuration and tied to the Δ-connected primary side of the transformer. This gives an additional factor-of-three increase from the line-to-neutral voltage of the PMSM to the line-to-line voltage on the transformer secondary side. A linear regulator at the output of the rectifier can then be used to achieve output voltage regulation.

Passive electronics, however, tend to be large and bulky. Transformers designed for use in the stator electrical frequency range of 10–20 kHz can be very large compared to the PMSM. The three-phase transformer used in [68] is eight times the volume of the PMSM stator. In addition, the transformer/diode bridge combination cannot provide power factor correction. Power factor correction is important at high speeds where the machine and transformer inductances will start to create a phase lag in the generator current relative to the generator voltage, reducing output power capability. At only 100 krpm the power factor for the transformer and diode bridge is 93.4%. At 300 krpm, the power factor drops to 90% while at 500 krpm the power

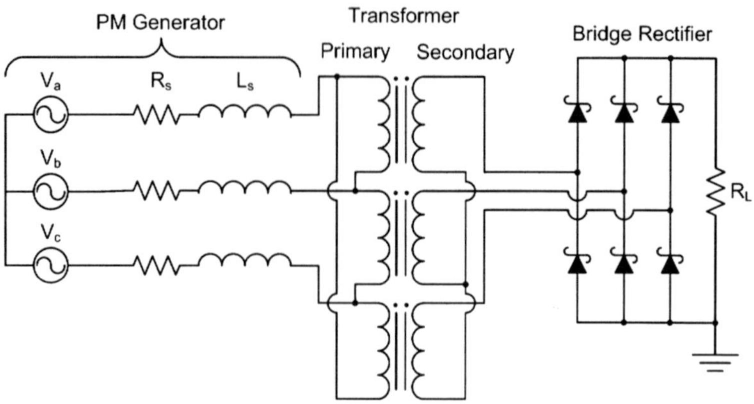

Fig. 8.26 Three-phase rectification circuit for DC power generation (Adapted from [64])

factor is only 82.6%. Clearly, as the speed rises, passive power electronics become a poor choice for power conversion.

Active power electronics can eliminate the transformer, diode bridge, and linear regulator, and provide all four functions (power factor correction, rectification, voltage step-up, and regulation) in a much more compact and efficient way. Active power electronics for a macroscale PMSM often requires speed and/or position sensing to determine the position of the magnetic poles on the rotor. This information is used to keep the stator currents in phase with the voltage to achieve maximum energy/power conversion. Hall effect sensors are sometimes used to directly measure the rotor position but are difficult to integrate into a microscale PMSM. "Sensorless" techniques to estimate the rotor position from current and terminal voltage measurements are also used in macroscale PMSMs but are computationally intensive, often requiring the use of high-performance microprocessors or digital signal processors. Real time processing of current and voltage information to determine the proper turn-on and turn-off times of the power transistors would be difficult achieve for the high speeds in the microscale PMSMs.

The active power electronics designed for microscale a PMSM uses a three-phase boost semi-bridge topology [72], shown in Fig. 8.27, operating in discontinuous conduction mode (DCM). Independent control of all three phases, as well as direct position, speed, current or voltage sensing is not needed to align the stator currents with the line-to-neutral voltages due to DCM nature of the circuit [72]. The three power MOSFETs are driven from the same control signal and are turned on/off at a constant switching frequency much higher than the stator electrical frequency. The converter turns the PMSM into a current source that can be controlled with a voltage control loop implemented using an off-the-shelf pulse-width-modulated (PWM) control IC. This topology provides near unity power factor and low-harmonic distortion without the need for large low-frequency passive magnetic components [73].

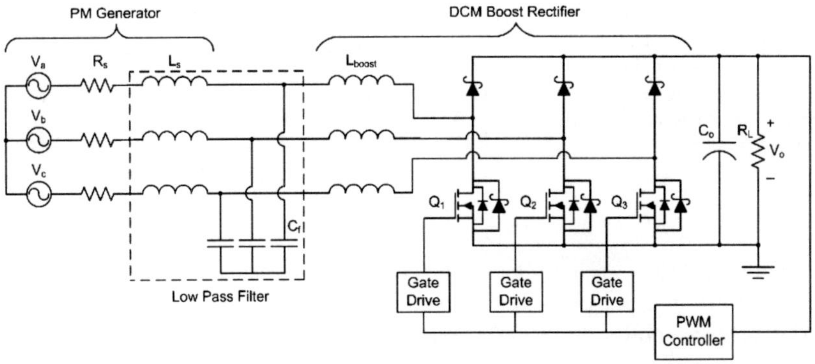

Fig. 8.27 DCM rectifier based on the boost semi-bridge topology (Adapted from [64])

The boost inductors are separate external inductors. The machine inductance and resistance, along with an external capacitor, act as a damped low-pass LC filter to attenuate the ripple current from the boost inductors so that the line voltage sees mainly a sinusoidal current at the line frequency. By using separate inductors, the high-frequency high-ripple currents are localized on the printed circuit board to minimize electromagnetic interference (EMI). In addition, the RMS stator currents are primarily due to the fundamental harmonic. This leads to low-harmonic distortion and reduced conduction loss in the stator windings. Even though the RMS currents are larger in the boost inductor, the resistance of the boost inductor is much smaller than the stator resistance (by design) so the conduction loss in the inductor is a small fraction of the overall conduction loss.

At the beginning of each switching period the three MOSFETs are turned on and the currents through the boost inductors rise to a value proportional to their respective open-circuit line-to-neutral voltages. The duty ratio determines the fraction of the switching period that the switches remain on. When the MOSFETs are turned off, the currents will flow through the diode bridge. If the currents through the boost inductor are positive, then current will flow through the high side (or boost) diode. If the currents are negative, they will flow through the free wheeling diode in parallel with the MOSFET.

The peak boost inductor current is proportional to the average value of the line-to-neutral voltage during the on-time of the MOSFET. This average value varies sinusoidally. Since the current in the boost inductor returns to zero each period, the variation in the average value of the current over a switching period will also be approximately sinusoidal, and will be in phase with the open-circuit voltage. For a DC output voltage much larger than the open-circuit voltage, such as is the case for microscale PMSMs, the time it takes for the boost inductor current to return to zero is much shorter than the on-time of the switch. Therefore, the average current over a switching period is almost independent of the output voltage leading to low-harmonic distortion of the current waveform [72].

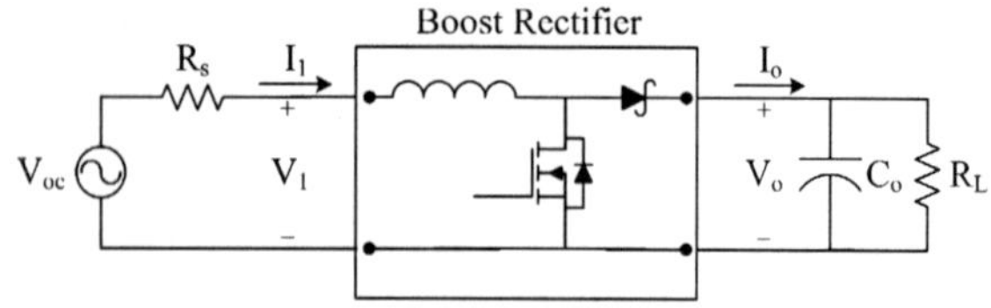

Fig. 8.28 Simplified model of the boost rectifier and generator. The input circuit is shown on a per-phase line-to-neutral basis (Adapted from [64])

Figure 8.28 shows a simplified model of the power generation system. The input side is shown on a per-phase line-to-neutral basis. Assuming an ideal lossless converter, the input power will equal the output power such that

$$\frac{3}{2}V_1 I_1 = V_o I_o = P_o, \tag{8.99}$$

where V_1 is the peak line-to-neutral voltage after the stator resistance and I_1 is the peak fundamental component of the current through a single boost inductor. The input power from the generator will be

$$\frac{3}{2}V_{oc} I_1 = \frac{3}{2}I_1^2 R_s + P_o. \tag{8.100}$$

Solving (8.100) for I_1, the required current in each phase to deliver P_o to the output is

$$I_1 \frac{V_{oc}}{2R_s}\left(1 - \sqrt{1 - \frac{P_o}{P_{max}}}\right). \tag{8.101}$$

where

$$P_{max} \equiv \frac{3V_{oc}^2}{8R_s}, \tag{8.102}$$

And P_{max} is the power that the three-phase generator can deliver to a matched resistive three-phase load. The input voltage to the power converter on a line-to-neutral basis will be

$$V_1 = V_{oc} - I_1 R_s = \frac{V_{oc}}{2}\left(1 + \sqrt{1 - \frac{P_o}{P_{max}}}\right). \tag{8.103}$$

In [74], the input and output currents are derived for the three-phase DCM boost rectifier as functions of input voltage, output voltage, duty ratio, boost inductance, and switching frequency. The terms in [74] are modified using the voltage after the machine resistance, V_1, rather than the input voltage, V_{oc}. The average output current for the boost rectifier is

$$I_o = \frac{1.46}{2L_{boost}f_{sw}} \frac{V_1^2 D^2}{(V_o - 1.67V_1)} \tag{8.104}$$

where D and f_{sw} are the PWM duty ratio and switching frequency, respectively. The per-phase input current to the converter can be found using (8.99) and (8.104), and is given by

$$I_1 = \frac{0.973}{2L_{boost}f_{sw}} \frac{V_o V_1 D^2}{(V_o - 1.67V_1)}. \tag{8.105}$$

To deliver 10 W at an output voltage of 12 V, I_o must be 833 mA. Under the assumptions in that $V_{oc} = 2.5$ V, and $R_s = 100$ mΩ, P_{max} equals 23.44 W. Using (8.103), V_1 is 2.196 V. Equation (8.104) can be used as guide to determine the desired boost inductor value and duty ratio. Rewriting (8.104),

$$\frac{L_{boost}}{D^2} = \frac{1.46}{2I_o f_{sw}} \frac{V_1^2}{(V_o - 1.67V_1)}. \tag{8.106}$$

A PWM switching frequency of 500 kHz is chosen to minimize frequency dependent losses. Since the output voltage is much larger than the input voltage, a large duty cycle can be chosen and still keep the boost inductor current discontinuous. A larger duty ratio typically leads to a smaller RMS value for the boost current, minimizing conduction loss. However, since the converter is not lossless, any duty ratio set by (8.106) will under predict the actual duty ratio. This is because the converter must draw extra power to overcome its own losses which it does by increasing the inductor current. The duty ratio is set to 0.5 which corresponds to a boost inductor value of 0.36 μH. The closest standard value available is 0.33 μH and the resulting nominal duty ratio is 0.57. PSpice simulations of the boost rectifier confirm that the inductor current remains discontinuous at maximum output power.

The peak current in the boost inductor will be

$$I_{pk} = \frac{V_1 D}{L_{boost}f_{sw}} = 7.6A. \tag{8.107}$$

The corresponding RMS current through the boost inductor is 3.8 A which is found from the simulation model. The D01813H-331 inductor from Coilcraft is a 0.33 μH surface-mount power inductor designed for high current applications. It is chosen as the boost inductor due to its high saturation (10 A) and RMS (7 A) current ratings, and its small size. The DC resistance of the inductor, R_{DC}, is only 4 mΩ. At a switching frequency of 500 kHz, the resistance of the inductor increases to 34 mΩ. The high-frequency resistance of the inductor windings is substantially larger than the DC resistance due to skin effect, and contributes a significant amount of loss in the inductor. This is expected due to the thickness of the wire used. In the future, the single thick wires can be replaced by multiple interposed thinner wires, or even Litz wire.

The PWM controller used for the boost rectifier is the UCC28C40 from Texas Instruments. It is intended for high-frequency power supplies for 12-V systems, and has a maximum switching frequency of 1 MHz. The operating supply current at 500 kHz is 4 mA. At a supply voltage of 12 V, the controller consumes 48 MW of power. The LM5112 gate driver IC from National Semiconductor is used to drive the power MOSFETS. The gate drive is capable of sourcing 3 A and sinking 7 A with a propagation delay of 25 ns. The selected power MOSFET (Q_1, Q_2, and Q_3 in Fig. 8.27) is the IRF7413Z from International Rectifier. It is a 30-V 13-A power MOSFET intended for notebook processor power applications. The diodes used in the bridge are 30-V 1-A DFLS130L surface-mount Schottky diodes from Diodes Inc. These diodes have very low forward voltages and junction capacitances.

The controller and gate drivers can be powered either from an external power supply or from the output voltage. To start up the converter, an external supply raises the input to 7 V, at which point the converter turns on and starts to bring up the output voltage. When the output voltage becomes larger than the external supply voltage, the converter powers itself up to 12 V. Two diodes provide an OR-ing of the two power sources.

The rectifier is shown in Fig. 8.29 along with a 3-turn-per-pole stator for size comparison. The rectifier power stage is 24 mm by 22.5 mm and has about the same surface area as the stator die. The PCB is 1.6 mm thick and most of the components are 1.75 mm in height or less with the boost inductors being 5 mm tall. The output filter capacitors and free wheeling diodes are placed on the bottom side of the power stage. The control section is 11 mm by 10 mm. The size of the control section could be made smaller by using EIA 0603 or 0402 sized packages for all the passives. This is not done here due to the challenge of soldering such small components.

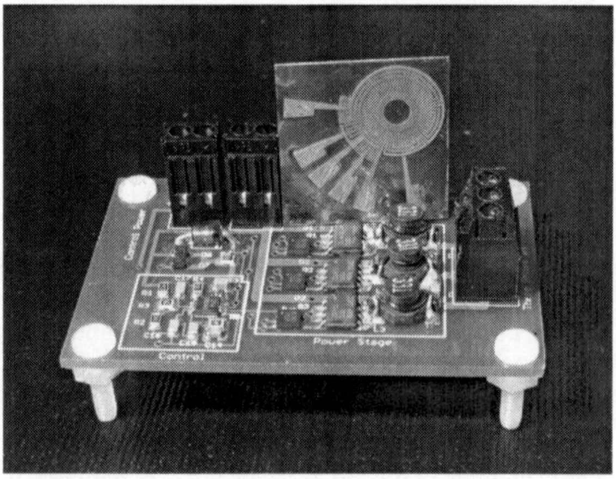

Fig. 8.29 Boost rectifier board shown with the 3-turn stator for size comparison (Adapted from [64])

Figure 8.30 shows the boost inductor current, stator current, and open-circuit voltage at a speed of 150 krpm. Both the boost inductor and stator currents are in phase with the open-circuit voltage. Figure 8.31 shows the boost inductor current,

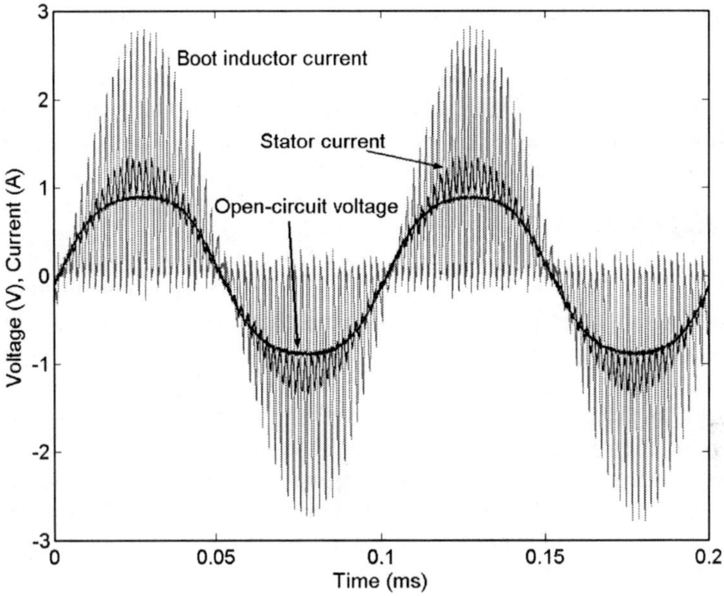

Fig. 8.30 Boost rectifier waveforms over two stator electrical periods at a generator speed of 150 krpm (Adapted from [64])

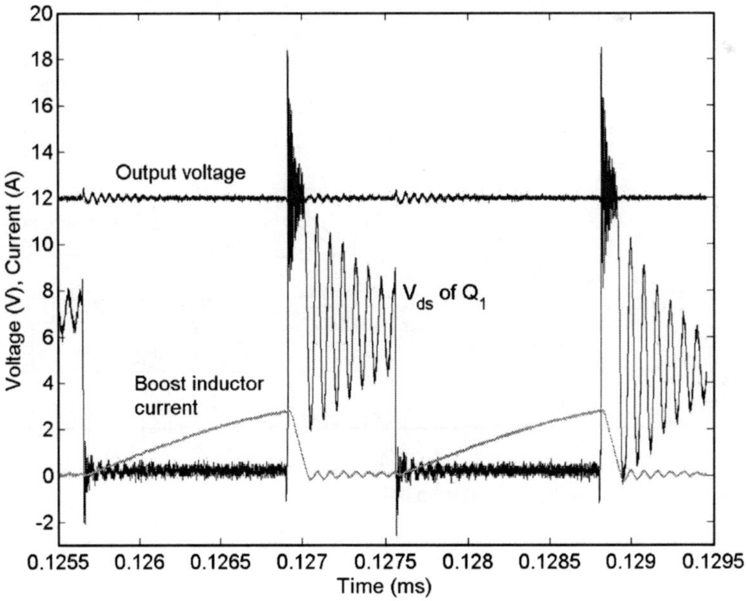

Fig. 8.31 Boost rectifier waveforms over two switching periods (Adapted from [64])

V_{ds} of Q_1 and the output voltage over a couple of switching periods. The ringing in the drain-to-source voltage and the inductor current is due to the LC tank created by the boost inductor and the output capacitance of Q_1. The output voltage also exhibits ringing when the MOSFET turns on and off. However, the magnitude is much smaller than the ringing on V_{ds} due to the output filter capacitor, C_o.

8.4.4 Experiments

ptA test stand has been developed to support spinning rotors in order to demonstrate electrical power generation while avoiding the design and fabrication complexities of integrated high-speed bearings [68]. The test stand consists of a high-speed, air-driven spindle to spin rotors with a controllable air gap over the surface of the stators, as depicted in Fig. 8.32. Powered by compressed nitrogen, the spindle provides rotational speeds in excess of 350 krpm. The rotor/shaft assembly is mounted in the spindle, and the rotation speed is measured with an optical shaft encoder or, in the case of electrical machine tests, using the frequency of the generated voltages. The stator is clamped to an xyz-micropositioner stage, having $\pm 5\ \mu m$ resolution in all three directions. Rotors are mounted to a shaft and inserted into the spindle, which is attached to an articulating arm providing translation and rotation in all directions.

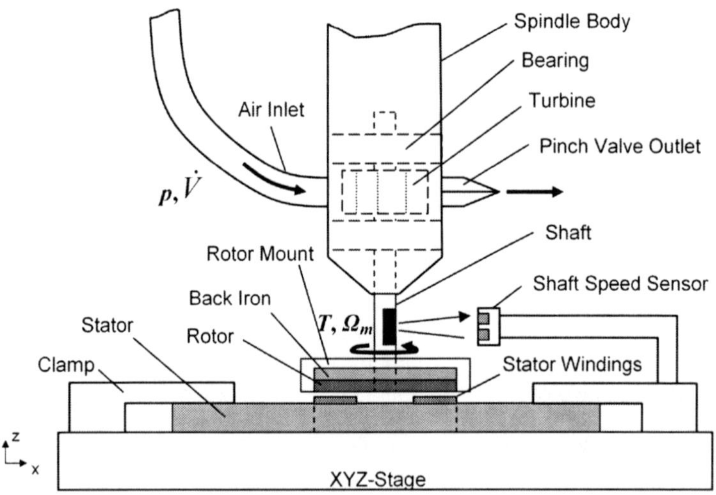

Fig. 8.32 Spinning rotor test stand, depicting the air-powered spindle spinning a magnetic rotor over the surface of the stator (Adapted from [64])

8.4.4.1 Electrical Characterization

The stators are first characterized in terms of their equivalent circuit parameters: phase resistance R_s, phase inductance L_s, and open-circuit voltage V_{oc} (Fig. 8.20).

The machines are then connected to power electronics for power generation tests, which are conducted at varying load resistances and speeds with the rotor–stator air gap set to 100 μm.

The stator has radial conductors with an average thickness of 188 μm. There is a layer of SU8 over the coils to protect them that is approximately 50 μm thick. The measured phase resistance is 145 mΩ which is 45% higher than predicted and is most likely due to variations in the thickness of the SU-8 layer between the top and bottom windings which can make the inner and outer end turns thinner than expected. The measured phase inductance is 0.34 μH, which is more than double the 0.138 μH predicted by the continuum model. This is because the continuum model does not take into account the leakage inductance from the inner or outer end turns. The leakage inductance of a macroscale machine is typically a small fraction of the magnetizing inductance [1]. In macroscale surface wound PM machines, the leakage can be on the same order of magnitude as the magnetizing inductance. In the microscale surface-wound PMSMs presented here, however, the leakage inductance dominates the overall stator inductance.

The permanent magnets used have an inner diameter of 5 mm and an outer diameter of 10 mm. The ideal square-wave magnetization with a B_r of 1 T for SmCo was seen to over-predict the experimental results in [65, 68, 69]. It is believed that the selective pulse magnetization process used to pattern the rotor magnetization does not fully saturate the material, resulting in a reduced remanence. Also, transition regions between poles are possible due to saturation of the magnetized heads at the edges of the magnetizer poles and the resulting leakage flux. A closer fit is obtained if the remanence used in the analytical model is reduced to 0.52 T. Further improvement, is made by introducing a 1.2 mm linear magnetization transition region, c_m, between north and south poles. This trapezoidal profile yields a more sinusoidal shape, and more likely represents the physical magnetization pattern.

The open-circuit voltage is shown as function of speed and air gap in Figs. 8.33 and 8.34 , respectively. The open-circuit voltage as function of speed is taken at an air gap of 100 and 250 μm while the voltage as a function of air gap is taken at 50, 100, and 150 krpm. The thickness of the SU8 layer is added to the air gap thickness. The RMS V_{oc} varies linearly with speed as expected and the open-circuit voltage is a decreasing function of air gap due to the exponential decay of rotor flux in the air gap. Notice that once B_r and c_m are set, the experimental measurements of V_{oc} agree very well with the values predicted by the analytical continuum model over a wide range of speeds and air gaps.

Power Generation Tests

Machine efficiency is an important consideration in the generator design. The power flow diagram for the generator is shown in Fig. 8.35. The total generator system efficiency, η_g, and electrical system efficiency, η_e, are defined as

$$\eta_g = \frac{P_{out}}{P_{mech}}, \tag{8.108}$$

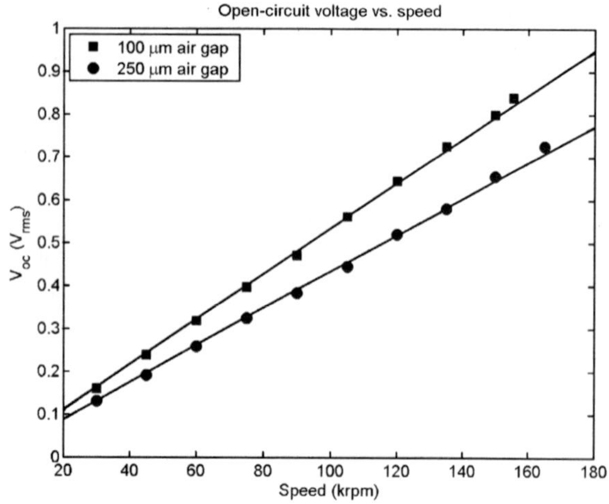

Fig. 8.33 Open-circuit voltages versus rotational speed at different air gaps. *Points* represent measurements; *curves* represent the analytical model (Adapted from [64])

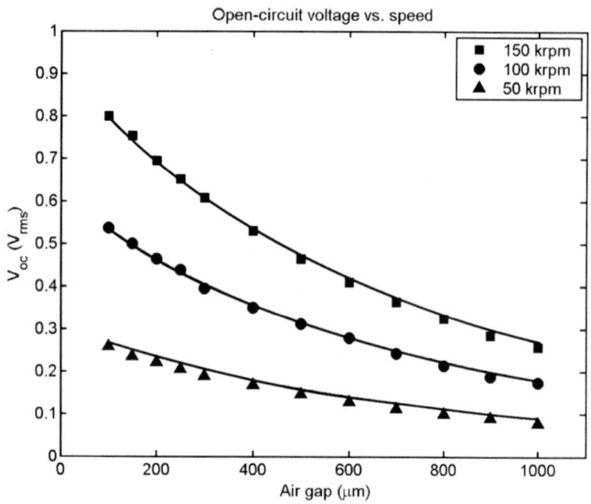

Fig. 8.34 Open-circuit voltages versus air gap at different speeds. *Points* represent measurements; *curves* represent the analytical model (Adapted from [64])

$$\eta_e = \frac{P_{\text{out}}}{P_{\text{elect}}}. \tag{8.109}$$

Direct experimental measurement of the total generator system efficiency is not possible because of the difficulty in measuring the mechanical input power, P_{mech}. Instead, the total input mechanical power is estimated by summing the proximity losses, core losses, and converted power; $P_{\text{mech}} = P_{\text{prox}} + P_{\text{core}} + P_{\text{elect}}$. The core losses P_{core}, and proximity losses P_{prox}, are computed from the analytical model. The core losses consists of both eddy-current and hysteresis losses. The converted

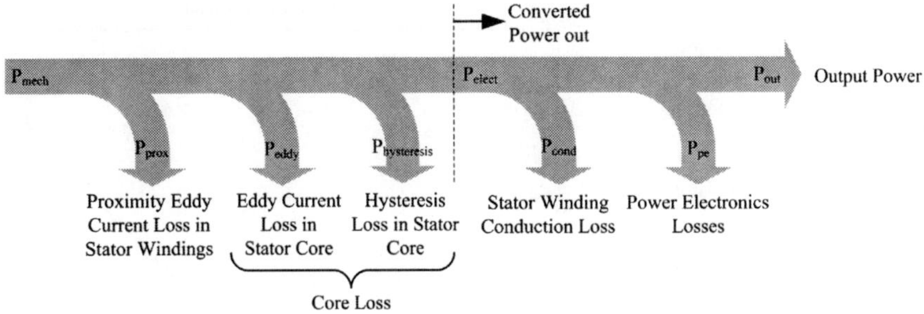

Fig. 8.35 Power flow diagram for generator system (Adapted from [64])

power P_{elect} is the power available from the open-circuit voltage in the equivalent circuit model, and is extracted from a PSpice model. P_{elect} consists of stator winding conduction loss P_{cond}, power electronics losses P_{pe}, and the power delivered to the load, P_{out}. The efficiency of the power electronics, η_{pe}, is

$$\eta_{\text{pe}} = \frac{P_{\text{out}}}{P_{\text{out}} + P_{\text{pe}}}. \tag{8.110}$$

The PMSM is tested using the transformer and diode bridge rectifier from [69]. Testing focuses on delivering maximum power to the load resistor as a function of speed. The output power versus speed is shown in Fig. 8.36. The load resistance is 37 Ω. At a rotor speed of 305 krpm, the PM generator delivers 8 W of DC power to the load resistor at an estimated electrical system efficiency of 49.2%. This corresponds to 16.27 W of mechanical-to-electrical power conversion. There is 7.46 W dissipated in the stator windings and 0.81 W dissipated in the transformer and diode bridge rectifier. The hysteresis losses in the stator core and the proximity eddy-current losses in the stator windings are 0.37 and 0.92 W, respectively. The stator core eddy-current losses are predicted to be 12.56 W. The resulting generator system efficiency, defined in Eq. (8.108) is only 26.6%. The measured values correspond well with a PSpice model of the generator and power electronics.

The dominant losses are the eddy current losses in the stator core which can be reduced significantly by using a laminated stator. The proximity eddy-current losses in the stator windings can be reduced by laminating the radial conductors. However, since the conduction losses are eight times as large as the proximity losses, it is better to leave the windings unlaminated to keep the winding resistance to a minimum.

Next, the PMSM is connected to the boost rectifier. To estimate the actual efficiency of the converter, the input current is measured using a Tektronix TCP202 current probe in conjunction with a Tektronix TDS3034B oscilloscope. The experimental input power is found by multiply the RMS values of the measured open-circuit voltage and stator currents by the power factor predicted by the PSpice model. This is done because the open-circuit voltage and stator currents could not be measured at the same time to compute the measured input power. The output

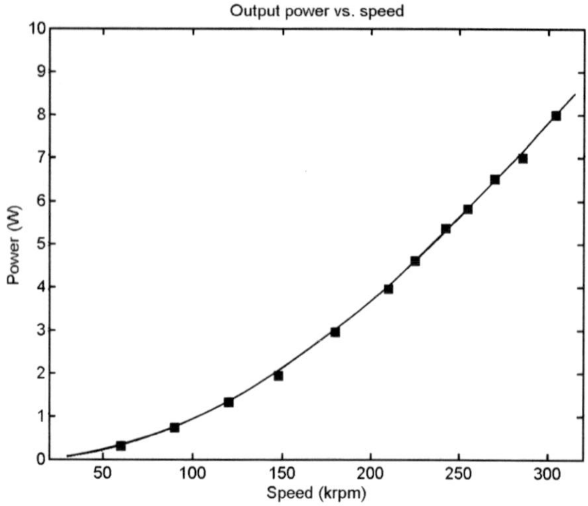

Fig. 8.36 DC output power across the load resistor for stator #3 vs. speed for a 37 Ω load at a 100 μm air gap. *Points* represent measurements; *curves* represent PSpice modeling (Adapted from [64])

power is measured in real time with a voltage probe across a load potentiometer, a current probe around a wire connecting the potentiometer to the rectifier and using the channel math function on the oscilloscope.

Figure 8.37 shows the output power at different speeds for both a self-sustaining converter and a converter with a 12-V external control power supply. Note that this is not the maximum output power at these speeds, but rather what can be achieved using the coarse pressure control knob on the nitrogen tank [64]. Also plotted in Fig. 8.37 is the input stator current required to achieve the output powers at various speeds. The externally powered boost rectifier demonstrates 5.47 W of power delivery at of 298 krpm, while the self-sustaining boost rectifier delivers 5.52 W at 303 krpm. These two data points were achieved by fixing the load resistance and then increasing the nitrogen pressure to achieve a 12-V output voltage. The measurements were then recorded and the pressure was decreased to slow down the rotor. Attempts were not made to adjust the load potentiometer to find the maximum power due to concerns for rotor structural integrity at these speeds.

The power electronics, electrical system, and generator system efficiencies for the self-sustaining boost rectifier are shown in Fig. 8.38. The power electronics efficiency varies between 40% at low speed and power, to 70% at high speed and power. This is because, as the speed increases, the output power delivered increases making the fixed losses in the control IC and gate drives a smaller fraction of the input power. At 303 krpm the self sustaining rectifier delivered 5.52 W at 69% power-electronics efficiency with an electrical system efficiency of 57%. Including the predicted core and proximity losses in the stator core of 13.72 W, the overall generator system efficiency is 23.6%. The self-sustaining rectifier has a generator system efficiency of only 11% at 150 krpm. The power factor for the output powers in Fig. 8.36 is between 98 and 99.5%.

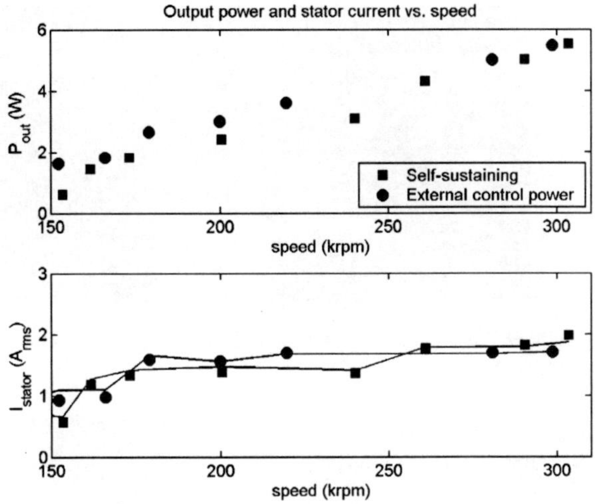

Fig. 8.37 Power and stator current versus speed for the boost rectifier with and without external control power. *Points* represent measurements; *curves* represent PSpice modeling (Adapted from [64])

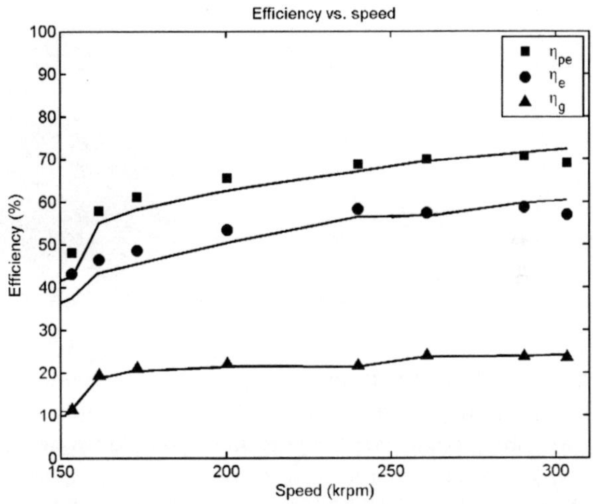

Fig. 8.38 Power electronics, η_{pe}, electrical system, η_e, and generator system, η_g, efficiencies versus speed for the self-sustaining boost rectifier. For η_{pe} and η_e, *points* represent measurements, and *curves* represent PSpice modeling. For η_g, *points* represent measurements/analytical model, and *curves* represent PSpice/analytical modeling (Adapted from [64])

To demonstrate the ability to power practical applications, the PMSM and rectifier is loaded with a series combination of six LEDs (two blue, two white, and two red). The generator is spun up to 150 krpm. A 9 V battery is connected temporarily to the control power terminal block and the converter boosts the output to 12 V. The test setup for this demonstration is shown in Fig. 8.39. Since the LEDS have a small viewing angle, the picture on the right shows the test setup from the top to better show the brightness of the LEDs. The control section of the rectifier consumes 0.48 W of power.

Figure 8.40 shows a self-contained microgenerator powering a PDA using only compressed gas [75]. The system integrates the PMSM, an air-turbine assembly

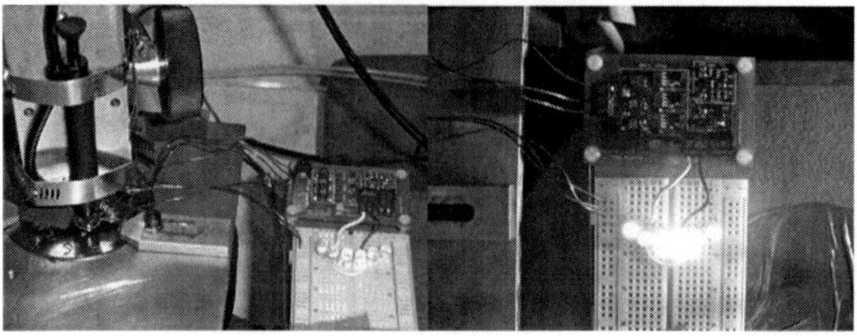

Fig. 8.39 Test setup for self-sustained operation. Boost rectifier is powering six LEDs (Adapted from [64])

Fig. 8.40 Standalone generator powering a PDA

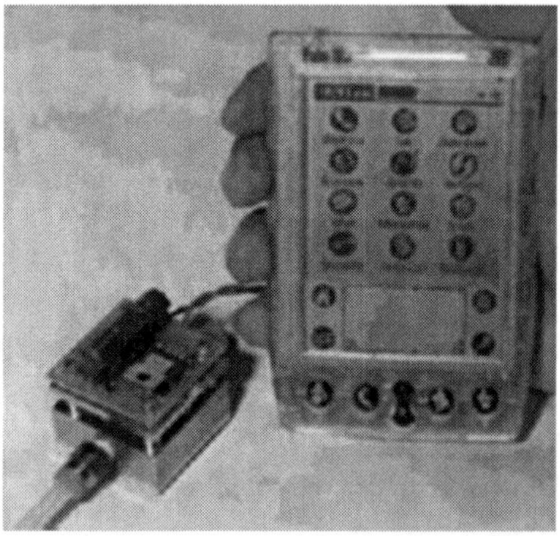

and power electronics into a compact, self-contained unit. The air-turbine assembly houses an off-the-shelf air-turbine in a conventionally milled stainless steel block. Two tubing connectors provide inlet and outlet connections to the air-turbine assembly. The rotor components are mounted on a steel shaft that fits into the air-turbine. Output wiring is soldered to the stator bond pads and routed to the power electronics board. The power electronics consist of three transformers for stepping up the generator voltages, a three-phase Schottky diode bridge for rectification and a linear regulator to provide a constant 5 V output under varying load conditions. The complete system measures 3.8 cm on a side, with a volume of 55 cm^3.

8.4.4.2 Comparison Between the Passive and Active Power Electronics

The passive and active power electronics are designed with different constraints and with different objectives. The goal of the passive power electronics is to maximize output power by efficiently stepping up and rectifying the generator's AC voltages with no constraints on the size of the circuit. In addition, output voltage regulation and power factor correction are not needed for initial low-speed power generation tests. With no size constraint, a very large transformer is used to make the passive electronics highly efficient. The boost rectifier, on the other hand, is designed with size being an important consideration. Unlike the passive circuit, the boost rectifier provides power factor correction and voltage regulation in addition to stepping up and rectifying the AC voltages from the PMSM. Its goal is to efficiently deliver power at a regulated DC voltage with high power factor in a compact size. The size of the boost-rectifier power stage is limited to the size of a stator die. The passive and active power electronics are shown in Fig. 8.41. The transformer in the passive circuit is much larger than the power stage of the boost rectifier, and the bridge rectifier in the passive circuit is larger than the control section of the boost rectifier. The black terminal blocks in Fig. 8.41 are used so that the generator can be easily connected and disconnected to the power electronics.

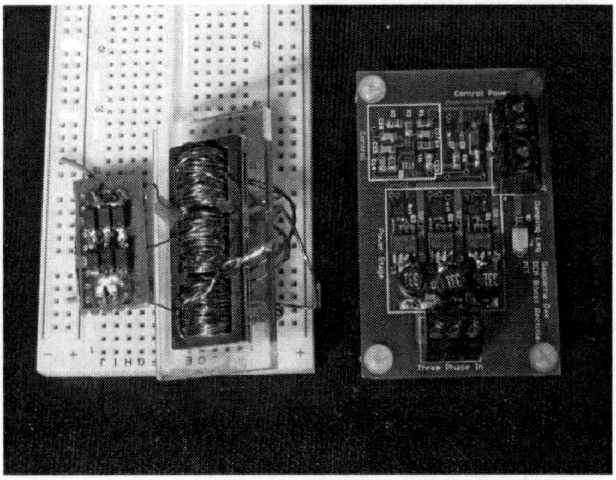

Fig. 8.41 Size comparison of the passive (*left*) and active (*right*) power electronics

Table 8.6 gives a breakdown of the performance of the PMSM with the passive and active power electronics. At 305 krpm, the PMSM with the transformer and diode bridge rectifier can deliver 8 W at an electrical system efficiency of 49.2% and overall generator system efficiency of 26.6%. The passive power electronics is 90.8% efficient, but the generator has significant conduction loss (7.46 W) which is an order of magnitude larger than the losses in the power electronics. The power factor for the passive circuit is 89.2%. At 303 krpm, the PMSM with the boost rectifier can deliver 5.52 W at 57% electrical system efficiency and 23.6% generator system efficiency. The power factor of 98.3% is very close to unity. The boost rectifier is 69% efficient and the power lost in the circuit (2.46 W) is actually larger than the stator conduction loss (1.71 W). In both cases, the 12+ W of eddy-current loss in the stator core leads to generator system efficiencies that are significantly lower than the electrical system efficiencies.

A better performance comparison may be between the transformer/diode bridge rectifier and the power stage of the boost rectifier. At 303 krpm, the power stage

Table 8.6 Performance breakdown of the first and second generation power electronics

Property	Transformer/diode bridge rectifier	Self-sustaining boost rectifier power stage	Self-sustaining boost rectifier
Speed (krpm)	305	303	303
P_{mech}(W)	30.12	23.4	23.4
P_{prox}(W)	0.92	0.91	0.91
P_{eddy}(W)	12.56	12.45	12.45
$P_{hysteresis}$(W)	0.37	0.36	0.36
P_{elect}(W)	16.27	9.68	9.68
P_{cond}(W)	7.46	1.71	1.71
P_{pe}(W)	0.81	1.98	2.46
P_{out}(W)	8	6	5.52
η_{pe}(%)	90.8	78	69
η_e (%)	49.2	62	57
η_g(%)	26.6	25.6	23.6
Power factor (%)	89.2	98.3	98.3
Volume (mm^3)	5017	1080	1190
Power density (MW/m^3)	1.59	5.56	4.64

delivers 6 W out, of which 0.48 W is fed back to power the control circuitry. Since both circuits provide voltage step-up and rectification, one can think of the 0.48 W of fixed loss as the cost of providing output voltage regulation and power factor correction. The electrical system efficiency in this case is 62%, and the power stage is 78% efficient. The generator system efficiency in this case is only 1% less than the PMSM with the passive circuit. Another important metric for power electronic circuits is power density. The frame of the transformer is 43 mm long, 25 mm wide, and 3 mm thick. The windings are cylinders that extend 4 mm beyond the thickness of the frame in each direction and are 10 mm long each. The volume of the transformer is 4833 mm^3. The six diodes are 2.9 mm by 4.6 mm by 2.3 mm each, for a bridge rectifier volume of 184 mm^3. The total volume of the passive power electronics is 5017 mm^3 and the power density is 1.59 MW/m^3 when delivering 8 W to the load. The power stage of the boost rectifier is 24 mm by 22.5 mm. The tallest components are the boost inductors which are 5 mm high. The other components are 1–1.75 mm in height. The bypass and filter capacitors on the bottom side of the PCB can easily fit in the empty spaces on the top side of power stage area. For a mean height of about 2 mm, the volume of the power stage is 1080 mm^3. The control section is 11 mm by 10 mm, and the average component height is 1 mm for a volume of 110 mm^3. The total power electronics volume is 1190 mm^3. At an output power of 5.52 W the boost rectifier has a power density of 4.64 MW/m^3, which is nearly three times the power density of the passive power electronics. Considering only the power stage of the boost rectifier, the power density is 5.56 M W/m^3 for 5.52 W of output power (6 W from the power stage). This is more than three and a half times larger than the passive power electronics.

8.4.5 Conclusions

This section presented the modeling, design, and testing of microscale generators and power electronics capable of generating and delivering Watt-level DC power to a load. The generators are three-phase 8-pole axial-air-gap PMSMs. The rotor consists of a 500-μm SmCo permanent magnet, and a Hiperco 50 core. The stator consists of 3-turn-per-pole windings electroplated on the surface of a 1-mm thick Moly Permalloy stator core. The generators were designed and fabricated using a combination of microfabrication, and precision machining and assembly.

PMSMs allow the use of both passive and active power electronics to deliver DC power. The passive power electronics consist of a three-phase Δ/wye-connected transformer and a three-phase diode bridge rectifier. The active power electronics consist of a three-phase semi-bridge boost rectifier operating in DCM. DCM operation provides power factor correction without the need for direct rotor position/speed or stator current/voltage sensing. In addition, DCM operation of the rectifier turns the generator/boost semi-bridge into a controlled current source so that only a single feedback loop is necessary to regulate the output voltage. The power stage of the rectifier uses the same area as a stator die.

One generator is able to deliver 8 W of DC power to a load resistor at 305 krpm using the transformer and diode bridge rectifier. The stator current is 5.36 A peak corresponding to a current density of $\sim 6\times 10^8$ A/m^2, which is two orders of magnitude larger than what is possible in a macroscale machine. This is possible in an MEMS device because of the intimate thermal contact between the windings and their surrounding materials. For an active machine volume of 138.4 mm^3, (10-mm outer diameter, 5-mm inner diameter, and 2.35-mm thickness), the 8-W output power corresponds to a power density of 57.8 MW/m^3. A typical 1000-MW turbine/generator with an active volume of 60 m^3 (8.4-m length, 3-m stator-core outer diameter) has a power density of 17 MW/m^3. With 16.27 W electrical power converted through the air gap at 305 krpm, the shear stress is 2.8 kN/m^2 using (8.40). This is much lower than the 180 kN/m^2 for the 1000-MW liquid-cooled turbine/generator. Despite the lower shear stress the microscale generator has more than three times the power density of a 1000-MW turbine generator due its much higher current densities and rotor tip speeds.

The estimated electrical and generator system efficiencies are 49.2 and 26.6%, respectively, when the generator is delivering 8 W to a load. The main difference between the two efficiency metrics is due to the eddy-current losses in the stator core. In fact, the 12.56-W loss due to eddy currents in the stator core is larger than either the output power or the sum of all the other losses in the system (9.56 W). This means that laminating the stator will be necessary to increase the overall generator system efficiency. The proximity eddy-current losses in the stator windings are on the order of 1 W at 305,000 rpm. At higher speeds, this will increase significantly and it may be necessary to laminate the radial conductors of the stator windings as well. However, because the PMSM is expected to operate at 300°C in a microengine, the conductivity of the stator windings will drop by more than a factor of two.

The resulting decrease in conductivity will partially offset the increased proximity eddy-current losses due to higher rotational speeds. The increase in conduction loss due to higher temperatures may require that the radial conductors be unlaminated to minimize the sum of the conduction and proximity losses.

The microscale generator, when connected to the self-sustaining boost rectifier, is able to deliver 5.52 W to a load resistor at a regulated output voltage of 12 V and at a speed of 303 krpm. The power electronics is 69% efficient and has a power factor of 98.3%. The electrical and generator system efficiencies are 57 and 23.6%, respectively, with the difference due to the large eddy-current losses in the stator core. The power density of the boost rectifier is 4.64 MW/m^3, which is three times the power density of the transformer/diode bridge, even though it is less efficient than the passive circuit. The generators are also able to power a string of LEDs and a PDA. These results prove that Watt-level power generation and delivery is achievable using microscale axial-flux PM machines and switch-mode power electronics, and that they are a viable compact power generation system for portable power applications.

8.5 Summary

For reasons discussed in Section 8.1.1, this chapter has focused on the electric induction machine and the permanent-magnet synchronous machine. It is difficult to make a conclusive comparison between the two because their characteristics are very different at the microscale. The advantages of the electric induction machine are at that it is CMOS compatible, and probably the simpler of the two to fabricate. Further, it can operate at very high temperatures if its rotor conductor is designed with temperature stability in mind. Its greatest disadvantages are the viscous losses incurred in its air gap, and the inconvenience and complexity of developing high-voltage and high-frequency power electronics. Its operation as a generator is also greatly influenced by parasitics. The advantages of the permanent-magnet synchronous machine are its power and torque density, which have so far exceeded that of the electric induction machine, and its insensitivity to parasitics. Additionally, because of its comparatively lower pole count, the permanent-magnet synchronous machine can operate with a much larger air gap, and correspondingly smaller viscous losses. Its greatest disadvantage is probably the increased complexity of its fabrication, and its inability to run at high temperatures.

References

1. A. E. Fitzgerald, C. Kingsley, and S. D. Umans, *Electric Machinery*, Sixth Edition, McGraw Hill, New York 2003.
2. H. H. Woodson and J. R. Melcher, *Electromechanical Dynamics: Parts I & II*, Wiley, New York 1968.
3. W. Leonhard, *Control of Electrical Drives*, Springer-Verlag, Berlin 1985.
4. B. K. Bose, *Power Electronics and AC Drives*, Prentice-Hall, Englewood Cliffs 1987.

5. H. A. Haus and J. R. Melcher, *Electromagnetic Fields and Energy*, Prentice Hall, Englewood Cliffs 1989.
6. F. Paschen, "Ueber die zum funkenübergang in luft, wasserstof und kohlensäure bei verschiedenen drucken erforderliche potentialdifferenz"; *Annal der Physik*, *37*, 69–96, 1889.
7. J. M. Meek and J. D. Craggs, *Electrical Breakdown of Gases*, John Wiley & Sons, New York 1978.
8. M. Zahn, "Solid, liquid, and gaseous electrical insulation", *Encyclopedia of Applied Physics*, *18*, 431–466, VCH Publishers, Weinheim 1997.
9. S. F. Bart, T. A. Lober, R. T. Howe, J. H. Lang, and M. F. Schlecht, "Design considerations for micromachined electric actuators"; *Sensors and Actuators*, *14*, 269–292, July 1988.
10. M. Mehregany, P. Nagarkar, S. D. Senturia, and J. H. Lang, "Operation of microfabricated harmonic and ordinary side-drive motors"; *Proceedings: IEEE Workshop on Micro Electro Mechanical Systems*, 1–8, Napa Valley, CA, February 12–14, 1990.
11. M. Mehregany, S. F. Bart, L. S. Tavrow, J. H. Lang, S. D. Senturia, and M. F. Schlecht, "A study of three microfabricated variable-capacitance motors"; *Sensors and Actuators*, *A-21,22,23*, 173–179, February–April 1990.
12. S. F. Bart, M. Mehregany, L. S. Tavrow, J. H. Lang, and S. D. Senturia, "Electric micromotor dynamics"; *IEEE Transactions on Electron Devices*, *39*, 566–575, March 1992.
13. C. H. Ahn, Y. J. Kim, and M. G. Allen, "A planar variable-reluctance magnetic micromotor with fully integrated stator and wrapped coils", *Proceedings: IEEE Workshop on Micro Electro Mechanical Systems*, 1–6, Fort Lauderdale, FL, February 7–10, 1993.
14. H. Guckel, T. R. Christenson, K. J. Skorbis, T. S. Jung, J. Klein, K. V. Hartojo, and I. Widjaja, "A first functional current excited planar rotational magnetic micromotor", *Proceedings: IEEE Workshop on Micro Electro Mechanical Systems*, 7–11, Fort Lauderdale, FL, February 7–10, 1993.
15. W. C. Young and R. G. Budynas, *Roark's Formulas for Stress and Strain*, 7TH Edition, p. 745, McGraw-Hill, New York 2002.
16. M. W. Layland, "Generalized electrostatic machine theory", *IEE Proceedings*, *116*, 403–405, 1969.
17. T. J. E. Miller, *Switched Reluctance Motors and their Control*, Magna Physics Publishing and Oxford University Press, London 1993.
18. D. A. Torrey and J. H. Lang, "Optimal-efficiency excitation of variable-reluctance motor drives", *IEE Proceedings Part B*, *138*, 1–14, January 1991.
19. D. E. Cameron and J. H. Lang, "The control of high-speed variable-reluctance generators in electric power systems", *IEEE Transactions on Industry Applications*, *29*, 1106–1109, November/December 1993.
20. Y. Tada, "Theoretical characteristics of generalized electret generator using polymer film electrets", *IEEE Transactions on Electrical Insulation*, *21*, 457–464, June 1986.
21. Y. Arakawa, Y. Suzuki, and N. Kasagi, "Micro seismic power generator using electret polymer film", *Proceedings: PowerMEMS Conference*, 187–190, Kyoto, Japan, November 28–30, 2004.
22. Y. Sakane, Y. Suzuki, and N. Kasagi, "Development of high-performance purfluorinated polymer electret", *Proceedings: IEEE Symposium on Electrets*, *13*, Tokyo, Japan, September 15–17, 2008.
23. W. Olthuis and P. Bergveld, "On the charge storage and decay mechanism in silicon dioxide", *IEEE Transactions on Electrical Insulation*, *27*, 691–697 April 1992.
24. H. Köşer, *Development of Magnetic Induction Machines for Micro Turbo Machinery*, Ph. D. Thesis, EECS Department, MIT, June 2002.
25. F. Cros, *Developpement D'une Micromachine Induction Magnetique - Developpement do Techniques de Microfabrication pour Micro-electroaimants*, Ph. D. Thesis, Ecole Doctorale de Toulouse, France, September 2002.
26. H. Köşer and J. H. Lang, "Magnetic induction micro machine – Part I: design and analysis"; *IEEE/ASME Journal of Microelectromechanical Systems*, *15*, 415–426, April 2006.

27. F. Cros, H. Köşer, M. G. Allen, and J. H. Lang, "Magnetic induction micro machine – Part II: fabrication and testing"; *IEEE/ASME Journal of Microelectromechanical Systems, 15*, 427–439, April 2006.
28. H. Köşer and J. H. Lang; "Magnetic induction micro machines – Part III: eddy currents and nonlinear effects"; *IEEE/ASME Journal of Microelectromechanical Systems, 15*, 440–456, April 2006.
29. J. R. Melcher, *Continuum Electromechanical Dynamics*, MIT Press, Cambridge.
30. L. G. Frechette, *Development of a Microfabricated Silicon Motor-Driven Compression System*, Ph. D. Thesis, AA Department, MIT, September 2000.
31. S. F. Nagle, *Analysis, Design and Fabrication of an Electric Induction Micromotor for a Micro Gas-Turbine Generator*, Ph. D. Thesis, EECS Department, MIT, October 2000.
32. S. F. Bart and J. H. Lang; "Electroquasistatic induction micromotors"; *Proceedings: IEEE Workshop on Micro Electromechanical Systems Workshop*, 7–12, Salt Lake City, UT, February 20–22, January 1989.
33. S. F. Bart and J. H. Lang; "An analysis of electroquasistatic induction micromotors"; *Sensors and Actuators, 20*, 97–106, November 1989.
34. L. G. Frechette, S. F. Nagle, R. Ghodssi, S. D. Umans, M. A. Schmidt, and J. H. Lang; "An electrostatic induction micromotor supported on gas-lubricated bearings"; *Proceedings: IEEE Workshop on Micro Electro Mechanical Systems*, 290–293, Interlaken, Switzerland, January 21–25, 2001.
35. C. Livermore, A. Forte, T. Lyszczarz, S. D. Umans, A. A. Ayon, and J. H. Lang; "A high-powerMEMS electric induction motor"; *IEEE/ASME Journal of Microelectromechanical Systems, 13*, 465–471, June 2004.
36. S. F. Nagle, C. Livermore, L. G. Frechette, R. Ghodssi, and J. H. Lang, "An Electric Induction Micromotor", *IEEE/ASME Journal of Microelectromechanical Systems, 14*, 1127–1143, October 2005.
37. J. L. Steyn, *A Microfabricated Electroquasistatic Induction Turbine Generator*, Ph. D. Thesis, AA Department, MIT, June 2005.
38. J. L. Steyn, S. H. Kendig, R. Khanna, T. M. Lyszczarz, S. D. Umans, J. U. Yoon, C. Livermore, and J. H. Lang; "Generating electric power with a MEMS electroquasistatic induction turbine generator"; *Proceedings: IEEE Workshop on Micro Electromechanical Systems*, 614–617, Miami, FL, January 30–February 3, 2005.
39. J. L. Steyn, S. H. Kendig, R. Khanna, T. M. Lyszczarz, S. D. Umans, J. U. Yoon, C. Livermore, and J. H. Lang, "A self-excited MEMS electroquasistatic induction turbine-generator"; *IEEE/ASME Journal of Microelectromechanical Systems, 18*, 424–432, April 09.
40. C. Kooy, "Torque on a resistive rotor in a quasi electrostatic rotating field"; *Applications of Scientific Research, 20*, 161–172, February 1969.
41. B. Bollee, "Electrostatic motors"; *Phillips Technical Review, 30*, 178–194, June–July 1969.
42. J. Ubbink, "Optimization of the rotor surface resistance of the asynchronous electrostatic motor"; *Applications of Scientific Research, 22*, 442–448, October 1970.
43. S. D. Choi and D. A. Dunn, "A surface-charge induction motor"; *IEEE Proceedings, 59*, 737–748, May 1971.
44. P. T. Krein and J. M. Crowley, "Harmonics affects in electrostatic induction motors"; *Electric Machines and Power Systems, 10*, 479–497, May–June, 1985.
45. E. R. Mognashi and J. H. Calderwood, "Asynchronous dielectric induction motor"; *IEE Proceedings A, 137*, 331–338, November 1990.
46. S. F. Bart, L. S. Tavrow, M. Mehregany, and J. H. Lang, "Microfabricated electrohydrodynamic pumps"; *Sensors and Actuators, A-21,22,23*, 193–197, February–April 1990.
47. S. Hardt and F. Schoenfeld, *Microfluidic Technologies for Miniaturized Analysis Systems*; Springer, New York 2007.
48. J. R. Melcher and M. S. Firebaugh, "Traveling wave bulk electroconvection induced across a temperature gradient"; *Physics of Fluids, 10*, 1178–1185, 1967.

49. T. Kamins, *Polycrystalline Silicon For Integrated Circuits And Displays*, 2nd Edition; Kluwer Academic Publishers, Dordrecht, 1998.
50. T. C. Nuegebauer, D. J. Perreault, J. H. Lang, and C. Livermore, "A six-phase multilevel inverter for MEMS electrostatic induction micromotors"; *IEEE Transactions on Circuits and Systems II: Express Briefs*, *51*, 49–56, February 2004.
51. A. M. Flynn and S. R. Sanders, "Fundamental limits on energy transfer and circuit considerations for piezoelectric transformers"; *IEEE Transactions on Power Electronics*, *17*, 8–14, January 2002.
52. C. C. Lin, R. Ghodssi, A. A. Ayon, D. Z. Chen, S. A. Jacobson, K. S. Breuer, A. H. Epstein and M. A. Schmidt, "Fabrication and characterization of a micro turbine/bearing rig"; *Proceedings: IEEE Micro Electro Mechanical Systems Workshop*, 529–533 Orlando, FL, 1999.
53. L. G. Frechette, S. A. Jacobson, F. F. Ehrich, R. Ghodssi, R. Khanna, C. W. Wong, X. Zhang, K. S. Breuer, M. A. Schmidt, and A. H. Epstein, "Demonstration of a microfabricated high-speed turbine supported on gas bearings"; *Proceedings: Solid-State Sensor and Actuator Workshop*, 43–47, Hilton Head, SC, June 2000.
54. R. Ghodssi, L. G. Frechette, S. F. Nagle, X. Zhang, A. A. Ayon, S. D. Senturia, and M. A. Schmidt, "Thick buried oxide in silicon (TBOS): an integrated fabrication technology for multi-stack wafer bonded MEMS processes"; *Proceedings: Tenth International Conference on Solid State Sensors*, 1456–1459, Sendai, Japan, June 1999.
55. B. Yen, *A Fully-Integrated Mult-Watt Permanent-Magnet Turbine/Generator*, Ph. D. Thesis, EECS Department, MIT, June 2008.
56. F. Herrault, *Microfabricated Air-Turbine and Heat-Engine-Driven Permanent Magnet Generators*, Ph.D. Thesis, Institut National des Sciences Appliquées, Toulouse, France, February 2009.
57. B. Wagner, M. Kreutzer, and W. Benecke, "Permanent magnet micromotors on silicon substrates"; *Journal of Microelectromechanical System*, *2*, 1, pp. 23–29, March 1993.
58. K.-P. Kamper, et al., "Electromagnetic permanent magnet micromotor with integrated micro gear box"; *Proceedings of the 5th International Conference on New Actuators* (Actuator '96), June 1996, pp. 429–432.
59. U. Berg, et al., "Series production and testing of a micro motor"; *Proceedings of the 6th International Conference on New Actuators* (Actuator '98), June 1998, pp. 552–555.
60. P.-A. Gilles, J. Delamare, O. Cugat, and J.-L. Schanen, "Design of a permanent magnet planar synchronous micromotor"; *Proceedings of the 35th Mtg. IEEE Industry Appl. Soc.*, *1*, October 2000, 223–227.
61. C. Yang, et al., "An axial flux electromagnetic micromotor"; *Journal of. Micromechanics and Microengineering*, *11*, 113–117, 2001.
62. A. S. Holmes, G. Hong, and K. R. Buffard, "Axial-flux permanent magnet machines for micropower generation"; *Journal of Microelectromechanical System*, *14*, 1, 54–62, February 2005.
63. H. Raisigel, et al., "Magnetic planar micro generator," in *Digest Tech. Papers Transducers '05 Conference*, Seoul, South Korea, June 10–14, 2005, 757–761.
64. S. Das, "Magnetic machines and power electronics for power MEMS applications," Ph.D. Thesis, Massachusetts Institute of Technology, Cambridge, MA, September 2005.
65. S. Das, D. P. Arnold, I. Zana, J. -W. Park, M. G. Allen, and J. H. Lang, "Microfabricated high-speed axial-air-gap multi-watt permanent-magnet generators – Part I: modeling"; *Journal of Microelectromechanical System*, *15*, 5, 1330–1350, October 2006.
66. M. Zahn, *Electromagnetic Field Theory: A Problem Solving Approach*, Kreiger, Malabar, 1979.
67. J. G. Kassakian, M. F. Schlecht, and G. C. Verghese, *Principles of Power Electronics*, Addison Wesley, Reading, 1991.
68. D. P. Arnold, S. Das, I. Zana, J. -W. Park, J. H. Lang, and M. G. Allen, "Microfabricated high-speed axial-air-gap multi-watt permanent-magnet generators – Part II: Design, Fabrication and Testing," *Journal of Microelectromechanical System*, *15*, 5, 1330–1350, October 2006.

69. D. P. Arnold et al., “Design optimization of an 8 W microscale, axial-flux, permanent-magnet generator”; *Journal of Micromechanics Microengineering, 16*, 290–296, 2006.
70. R. M. Bozorth, *Ferromagnetism*, IEEE Press, New York, Reissue, 1993.
71. “MMPA Standard No. 0100-00, Standard Specifications for Permanent Magnet Materials,” published by International Magnetics Association (formerly Magnetic Materials Producers Association), Available: http://www.intl-magnetics.org/pdfs/0100-00.pdf.
72. E. H. Ismail and R. Erickson, “Single-Switch 3-Phase PWM Low Harmonic Rectifiers”; *IEEE Transaction on Power Electron*, *11*, 2, 338–346, March 1996.
73. A. R. Prasad, P. D. Ziogas, and S. Manias, “An active power factor correction technique for three-phase diode rectifiers”; *IEEE Transaction on Power Electron*, *6*, 83–92, January 1991.
74. D. Simonetti, J. L. Vieira, and G. Sousa, “Modeling of the high-power-factor discontinuous boost rectifiers”; *IEEE Transaction on Industrial Electronics*, *46*, 4, 788–795, August 1999.
75. D. P. Arnold, P. Galle, F. Herrault, S. Das, J. H. Lang, and M. G. Allen, "A self-contained, flow-powered microgenerator system"; Technical Digest 5th International Workshop Micro Nanotechnology For Power Generation and Energy Conversion Apps. (PowerMEMS 2005), Tokyo, Japan, 113–115, November 2005.

Chapter 9
Microcombustors for Rotating Machinery

C.M. Spadaccini and I.A. Waitz

9.1 Introduction

To drive microscale rotating machinery such as a microengine, motor-compressor, or generator, energy must be added to the system. This can take many forms depending on the application, however, for an engine intended to be a primary power source; this is usually thermal energy. Adding heat to an engine is most commonly achieved via a *combustion system* where a fuel is mixed with an oxidizer and burned. Due to the high-energy density of liquid hydrocarbons, they are the most common fuels and are mixed and burned with air. For a microengine, this system is termed the *microcombustor*.

Engines with rotating machinery usually operate on the Brayton cycle which is most commonly associated with gas turbine engines (a notable exception is the Wankel rotary engine which operates on an Otto cycle). An ideal Brayton cycle consists of an isentropic compression step (1-2), followed by constant pressure heat addition (2-3), isentropic expansion (3-4), and heat rejection (4-1). In terms of gas turbine components, these individual processes correspond to the compressor (1-2), combustor (2-3), turbine (3-4), and exhaust (4-1) as shown in Fig. 9.1.

In this chapter, we are primarily concerned with the miniaturization and performance of the constant pressure heat addition step. A typical combustor is a flow through device (hence the constant pressure) where high-pressure air from the compressor enters a chamber, mixes with fuel which is injected from a separate source, ignites and burns. This raises the temperature of the working fluid as well as the enthalpy. The hot combustion products then exhaust from the chamber and enter a turbine where the gas expands and work is extracted. Figure 9.2 shows a conventional combustor found in a large scale gas turbine engine.

At the microscale, combustion for the purpose of driving rotating machinery must satisfy similar requirements. It should accept high-pressure air from a microcompressor, inject and mix fuel, burn the mixture efficiently, and provide high enthalpy

C.M. Spadaccini (✉)
Center for Micro and Nano Technology, Engineering Technologies Division, 7000 East Avenue, Livermore, CA 94550, USA

J.H. Lang (ed.), *Multi-Wafer Rotating MEMS Machines*, MEMS Reference Shelf,
DOI 10.1007/978-0-387-77747-4_9,

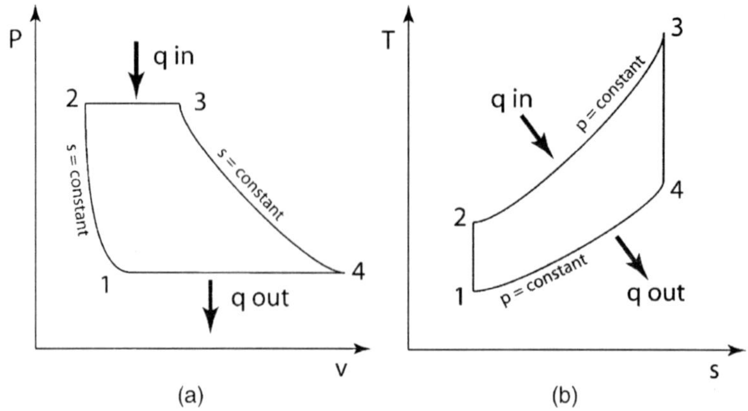

Fig. 9.1 Ideal Brayton cycle; (**a**) pressure–volume coordinates and (**b**) temperature–entropy coordinates

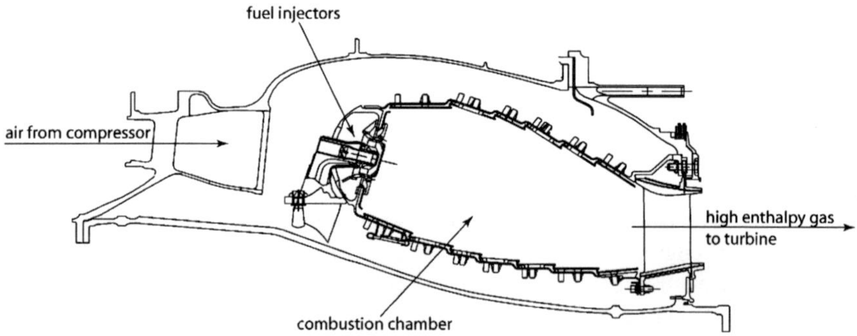

Fig. 9.2 Conventional gas turbine combustor. (Reprinted with permissions from Pratt & Whitney Aircraft Engines)

fluid to the turbine for work extraction; however, there are additional constraints due to the small scale. These include dealing with very short time-scales, increased heat loss via fluid-structure coupling, and fabrication and materials issues.

All of this will be discussed in this chapter. We will begin by presenting some basic combustion concepts which are not unique to microsystems. This includes several points of nomenclature and the basics of combustion physics and chemistry. Next, the challenges specific to combustion in microengines will be presented. This will be followed by a discussion of the two primary strategies for operating a micro-combustor; homogenous gas-phase systems and heterogeneous catalytic systems. This will include both modeling and experimental results for fabricated devices designed for integration with a rotating structure as well as some more generally applicable conclusions. A specific section on the fabrication of these devices is then presented and followed by a conversation regarding future trends and concepts.

9.2 Basic Combustion Concepts

Before discussing issues pertaining specifically to microcombustors, the reader should be familiar with some basic combustion concepts. These include the fundamental chemical equations of combustion reactions, the concept of equivalence ratio (normalized fuel-air ratio), reaction rate and Arrhenius expressions, and various definitions of efficiency.

9.2.1 Combustion Chemistry

Combustion consists of a chemical reaction between a fuel (typically hydrogen or a hydrocarbon species) and an oxidizer leading to heat release. In general, the overall chemical equation for complete combustion of hydrocarbons with air can be written as

$$C_xH_y + \left(x + \frac{y}{4}\right)(O_2 + 3.76N_2) \rightarrow xCO_2 + \frac{y}{2}H_2O + 3.76\left(x + \frac{y}{4}\right)N_2. \quad (1)$$

Typical fuels for conventional gas turbines are long chain kerosene-based liquids containing a mixture of molecules with 12–15 carbon atoms in each molecule. For the microcombustors discussed here and in most literature sources, gaseous fuels such as hydrogen (H_2), ethylene (C_2H_4), and propane (C_3H_8) are most common [1–6]; however, more recent work has moved in the direction of long chain liquid fuels [7]. Choice of fuel often depends on fuel storage constraints, reaction rates, and emissions requirements.

The combustion chemistries for hydrogen and propane with air respectively are shown below.

$$H_2 + \frac{1}{2}O_2 + 1.88N_2 \rightarrow H_2O + 1.88N_2 \quad (2)$$

$$C_3H_8 + 5O_2 + 18.8N_2 \rightarrow 3CO_2 + 4H_2O + 18.8N_2 \quad (3)$$

The only products of complete combustion are water and carbon dioxide (in the ideal case nitrogen is inert); however, in reality this is not the case. Trace quantities of pollutants such as oxides of nitrogen (NO_x) and carbon monoxide (CO) are usually present as well as some unburned hydrocarbons. Incomplete combustion may occur if the reactants are not well mixed or if there is insufficient residence time for the combustion to be completed (as is often the case with microcombustors). In these scenarios, there will be significant quantities of reactants and other chemical species remaining in the exhaust gas stream.

Although combustion chemistries can be written generally as single chemical equations, there are many individual subreactions which combine to give the overall reaction. These sets of chemical equations are referred to as the reaction mechanism.

9.2.2 Equivalence Ratio

If propane and air, for example, are provided to a combustion system in a 23.8:1 molar ratio, the mixture is considered stoichiometric. In this case, all fuel and air are consumed in the reaction provided there is adequate time. If there is excess oxidizer, the reaction is considered to be *lean*, whereas with excess fuel, it is *rich*. This ratio is referred to as the fuel/air ratio and is usually given on a mass basis which can be obtained by scaling the molar ratio by the molecular weights of the constituents. In a combustor, the mass-based fuel/air ratio can be found directly from measurements of the fuel and air mass flow rates.

$$\mathrm{fa} = \frac{\dot{m}_{\mathrm{fuel}}}{\dot{m}_{\mathrm{air}}} \tag{4}$$

The fuel/air ratio can be normalized by the stoichiometric ratio to yield a quantity known as equivalence ratio.

$$\phi = \frac{\mathrm{fa}_{\mathrm{actual}}}{\mathrm{fa}_{\mathrm{stoichiometric}}} \tag{5}$$

This is a more commonly used measure of mixture ratio. An equivalence ratio of $\phi=1$ indicates that the fuel/air mixture is stoichiometric. When $\phi<1$ the mixture is lean, and if $\phi>1$ it is rich. Combustion reactions typically peak in temperature around an equivalence ratio of unity and are reduced the more lean or rich the mixture. This is due primarily to the unreacted constituents absorbing some of the heat released from the reaction. The theoretical temperature achieved with an adiabatic boundary condition (all released heat is used to heat the products) is known as the adiabatic flame temperature and is a useful reference point.

9.2.3 Kinetics

Chemical kinetics is the study of the rate at which a reaction occurs. Generally, the conditions that can impact reaction rate include reactant concentration, temperature, and the presence of a catalyst. This rate can be expressed as the rate of change in concentration of the reacting substances (a decrease for reactants and an increase for products) per unit time. The rate of change can also be equated to a rate constant multiplied by the concentrations of the reactant species raised to the power of their respective stoichiometric coefficients. For example, an arbitrary reaction may be written as

$$A + 2B \rightarrow C + D \tag{6}$$

where the reaction rate is,

$$-R = \frac{\mathrm{d}[A]}{\mathrm{d}t} = -k\,[A]\,[B]^2 \tag{7}$$

with brackets denoting concentration and k being the rate constant.

The temperature dependence of reaction rate is contained within the rate constant k. This is embodied in the Arrhenius expression for the rate constant which is generally given as

$$k = A\exp\left(\frac{-E_a}{RT}\right) \tag{8}$$

where E_a is the activation energy for the reaction, R is the gas constant, T is the temperature, and A is the kinetic pre-exponential factor. The pre-exponential factor accounts for molecular collisions and steric factors, and along with the activation energy, it is usually obtained empirically. Typically, the constants are extracted from a linear fit to empirical data on a ln(k) vs. (1/T) plot where the slope of the fit is ($-E_a/R$). Collision theory and more advanced techniques such as transition state theory can be used to calculate pre-exponentials and rate constants; however, these methods tend to over-predict reaction rates.

The activation energy represents an energy barrier which must be overcome in order to begin progress from reactants to products. This energy is typically provided in the form of thermal energy and once it is provided, the reaction rapidly proceeds to completion. Figure 9.3 shows energy as a function of reaction coordinate for a typical exothermic combustion reaction. With a catalyst present, the activation energy is reduced while all other aspects of the reaction remain unchanged.

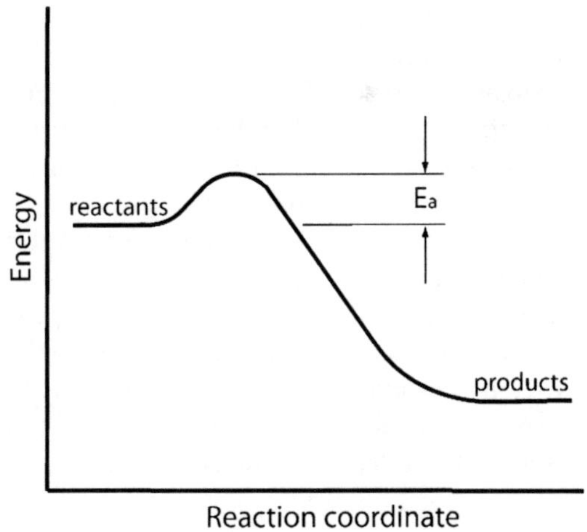

Fig. 9.3 Energy as a function of reaction coordinate for a generic exothermic reaction

9.2.4 Microcombustor Efficiency Definitions

For the purpose of this discussion, overall combustor efficiency will be based on an enthalpy balance across the combustor and is defined as:

$$\eta_c = \frac{(\dot{m}_a + \dot{m}_f)h_2 - \dot{m}_a h_1}{\dot{m}_f h_f} \tag{9}$$

where station (1) refers to the combustor inlet and station (2) refers to the combustor exit. The combustor efficiency can be written as the product of two sub-efficiencies; a chemical efficiency, and a thermal efficiency. These two parameters can be written as,

$$\eta_{\text{chem}} = \frac{[(\dot{m}_{\text{a}} + \dot{m}_{\text{f}})h_2 - \dot{m}_{\text{a}}h_1] + Q_{\text{loss}}}{\dot{m}_{\text{f}}h_{\text{f}}} = \frac{\text{total enthalpy released}}{\text{maximum enthalpy release possible}} \tag{10}$$

$$\eta_{\text{therm}} = \frac{(\dot{m}_{\text{a}} + \dot{m}_{\text{f}})h_2 - \dot{m}_{\text{a}}h_1}{[(\dot{m}_{\text{a}} + \dot{m}_{\text{f}})h_2 - \dot{m}_{\text{a}}h_1] + Q_{\text{loss}}} = \frac{\text{enthalpy rise of fluid}}{\text{total enthalpy released}} \tag{11}$$

where Q_{loss} is the heat lost to the structure. Dividing the combustor efficiency into two parts like this is not typical for large scale combustors, but is useful for micro-combustors because of the significant role of heat loss in their performance.

9.2.5 Combustion Strategies

There are two fundamental combustion techniques; homogeneous gas-phase combustion and heterogeneous catalytic combustion. Both will result in the heat release defined by the chemical mechanism and its associated thermodynamics. The primary difference between the two methods is implementation.

Homogenous gas-phase combustion is simply the mixing and reacting of the fuel and oxidizer species while both are gaseous. This is the most commonly encountered combustion process. Even if the fuel is introduced as a liquid, it is first vaporized before reacting with the oxidizer.

Heterogeneous catalytic combustion involves the use of a solid-phase catalyst. Typically, for hydrocarbon reactions, this is a noble metal with platinum and palladium being the most common. In this type of process, the solid catalyst surface provides local sites at which the fuel and oxidizer species will bind and react followed by desorption of products as shown in Fig. 9.4. The reaction occurring at these sites propagates at a significantly faster rate than in the gas-phase. This is the result of the surface site lowering the activation energy of the reaction; other than

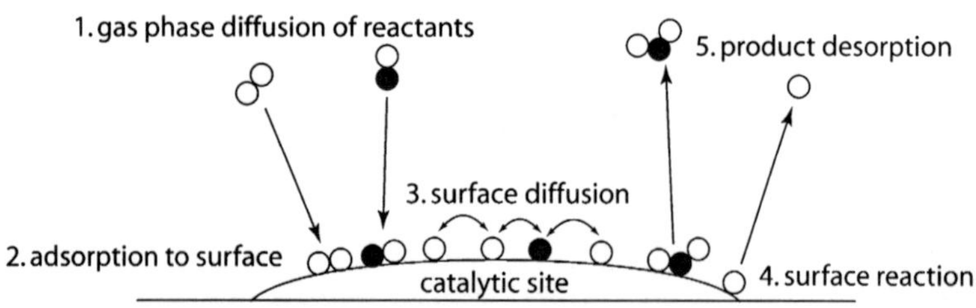

Fig. 9.4 Schematic of catalytic combustion process

this, the catalyst does not change the reaction in any other way. The products generated and the total heat released is identical to the gas-phase reaction. The addition of a catalyst adds complexity to the system and its primary limitations include:

1. diffusion of the fuel and oxidizer species to the active surface sites;
2. ensuring adequate catalyst surface area while minimizing pressure loss;
3. start-up and stability of the reaction process;
4. heat loss through the solid catalyst surface;
5. and material cost.

The physical phenomena occurring at the catalyst surface which cause the increased reaction rates are not well understood and constitute a fertile field of research.

9.3 Microcombustor Challenges

There are fundamental concepts and challenges which are unique to a microcombustor being designed for integration with rotating machinery. These include residence time constraints, material constraints, fluid-structure thermal coupling, the overall engine (or other rotating apparatus) design space, and fabrication and packaging. These concepts and challenges are outlined below.

9.3.1 Residence Time Constraints

Amongst the most fundamental physical restrictions of a microcombustion system is the significantly reduced residence time. For energy conversion applications, such as an engine or other rotating apparatus, power density is a critical metric. Using the gas turbine combustor as a model system, the differences in power density spanning length scales become evident. As shown in Table 9.1, the high power density of a microcombustor directly results from high mass flow per unit combustor volume. Chemical reaction times do not scale with mass flow rate or combustor volume, only temperature, pressure, and species concentrations. As a result, the realization of high power density is contingent upon completing the combustion process within a significantly shorter combustor through-flow time.

This fundamental time constraint can be quantified in terms of a homogeneous Damköhler number; the ratio of the residence time to the characteristic chemical reaction time.

$$Da_{\mathrm{h}} = \frac{\tau_{\mathrm{residence}}}{\tau_{\mathrm{reaction}}} \tag{12}$$

To ensure complete and efficient combustion, a Da_{h} of order unity or greater is required. A microcombustor designer should either increase the flow residence time

Table 9.1 Comparison of operating parameters and requirements for a micro gas turbine engine combustor with those estimated for a conventional GE90 gas turbine combustor [2]

	Conventional combustor	Microcombustor
Length	0.2 m	0.001 m
Volume	0.073 m^3	6.6×10^{-8} m^3
Cross-sectional area	0.36 m^2	6×10^{-5} m^2
Inlet total pressure	37.5 atm	4 atm
Inlet total temperature	870 K	500 K
Mass flow rate	140 kg/s	1.8×10^{-4} kg/s
Residence time	~7 ms	~0.5 ms
Efficiency	>99%	>90%
Pressure ratio	>0.95	>0.95
Exit temperature	1800 K	1600 K
Power density	1960 MW/m^3	3000 MW/m^3

Note: residence times are calculated using inlet pressure and an average flow temperature of 1000 K.

or decrease the chemical reaction time in order to ensure this constraint is satisfied; however, this is often difficult or even impossible when the system geometry and thermo-fluid requirements are considered.

The characteristic combustor residence time is given by estimating the bulk flow rate through the combustor volume

$$\tau_{\text{residence}} \approx \frac{\text{volume}}{\text{volumetric flow rate}} = \frac{VP}{\dot{m}RT} \tag{13}$$

and can be increased by increasing the size of the combustion chamber, reducing the mass flow rate, or increasing the operating pressure. Chemical reaction times can be approximated using an Arrhenius type expression.

$$\tau_{\text{reaction}} \approx \frac{[\text{fuel}]_0}{A[\text{fuel}]^a[\text{O}_2]^b e^{-E_a/RT_0}} \tag{14}$$

It is clear from Eq. (14) that reaction time is primarily a function of species concentrations and temperature.

Since high power density requirements mandate high mass flow rates through small chamber volumes, the mass flow rate per unit volume cannot be reduced without compromising device power density. Hence, there is a basic tradeoff between power density and flow residence time.

$$\text{Power density} \propto \frac{\dot{m}}{V} \propto \frac{\dot{m}_f LHV}{V} \propto \frac{\rho}{\tau_{\text{residence}}} \tag{15}$$

For a given operating pressure (and thus density), and assuming a Da_{h} of unity, reducing the chemical reaction time, and thus required residence time, is the only means of ensuring complete combustion without compromising the high power density of the device.

The more complex catalytic combustors depend upon yet another time-scale in addition to chemical reaction time and residence time. Although the reaction time and residence time are still critical, the diffusion time required for species to reach the active catalytic surface must also be accounted for. The time required for a fuel and oxidizer molecule to come into contact with the catalytic sites is often the rate limiting time-scale, especially since the chemical reaction times are greatly reduced in these systems.

As a result there are several non-dimensional time-scale ratios that can be examined to determine which phenomenon governs overall performance. Combustor residence time and reaction time can be estimated by the same means used in the gas-phase case provided there are catalytic reaction mechanisms available in the literature. With the additional diffusion time-scale, a second Damköhler number can be defined. This parameter is the ratio of the diffusion time to the reaction time and indicates the relative rates of these two processes.

$$Da_2 = \frac{\tau_{\text{diffusion}}}{\tau_{\text{reaction}}} \tag{16}$$

Finally, there is a third ratio of time-scales to consider. The ratio of diffusion time to residence time is known as the Peclet number. For large Peclet numbers (>>1) reactants do not have time to diffuse to the active surface and simply pass through the combustor unreacted. A device operating under these conditions is said to be diffusion limited.

$$Pe = \frac{\tau_{\text{diffusion}}}{\tau_{\text{residence}}} \tag{17}$$

For low Peclet numbers (<<1), the chemical reaction may be controlling the combustion process and the two Damköhler numbers must be examined. If they are both <<1, then the device is kinetically limited or reaction controlled.

As discussed previously, high power density requires high mass flow rates through small chamber volumes. At the small volumes of interest, residence time will be correspondingly low while catalytic reaction rates are generally fast (short time-scale). As a result these devices will likely be diffusion limited and Peclet number is the governing non-dimensional parameter. Power density can be rewritten as

$$\text{Power density} \propto \frac{\rho Pe}{\tau_{\text{diffusion}}}. \tag{18}$$

Hence, there is a basic tradeoff between power density and diffusion time. Noting that Peclet number can also be written as

$$Pe = \frac{Lv}{D} \tag{19}$$

where L is reactor length and v is velocity, power density can again be re-written.

$$\text{Power density} \propto \frac{\rho L v}{D\tau_{\text{diffusion}}} \tag{20}$$

For a given operating pressure and temperature (thus density), and reactant species, reducing diffusion times will increase the power density of the device and improve performance. The easiest means of reducing diffusion timescales for a given fuel-air mixture is by increasing catalyst surface area-to-volume ratio.

9.3.2 Heat Transfer and Fluid-Structure Coupling

Energy loss due to heat transfer at the walls of the combustion chamber in a large-scale flame combustor is typically neglected; however, this is an important factor in a microcombustor. The surface area-to-volume ratio for these combustors is approximately 500 m^{-1}; two orders of magnitude larger than that of a typical scale combustor.

Waitz et al. [5] have shown that the ratio of heat lost to that generated scales with the hydraulic diameter:

$$\frac{Q_{\text{loss}}}{Q} \propto \frac{1}{d_{\text{h}}^{1.2}}. \tag{21}$$

The hydraulic diameter of a microcombustor is hundreds of times smaller than that of a conventional combustor and is approximately millimeters or less. Therefore, the ratio of heat lost to that generated may be as much as two orders of magnitude greater in a microsystem.

The effect of this large surface heat loss on combustion is two-fold. First, large thermal losses have a direct impact on overall combustor efficiency making conventional efficiencies of greater than 99% unattainable. Second, the heat loss leads to lower reaction temperatures which can increase kinetic times and narrow flammability limits exacerbating residence time constraints.

In a catalytic device, the heat loss problem becomes worse. The reaction and heat release process occurs on the solid surface of the catalyst, so there is a direct conduction path through the structure to the exterior of the device. In the case of a gas-phase microcombustor, there is a thermal boundary layer between the hot reacting gases and the wall helping to limit heat loss. In a catalytic device, the thermal boundary layer acts to isolate the hot wall from the relatively cooler free-stream flow. In addition, surface area-to-volume ratios are likely to be even higher in a catalytic combustor due to the need to fill the chamber with a large quantity of the catalyst itself.

9.3.3 Material Constraints

Choice of materials for microcombustors is often limited due to the available fabrication techniques. Typically, deep reactive ion etching (DRIE) in silicon wafers and aligned fusion bonding are used to generate 2-D extruded geometries which serve as combustion chambers, fuel/air mixers, fuel injectors, and other associated components. Consequently, many microcombustion systems are silicon-based.

There are several material constraints imposed upon a silicon microcombustion system. The most critical requirement is a wall temperature limit of less than 950 K. At temperatures above this, silicon begins to soften and lose its structural integrity. However, high surface heat transfer and the high thermal conductivity of silicon are beneficial in this case. Combustor wall temperatures can be kept below the 950 K requirement by conduction of heat through the structure to the ambient.

It is also important to note that silicon-based microcombustors are typically isothermal. This is a result of the combination of the high thermal conductivity of silicon and the short conduction paths in a microsystem. This can be quantified by the heat transfer Biot number

$$Bi = \frac{hL}{k} = \frac{\text{resistance to conductive heat transfer}}{\text{resistance to convective heat transfer}} \tag{22}$$

which is typically << 1 for these systems.

Material constraints can be even more severe in a catalytic microcombustor. First, the catalyst material itself can be problematic. For noble metals such as platinum (a typical hydrocarbon/air catalyst), agglomeration of thin layers will occur at elevated temperatures thus greatly reducing the surface area available for catalysis. This phenomenon has been shown to occur at temperatures greater than 1000 K [8, 9]. This is problematic during the high temperature combustion process as well as during any fabrication and processing steps which require elevated temperatures (such as wafer bonding and packaging). Furthermore, catalyst surface area can be lost via diffusion of the catalyst material into its substrate at elevated temperature.

9.3.4 Design Space

While residence time, heat transfer, and material constraints are important, often the most limiting constraints are those imposed by the engine thermodynamic cycle. Figure 9.5 shows a typical design space for a hydrogen-fueled microscale combustor. Equivalence ratio is plotted on the x-axis versus heat rejected on the y-axis. A turbine inlet temperature between 1600 and 1800 K is required for the cycle proposed for this particular engine. In Fig. 9.5 the flame stability limit bounds the stable operation of a non-adiabatic perfectly stirred reactor. The thermal stress constraint indicates the material limits of an all silicon structure. The design space is further bounded by the desire to burn lean and the flammability limits of hydrogen–air combustion.

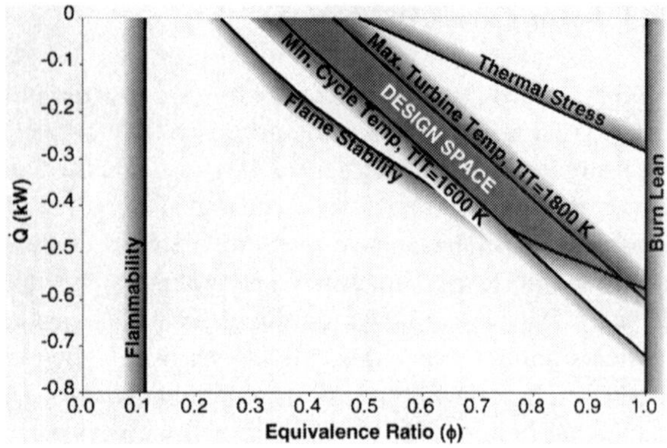

Fig. 9.5 Microcombustor design space [2]

9.4 Homogeneous Gas-Phase Microcombustor Development

The development of homogeneous gas-phase microcombustors was initiated in the mid 1990s. Preliminary experiments were with a simple metal flame tube geometry and eventually evolved to a six silicon wafer, axisymmetric, dual-zone device designed for integration with rotating structures to form a complete engine. This section will review some of this early work and discuss the details of later devices which were intended to be part of an engine system. General conclusions regarding gas-phase microcombustors will also be presented.

9.4.1 Review of Early Microcombustor Work

Combustion systems for microengines with rotating machinery were initially investigated by Waitz and Tzeng [5, 10]. A flame tube was used to map the flammability boundaries of hydrogen–air mixtures and to establish the conditions under which a stable flame can be sustained inside a small diameter tube. The burning of low equivalence ratio hydrogen–air mixtures in a millimeter-scale tube was demonstrated and found to be limited by heat loss.

Using the knowledge gained from the flame tube experiments, the first microcombustor compatible with a microengine system was developed. This combustor was 0.13 cm^3 and conventionally machined out of steel. Premixed hydrogen–air combustion was stabilized and the results from this device laid the foundation for the development of the first silicon microfabricated combustors [10].

Mehra and Waitz went on to develop this first silicon microcombustor for integration with an engine geometry [11]. This three-wafer combustor, shown in Figs. 9.6 and 9.7, was 0.066 cm^3 in volume and was designed to operate on a lean, premixed combustion strategy.

The combustor was tested over a range of equivalence ratios spanning from 0.4 to 1.6 for a fixed mass flow rate of 0.045 g/s and atmospheric pressure. This resulted

Fig. 9.6 Schematic of three-wafer microcombustor [11]

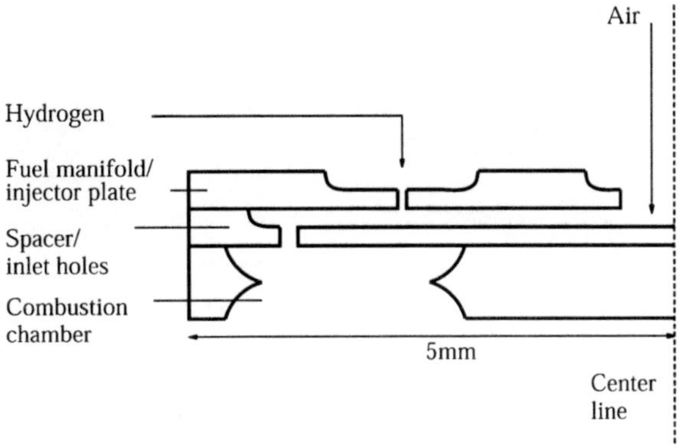

in a residence time of approximately 0.5 ms. For premixed hydrogen–air operation, exit gas temperatures in excess of 1800 K were achieved with combustor efficiencies of up to 70%.

Low combustor efficiencies were attributed to poor thermal isolation of the device. Chemical conversion was estimated to be nearly 100% while heat transfer from the combustion chamber to the surroundings lowered the overall efficiency. The power density of the device was approximately 1200 MW/m^3. The device also maintained its structural integrity over tens of hours of operation [11].

Mehra went on to design and develop a baseline six silicon wafer microcombustor based on the engine layout shown in Figs. 9.8 and 9.9. Several flame holding geometries, fuel injections schemes, and fuels were evaluated in this device.

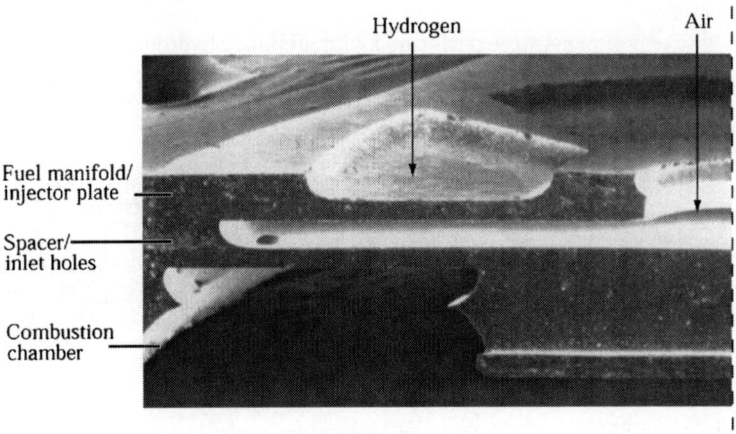

Fig. 9.7 Scanning electron micrograph (SEM) of three-wafer microcombustor [11]

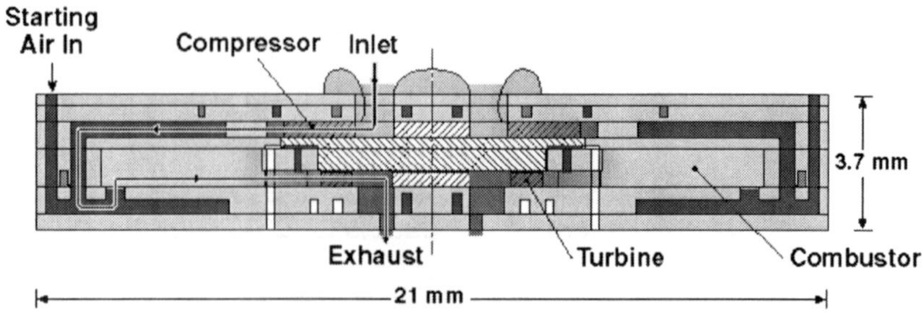

Fig. 9.8 Baseline micro gas turbine engine schematic

Fig. 9.9 3-D schematic of micro gas turbine engine

9.4.2 Baseline Six Silicon Wafer Gas-Phase Microcombustor

A baseline axisymmetric microcombustor fabricated from six silicon wafers is shown in Figs. 9.10 and 9.11. It includes a 191 mm^3 radial inflow combustion chamber, various flame holding chamber inlet geometries, a recirculation jacket which serves as both a thermal management system and fuel/air mixer, and several sets of fuel injectors.

Baseline experiments were performed with premixed hydrogen–air in this microcombustor. Figure 9.12 shows a plot of combustor exit temperature for constant

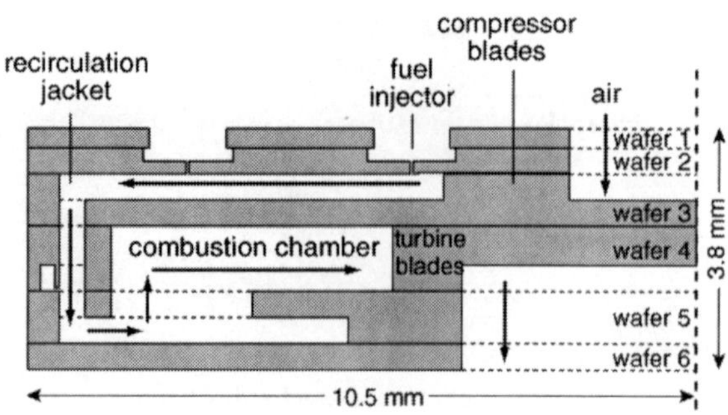

Fig. 9.10 Schematic of baseline six-wafer microcombustor [11]

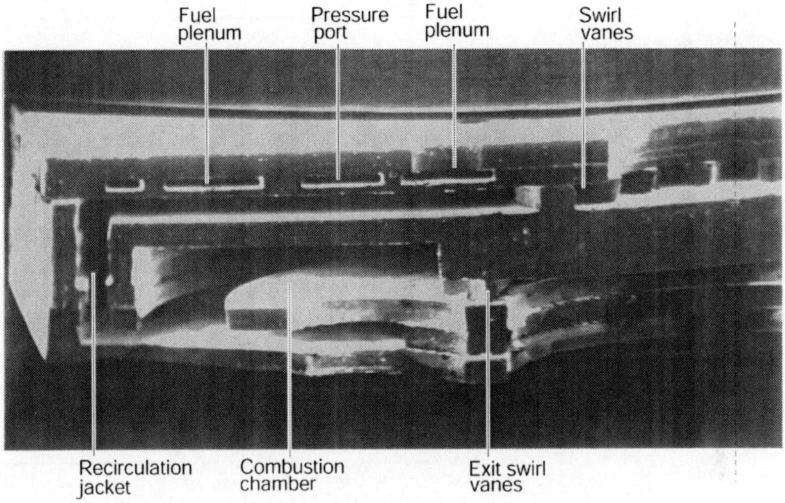

Fig. 9.11 SEM of baseline six-wafer microcombustor [11]

equivalence ratio while Fig. 9.13 shows overall combustor efficiency. Mass flow rate was controlled by varying the inlet pressure with a fixed exit area exhausting to atmospheric pressure.

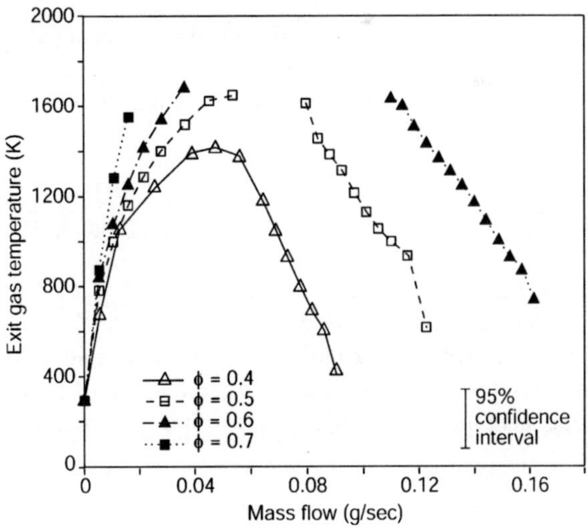

Fig. 9.12 Exit gas temperature versus mass flow rate for baseline six-wafer microcombustor [11]

The combustor produced exit gas temperatures in excess of 1600 K for a mass flow rate of 0.11 g/s and an operating pressure of 1.13 atm. The overall efficiency in this case exceeded 90% and the power density was approximately 1100 MW/m^3. The break in the $\phi = 0.5$ and $\phi = 0.6$ exit temperature and efficiency curves is due

Fig. 9.13 Overall combustor efficiency versus mass flow rate for baseline six-wafer microcombustor [11]

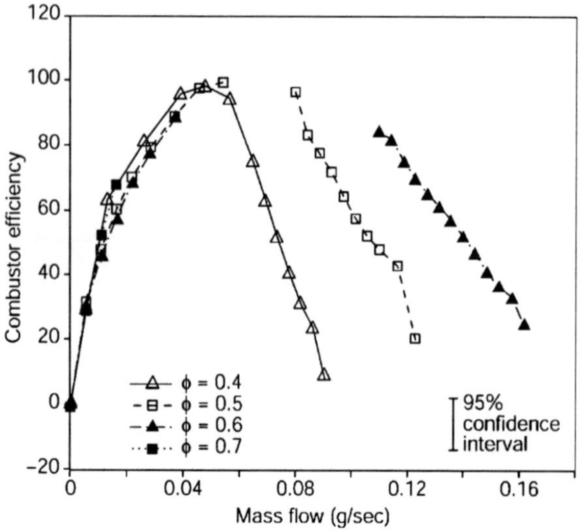

to a lack of measurement capabilities for gas temperatures over 1600 K. The $\phi = 0.7$ curve terminates at a mass flow rate of 0.015 g/s due to upstream burning in the recirculation jacket which can result in overheating and destruction of the device [2, 4, 11].

The blowout characteristics of this microcombustor can be explained by examining the change in Da_{h} over a constant equivalence ratio operating line. Using Eq. (13) for residence time, and a simple one step mechanism for hydrogen–air combustion

$$\tau_{\text{reaction}} = \frac{[\mathrm{H_2}]_0}{A[\mathrm{H_2}]^2[\mathrm{O_2}]\mathrm{e}^{-E_a/RT}}, \tag{23}$$

with $E_a = 10950$ cal/mole and $A = 1.62 \times 10^{18}$ (cm^3/mole)^2 · sec, an estimate of Da_{h} can be obtained [11]. Figure 9.14 shows a plot of reaction time, residence time, and Da_{h} versus mass flow rate. Da_{h} rapidly drops below unity as a result of the residence time decreasing more quickly than the reaction time. This results in chemical inefficiency at high mass flow rates and flame blowout.

9.4.3 Recirculation Zones and Flame Stability

In a gas-phase combustor, fluid recirculation zones and eddies are useful in stabilizing flames. Two types of combustion chamber inlets were utilized in this device to understand the impact of flame holding recirculation zones. The first was a simple annular inlet 1.2 mm wide while the second was an array of 60 slots 2.2 mm

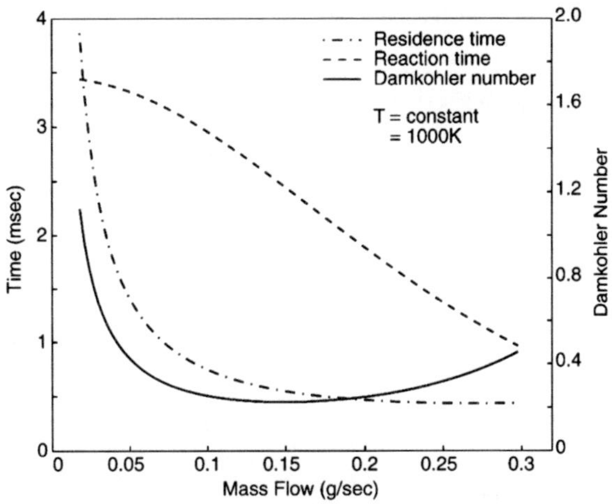

Fig. 9.14 Reaction time, residence time, and Da_h versus mass flow rate for baseline six-wafer microcombustor [11]

long intended to create multiple small recirculation zones for more rapid and uniform ignition of the incoming flow. SEMs of both types of inlets are shown in Fig. 9.15 [11].

Combustor efficiency for premixed hydrogen–air in the slotted-inlet geometry is plotted in Fig. 9.16. Performance is similar to that of the annular-inlet combustor (Figs. 9.12 and 9.13); however, the slotted-inlet device produced higher exit temperatures and efficiencies in the high mass flow, kinetically limited regime of the operating line. This can be attributed to the presence of multiple recirculation zones at the inlet which more rapidly and uniformly ignite the incoming mixture.

Fig. 9.15 SEMs of inlet geometries [11]

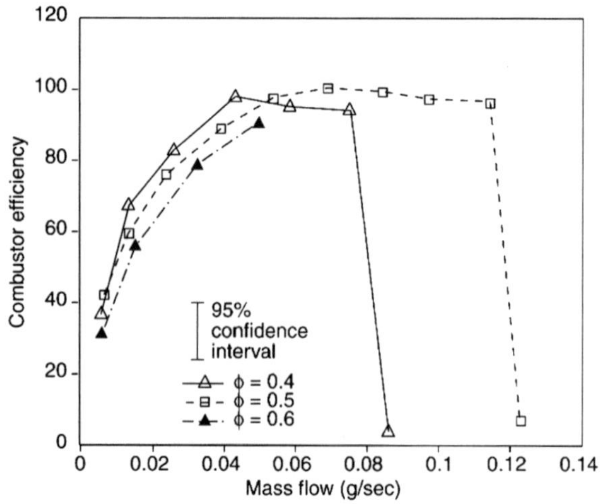

Fig. 9.16 Overall combustor efficiency versus mass flow rate for a six-wafer combustor with slotted inlets [11]

The sharp drop in performance for the slotted combustor was attributed to the rapid extinction of these multiple small ignition zones. With the annular inlet geometry, the single large ignition zone continues burning over a broader range of mass flow rates, but does not ignite the incoming reactants as uniformly and completely [11].

Without internal measurements or flow visualization, numerical simulations must be used to understand this performance difference. Figure 9.17a shows temperature contours generated from computational fluid dynamics (CFD) software, for the combustor with the annular inlet. In Fig. 9.17b contours of temperature are plotted for a 2-D cut through the combustor inlet in the θ–z plane. These contours show a temperature gradient from the bottom of the combustion chamber to the top and indicate the primary ignition zone is in the upper right corner of the combustor. Figure 9.18a, b shows the same contour plots for a combustor with a slotted inlet. There are significantly higher temperatures in the lower regions of the combustion chamber and higher overall temperature is achieved. The shape of the contours also indicates there are small, hot ignition zones near the slotted inlet [1, 2].

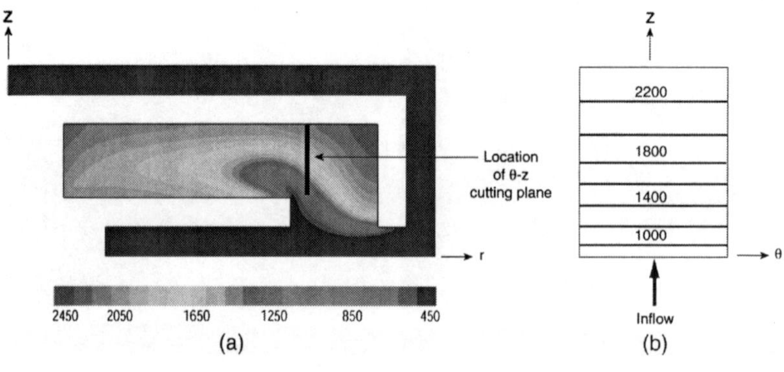

Fig. 9.17 Temperature (K) contours for annular inlet six-wafer microcombustor [1, 2]

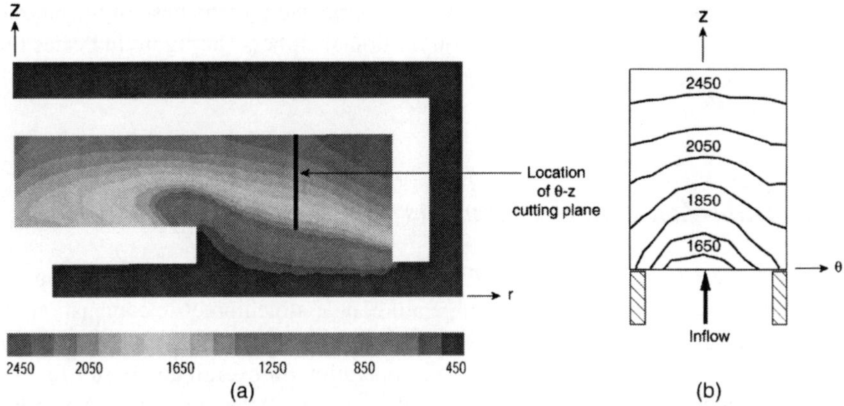

Fig. 9.18 Temperature (K) contours for slotted inlet six-wafer microcombustor [1, 2]

9.4.4 Fuel Injection and Fuel/Air Mixing

The baseline device was also used to evaluate fuel injection schemes and required fuel air mixing lengths. Three sets of fuel injectors were designed into the device at different locations within the recirculation jacket. By comparing overall combustor efficiencies for each set of fuel injectors to pre-mixed operation, injectors performance and required mixing length can be observed. Pre-mixed operation consisted of fuel/air injection through the primary air inlet only. Figure 9.19 plots overall efficiency for each set of fuel injectors versus equivalence ratio at a constant mass flow rate of 0.045 g/s. The injectors were located at radii of 4.8 and 8.0 mm into the

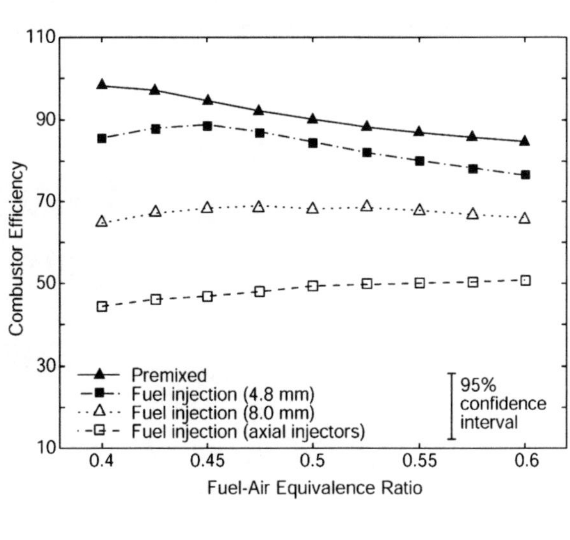

Fig. 9.19 Overall combustor efficiency for each set of fuel injector versus equivalence ratio [11]

recirculation jacket while a set of sidewall injectors were at the base of the recirculation jacket just upstream from the combustion chamber. The figure indicates that performance decreases as the injectors get closer to the combustion chamber due to the decreased mixing [11].

9.4.5 Dual-Zone Microcombustors

Typical large scale combustors utilize a multi-zone combustion strategy. An initial primary zone consists of high-temperature, near-stoichiometric combustion followed by dilution jets of air to form a secondary zone with a leaner mixture and lower temperature. Such a strategy enables operation for overall combustor fuel–air mixtures that are leaner (exit temperatures that are lower) than those that can sustain steady combustion (they are outside the flammability envelope); a high-temperature region is used to sustain the flame followed by subsequent dilution.

A multi-zone microcombustor which operates in a similar fashion may be beneficial for overall performance. This was fabricated by adding a series of holes through the inner wall to connect the upper recirculation jacket to the combustion chamber allowing inlet air to bleed into the combustion chamber. This dilution air serves two purposes; first, it splits the combustion chamber into the two zones, and second, it dilutes the hot combustion products reducing their temperature to the required outlet temperature. In the primary zone, the fuel–air mixture is burned near stoichiometric conditions. By burning at higher temperature, kinetic rates increase and reaction time decreases, effectively increasing Da_{h}. The dilution jets also serve to strengthen recirculation zones and improve flame stability. Fuel injection occurs just downstream of the dilution holes in the cooling jacket. A three-dimensional, adiabatic, reacting flow CFD solution shows the effect of the two zones in Fig. 9.20.

In operation, the dual-zone combustor exhibited slightly lower exit gas temperatures and efficiencies than the baseline six-wafer microcombustor. However, its

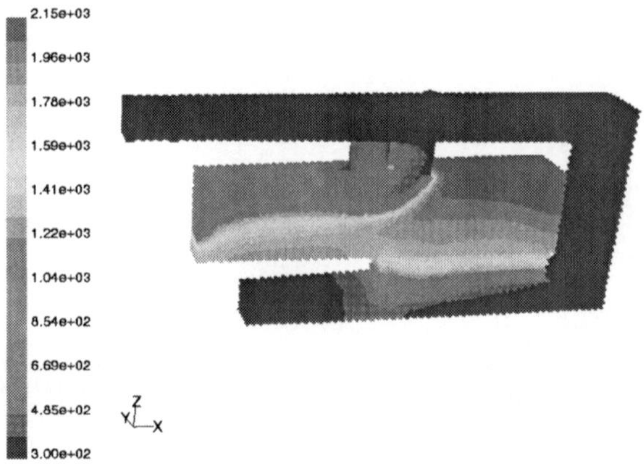

Fig. 9.20 CFD solution and temperature (K) contours for a dual-zone microcombustor [1, 2]

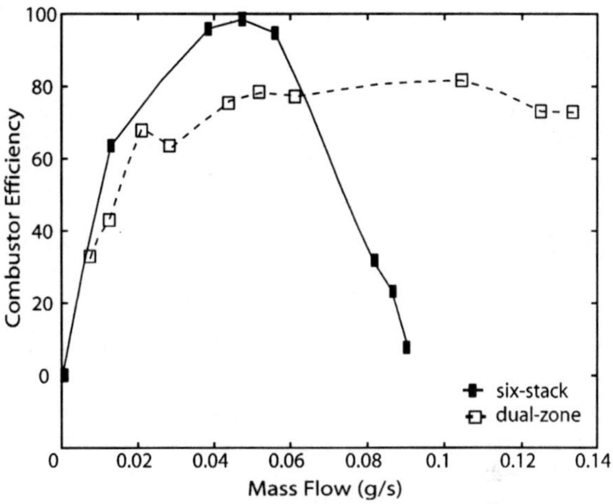

Fig. 9.21 Comparison of dual-zone to baseline microcombustor for hydrogen–air operation [1, 2]

operating range was much wider for hydrogen–air mixtures. Combustion was stabilized at overall equivalence ratios as low as 0.2 and mass flow rates as high as 0.2 g/s. Figure 9.21 compares the dual-zone combustor efficiency to the baseline design for similar conditions [1, 2]. The significant extension in operating range is largely due to strengthened recirculation zones and the hot, stable primary zone. For the same overall efficiency, the dual-zone combustor achieved a 100% increase in mass flow rate before blowout. The ability to operate at low overall equivalence ratio is a result of a higher local equivalence ratio in the primary zone.

The non-uniform exit temperature profile generated by the dual-zone combustor (as shown in CFD solutions) may be beneficial for turbine cooling. In the case of a silicon microengine system with a single rotating disk (compressor on one side with a turbine on the other), heat transfer from the turbine to the compressor can degrade compressor performance by as much as 20–30%. Reducing the heat load on the turbine would reduce the heat transferred to the compressor and improve overall engine performance. The dual-zone combustor's non-uniform temperature profile has the potential, if properly tailored, to provide a cooler layer of gas over the turbine disk while still maintaining a high enthalpy flow over the blades for work extraction. A detailed analysis of this possibility can be found in [1].

9.4.6 Hydrocarbon-Air Operation

Hydrocarbon fuels were tested in both the baseline and dual-zone devices. For a microengine to be practical, fuels such as ethylene (C_2H_4) and propane (C_3H_8) will be required due to their high-energy densities and favorable storage properties. Results from combustion tests for these two fuels in the six-wafer combustor are shown in Figs. 9.22 and 9.23. The ethylene–air mixture achieved maximum

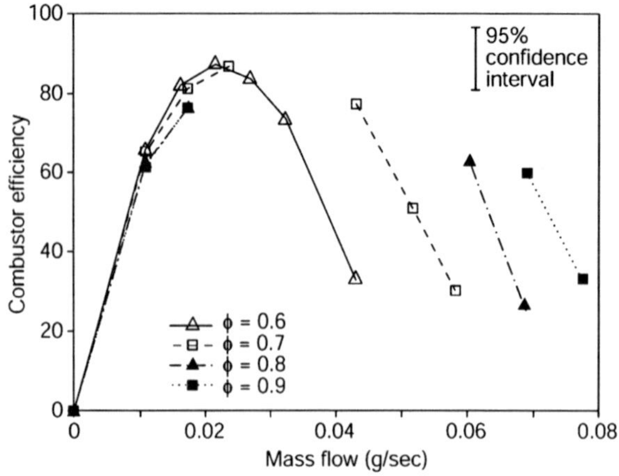

Fig. 9.22 Overall combustor efficiency for baseline six-wafer microcombustor with ethylene/air mixtures [1, 11]

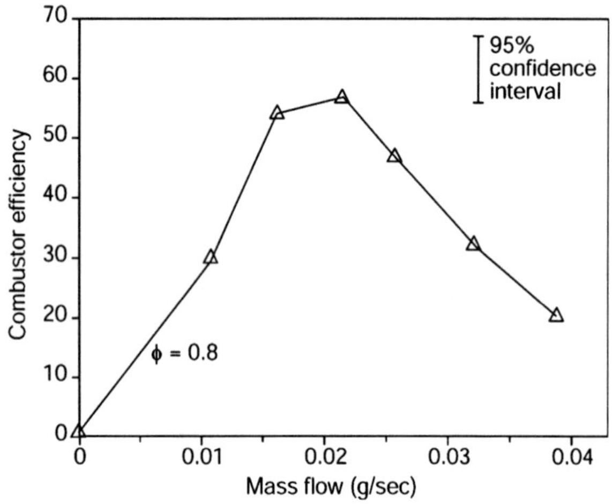

Fig. 9.23 Overall combustor efficiency for baseline six-wafer microcombustor with propane/air mixtures [1, 11]

power density at an equivalence ratio of 0.9 and ~1 atm pressure, with exit temperatures exceeding 1400 K and an overall efficiency of 60%. The estimated power density for these conditions is ~500 MW/m^3. The reaction rates of propane are significantly lower than ethylene and as a result, combustion could only be stabilized in the device at an equivalence ratio of 0.8. Exit temperatures of 1200 K were reached at ~1 atm with an efficiency of 55%. This corresponds to a power density of ~140 MW/m^3. Nominal residence times for ethylene–air and propane–air combustion in this device were 1.6 and 2.8 ms, respectively. Typical reaction times are 0.5–1 ms for ethylene–air and 1–2 ms for propane–air whereas hydrogen–air reaction times are approximately 0.2 ms.

The dual-zone combustor did not provide the expected broader range of mass flow for hydrocarbon-fueled operation. This is believed to be due primarily to inadequate mixing lengths (premixed operation is not feasible in this device). Table 9.2 summarizes the maximum power densities achieved for both devices with various fuels [1].

Table 9.2 Maximum power densities and efficiencies for microcombustors [1]

Fuel	Baseline microcombustor		Dual-zone microcombustor	
H_2	1150 MW/m^3	>99%	1400 MW/m^3	92%
C_2H_4	500 MW/m^3	85%	100 MW/m^3	38%
C_3H_8	140 MW/m^3	50%	75 MW/m^3	29%

9.4.7 Operating Space

Figure 9.24 plots the empirically identified operating space for the baseline six-wafer and the dual-zone microcombustors operating on hydrogen–air mixtures. The operating boundaries include:

1. The lean blowout limit.
2. The upstream burning limit where combustion in the cooling jacket occurs.
3. The structural limit.
4. The 1600 K T_{t4} microengine cycle limit (assuming 80% combustor efficiency) for a nominal engine system design.

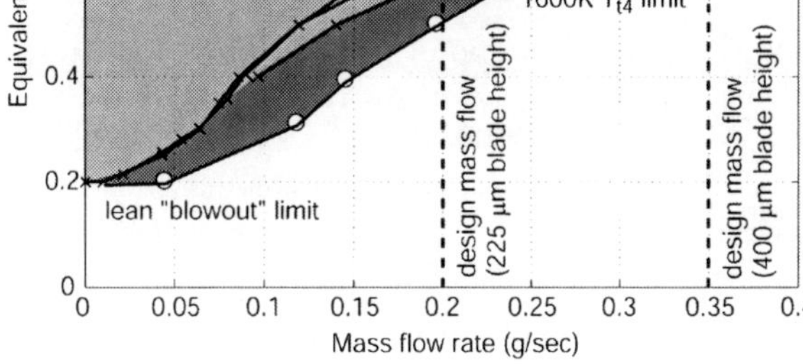

Fig. 9.24 Empirically determined operating space for six-wafer and dual-zone microcombustors [1]

A narrow operating corridor exists between the lean blowout limit and the upstream burning limit. Two nominal design mass flow rates are also shown at approximately 0.2 and 0.35 g/s and indicate the difficulty in achieving the desired conditions for a microengine system. The dual-zone combustor has a broader operating space due to the stable, hot primary zone. The most notable features of this operating space for the dual-zone device are the lower lean blowout limit and the absence of an upstream burning limit due to non-premixed operation.

Microcombustor performance is a function of both chemical and thermal losses. Performance can be quantified in terms of non-dimensional power density, defined as power density normalized by the maximum possible power density at those conditions (the latter being a function of mass flow rate and fuel type). If it is assumed that there is adequate fuel–air mixing and rapid, uniform ignition, this quantity is a function of only two parameters; Damköhler number and non-dimensional heat loss. Damköhler number can be estimated as previously described using exit gas temperature for residence time calculations and Chemkin III (a reaction kinetics software package) with detailed reaction mechanisms from the literature for chemical time approximations. Heat loss can be cast in non-dimensional terms by normalizing the heat lost from the device by the maximum possible heat that can be generated at a given operating condition. The heat lost was estimated using a 1-D heat transfer model (which is an adequate approximation in this case due to the low Biot number and isothermal nature of the structure).

Figure 9.25 plots this non-dimensional power density for several microcombustors in three-parameter space. The plot includes multiple fuels as well (hydrogen,

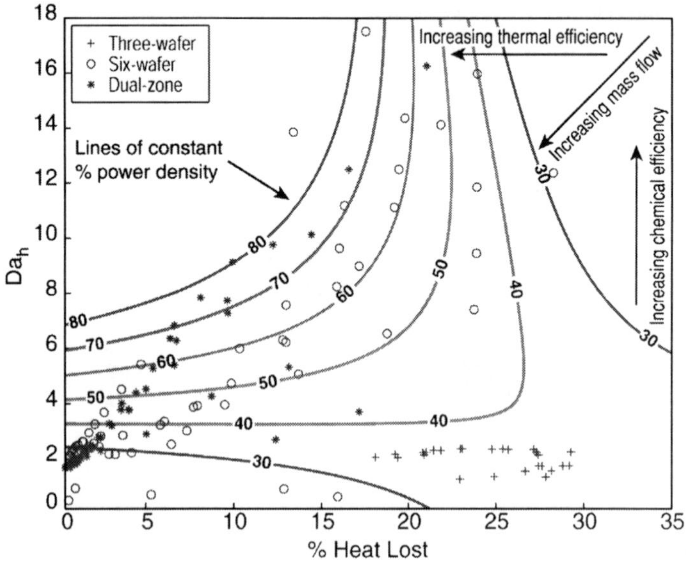

Fig. 9.25 Non-dimensional operating space for six-wafer and dual-zone microcombustors [1, 2]

ethylene, and propane). The heat loss parameter and homogenous Damköhler number are on the x- and y-axes, respectively. A second order, least squares surface fit was used to generate contours of non-dimensional power density (the fit explains 65% of the variance in the data) [1, 2].

Thermal efficiency in these devices is inversely proportional to heat loss while chemical efficiency is directly proportional to Damköhler number. Hence, optimum performance is achieved at low levels of heat loss and high Damköhler number as indicated in Fig. 9.25. In the low mass flow region of the space, performance is limited by heat loss, while at high mass flow, Damköhler number falls and performance is limited by chemical inefficiency. In the area of moderate mass flow, the center of the operating space, the highest power densities are achieved.

9.4.8 Suggested Design Methodology

The non-dimensional operating space of Fig. 9.25 can also be useful in the design process. For example, a good estimate of the volume required for a propane–air microcombustor can be obtained. If operating conditions such as a 1600 K exit gas temperature, 1 atm pressure, a mass flow rate of 0.15 g/s, and an equivalence ratio of $\phi = 0.8$ are required, the non-dimensional power density can be computed and is found to be approximately 60%. The heat loss from the device can be assumed to be approximately 5% based on data for similar devices or can be estimated using a heat transfer model. From Fig. 9.23, the desired Damköhler number can be found and is approximately 5.5. The chemical time can be computed using Chemkin III or any other suitable technique and the desired residence time found from the Damköhler number. This results in a residence time of 1.44 ms and a volume of 950 mm^3, or about a five-fold increase in volume compared to the gas-phase microcombustors presented here. Table 9.3 shows estimated combustor volumes required for a series

Table 9.3 Design parameters for a range of microengine designs

Parameter	Units	Engine #1	Engine #2	Engine #3	Engine #4	Engine #5
$\dot{m}$	(g/s)	0.5	0.6	0.5	1.15	0.7
T_{inlet}	K	300	300	300	300	300
P_{inlet}	atm	1	1	1	1	1
η_{comp}	%	0.75	0.78	0.78	0.78	0.78
π_{comp}	n/a	2.5	2.9	3.0	3.9	4.5
T_{t4}	K	1600	1600	1700	1700	1800
η_{burner}	%	90	90	90	90	95
π_{burner}	n/a	0.95	0.95	0.98	0.98	0.98
fuel	n/a	C_3H_8	C_3H_8	C_3H_8	C_3H_8	C_3H_8
ϕ	n/a	0.55	0.55	0.55	0.58	0.63
Da_{h}	n/a	8.5	8.5	8.5	8.5	9.0
τ_{chem}	ms	0.3	0.3	0.3	0.2	0.1
τ_{res}	ms	2.55	2.55	2.55	1.7	0.90
Volume	mm^3	2310	2390	2046	2413	715
%increase	%	1210	1250	1070	1260	375

of other engine designs. The technique described above was used and all results are compared to the baseline six-wafer microcombustor design.

It is important to note that these results are for design point operation only and operating line issues must be kept in mind. Namely, the combustors will likely reach a minimum Damköhler number at which blowout may occur, prior to reaching the design mass flow rate. To account for this, a gas-phase microcombustor volume should be sized for the worst case off-design operating condition.

9.5 Heterogeneous Catalytic Microcombustor Development

Thus far, we have focused primarily on homogeneous gas-phase combustion and have an understanding of the combustor volume required for a practical device given requirements for mass flow, exhaust gas temperature, and fuel type. Experimental results and the design guidelines presented in Section 9.4.8 suggest that a microcombustor operating with a more practical hydrocarbon fuel such as propane will need to be at least five times larger than a device operating with hydrogen. The difference is attributable to the difference in chemical kinetic rates of the two fuels.

For most practical applications of microcombustors, the goal is to operate with storable liquid hydrocarbon fuels such as propane, butane, or even more common logistics fuels such as JP8. To utilize these fuels without dramatically increasing combustor volume or temperature, chemical conversion rates must be increased significantly. Surface catalysis is a means of increasing reaction rates. Hydrocarbon–air mixtures reacting on noble metal catalysts are known to be faster than those occurring in the gas-phase. For this basic reason, catalytic microcombustors have been developed.

A typical large-scale catalytic combustor consists of a set of parallel channels coated with an active catalyst material. Common catalyst materials include platinum and palladium as well as more exotic and proprietary metal oxides. The fuel and air flow through the channels and diffuse to the active surface. When a fuel molecule and an oxygen molecule reach the surface they are adsorbed and react, releasing heat. The products of the reaction are then desorbed and flow out of the channel. The heat that has been released from the reaction is transferred both conductively through the solid and convectively to the fluid. It is also possible for the remaining fuel–air mixture to get hot enough to ignite in the gas-phase. Figure 9.26 shows a schematic of a typical catalytic combustor.

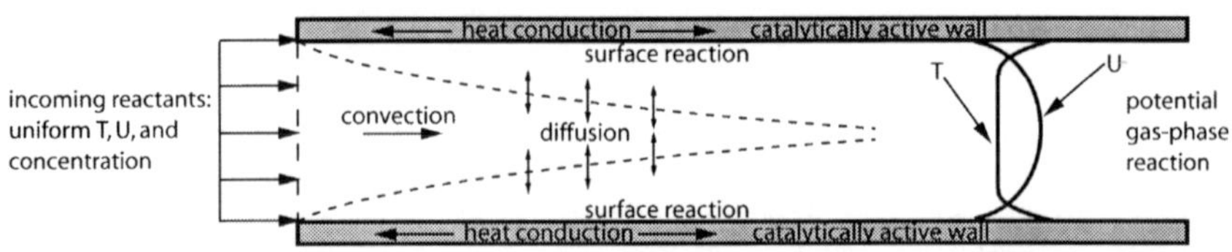

Fig. 9.26 Schematic of a typical catalytic combustor [12]

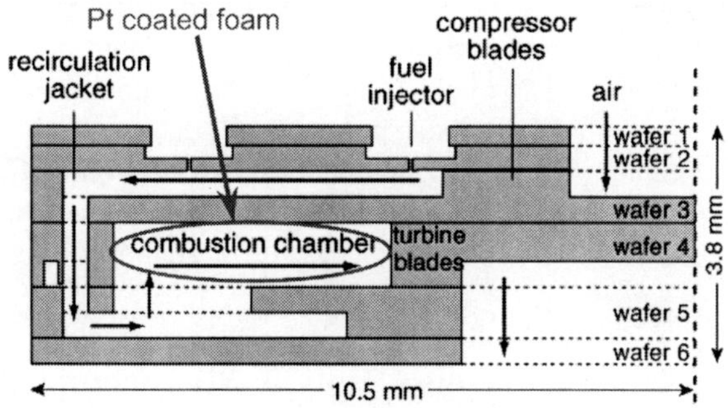

Fig. 9.27 Baseline six-wafer microcombustor to be filled with catalytic (platinum) foam [1, 3]

Figure 9.27 shows the gas-phase microcombustor geometry and indicates that the combustion chamber is filled with a porous substrate coated with platinum which will act as the active catalytic surface. This was chosen due to fabrication considerations and a desire to maintain a similar geometry to the gas-phase device for comparison purposes.

9.5.1 Ignition Techniques and Characteristics

To achieve auto-thermal operation of a catalytic microcombustor the catalyst has to be heated to a suitable ignition temperature. This can be accomplished by either:

1. Beginning with hydrogen–air operation (which will catalytically ignite at lower temperatures) to pre-heat the catalyst. The fuel can then be converted to a hydrocarbon fuel once the catalyst has reached adequate temperature (typically > 500 K).
2. Externally heating the entire microcombustor and begin flow of the hydrocarbon–air mixture when the desired temperature is achieved.
3. Heating the catalytic element directly by resistive heating while flowing the fuel–air mixture. This requires that the catalyst itself be fabricated with electrical contacts.

Typically, a catalytic combustor will exhibit an ignition/extinction hysteresis similar to that shown by Schmidt and co-workers [13, 14]. Figure 9.28 indicates this for the six-wafer catalytic microcombustor fitted with catalytic foam of 95 and 88.5% porosity where porosity is defined as open volume divided by total volume;

$$\alpha = \frac{V_{\text{open}}}{V_{\text{total}}}. \tag{24}$$

The combustor was brought to ignition temperature via an external electrical heater while a hydrogen–air mixture was flowing through the device. Wall

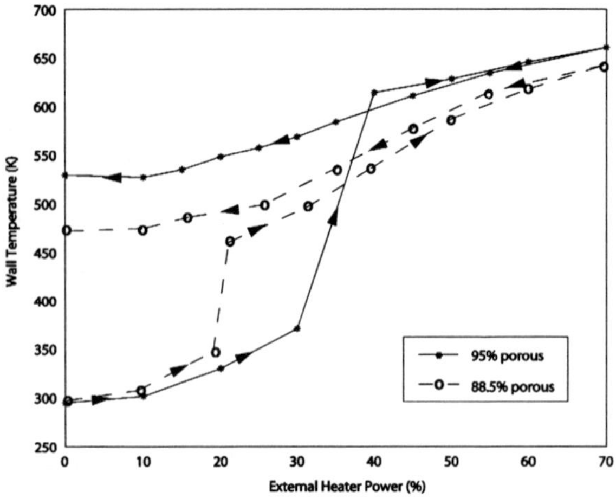

Fig. 9.28 Ignition characteristics for a catalytic microcombustor [1, 12]

temperature is plotted against heater power and ignition occurs at a heater power of approximately 20–30% (around 80–100°C). Upon removal of the external heater the combustor continued to operate auto-thermally resulting in the hysteresis shown in the figure.

9.5.2 Performance of a Baseline Catalytic Microcombustor

The baseline six-wafer catalytic device contained a 95% porous foam material coated with platinum as the active surface. For propane–air operation, plots of exit gas temperature and overall combustor efficiency and a range of mixture ratios are shown in Figs. 9.29 and 9.30, respectively. The maximum exit gas temperature achieved was approximately 850–900 K over a wide range of flow rates, and

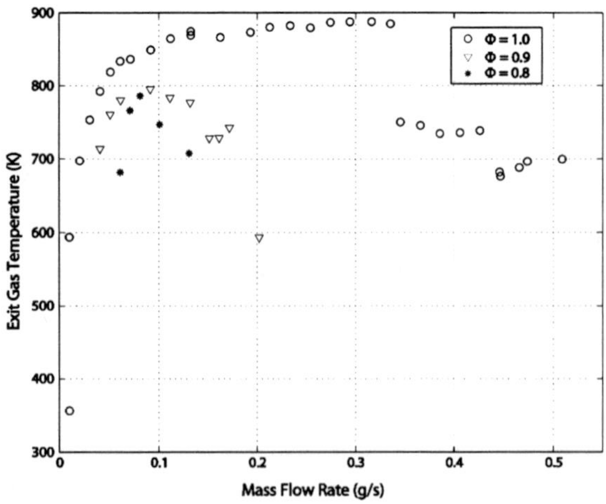

Fig. 9.29 Exit gas temperature versus mass flow rate for a catalytic microcombustor with a 95% porous platinum catalyst and propane–air mixtures [1, 3]

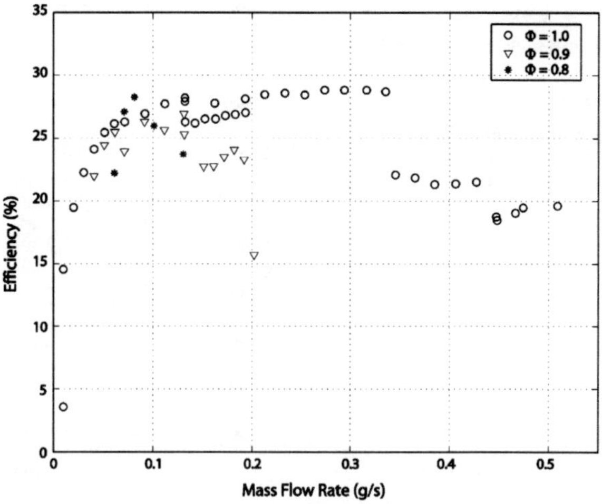

Fig. 9.30 Overall combustor efficiency versus mass flow rate for a catalytic microcombustor with a 95% porous platinum catalyst and propane–air mixtures [1, 3]

overall efficiencies were up to 30%. The reduction in performance at 0.35 g/s was a result of limitations in the experimental setup, not the device itself. Although these temperatures and efficiencies are quite low, the mass flow rates achieved were in excess of 0.4 g/s, well beyond those measured for the baseline gas-phase device. The maximum power density achieved was approximately 1050 MW/m^3, which is a 7.5-fold increase over gas-phase propane–air power densities, and about 95% of that for gas-phase hydrogen–air mixtures [1, 3].

The overall combustor efficiency can be broken down into its thermal and chemical components. A wall temperature measurement combined with a 1-D heat transfer model reveal that although heat loss is greater than in the gas-phase case, losses are dominated by chemical inefficiency indicating that most of the reactants pass through the device without contacting the active surface (Fig. 9.31).

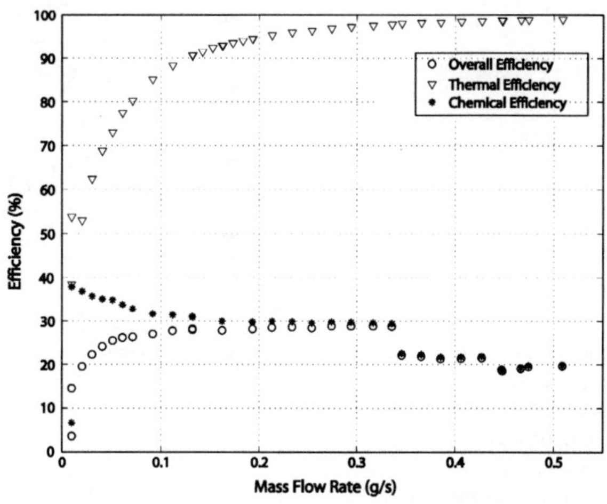

Fig. 9.31 Efficiency break down for a catalytic microcombustor with a 95% porous platinum catalyst and a propoane–air mixture at a ϕ=1.0 [1, 3]

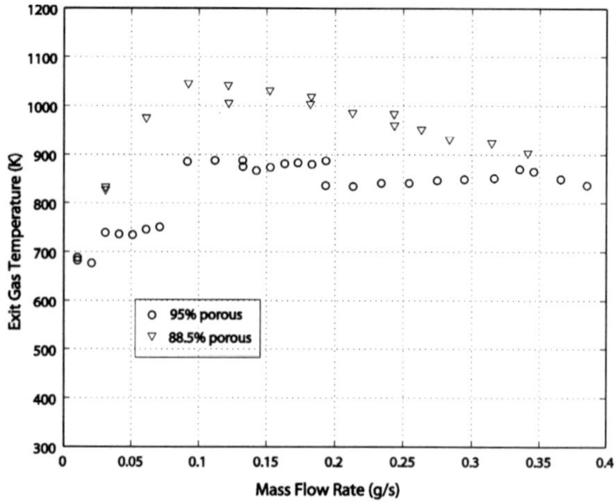

Fig. 9.32 Exit gas temperature versus mass flow rate; comparison of a devices with 95% and 88.5% porous catalyst material burning propane–air mixtures at ϕ=1.0 [1, 3]

Figure 9.32 shows exit gas temperature and overall combustor efficiency comparisons for two devices with different porosity catalysts; 95 and 88.5%. Both are operating at a stoichiometric equivalence ratio. The 88.5% porous catalytic combustor (higher surface area) achieved exit gas temperatures over 150 K higher than the 95% porous device. This corresponds to an approximately 2–10% higher efficiency over a range of mass flow rates and is shown in Fig. 9.33. Low-order modeling and analysis described in the following sections will further explain these trends.

9.5.3 Diffusion and Time-Scales

To determine which phenomena control the combustion process in a catalytic micro-combustor, relevant physical time-scales can be evaluated. These include reaction time, residence time, diffusion time of the fuel species, and diffusion time of the

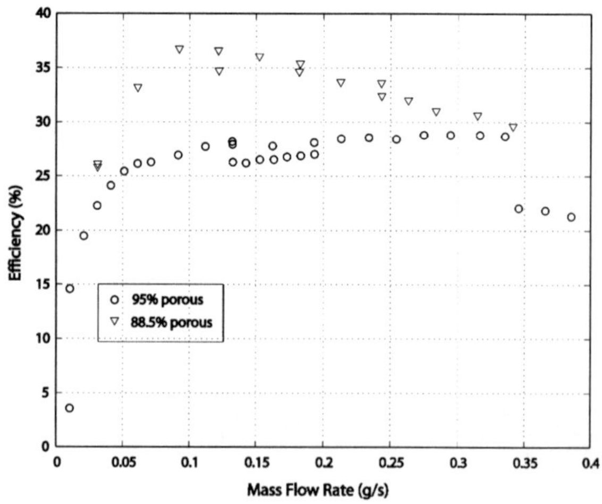

Fig. 9.33 Overall combustor efficiency versus mass flow rate; comparison of a devices with 95 and 88.5% porous catalyst material burning propane–air mixtures at ϕ=1.0 [1, 3]

oxidizer. Residence time can be estimated from the volume, mass flow rate, pressure, and bulk gas temperature as in the gas-phase case (Eq. 13). Reaction rate can be obtained from the appropriate rate expression. For a propane–air reaction on a platinum catalyst, one possible mechanism is [15]

$$-R_{C_8H_8} = -k_S\,[C_3H_8] \tag{25}$$

where the rate constant is

$$k_S = 2.4 \times 10^5 \exp\left(\frac{-1.08 \times 10^4}{T}\right). \tag{26}$$

Molecular diffusion coefficients for propane and oxygen diffusion through air can be obtained from the Fuller correlation [15];

$$D_{AB} = \frac{1.013 \times 10^{-2} T^{1.75} \left(\frac{1}{M_A} + \frac{1}{M_B}\right)^{1/2}}{P\left[\left(\sum v_i\right)_A^{1/3} + \left(\sum v_i\right)_B^{1/3}\right]^2}. \tag{27}$$

M_A and M_B are the molecular weights of the species and v_i is the diffusion volume of each atom in the molecule. Diffusion volumes for carbon, hydrogen, and oxygen are 16.5, 1.98, and 5.48 respectively.

Using estimates of these various time-scales, non-dimensional parameters can be calculated and used to determine the governing physics. These parameters and their approximate values for a catalytic microcombustor are summarized below in Table 9.4. From this simplified time-scale analysis, it is clear that diffusion of reactants to the surface is the controlling phenomenon. The second Damköhler number indicates that the surface reaction rate is much faster than the rate of diffusion and the large Peclet number reveals that reactants can flow through the device without ever coming into contact with the catalytic surface. The longest time-scale in the system is the diffusion time.

Since the above analysis indicates that the process is diffusion controlled, a closer examination of Peclet number is useful. If a tubular plug flow reactor is assumed, the Peclet number can be estimated and shown to be a strong function of geometry. For a given set of flow conditions (pressure, temperature, and mass flow rate), Peclet number can be calculated for a range of diameters (or pore sizes). Figure 9.34

Table 9.4 Summary of non-dimensional parameters [1, 3]

	Non-dimensional parameter	Range
Damköhler number for residence time	$Da_h = \frac{\tau_{residence}}{\tau_{reaction}}$	~1–5
Damköhler number for diffusion time	$Da_2 = \frac{\tau_{diffusion}}{\tau_{reaction}}$	~5–20
Peclet number	$Pe = \frac{\tau_{diffusion}}{\tau_{residence}}$	~1–20

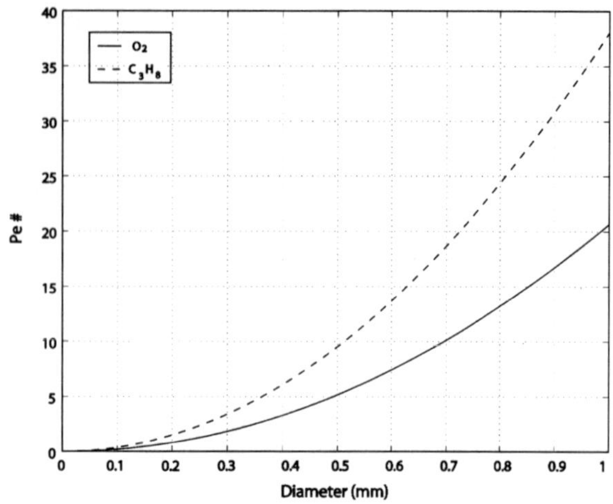

Fig. 9.34 Peclet number versus diameter for oxygen and propane [1, 3]

shows this Peclet number for P=2 atm, T=1000 K, and a mass flow rate=0.3 g/s for both reactants; propane and oxygen. For diameters greater than ~100 μm, the Peclet number is larger than unity and diffusion is critical. It is also important to note that propane diffuses more slowly than oxygen due to its larger molecular size resulting in a larger Peclet number. This indicates that propane diffusion to the active surface is the controlling phenomenon.

9.5.4 Pressure Loss

For many systems such as a microengine, minimizing pressure loss through the device is critical to the thermodynamic cycle. The addition of a porous catalyst material or a monolithic catalyst support will significantly increase the pressure loss of the system. This effect must be minimized and optimized with the benefits of the catalytic surface.

Pressure loss through a porous substrate can be estimated from the Ergun equation for Reynolds numbers less than 300 [15];

$$-\frac{\mathrm{d}P}{\mathrm{d}z} = \frac{v}{L}\frac{(1-\alpha)}{\alpha^3}\left[\frac{150\mu\,(1-\alpha)}{L} + 1.75\rho v\right], \tag{28}$$

where μ is the viscosity, v is velocity through the porous media, and l is a characteristic length scale, usually the width of a foam fiber or diameter of a particle in a packed bed.

9.5.5 Isothermal, Porous, Plug Flow Reactor Model

In many cases, catalytic microcombustors can be modeled as isothermal plug flow reactors. The equations developed here have been derived from a control volume analysis of a fluid element in a plug flow reactor. These equations, coupled with the isothermal assumption, yield trends in fuel conversion and bulk gas temperature rise through a reactor as a function of flow conditions and geometry. Figure 9.35 shows the control volume element used for these derivations.

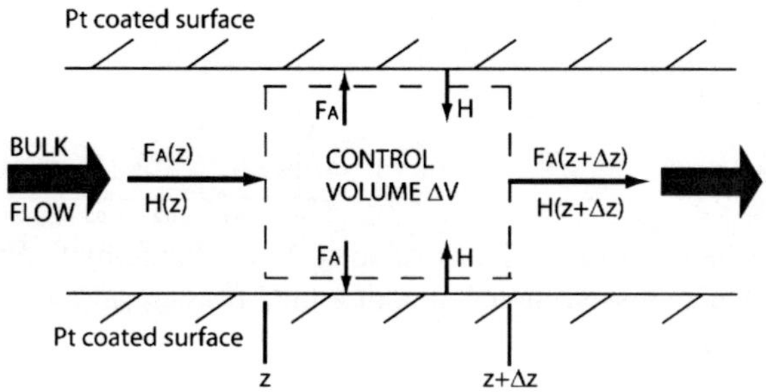

Fig. 9.35 Control volume for a fluid element in a catalytic combustor [1]

The Peclet number analysis indicated that the diffusion of fuel species is the governing phenomenon, therefore, only the mass transport and consumption of the fuel is taken into account in subsequent derivations. Homogeneous gas-phase reactions will be neglected and the catalytic reaction mechanism shown in Eqs. (25) and (26) will be utilized.

A gas-phase mole balance can be written as

$$\begin{bmatrix}\text{moles transported}\\ \text{into volume}\end{bmatrix} - \begin{bmatrix}\text{moles transported}\\ \text{out of volume}\end{bmatrix} - \begin{bmatrix}\text{moles reacted}\\ \text{in gas phase}\end{bmatrix} - \begin{bmatrix}\text{moles transported}\\ \text{to Pt surface}\end{bmatrix} = 0 \tag{29}$$

The third term, moles reacted in gas-phase, is equal to zero because homogeneous reactions are being neglected. The first two terms can be written as a molar flow rate of fuel species (propane in this case).

$$\begin{bmatrix}\text{moles transported}\\ \text{into volume}\end{bmatrix} - \begin{bmatrix}\text{moles transported}\\ \text{out of volume}\end{bmatrix} = (F_{\text{fuel}})_Z - (F_{\text{fuel}})_{Z+\Delta Z} = -\Delta F_{\text{fuel}} \tag{30}$$

The fourth term, which describes the transport of fuel species out of the control volume to the catalyst surface, can be written as

$$\begin{bmatrix}\text{moles transported}\\ \text{to Pt surface}\end{bmatrix} = k_{\text{m}} C_{\text{b}} \left(Y_{\text{fuel,b}} - Y_{\text{fuel,S}}\right) \Delta A_{\text{S}} \tag{31}$$

where k_{m} is a mass transport coefficient, C_{b} is the molar density, $Y_{\mathrm{fuel,b}}$ is the mole fraction of fuel in the bulk gas stream, $Y_{\mathrm{fuel,S}}$ is the mole fraction of fuel at the surface, and ΔA_{S} is the incremental surface area. The transport of fuel species is driven by the concentration gradient.

Combining all of the terms from above, the expression for the gas-phase mole balance over the control volume becomes

$$\Delta F_{\mathrm{fuel}} + k_{\mathrm{m}} C_{\mathrm{b}} \left(Y_{\mathrm{fuel,b}} - Y_{\mathrm{fuel,S}} \right) \Delta A_{\mathrm{S}} = 0. \tag{32}$$

Noting that

$$\Delta F_{\mathrm{fuel}} = \Delta \left(C_{\mathrm{b}} v Y_{\mathrm{fuel,b}} \right) \tag{33}$$

where v is the bulk gas velocity. Normalizing by the incremental volume where A_{c} is the cross-sectional area of a fluid passage and a_{v} is the surface-area-to-volume ratio

$$\Delta V = A_{\mathrm{C}} \Delta z = \Delta A_{\mathrm{S}} / a_{\mathrm{v}} \tag{34}$$

the expression can be written as

$$\frac{\Delta \left(C_{\mathrm{b}} v Y_{\mathrm{fuel,b}} \right)}{\Delta z} + a_{\mathrm{v}} k_{\mathrm{m}} C_{\mathrm{b}} \left(Y_{\mathrm{fuel,b}} - Y_{\mathrm{fuel,S}} \right) = 0. \tag{35}$$

Finally, assuming that the control volume is a differential element, the following first-order differential equation is obtained.

$$C_{\mathrm{b}} v \frac{\mathrm{d}Y_{\mathrm{fuel,b}}}{\mathrm{d}z} + a_{\mathrm{v}} k_{\mathrm{m}} C_{\mathrm{b}} \left(Y_{\mathrm{fuel,b}} - Y_{\mathrm{fuel,S}} \right) = 0 \tag{36}$$

The surface area-to-volume ratio can be written in terms of porosity and foam fiber thickness as

$$a_{v} = \frac{4 (1 - \alpha)}{w}. \tag{37}$$

A similar molar balance can be conducted at the catalyst surface. Assuming that all fuel species, which come into contact with the catalyst react, the moles transported to the surface must equal the moles reacted on the surface.

$$\begin{bmatrix} \text{moles transported} \\ \text{to Pt surface} \end{bmatrix} = \begin{bmatrix} \text{moles reacted} \\ \text{on Pt surface} \end{bmatrix} \tag{38}$$

The transport term is the same as in the gas-phase mole balance. The surface reaction term can be written as

$$\begin{bmatrix}\text{moles reacted}\\ \text{on Pt surface}\end{bmatrix} = (-R_{\text{fuel}})_{\text{S}}\,\Delta A_{\text{S}}\,(1-\alpha) \tag{39}$$

where $(-R_{\text{fuel}})_{\text{S}}$ is the reaction rate obtained from the mechanism in Eqs. (25) and (26) and ΔA_{S} is again the incremental surface area. The mole balance can then be written as

$$k_{\text{m}}C_{\text{b}}\left(Y_{\text{fuel,b}} - Y_{\text{fuel,S}}\right)\Delta A_{\text{S}} = (-R_{\text{fuel}})_{\text{S}}\,\Delta A_{\text{S}}\,(1-\alpha)\,. \tag{40}$$

Canceling the surface area results in the algebraic equation

$$k_{\text{m}}C_{\text{b}}\left(Y_{\text{fuel,b}} - Y_{\text{fuel,S}}\right) = (-R_{\text{fuel}})_{\text{S}}\,(1-\alpha)\,. \tag{41}$$

The third and final equation (second ordinary differential equation) can be obtained by performing a gas-phase energy balance across the fluid element shown in Fig. 9.35. In general terms, this energy balance is

$$\begin{bmatrix}\text{enthalpy rise from}\\ \text{heat addition}\end{bmatrix} = \begin{bmatrix}\text{heat added}\\ \text{from surface}\end{bmatrix} + \begin{bmatrix}\text{heat added from}\\ \text{gas - phase reaction}\end{bmatrix}. \tag{42}$$

The gas-phase reaction term is zero because only catalytic reactions are being considered and the enthalpy rise (left hand side of the expression) can be written as

$$\begin{bmatrix}\text{enthalpy rise from}\\ \text{heat addition}\end{bmatrix} = \dot{m}C_{\text{p}}\Delta T_{\text{b}} \tag{43}$$

where $\dot{m}$ is the mass flow rate, C_{p} is the specific heat of the gas, and T_{b} is the bulk gas temperature. The heat generated from the surface reaction is transferred to the gas stream and can be expressed as

$$\begin{bmatrix}\text{heat added}\\ \text{from surface}\end{bmatrix} = \Delta A_{\text{S}}h\,(T_{\text{S}} - T_{\text{b}}) \tag{44}$$

where h is the heat transfer coefficient and T_{S} is the catalyst surface temperature. All other variables are the same as in previous sections. The energy balance then becomes

$$-\,\dot{m}C_{\text{p}}\Delta T_{\text{b}} + \Delta A_{\text{S}}h\,(T_{\text{S}} - T_{\text{b}}) = 0. \tag{45}$$

Again normalizing by ΔV and taking the control volume to be a differential element, an ordinary differential equation is obtained.

$$-\rho C_p v \frac{dT_b}{dz} + a_v h (T_S - T_b) = 0 \tag{46}$$

Analytical solutions to this system of three equations can be found or an appropriate numerical method can be applied. In the case of a non-isothermal structure, a fourth energy balance equation across a differential element in the wall, accounting for conduction in the solid, would be required.

Correlations for gases in packed bed reactors can be used to approximate the heat and mass transfer coefficients in porous substrates. These transport coefficients are given in terms of non-dimensional j factors; j_D for mass transfer and j_H for heat transfer [15].

$$j_D = \frac{k_m}{v} Sc^{2/3} \tag{47}$$

$$j_H = \frac{h}{C_p \rho v} \Pr^{2/3} \tag{48}$$

These j factors can be estimated using the following correlations [15]:

$$j_D = j_H = 0.91 \mathrm{Re}_j^{-0.51} S_f \quad \mathrm{Re}_j < 50 \tag{49}$$

$$j_D = j_H = 0.91 \mathrm{Re}_j^{-0.41} S_f \quad \mathrm{Re}_j > 50. \tag{50}$$

S_f is a shape factor and for this analysis a value of 0.79 was used (shape factor for ring-type structures) [15]. The Reynolds number used here can be found from

$$\mathrm{Re} = \frac{\rho v w}{6(1-\alpha)\mu S_f}. \tag{51}$$

A typical set of boundary and flow conditions required to utilize this model is shown below in Table 9.5.

Table 9.5 Typical boundary and flow conditions for reactor inlet in 1-D isothermal, porous, plug flow model

Parameter	Value
T_b	500 K
T_S	1000 K
$Y_{fuel,b}$ (from ϕ)	0.04 (ϕ=1.0)
$Y_{fuel,S}$	0.00
Pressure	2 atm
Mass flow	0.3 g/s

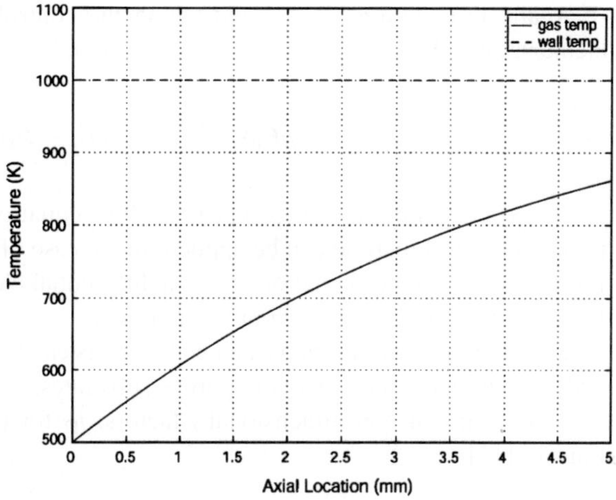

Fig. 9.36 Axial temperature profile along porous media plug flow reactor [1, 3]

Figures 9.36 and 9.37 show temperature and fuel concentration profiles, respectively, for the conditions listed above in Table 9.4. A substrate porosity of 95% and an average foam fiber thickness of 90 μm were also used.

With this model, the effect of key design variables such as porosity and surface area can be further examined. For constant flow conditions, the model can produce fuel conversion profiles for various porosity and surface area materials. Figure 9.38 shows trends in fuel conversion for increasing surface area-to-volume ratios using the same conditions. The profiles indicate that higher surface area-to-volume ratios (usually a result of lower porosity for a constant fiber thickness) will significantly improve fuel conversion.

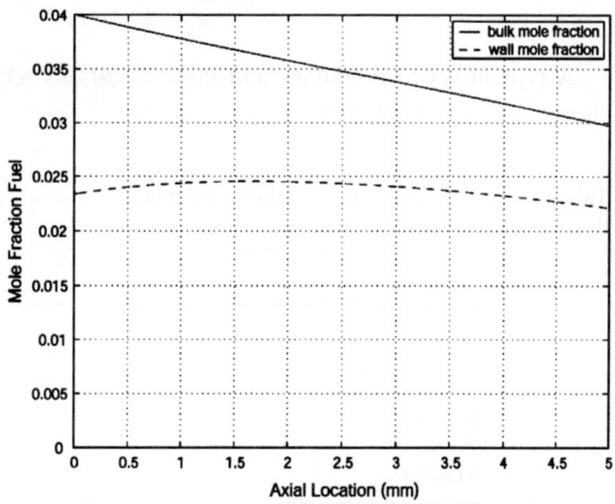

Fig. 9.37 Axial fuel concentration profile along porous media plug flow reactor [1, 3]

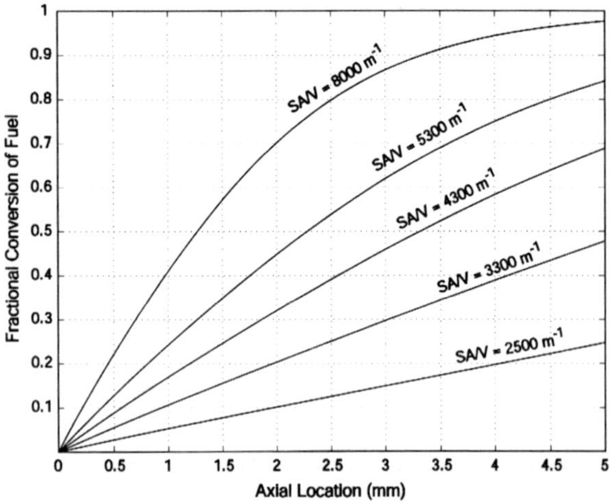

Fig. 9.38 Fuel conversion profiles for various surface area-to-volume ratios in porous media plug flow reactor [1, 3]

9.5.6 *Catalytic Microcombustor Operating Space*

The 1-D isothermal, porous, plug flow model can also be used to visualize a catalytic microcombustor's operating space. Figure 9.39 shows lines of constant power density on a plot of total pressure loss through the device versus catalyst temperature for an equivalence ratio of unity. The maximum power density is obtained from the exit gas temperature predicted by the model. Figure 9.39 indicates that at higher pressure loss (higher surface area-to-volume ratio) and higher catalyst temperatures, combustor performance improves. It is also clear that by relaxing the pressure loss constraint

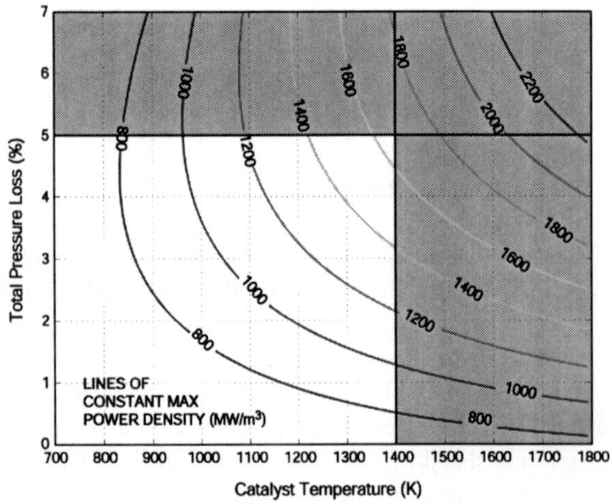

Fig. 9.39 Operating space for catalytic microcombustor; lines of constant power density [1, 3]

and utilizing a catalyst which can survive at higher temperatures, the available operating space will become larger. A representative constraint of less than 5% total pressure loss is indicated by the black line and gray shaded area. The shaded area above a 1400 K catalyst temperature represents an approximate failure temperature for a thin film of noble metal catalyst.

The catalytic microcombustor operating space can be viewed more generally by plotting non-dimensional parameters. Figure 9.40 shows lines of constant combustor efficiency on a chart with Peclet number versus thermal efficiency. As Peclet number decreases, combustor efficiency increases due to the time available for diffusion of the reactant species to the active surface. However, as heat is lost from the system (lower thermal efficiency), the overall efficiency decreases.

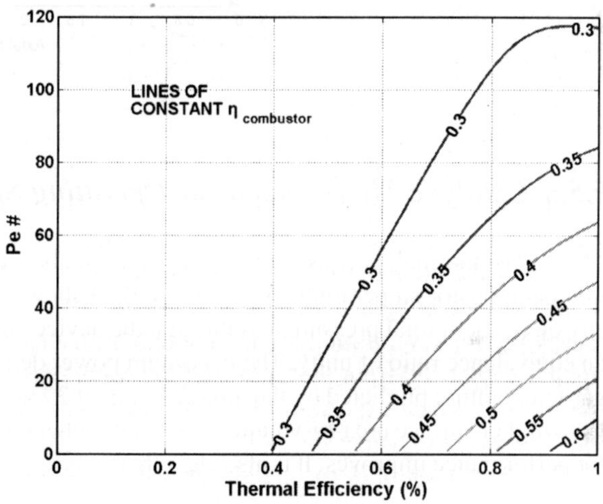

Fig. 9.40 Non-dimensional operating space; Peclet number versus thermal efficiency [3]

9.6 Fabrication and Packaging

9.6.1 Basic Techniques

Fabrication of the silicon microcombustors presented here was enabled by three fundamental technologies; photolithography, DRIE, and aligned fusion wafer bonding. Devices consist of 2-D extruded geometries formed by photolithography and DRIE on single crystal silicon wafers. Multiple wafers with various structures are then aligned and fusion bonded to form a quasi three-dimensional device. Up to ten microcombustors were batch processed on a set of 4 inch wafers. The microcombustors consisted of six silicon wafer levels, 11 photomasks, and 10 deep etches. Figure 9.41 shows a schematic of the fabrication and assembly.

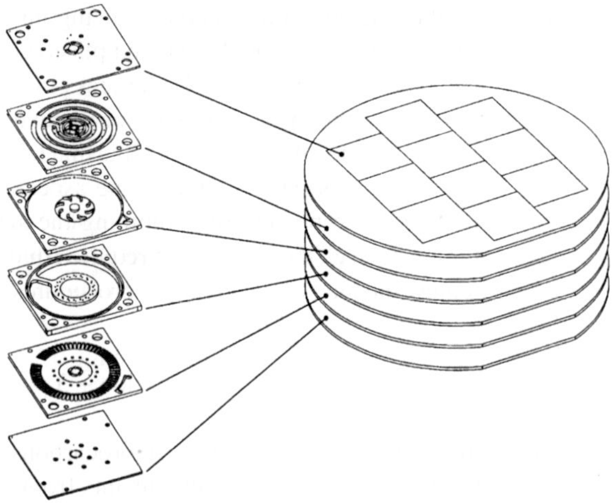

Fig. 9.41 Schematic of typical microcombustor wafer assembly

Photolithography consists of coating a wafer with a layer of photosensitive polymer called photoresist (in this case, positive photoresist is used). For the microcombustors of interest here, this layer is typically ~20 μm and is considered to be very thick. The optical template used to define the 2-D features, which are to be etched, is the photomask. This mask is aligned with the coated wafer and exposed to an ultra-violet light source for up to 3 min. After exposure, the wafer is soaked in developer for several minutes and the photoresist areas which were exposed to the light are removed from the surface of the wafer exposing a pattern. The remaining photoresist protects the wafer while the exposed silicon is etched. A schematic of this process is shown in Fig. 9.42.

The formation of deep, straight-walled structures in silicon is based on Bosch's patented technique for time-multiplexed deep etching (TMDE) [16]. The high aspect ratio, complex fluid paths required for a microcombustor and microengine can be

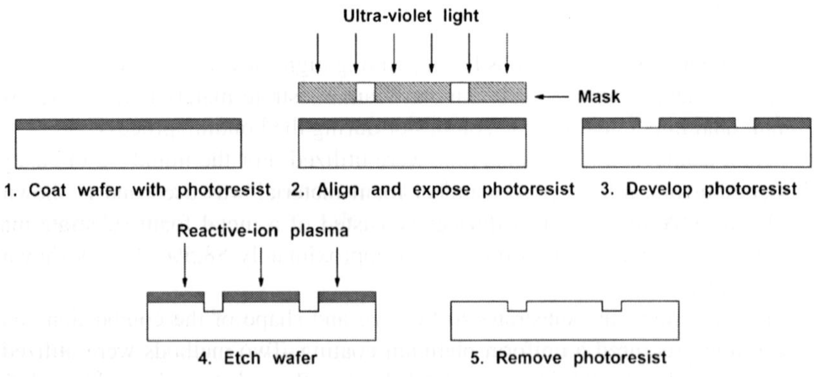

Fig. 9.42 Schematic of photolithography process [11]

micromachined to the necessary tolerances using this method. The anisotropic etch uses a time-multiplexed inductively coupled plasma of SF_6 as the etchant and C_4F_8 as the sidewall passivating polymer. The process involves alternating between etching and deposition of the protective passivating layer. The C_4F_8 protects the sidewalls from etching while allowing the SF_6 plasma to attack the lower, perpendicular surface. This technique is capable of attaining aspect ratios in excess of 20.

Wafer bonding is the process of combining the six individual wafers to form a single, multilevel device. This procedure requires that the wafers have minimal bow and are extremely clean. Typical cleaning procedures are listed below [17]:

1. 10 min Pirahna (1 part hydrogen peroxide + 3 parts sulfuric acid) dip to remove photoresist
2. 1 h plasma ashing to remove excess fluorocarbons from the DRIE
3. 10 min Pirahna dip to remove contaminants from asher
4. 10 min hydrofluoric acid dip to remove protective layer of silicon dioxide and any other defects/particles which have adhered to the surface
5. RCA cleaned to ensure surfaces are pristine [18]

The single crystal silicon wafers are then aligned and contacted. Van der Waals forces cause the wafers to weakly adhere. After applying some pressure (approximately 4 atm) and annealing the stack in a nitrogen tube furnace at 1100°C for 1 h, the individual silicon wafers inter-diffuse forming a single, robust silicon piece. Figure 9.43 shows a schematic of the individual combustor layers before and after contacting and bonding. Figure 9.44 shows an infrared image of a bonded six-wafer stack. Optical fringe patterns in the image indicate poorly bonded areas. A more comprehensive treatment of microfabrication for rotating structures and microengine components can be found in Chapter 4.

9.6.2 Catalytic Components

There are many possible methods for implanting high surface area catalyst materials into a microdevice. In this case, a metal foam substrate material was coated with platinum and integrated into the combustor during the bonding process.

Several catalyst substrate materials were utilized. For the initial set of catalytic microcombustors, a 95% porous nickel foam material was used and is shown in Fig. 9.45a, b. A second set of devices consisted of a metal foam substrate material made of FeCrAlY with a porosity of approximately 88.5%. This is shown in Fig. 9.46a, b.

After machining the substrates to the size and shape of the combustion chamber, the foam required a uniform platinum coating. Two methods were utilized to produce these layers. The first involved dipping the substrate in a chloroplatinic acid solution while the second consisted of using ionic plasma deposition. The first

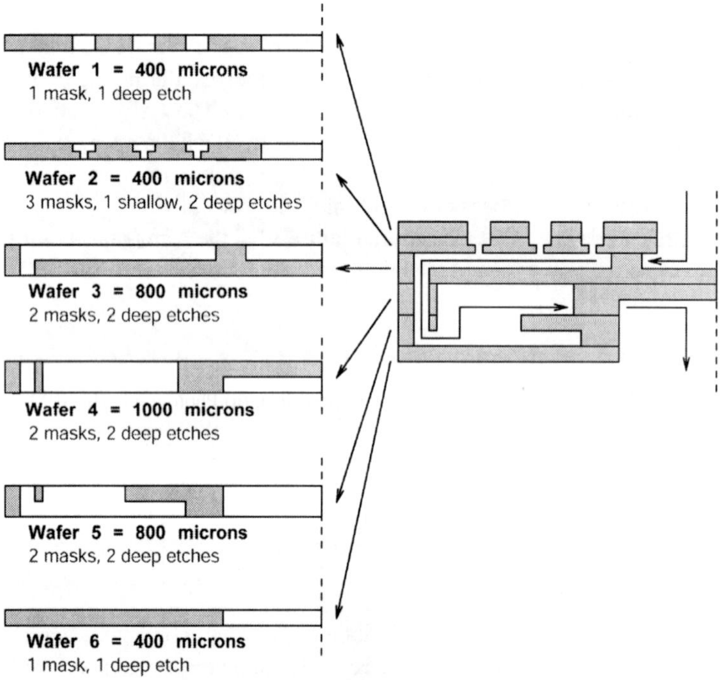

Fig. 9.43 Schematic of microcombustor layers pre- and post-bonding [11]

Fig. 9.44 Infrared image of bonded six-wafer microcombustor stack [1]

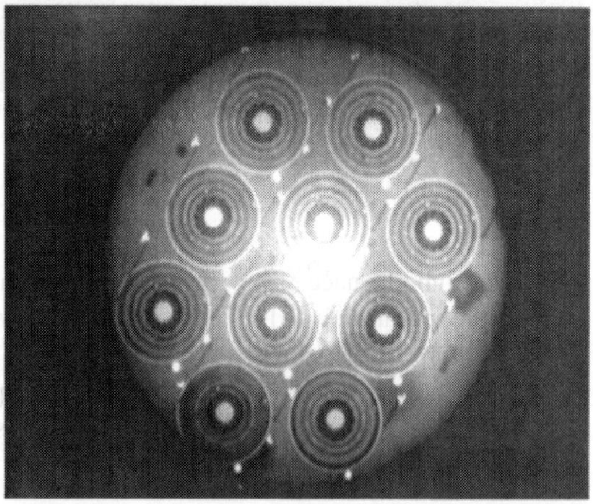

technique was performed on the 95% porous nickel substrate and the second on the FeCrAlY.

Unsupported metal catalysts can be deposited onto a substrate using a solution of metal salt. In this procedure, the metal salt (H_2PtCl_6 in this case) is dissolved in deionized water. The substrate pieces are soaked in this solution then placed

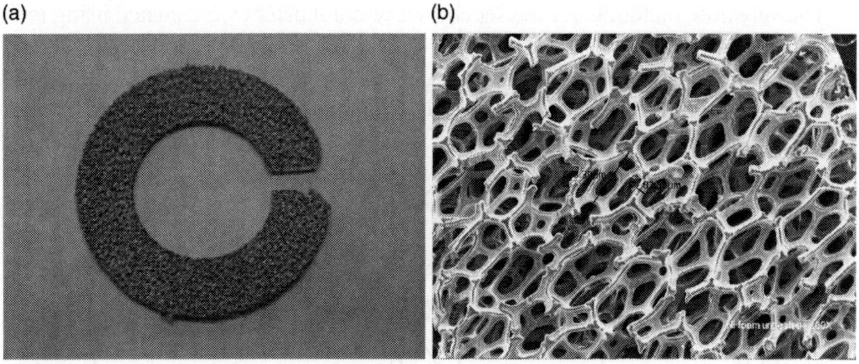

Fig. 9.45 Nickel foam substrate material (∼95% porosity); (**a**) Photograph; (**b**) SEM (50×) [1]

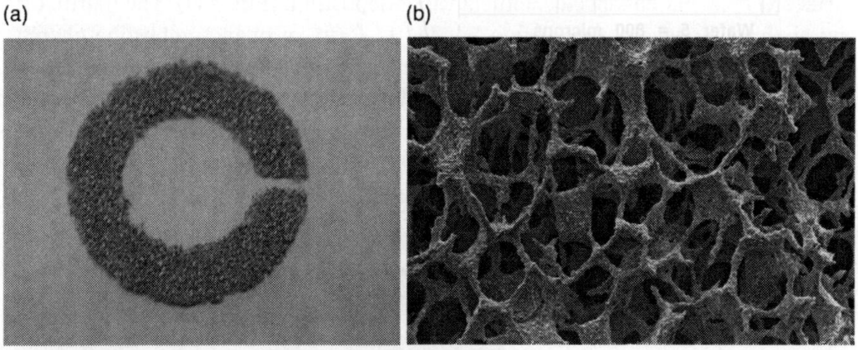

Fig. 9.46 FeCrAlY foam substrate material (∼88.5% porosity); (**a**) Photograph; (**b**) SEM (50×) [1]

in a small tube furnace. The water evaporates and the remaining compound reacts with hydrogen at elevated temperature to remove Cl in the form of HCl, leaving only platinum on the surface. The nickel foam substrates were coated using this technique and typically increased in weight by approximately 3–5% creating films approximately 2–3 μm thick. The procedure is listed below.

1. Start with *dihydrogen hexochloroplatinate* salt (H_2PtCl_6).
2. Mix approximately 2.5 g of H_2PtCl_6 with about 20 ml of deionized water to form H_2PtCl_6–H_2O (∼10% Pt-salt solution).
3. Soak nickel foam pieces in solution for approximately 1 h.
4. Insert pieces into tube furnace.
5. Flow approximately 100 sccm of N_2 at >100°C for ∼2 h.
6. Flow approximately 100 sccm of H_2 at ∼300°C for ∼3 h.

The substrate materials for the second set of combustors were coated using ionic plasma deposition. This is a proprietary process developed by Ionic Fusion Corporation and involves propelling atoms of a given material by ionic acceleration and ballistically impregnating a substrate. It is a high-energy, low-temperature process that provides good penetration into porous substances [19]. The foam pieces were first coated with 2–3 μm of Al_2O_3 to act as a diffusion barrier followed by 2–3 μm of platinum. The substrates increased in weight by approximately 4%.

9.6.3 Catalytic Microcombustor Assembly and Bonding

Incorporating the foam catalytic components into the microdevices, required that they be integrated into the bonding procedure. To protect the bonding surfaces from the catalytic inserts, a "shield wafer" was fabricated and utilized. This wafer was etched through with the shape of the inserts and silicon dioxide was deposited on its surface via plasma-enhanced chemical vapor deposition (PECVD). The fourth, fifth, and sixth wafer levels were fusion bonded to form an open combustion chamber prior to inserting the catalyst. The shield wafer was then aligned and contacted with these three levels. While in contact, the catalytic pieces were placed into the bonded 4-5-6 wafer stack through the shield wafer. Finally, the protective shield wafer was carefully removed. The PECVD oxide prevents the shield wafer from adhering to the 4-5-6 stack. Insertion of the catalytic pieces can be seen in Fig. 9.47 along with the assembly and bonding process in Fig. 9.48.

Fig. 9.47 Catalytic pieces being inserted into the six-wafer microcombustor during fabrication [1]

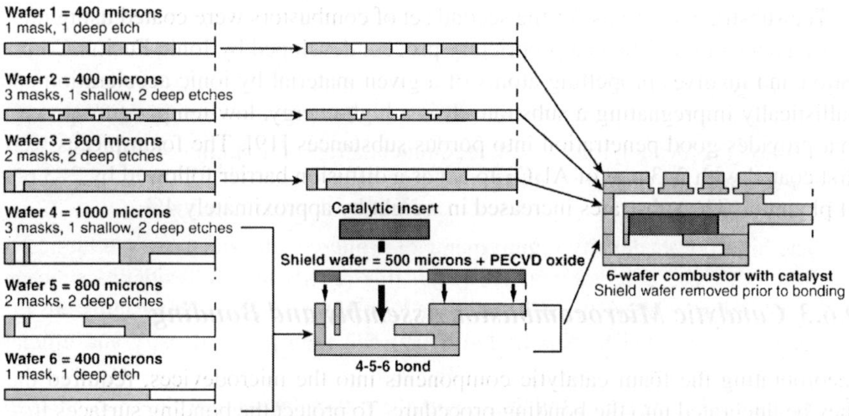

Fig. 9.48 Fabrication, assembly, and bonding process for six-wafer catalytic microcombustor [1]

9.6.4 Packaging

Prior to testing, the microcombustors were packaged into a suitable test rig. In order to connect the device's microscale fluid channels to a macro-scale feed system, a glass bead interconnect scheme was developed by Mehra [11]. Small diameter Kovar tubing was hermetically sealed to the silicon with glass beads and brazed to a larger kovar plate for connection to conventional fittings. The kovar–glass–silicon system provided the best thermal expansion matching available. A fully packaged device is shown in Fig. 9.49. More details of device packaging can be found in Chapter 3.

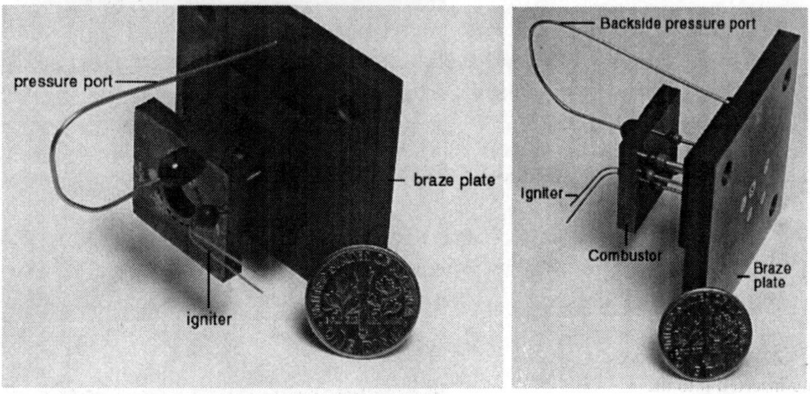

Fig. 9.49 Fully packaged microcombustor [2, 11]

9.7 Summary

9.7.1 Design Guidelines

The following design trades and recommendations for homogeneous gas-phase microcombustion systems should be considered:

1. The high power density requirements of a microcombustor mandate high mass flow rates per unit volume. However, this results in low residence time and Damköhler numbers of order unity or less. Microcombustion systems are fundamentally limited by this trade-off between high power density and efficient combustion.
2. The large surface area-to-volume ratio of a microcombustor makes non-adiabatic operation intrinsic to its design. The coupling between heat transfer and chemical kinetics is a critical element of the design process. Thermal management strategies such as the recirculation jacket are effective solutions. While keeping the inner wall cool, it acts as a fuel–air pre-heater to improve kinetic rates, offsetting the effect of lower wall temperatures.
3. Recirculation zones are critical for rapid and uniform ignition of incoming reactants. However, if these ignition zones are too large, the effective volume of the combustion chamber will be significantly reduced resulting in decreased residence time and Da_{h}.
4. Fuel injection for non-premixed operation should occur as far upstream of the combustion chamber as possible to provide adequate mixing length.
5. It is not sufficient to design a microcombustor exclusively for design point operation. Operating line issues need to be considered.

Design recommendations for heterogeneous catalytic microcombustion systems include the following:

1. The high power density catalytic microcombustors presented here are diffusion controlled. A designer should seek to approach the high-temperature reaction-controlled regime to maximize performance. This can be achieved by implementing the following:
   a. Utilize the most thermally durable catalytic materials to achieve high-temperature operation.
   b. Utilize the highest surface area-to-volume ratio substrate material available which does not violate the system pressure loss constraint.
   c. Relax the total pressure loss constraint as far as the thermodynamic cycle will permit.
2. Searching for a more active catalytic material is not required unless:
   a. The ignition transient is of concern.
   b. The overall design lies in the reaction-controlled regime.

3. High surface area-to-volume ratio catalyst materials result in increased heat loss from the system. As a result, thermal management is more critical than in the gas-phase case. A materials solution such as a thermal barrier combined with a concept similar to the recirculation jacket may mitigate these losses.

9.7.2 Recommendations for Future Work

Use of catalytic elements was successful in significantly increasing power density and mass flow range for hydrocarbon-fueled devices. However, the exit gas temperatures achieved were significantly lower than that required for a useful thermodynamic cycle. Attaining these temperatures in a catalytic device of this kind is unlikely because the catalyst wall temperatures required would be high enough to cause significant agglomeration and degradation of the catalyst layer and would even approach the softening points of both the catalyst and some substrate materials. In addition, to accomplish this in a relatively small volume, the total pressure loss would be significant and most likely greater than any constraint imposed by an engine cycle. Although the gas-phase devices were capable of achieving high exit gas temperatures and efficiencies with hydrogen-air mixtures, their mass flow rate capability was significantly less than would be necessary and was even worse with hydrocarbon fuels.

As a result, a strategy for achieving both high temperature and high mass flow rate operation in a minimum volume may be to combine the two devices into a "hybrid microcombustor." This is not a new concept and is commonly used in conventional-scale catalytic combustors, often referred to as a "staged catalytic combustor."

For this type of device, the entrance region of the combustion chamber would consist of a porous substrate or microchannel type geometry coated with a catalyst film. The first stage would ignite the fuel–air mixture and bring it to some mid-range temperature ($\sim$1000 K) with minimal total pressure loss. A second stage of open volume for homogeneous combustion would follow. The exit conditions of the catalytic section would serve as the inlet to the gas-phase section, which would burn the remaining fuel–air mixture. The reactions in the second stage would proceed quickly due to the already very high inlet gas temperature and would likely achieve temperature in excess of 1600 K in a minimum total volume. In conventional-scale catalytic combustors, this section is often referred to as the "burnout section." The burnout section may also include additional fuel injection to raise the local equivalence ratio. This may be required as a result of the reduction in mixture ratio due to the upstream catalytic combustion process.

One final subject of potential future work is to include the use of liquid fuels such as kerosene and JP8 in microcombustor development. These types of fuels are the most commonly used hydrocarbons in gas turbine engines. For potential end users, these are the most practical fuels and are often referred to as "logistics fuels." A microcombustor, which was capable of burning liquid fuels, would provide a micro-engine with a significant advantage over competing microscale power technologies that require more exotic fuels. The catalytic microcombustors discussed here may be a first step toward achieving this and work in this area has already begun [7].

References

1. Spadaccini C M 2004 Combustion systems for power-MEMS applications. Ph.D. thesis, Massachusetts Institute of Technology
2. Spadaccini C M, Mehra A, Lee J, Zhang X, Lukachko S, and Waitz I A 2003 High power density silicon combustion systems for micro gas turbine engines. *ASME Journal of Engineering for Gas Turbine and Power*, Vol. 125, pp. 709–719
3. Spadaccini C M, Peck J, Waitz I A 2006 Catalytic combustion systems for micro-scale gas turbine engines. *ASME Journal of Engineering for Gas Turbines and Power*, Vol. 129, pp. 49–60
4. Mehra A, Zhang X, Ayon A A, Waitz I A, Schmidt M A, and Spadaccini C M, 2000 A six-wafer combustion system for a silicon micro gas turbine engine. *IEEE/ASME Journal of Microelectromechanical Systems*, Vol. 9, No. 4, pp. 517–527
5. Waitz I A, Gauba G, and Tzeng Y-S 1998 Combustors for micro gas turbine engines. *ASME Journal of Fluids Engineering*, Vol. 20, pp. 109–117
6. Arana L R 2003 High-temperature microfluidic systems for thermally-efficient fuel processing. Ph.D. thesis, Massachusetts Institute of Technology
7. Peck J J 2008 Development of a liquid-fueled micro-combustor. Ph.D. thesis, Massachusetts Institute of Technology
8. Maskell W C, Sammes N M, and Steele B C H 1987 Agglomeration of thin platinum films on a zirconia substrate. *Journal of Physics D: Applied Physics*, Vol. 20, pp. 99–104
9. Wang J, Zhao D G, Sun Y P, Duan L H, Wang Y T, Zhang S M, Yank H, Zhou S, and Wu M 2003 Thermal annealing behaviour of Pt on n-GaN schottky contacts. *Journal of Physics D: Applied Physics*, Vol. 36, pp. 1018–1022
10. Tzeng, Y-S 1997 An investigation of microcombustion thermal phenomena. Masters Thesis, Massachusetts Institute of Technology
11. Mehra A 2000 Development of a high power density combustion system for a silicon micro gas turbine engine. Ph.D. thesis, Massachusetts Institute of Technology
12. Spadaccini C M and Waitz I A 2007 Emerging Topics: Microcombustion, *Comprehensive Microsystems,* Gianchandani Y, Tabata O, and Zappe H, Eds., Elsevier Science Publishing
13. Goralski C T Jr and Schmidt L D 1996 Lean catalytic combustion of alkanes at short contact times. *Catalysis Letters*, Vol. 42, pp. 15–20
14. Veser G and Schmidt L D 1996 Ignition and extinction in catalytic oxidation of hydrocarbons. *AIChE Journal*, Vol. 42, pp. 1077–1087
15. Hayes R E and Kolaczkowski S T 1997 *Introduction to Catalytic Combustion*, Gordon and Breach Science Publishers, Canada
16. Robert Bosch Gmbh, Patents 4855017 and 4784720 (USA), and 4241045C1 (Germany)
17. Miki N, Zhang X, Khanna R, Ayon A A, Ward D, and Spearing S M 2003 Multi-stack silicon-direct wafer bonding for 3D MEMS manufacturing *Sensors and Actuators A: Physical*, Vol. 103, pp. 194–201
18. Kern W, and Puotinen D A, 1970 Cleaning solutions based on hydrogen peroxide for use in silicon semiconductor technology *RCA Review*, Vol. 31, pp. 187–206
19. Ionic Fusion Corporation, 2003 corporate literature, Longmont, CO

Index

J.H. Lang (ed.), *Multi-Wafer Rotating MEMS Machines*, MEMS Reference Shelf,
DOI 10.1007/978-0-387-77747-4, © Springer Science+Business Media, LLC 2009

Lightning Source UK Ltd.
Milton Keynes UK
02 January 2009

148025UK00007B/104/P

9 780387 777467